Springer Series in Solid and Structural Mechanics

Volume 5

Series editors

Michel Frémond, Rome, Italy
Franco Maceri, Rome, Italy

More information about this series at http://www.springer.com/series/10616

Mario Como

Statics of Historic Masonry Constructions

Second Edition

Springer

Mario Como
Department of Civil Engineering
University of Rome Tor Vergata
Rome
Italy

ISSN 2195-3511 ISSN 2195-352X (electronic)
Springer Series in Solid and Structural Mechanics
ISBN 978-3-319-24567-6 ISBN 978-3-319-24569-0 (eBook)
DOI 10.1007/978-3-319-24569-0

Library of Congress Control Number: 2015958658

Springer Cham Heidelberg New York Dordrecht London
© Springer International Publishing Switzerland 2013, 2016
This work is subject to copyright. All rights are reserved by the Publisher, whether the whole or part of the material is concerned, specifically the rights of translation, reprinting, reuse of illustrations, recitation, broadcasting, reproduction on microfilms or in any other physical way, and transmission or information storage and retrieval, electronic adaptation, computer software, or by similar or dissimilar methodology now known or hereafter developed.
The use of general descriptive names, registered names, trademarks, service marks, etc. in this publication does not imply, even in the absence of a specific statement, that such names are exempt from the relevant protective laws and regulations and therefore free for general use.
The publisher, the authors and the editors are safe to assume that the advice and information in this book are believed to be true and accurate at the date of publication. Neither the publisher nor the authors or the editors give a warranty, express or implied, with respect to the material contained herein or for any errors or omissions that may have been made.

Printed on acid-free paper

Springer International Publishing AG Switzerland is part of Springer Science+Business Media (www.springer.com)

To Ida

Preface

Masonry constructions are the majority of the buildings in Europe's historic centers and the most important monuments in its architectural heritage. Given the age of much of these constructions, the demand for safety assessments and restoration projects is pressing and constant. Nevertheless, there is a lack of a widely accepted approach to studying the statics of masonry structures. Simple linear elastic models, which form the foundation of common structural analyses, cannot in fact be applied to masonry, because of its inherent, widely differing response to tension and compression.

The ingenious Heyman no-tension model well interprets the masonry behavior and is widely used and was fruitfully applied in analyzing the statics of systems of arches. However, completely different assumptions are commonly used for other types of masonry structures in other contexts, for example, strength evaluations of masonry buildings under seismic forces, which is rather perplexing, given that a masonry arch, a vault, and a building wall are all still made of the same material. Moreover, most masonry studies approach strength evaluations of structures through Limit Analysis, forgoing any study of the construction's actual state.

This book aims to fill these gaps in the study of masonry structures by formulating a new, comprehensive, unified theory of statics of masonry constructions extending the Heyman model to the analysis of the masonry continuum. The book features complete mathematical derivation of all the given results and, through an interdisciplinary approach combining engineering, architecture, and a bit of history, advances from the simple to the complex, while striving, above all, for clarity.

The book is the result of 30 years of research and professional experience. It is divided into nine chapters, each of which begins with historical notes and an introduction highlighting the main aspects of the topics covered.

The strength and deformability of masonry materials are addressed in the first chapter. The second chapter deals with the deformation and equilibrium of masonry solids. The kinematics of strains and crackings, as well as internal stress states are analyzed. The fundamental concepts of admissible equilibrium and the parameters governing collapse strength are examined in detail to highlight the strict relation between structural geometry and strength. The notion of minimum thrust is then introduced—an aspect of masonry structural behavior that extends the field of application of Limit Analysis to include the study of the actual stress states of masonry constructions. The third and fourth chapters examine the static behavior of the main basic masonry structures, such as arches and vaults.

By way of example, static analyses are conducted on a number of renowned examples from the world's architectural heritage, such as ancient Mycenaean domes, the Roman Pantheon, the large cross vaults of the Baths of Diocletian, and the domes of Santa Maria del Fiore in Florence and Saint Peter's in Rome. The fifth chapter turns to a detailed analysis of the statics of the Roman Colosseum and examines the reasons for its actual state of damage. The sixth chapter describes and analyzes the statics of cantilevered stairways, a typical element whose structural behavior is still somewhat unknown. Chapter seven then takes up the structural analysis of walls, piers, and towers under vertical loads. The stability of such structures is heavily affected by the nonlinear interactions between the destabilizing effects of the axial loads and masonry's no-tension response. The instability of towers, leaning towers in particular, is addressed in a specific section of the chapter. In this regard, a detailed stability analysis was conducted at the famous Leaning Tower of Pisa, which has recently undergone a successful restoration work. The eighth chapter then analyzes the statics of Gothic cathedrals, with particular reference to analysis of their resistance to wind actions. The 1294 collapse of the Beauvais cathedral is also examined in-depth. The last chapter deals with the seismic behavior of historic masonry buildings and crucial issues regarding their conservation. The latter part of the chapter regards, in particular, the analysis of the transmission of seismic forces between the various constituents of a building, together with the out-of-plane and in-plane strengths evaluations of multi-story walls with openings.

The book is addressed especially to researchers, engineers, and architects operating in the field of masonry structures and of their consolidation and restoration, as well as to students of civil engineering and architecture. It is, for the most part, an English translation of a recent Italian book of mine "Statica delle Costruzioni Storiche in muratura." The English edition has, however, been revamped to address some new questions and, hopefully, improve on the original.

Many thanks go to colleagues Michel Frémond and Franco Maceri for their precious encouragement to prepare the book. Many thanks also go to Anthony Cafazzo, English Lecturer at the University of Pisa, who insightfully and patiently assisted me in revising the text.

I would also like to thank all the graduate and postdoctoral students, researchers, visiting scholars, external collaborators, and students, who attended my courses at the Faculty of Engineering at the University of Rome Tor Vergata—all of whose contributions have been duly noted—for their invaluable assistance in the various research studies without which this book would not have been possible.

Rome
January 2012

Mario Como

Preface to the Second Edition

The interest accomplished in the first edition together with the need to improve the text with new developments, widening, and revisions, due to the recent research achievements on the subject matter, is the motivation of this second edition of the book.

A new section has been added to the first chapter, analyzing new test results for masonry strength under inclined compression with respect to the joint direction; this is a subject of great importance for the evaluation of the seismic strength of masonry walls. Within the second chapter, dedicated to the Fundamentals of Statics of Masonry Structures, a new Limit Analysis of elastic no-tension one-dimensional systems has been included; this is a very useful tool, for instance, in the strength analysis of masonry walls reinforced with steel ties. The third, fourth, fifth, and eighth chapters are substantially unchanged, except for some additions concerning the construction of the Brunelleschi Dome in Florence and the inclusion of a new section dealing with the thrust evaluation of round cross vaults, then applied to the vaults of the Diocletian Baths in Rome. In the sixth chapter the study of the effect of the inclined cracking on the buttresses and leaning towers static behaviors has been included. This study has been very useful for the analysis concerning the strength assessment of Gothic Cathedrals under side wind, a topic which has been revised and developed within the seventh chapter. The ninth chapter, dealing with the seismic analysis of masonry buildings—a topic in which the current research has produced new remarkable results—has received the most important revisions and widening.

All these developments have been obtained thanks to the precious teamwork with Simona Coccia and Fabio Di Carlo of the Department of Civil Engineering and Computer Science at the University of Tor Vergata in Rome. To them, I address my grateful thanks.

Reflections and ideas on the topic were also triggered from the fruitful discussions that I had with the students of the Doctorate course in "Restoration of Historic and Contemporary Buildings," held at the D.I.C.A.T.A. of the University of Brescia. Also, to all these students my thanks are directed.

Rome
July 2015

Mario Como

Contents

1 Masonry Strength and Deformability 1
  1.1 Brief Notes on the History of Masonry Constructions 1
  1.2 The Masonry of Historic Buildings 4
  1.3 Compression Strength of Brick and Stone Elements 7
    1.3.1 Bricks 7
    1.3.2 Stone Blocks 8
  1.4 Mortars 10
    1.4.1 Binders 10
    1.4.2 Aggregates 11
    1.4.3 Mortars of Lime 11
  1.5 Tests on Rock and Mortar Specimens 12
    1.5.1 Tests on Rock Specimens 13
    1.5.2 Uniaxial Compression Tests on Mortar Specimens ... 15
    1.5.3 Stress Strain Diagrams for Stone and Mortar Materials 16
  1.6 A Triaxial Failure Criterion for Stone Materials 17
    1.6.1 Preliminary Considerations 17
    1.6.2 Porosity Effects: Micro-Macro Stress States 18
    1.6.3 Micro-Macro Failure Condition: Reasons of the Different Tensile and Compression Strengths 21
    1.6.4 Valuation of the Pores Shape Irregularity Factor 24
    1.6.5 Failure Interaction Domains 25
  1.7 Masonry Compression Strength 30
    1.7.1 Features of Compressed Masonry Failure 30
    1.7.2 Valuation of Masonry Compression Strength 30
  1.8 Masonry Tensile Strength 36
  1.9 Masonry Shear Strength 37
  1.10 Masonry Strength and the Variable Course Inclinations 40

	1.11	Masonry Deformations	43
		1.11.1 Masonry Elastic Modulus	43
		1.11.2 Masonry Deformation at the Onset of Blocks Failure	46
		1.11.3 Stress–Strain Diagram of the Compressed Masonry	47
		1.11.4 Creep Deformation of Mortar	49
		1.11.5 The Concept of Memory in Constitutive Creep Models	50
		1.11.6 Mortar Shrinkage	52
	References		53
2	**Fundamentals of Statics of Masonry Solids and Structures**		**55**
	2.1	Introduction	55
	2.2	No-Tension Masonry Models	56
		2.2.1 No-Tension Assumption	56
		2.2.2 The Problem of Elastic Compressive Strains	57
	2.3	The Rigid No-Tension Model	61
		2.3.1 The Heyman's Assumptions	61
		2.3.2 The Unit Resistant Masonry Cell	62
		2.3.3 Properties of the Rigid No Tension Material	67
	2.4	The Masonry Continuum	69
		2.4.1 Compatibility Conditions on Loads	71
		2.4.2 Compatibility Conditions on Stresses	71
		2.4.3 First Consequence of the No-Tension Assumption	72
		2.4.4 Impenetrability Condition on the Displacement Fields	74
		2.4.5 Compatibility Conditions on Strains and Detachments	74
		2.4.6 The Boundary of the Cracked Body	76
		2.4.7 Coupled Conditions on Stresses and Strains and on Stress Vectors and Detachments	78
		2.4.8 Specifications for One-Dimensional Masonry Systems	79
		2.4.9 Indeformable Masonry Structures	80
	2.5	Equilibrium and Compatibility	81
		2.5.1 Principle of Virtual Work	81
		2.5.2 Variational Inequality for the Existence of the Admissible Equilibrium State	86
		2.5.3 No-Existence of Self-Equilibrated Stress Fields in Deformable Structures	89
		2.5.4 Indeformable Structures: Statically Indeterminate Behaviour	90
		2.5.5 Admissible Equilibrium in One-Dimensional Systems	90

		2.5.6	Admissible Equilibrium of Elastic No Tension	
			One-Dimensional Systems	93
		2.5.7	Weight and Live Loads	93
		2.5.8	Mechanism States	95
	2.6	Collapse State.	97
		2.6.1	Definitions	97
		2.6.2	The Static Theorem	99
		2.6.3	The Kinematic Theorem.	100
		2.6.4	Uniqueness of the Collapse Multiplier	102
		2.6.5	Indeformable Systems: Lack of Collapse	102
		2.6.6	Collapse State for the Elastic No Tension Systems . . .	103
	2.7	Incipient Settled States.	104
		2.7.1	Definitions	104
		2.7.2	Features of Incipient Settled States	106
		2.7.3	Statically Admissible Thrusts: The Static Theorem	
			of Minimum Thrust.	107
		2.7.4	Kinematically Admissible Thrusts:	
			The Kinematic Theorem of the Minimum Thrust.	109
		2.7.5	Uniqueness of the Settlement Multiplier.	110
		2.7.6	The Class of the Statical and Kinematical	
			Settlement Multipliers	110
		2.7.7	An Application of the Kinematic Approach.	
			A Minimum Thrust Assessment	113
		2.7.8	Masonry Structures at Their Actual State	116
	2.8	Geometry and Strength: The Theory of Proportions	
		of the Past Architecture	117
	Appendix	124
	References.	131
3	**Masonry Arches**	133
	3.1	Definitions and History	133
	3.2	The Birth of the Statics of the Arch and Its Evolution	138
	3.3	Internal Equilibrium in the Arch	152
		3.3.1	Shear Force in Arches	153
	3.4	Limit Analysis	154
	3.5	Minimum and Maximum Thrust	156
		3.5.1	Effects of the Elastic Deformation on the Thrust	
			of the Arch	156
		3.5.2	Cracking	159
		3.5.3	Minimum Thrust State.	160
		3.5.4	Minimum Thrust in the Round Arch	161
		3.5.5	Minimum Thrust in the Depressed Arch.	162
	3.6	Coupled Systems of Arches of Different Spans.	164
	3.7	Masonry Arches Loaded by Horizontal Forces	165

	3.8	Some Experimental Results and Comments on Failure Tests of Masonry Arches.............................	170
		3.8.1 Test Description	170
		3.8.2 Comments on Test Results..................	172
	References...		174
4	**Masonry Vaults: General Introduction**....................		177
	4.1	Brief Historical Notes	177
	4.2	The Implemented Static Approach....................	182
	References...		184
5	**Masonry Vaults: Domes**		185
	5.1	Some Recalls on Membrane Equilibrium of Rotational Shells...............................	185
	5.2	Meridian Cracking: Definitive Stress State in Masonry Domes...	192
		5.2.1 From the Membrane to the Cracked State...........	192
		5.2.2 Safety Check of Domes: Static and Kinematic Approaches	194
		5.2.3 Minimum Thrust for the Hemispherical Dome with Constant Thickness	196
		5.2.4 Domes of More Complex Shape: The Kinematic Approach	207
	5.3	Mycenaean Tholos	208
		5.3.1 Description and Historical Notes................	208
		5.3.2 Statics of the Mycenaean Tholos................	210
	5.4	Roman Concrete Vaults: Do They Push on Their Supports?...	213
	5.5	The Pantheon.................................	213
		5.5.1 Introductive Notes.........................	213
		5.5.2 Structural Aspects of the Temple................	214
		5.5.3 Thrust of the Dome........................	219
	5.6	Brunelleschi's Dome in Florence.....................	222
		5.6.1 A Brief Account of the Cathedral's Construction......	222
		5.6.2 The Supporting Pillars......................	223
		5.6.3 The Drum	223
		5.6.4 The Dome	226
		5.6.5 Crack Patterns in the Dome	234
		5.6.6 Thrust of the Dome........................	237
		5.6.7 Loads Transmitted to the Pillars	239
		5.6.8 Stresses at the Base of the Pillar................	244
	5.7	St. Peter's Basilica Dome by Michelangelo: The Static Restoration by Poleni and Vanvitelli	246
		5.7.1 Dome Geometry	246
		5.7.2 Damage to the Dome: Early Studies Assessing Its Safety.........................	250

		5.7.3	The Three Mathematicians' and Poleni's Differing Opinions	251
		5.7.4	Poleni and Vanvitelli's Restoration Works	254
		5.7.5	Further Considerations on the Heated Debate	256
		5.7.6	Minimum Thrust in the Dome	257
		5.7.7	Comparisons: St. Peter's Lower Static Efficiency	265
		5.7.8	Checking Safety of the Drum/Attic/Buttresses System.................................	265
		5.7.9	Insertion of Iron Hoops	269
		5.7.10	Conclusions	270
	References...			271
6	**Masonry Vaults: Barrel Vaults**			273
	6.1	Introduction		273
	6.2	Membrane Stresses in Cylindrical Vaults		274
	6.3	Transition from the Uncracked to the Cracked State. The No-Tension Model of the Barrel Vault		279
	6.4	Systems Made up of Vaults and Walls...................		280
		6.4.1	The Barrel Vault with Side Walls	280
		6.4.2	Tie Rod Reinforcing Vault/Walls Systems	287
	References...			289
7	**Masonry Vaults: Cross and Cloister Vaults**			291
	7.1	Geometric Generation of Cross and Cloister Vaults..........		291
	7.2	Surface Areas and Weights of Webs and Lunes		293
	7.3	Historical Notes on Cross Vaults.......................		297
	7.4	Statics of Cross Vaults...............................		299
		7.4.1	Initial Membrane Stresses.....................	299
		7.4.2	Transition from the Uncracked to the Cracked State	302
		7.4.3	The Definitive Sliced Model....................	304
		7.4.4	Are Ribs Necessary for Cross Vault Equilibrium?....	326
		7.4.5	The Cross Vaults of the Diocletian Baths in Rome ...	327
	7.5	Cloister Vaults		332
		7.5.1	Initial Membrane Stresses.....................	333
		7.5.2	Cracking	342
		7.5.3	Definitive Resistant Model....................	343
	References...			349
8	**The Colosseum**...			351
	8.1	The Original Colosseum Structure.......................		351
	8.2	Static Analysis of the Colosseum's Original Configuration....		358
		8.2.1	Pier Stresses..............................	358
	8.3	Limit Analysis		365
		8.3.1	Preliminary Remarks........................	365
		8.3.2	The Collapse Load	370

	8.4	Damage and Subsequent Repairs.	372
	8.5	Possible Causes of the Damage.	377
		8.5.1 Seismic Excitability of the Monument: Effects of Soil-Structure Interactions	377
		8.5.2 Seismic Strength of the Monument	381
		8.5.3 The Dismantling Hypothesis.	385
		8.5.4 Conclusions	387
	References.		388
9	**Masonry Stairways**		391
	9.1	Geometrical Features of Masonry Stairs: Cantilevered Stairs.	391
	9.2	Brick Layout	393
	9.3	Other Types of Stairs	394
	9.4	Paradoxical Static Behavior of Cantilevered Masonry Stairs	395
	9.5	Numerical Investigations on the Statics of a Single Cantilevered Masonry Flight.	395
	9.6	Resistant Model of the Horizontal Flight of Stairs with Side Landings.	399
	9.7	Determination of the Horizontal Forces P_i	403
	9.8	An Inclined Flight of Stairs	404
	9.9	Cantilevered Stairways as a System of Flights and Landings	405
	References.		407
10	**Piers, Walls, Buttresses and Towers.**		409
	10.1	Introduction	409
	10.2	Piers	410
		10.2.1 Strength of Masonry Piers Under Eccentric Axial Loads: Mechanical Aspects of the Problem	410
		10.2.2 Differential Equation of the Inflexion of an Eccentrically Loaded Cracked Pier	411
		10.2.3 Collapse Load	416
		10.2.4 Pier Strength with Variable Load Eccentricity.	418
		10.2.5 Influence of the Pier Weight.	419
		10.2.6 The Use of Nonlinear Programs in Stability Analysis of Piers	420
		10.2.7 Influence of Mortar Creep on the Behavior of an Eccentrically Loaded Pier	422
	10.3	Building Walls.	426
		10.3.1 Introductory Remarks	426
		10.3.2 Crack Patterns in Buildings Under Vertical Loads.	428
		10.3.3 Stresses Due to Vertical Loads	430

	10.4	Buttresses	439
		10.4.1 Geometry of the Detachment Inclined Crack (Ochsendorf)	440
		10.4.2 Buttress Side Strength	441
	10.5	Towers	445
		10.5.1 Introductory Remarks	445
		10.5.2 Crack Patterns in Masonry Towers	446
		10.5.3 Plastic Model of the Tower Foundation	448
		10.5.4 Stability of Leaning Towers	454
		10.5.5 Counter Weights to Stabilize Leaning	458
		10.5.6 Evolution of Tower Tiling	459
		10.5.7 The Leaning Tower of Pisa	466
		10.5.8 Stability of Other Leaning Towers	469
		10.5.9 Cracking of Leaning Towers. Heyman Collapse Analysis	469
	Appendix		477
	References		479
11	**Gothic Cathedrals**		**481**
	11.1	Introduction and Some Historical Notes	481
	11.2	Brief Notes on the Construction Techniques	482
	11.3	Relevant Static Problems	487
	11.4	Transverse Wind Strength	487
		11.4.1 Wind Action on the Transverse Segment of the Cathedral: The Assumed Typology of the Cathedral of Notre Dame D0Amiens	487
		11.4.2 Dead Loads: Vertical Forces and Horizontal Vault Thrusts	492
		11.4.3 Valuation of the Lateral Wind Strength of the Cathedral: The Assumed Mechanisms	498
		11.4.4 Conclusion	513
	11.5	The Failure at Beauvais at the 1284	513
		11.5.1 Introductory Notes	513
		11.5.2 Thrust of the Cross Vault Spanning the Choir	517
		11.5.3 Loads Acting on the Piers	518
		11.5.4 Creep Buckling of the Piers	521
		11.5.5 Conclusions	523
	References		524

12	**Masonry Buildings Under Seismic Actions**	525
	12.1 Introduction	525
	12.2 The Masonry Panel Under Horizontal Forces: Strength and Ductility	527
	12.2.1 Limit Strength in the Framework of the Rigid in Compression, no-Tension Model..............	527
	12.2.2 Panel Side Strength via Plastic Analysis...........	533
	12.2.3 Panel Ductility in the Framework of the Elastic-Plastic Model	536
	12.2.4 Geometrical Ductility in the Framework of the no-Tension Panel Model.................	538
	12.2.5 The Problem of the Comparison with Test Results ...	541
	12.2.6 Behavior of the Panel Under Alternating Thrust Actions	542
	12.3 Some Recalls of Earthquake Engineering: The Elastic and the Elasto-Plastic Simple Oscillator.................	543
	12.3.1 The Elastic Bilinear Oscillator Representative of the Masonry Behavior....................	547
	12.3.2 The Structure Coefficient Q for Masonry Buildings...	549
	12.4 Seismic Resistant Structure of Masonry Buildings: Active and Inactive Walls	551
	12.5 Distribution of Seismic Forces Along the Height...........	552
	12.6 Design for Seismic Loadings	555
	12.7 Seismic Failure Modes...............................	557
	12.8 Out-of-Plane Strength of Inactive Walls..................	558
	12.8.1 Unreinforced Wall. Collapse of the First Mode......	558
	12.8.2 Walls Reinforced Against Out of Plane Actions	564
	12.8.3 Force Transmission from Inactive to Active Walls....	564
	12.8.4 Out of Plane Limit Strength of Fastened Walls	567
	12.9 In-Plane Strength of Multi-storey Masonry Walls with Openings	574
	12.9.1 The Different Models of Walls	574
	12.9.2 Seismic Strength of Multi-storey Masonry Walls with Openings and Large Architraves	575
	12.9.3 Seismic Strength of the Multi-storey Wall with Openings and Thin Architraves	592
	References......................................	617
Erratum to: Masonry Vaults: Cross and Cloister Vaults		E1

Chapter 1
Masonry Strength and Deformability

Abstract This chapter deals with the strength and deformability of masonry materials composing the structure of the so called historic constructions. After some historical introductory notes, special attention has been given to the analysis of various strength features of these materials and of their components, as bricks, stone blocks, and mortars. The common peculiarity of all the stone materials, a strength in tension much lower than in compression, is analyzed in detail and a suitable tri-axial failure criterion is thoroughly discussed. These results are then applied to the strength evaluation of uniaxial compression strength of the masonry, composed by regular patterns of blocks and mortar courses, as function of the geometry and strength properties of its components. The study of the masonry deformations, both the instantaneous as the delayed, ends the chapter.

1.1 Brief Notes on the History of Masonry Constructions

Masonry constructions, whose oldest examples date back to about 8000 years ago, developed during the beginnings of the earliest urban civilizations, when more ancient techniques employing building materials such as wood, straw and hides were gradually replaced by more advanced technologies, enabling the construction of stronger, longer-lasting structures.

Initially, masonry walls were built by setting large rough-hewn stones one on the other, dry, without mortar, to form so-called Cyclopean masonry. During the Classic Age regularly shaped stone blocks with smooth outer faces were used to build walls or piers, still without the use of mortar. This technique was utilized in the construction of many of the temples of the Athens Acropolis and later the Roman Colosseum. Because of the scarcity of suitably hard rocks in the Mesopotamian area, the societies there developed techniques to produce artificial building blocks. Initially, bricks were sun-baked, friable and unreliable over time. The use of kilns to harden the clay developed later. This allowed producing more resistant elements—fired bricks,—a technique still in widespread use today.

The use of binders, substances that set and harden, in masonry construction is also an ancient technique. Over the course of history, various materials have been used as binders. The first mortars were made by mixing mud and clay. Then, the ancient Egyptians added gypsum as binder, while the ancient Persians used bitumen. The discovery of lime by the Etruscans was the last fundamental turning point in the evolution of masonry. It was discovered that limestone, when burnt and combined with water, produced the lime that would harden with age. The mixture of lime with pozzolana, a volcanic ash that reacts with calcium hydroxide in the presence of water, improved the quality of mortars, which would set under water. Historically, constructions with pozzolanic mortar first appeared in Greece, though it was the Romans (Choisy 1873; Giuliani 1995) who developed this technique to its full potential. Over time, they defined a number of different types of *opus* (literally 'work') used at different times in different structures. These *opera* were remarkable for the construction procedures used and the different geometries of the masonry patterns achievable:

- *opus caementicium*: a construction technique using aggregates, water and a binding agent. The aggregate, rubble of broken fragments of uncut stones or fist-sized tuff blocks (*caementa*), was mixed to lime and pozzolana mortar (Choisy 1873);
- *opus incertum*, a crude masonry made up of irregularly shaped, uncut (or 'undressed') stones randomly inserted into a core of *opus caementicium*;
- *opus quadratum*, facings built with cut stone blocks laid in regular horizontal courses;
- *opus testaceum*, or *latericium*, brick-faced masonry with kiln-backed bricks, which prevailed throughout the Imperial Age;
- *opus reticulatum*, a Roman decorative design using small square slabs of stone or small bricks embedded into a regular, tightly knit diamond pattern;
- *opus mixtum*, masonry of reticulated material reinforced and/or intersected by brick bands or interlocked with bricks;
- opus *vittatum,* oblong (occasionally square) tuff blocks intersected by one or more brick bands at more or less regular intervals.

Typical Roman masonry walls were usually quite thick and made up of an inner rubble core of *opus caementicium* and two outer facings. In particular, a wall, or pier, made with *opus quadratum* had facings of large bricks placed along horizontal courses. In the Imperial Age, brick facings were built using square-shaped bricks (*opus testaceum*), as in the Baths of Diocletian. Wall facings were otherwise built using *opus reticulatum, opus vittatum* or *opus testaceum* (Fig. 1.1).

Dead loads tend to pull the walls horizontally apart, causing vertical cracking. Roman masons deviced a method to connect the facings and the inner rubble core. They sawed square bricks diagonally and laid these triangular half-bricks in the core with their hypotenuses outward to create a toothed bonding surface to the facings. Figure 1.1 illustrates this technique, showing the structure of a wall built *with opus testaceum*. Although a large variety of bricks were produced, they came in three main sizes:

1.1 Brief Notes on the History of Masonry Constructions

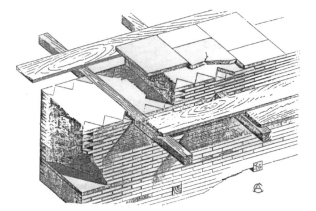

Fig. 1.1 An *opus testaceum* masonry wall (from Choisy 1873)

- *bessales*, 8 in. (19.7 cm) square;
- *sesquipedales*, 1.5 ft (44.4 cm) square;
- *bipedales*, 2 ft (59.2 cm) square.

Constructing a wall able to sustain loads and eventual settling of the foundation without severe damage was a difficult task. Greek architects first recognized the benefits of laying blocks with staggered vertical joints to achieve more compact walls (Fig. 1.2) (Giuffrè 1990).

This technique also defined the positioning of the bricks within the wall's thickness.

Initially, in walls laid according to Etruscan methods, some discontinuities occurred along the courses and some blocks had to be shaped differently from the others, as can be seen in some examples of walls built in ancient Etruscan towns and later in Rome (underground reservoirs, terracing walls, and temple podiums) (Fig. 1.3a, c). The Greeks later solved this problem (Sparacio 1999) by laying blocks in alternating longitudinal and transverse rows (Fig. 1.3b). Finally, the Roman fashion, shown in Fig. 1.3d, enjoyed widespread application.

The disastrous economical conditions ensuing in Europe after the fall of the Roman Empire made it necessary for Romanesque builders to reduce transport costs and thus to use materials that were easily available locally, such as the marl from nearby quarries. Moreover, using small elements simplified loading and unloading and, at the same time, reduced the amount of mortar needed. It thus became very common to build walls with small-sized blocks of tuff or bricks and mortar. Such simple building procedures continued throughout the Middle Ages and into the Renaissance (Morabito 2004).

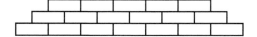

Fig. 1.2 Isodomic pattern of a masonry facing

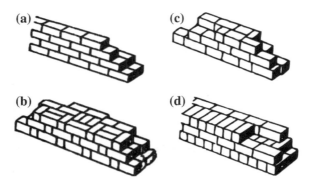

Fig. 1.3 First Greek and Romans patterns of the masonry texture: **a** isodomic Greek system; **b** Greek system with alternating stretchers and headers; **c** archaic Roman system; **d** Roman system with courses of stretchers and headers

1.2 The Masonry of Historic Buildings

A wide variety of types of masonry are present in historic buildings. Except for low-cost housing, whose walls were usually built with stone rubble, historic masonry is composed of mortar-cemented parallelepiped-shaped elements, usually bricks, whose standard size is 5.5 cm × 12 cm × 25 cm, though other types of elements are also common. Thus, according to the elements used, masonry can be subdivided in:

- *regular brickwork*: constructed with brick elements laid with mortar in horizontal courses with staggered vertical joints (Fig. 1.4).
  In this arrangement the bricks are named according to their placement in the wall. *A stretcher* is a brick laid horizontally flat, with its long side exposed on the outer face of the wall. *A header* is a brick laid flat across the wall's width with its short end exposed. Bricks may be laid in a variety of patterns, or bonds, of alternating headers and stretchers. Thinner walls are made using a stretcher bond, also known as a running bond, with stretchers forming the entire thickness of the wall, i.e. 5.5 cm (excluding the plaster or stucco facing). Others walls are constructed with a single row of stretchers, so that the wall is as thick as the brick head, 12 cm. There are many other types of bonds that use two or three headers in different alternating configurations with stretchers.
- *regular brickwork with squared stone blocks*: built with tuff blocks bound by horizontal mortar and vertically staggered joints, as in *regular brickwork* (Fig. 1.5). Thick walls may present an inner and outer tuff facing over an internal rubble core.

1.2 The Masonry of Historic Buildings

Fig. 1.4 Regular brickwork

Fig. 1.5 Masonry built with tuff blocks

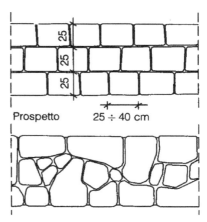

- *brickwork with mixed stone and brick*: come in two different types. In the first, called *edged masonry,* the bricks are arranged in horizontal courses along the entire thickness of the wall at varying distances (80–160 cm) between the stone masonry. In the second, *mixed masonry with bricks,* single bricks are laid in various places to level the stone planes (Fig. 1.6).
- *ordinary brickwork with huddled stones*: obtained by mortaring irregularly shaped elements, such as chunks of bricks or stones, along roughly horizontal planes in such a way as to reduce the spaces between them (Fig. 1.7). Such masonry, used frequently to build homes in small historical communities in southern Italy, are particular vulnerable to earthquakes.

The Italian Department of Civil Defense (2002) has issued its own classification of masonry, according to which there are five classes, each with a number of subclasses:

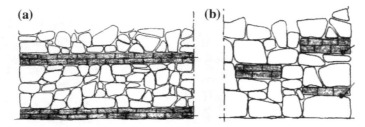

Fig. 1.6 Two examples of masonry with a mix of stones and bricks. **a** edged masonry; **b** mixed masonry with bricks

Fig. 1.7 Masonry built with huddled stones and mortar

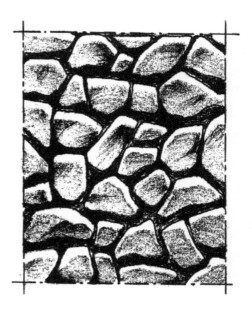

(a1) *rounded stone masonry:* built with small- or medium-sized cobblestones laid randomly or in ordered patterns (i.e. bonds):

- with neither courses nor regular bonding pattern;
- without courses, but presenting an orderly bonding pattern;
- stones with brick courses;

(a2) *rough stone masonry*: generally built with irregularly-shaped, undressed elements of varying sizes, such as chunks of brick or stone:

- with neither stone courses nor regular bonding pattern;
- without stone courses, but with an orderly bonding pattern;
- coursed with flat interlocking tiles and stones;
- with brick courses;

1.2 The Masonry of Historic Buildings

(b1) *masonry with ribbon-like stones:* built with rocks that tend to split along horizontal planes:

- without courses;
- with courses;

(b2) *semi-regular masonries*: built with semi-finished medium-sized elements:

- semi-finished limestone in courses;
- semi-finished limestone without courses;

(c1) *squared stone masonry*: made of "dressed" or worked stones, also called *ashlars*.

- tuff ashlars without courses;
- tuff ashlars with courses.

1.3 Compression Strength of Brick and Stone Elements

1.3.1 Bricks

Bricks, or masonry units, are made of clay, shale, soft slate, and calcium silicate. Bricks are generally manufactured by extrusion. Masonry units come in standard sizes of: 5.5 cm × 12 cm × 25 cm. The compression strength of fired bricks is about 200–250 kg/cm^2. However, when poorly fired, bricks may exhibit severely reduced compression strength, as low as 50 kg/cm^2. Standard compression tests are performed by cutting a brick in half and then gluing the two parts together with cement paste. These glued interfaces reproduce the effects of the mortar joints present in masonry. Four wet and four dry samples are then placed under platens of a so-called 'universal testing machine', which applies a compression load at a preset rate. The standard compression strength of a unit is obtained via the relation

$$f_b = f_{bm}(1 - 1.64\delta) \tag{1.1}$$

where f_{bm} is the mean strength of the three most consistent results, and $\delta = s/f_{bm}$ is the variation factor, with s the root-mean-square deviation. The tensile strength is assumed to be equal to about 1/10 the compression strength. Another compression test is performed by placing a single prism-shaped brick specimen directly between the platens of the universal testing machine and evaluating the corresponding compression strength. The failure pattern in this case is the so-called *hourglass* mode, typical of the failure of concrete specimens. Alternatively, before the test, the platens of the machine are treated with wax or stearic acid. In this case, the specimen, which can freely expand laterally during the test, breaks through vertical cracking under lower compression stresses.

1.3.2 Stone Blocks

As discussed above, squared elements, hewn from stone quarried in many different sites, have been used in masonry constructions for centuries. The mechanical features of these stone elements thus depend heavily on the source of the rock.

1.3.2.1 Strength of Stone Materials

Table 1.1 shows a classification of stones into five types according to the compression strength of undamaged rock samples. Few rocks belong to *class* A, the most notable examples being quartzite and basalt. *Class* B includes most magmatic rocks, the more resistant metamorphic rocks and few sedimentary rocks, as well as most lime stones and dolomites.

Class C comprises many argillites, marls and metamorphic rocks, such as shale. *Classes D* and *E* include many porous rocks, such as brittle sandstones, tuff, halite, etc. Another, simpler classification subdivides rocks into *soft, medium hard* and *hard*. Tuff, of both volcanic and sedimentary origins, are *soft* rocks. Sandstone, limestone, travertine are *medium-hard* rocks. Dolostone, trachyte, porphyry, gneiss, granite, basalt are classified as *hard*.

Table 1.2 gives the corresponding values of the uniaxial compression strengths f_c and the elastic modulus E_e, this latter measured as the tangent modulus on the σ–ε diagram at 50 % compression strength. Rocks used in constructions are mainly those designated as B, C, D, E. For instance, travertine was used to build the piers and perimeter arcades of the Rome Coliseum. The Milan cathedral was instead built of hewn marble blocks from quarries near Lake Maggiore in northern Italy. Hard sandstone was the main building material used for many Gothic cathedrals, and tuff is widespread in many types of historic architecture.

1.3.2.2 Tuff Blocks

The term tuff derives from the Latin name, *tuphos,* which was originally used to indicate both a pyroclastic rock formed by slow consolidation of volcanic materials,

Table 1.1 Classification of stones according to compression strength f_c

Class	Strength	fc [MPa]
A	Very high	>225
B	High	225 ÷ 112
C	Mean	112 ÷ 56
D	Low	56 ÷ 28
E	Very low	<28

1.3 Compression Strength of Brick and Stone Elements

Table 1.2 Density, elastic modulus and compression strength of some rocks

	Density (g/cm^3)	Compression strength (kg/cm^2)	Elastic Modulus (kg/cm^2 × 10^5)
Igneous rocks			
Granite, syenite	2.6–2.8	1600–2400	5–6
Diorite, gabbroid	2.8–3.0	1700–3000	8–10
Porphyry, quartz	2.6–2.8	1800–3000	5–7
Basalt	2.9–3.0	2000–4000	9–12
Pumice	0.5–1.1	50–200	1–3
Sedimentary Rocks			
Soft limestone	1.7–0.6	200–900	3–6
Compact limestone	2.7–2.9	800–1900	4–7
Dolomite	2.3–2.8	200–600	2–5
Metamorphic Rocks			
Gneiss	2.6–3.0	1600–2800	3–4
Shale	2.7–2.8	900–1000	2–6
Marble	2.7–2.8	1000–1800	4–7
Quartzite	2.6–2.7	1500–3000	5–7

such as lapillus, ash and sand, as well as sedimentary rocks, such as *Apulia* tuff. Actually, tuff properly refers only to the volcanic rock types, while the term *tufa* should be reserved for the sedimentary type. Both are considered soft rocks and were used without distinction in historic buildings.

Tuff is frequently used in constructions because it is light and soft, and therefore easily worked. It is also quite porous, and thus its density is low compared to other rock materials, such as limestones, shales and so forth, though it nonetheless offers fairly high compression strength. Standard tuff block dimensions are 30 cm × 40 cm × 13 cm. Thus, building with tuff blocks yields wall thicknesses ranging from 30 to 40 cm, or multiples thereof. Tuff blocks can also be laid together with bricks because their bases are about the same size (13 cm). Some mechanical parameters of tuff:

- Poisson's ratio: $v = 0{,}15$;
- Elastic modulus: 30,000–150,000 kg/cm^2;
- Unit weight: (volcanic tuff) 1,100–1,700 kg/m^3;
- Compression strength: $\sim$40–50 kg/cm^2.
- Tensile strength: $\sim$1/15 of compression strength.

1.4 Mortars

Mortar is a workable paste used to bind masonry blocks together and fill the gaps between them. Mortar becomes hard when it sets, resulting in a rigid aggregate structure. Modern mortars are typically made from a mixture of sand, a binder such as cement or lime, and water.

1.4.1 Binders

Binders used in mortar preparation are:

- gypsum;
- lime;
- hydraulic lime;
- cement.

Gypsum, the oldest binder, was used in the first Egyptian pyramids. It is present in alabaster, a decorative stone used in Ancient Egypt. It is obtained by baking the gypsum stone, made up of calcium sulfate, at a temperature of 110–200 °C, after which the stone turns to powder. Mixed with water, the powder hardens rapidly, though it has very low strength. Calcium oxide is the main component of lime. Lime is produced through a two-step process: firing and slaking. First the limestone is burnt at a temperature of 850–900 °C, to produce so-called *quicklime*, which is able to absorb large quantities of water. The quicklime is then combined with water and crushed into powder, giving rise to *slaked lime* or calcium hydroxide. The slaked lime is then used to produce either simple lime mortar, by mixing it with sand and more water, or hydraulic mortar, by mixing it with pozzolana. Simple lime will only set in contact with the air. Hydraulic limes, which will instead even set in water, are made from marly limestone or mixtures of limestone and clayey materials Cement is made by grinding together its main raw materials, which are (a) argillaceous, such as clay and shale, and (b) calcareous, such as limestone, chalk and marls. The mixture is then burnt in rotary kilns at temperatures between 1400 and 1500 °C to form clinkers. These are ground to a powder and mixed with gypsum to create the gray flour-like substance known as cement. When water is added to cement, a chemical process occurs as it hydrates, allowing it to harden anywhere, even under water. Cement, patented in 1824 by Joseph Aspdin in the UK, was called *Portland* because this artificial stone resembled the Portland stone. As cement began to be used only towards the end of the 19th century, cement mortars are generally not found in the masonry of historic buildings.

1.4.2 Aggregates

Aggregates are classed as fine or coarse. Sands are used as fine aggregates, while gravel or crushed rocks represent coarse aggregates. Sand, whose grain dimensions range from 0.5–1 mm, is generally used to prepare masonry mortars. Sand is the mineral skeleton of the mortar: it increases the volume of the paste and facilitates penetration of carbon dioxide within the mixture to improve setting. Moreover, sand reduces shrinkage and the consequent cracking that may occur during setting and hardening of the paste. Romans used *Caementa*, irregular pieces of stone or brick, as aggregate in preparing *opus caementicium* masonry.

1.4.3 Mortars of Lime

Mixtures of lime, water and sand form the mortar paste, which sets and hardens.

1.4.3.1 Roman Mortars

Roman mortar contained pozzolana, a volcanic ash that added a useful property lacking in the simple lime mortars used by the Greeks: *hydraulicity*, that is, the ability to set underwater. o the material called *pulvis puteolanus*, discovered in ancient times in the Bay of Naples. Pozzolana was also produced in the volcanic districts to the south of Rome, where it was termed *harena fossicia*, a volcanic sand with similar water-setting features to *pulvis puteolanus*, though it was less effective in practice than this latter. Table 1.3 gives the compositions of some Roman mortars quoted in Vitruvius' treatise *De Architectura*. Table 1.3 also gives the proportions for producing *cocciopesto*, or *opus signinum*, a mortar made with crushed terra-cotta. *Cocciopesto* is the material most commonly used to line cisterns and to protect the extrados of vaults exposed to the elements.

Table 1.3 Roman mortars (Adam 1988)

Binder	Aggregates	Water (%)
1 part lime	3 parts *harena fossicia* (Vitruvius, II, V, 5)	15–20
1 part lime	2 parts *river sand* (Vitruvius, II, 5,6)	15–20
1 part lime	1 part *terra cotta* (Vitruvius, II, V, 7)	15–20
1 part lime	2 parts *pulvis puteolanus* (Vitruvius, V, XII, 8–9, sea works)	15–20

Table 1.4 Composition of mortars according to Italian building codes

Class	Mortar	Cement	Simple lime	Hydraulic lime	Sand	Pozzolana
M_4	Hydraulic	–	–	1	3	–
M_4	Pozzolanic	–	1	–	–	3
M_4	Composite	1	–	2	9	–
M_3	Composite	1	–	1	5	–
M_2	Cementitious	1	–	0.5	4	–
M_1	Cementitious	1	–	–	3	–

1.4.3.2 Mortars of Historic Masonries

The mortars present in the masonry of historic buildings are as a rule composed of simple or hydraulic limes. They can be subdivided in:

- simple mortars;
- hydraulic mortars;
- composite mortars.

Italian building codes provide for dividing mortars into the types: *cementitious*, classified as M_1 and M_2, according to the cement content; *composite*, indicated as M_3, containing both lime and cement; and *hydraulic* or *pozzolanic*, indicated as M_4, containing only hydraulic lime or lime and pozzolana. For instance, a mix designated as M_1 has 3 parts sand by volume and 1 part cement, while an M_4 mix has 3 parts sand by volume and 1 part hydraulic lime (Table 1.4).

Simple limes were in widespread use in the past because of the efficiency and easy workability of quicklime. They harden slowly and weaken in the air. Their compression strength is very low, about 5 kg/cm^2.

Hydraulic mortars are prepared with mixtures of hydraulic limes, water and sand. The standard composition of a hydraulic mortar is given in Table 1.4: three parts sand and 1 part hydraulic lime by volume. The compression strength of hydraulic mortar is about 25 kg/cm^2, lower than that of composite mortars (about 50 kg/cm^2), and much lower than that of cementitious mortars (at least 120 kg/cm^2). As a rule, the strength of mortar is less than that of concrete.

1.5 Tests on Rock and Mortar Specimens

The most common experiments carried out on specimens of rocks or mortars, are:

(a) uniaxial tension and compression tests
(b) multi-axial tests
(c) torsion tests;
(d) flexural tests.

1.5 Tests on Rock and Mortar Specimens

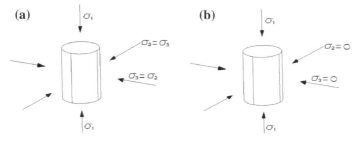

Fig. 1.8 Triaxial tests: **a b** uniaxial test

1.5.1 Tests on Rock Specimens

Cylindrical specimens are used. In multi-axial testing, three stresses (σ_1, σ_2, σ_3) and three strains (ε_1, ε_2, ε_3) are measured. Usually, σ_1 is the major principal stress, generally vertical, and $\sigma_2 = \sigma_3$ the intermediate lateral stresses (Fig. 1.8).

In a uniaxial tension test, a tensile stress σ_1 is applied by means of pincers or a metal plate glued to the specimen. There are also indirect splitting tests, such as the so-called Brazilian and Flexural tests. In a standard uniaxial compression test a load is applied to a cylindrical specimen by means of steel loading platens. The friction strength between the rigid platens of the testing machine and the specimen heads prevents lateral expansion of the specimen. Shears (Fig. 1.9) are thus superimposed on the vertical compression causing a three-dimensional stress state leading to cracks splitting the specimen diagonally). Figure 1.10 shows the typical *hourglass*

Fig. 1.9 Shear friction stresses

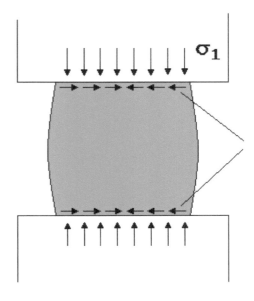

Fig. 1.10 Hourglass failure

failure mode of a concrete specimen obtained through a standard compression test. Mortar specimens exhibit the same behavior.

Figure 1.11 presents the results obtained by Brown testing marble prisms under biaxial compression at a constant ratio σ_2/σ_1. The strengths in the diagram are presented as ratios σ_2/σ_c and σ_1/σ_c, where σ_c is the corresponding uniaxial

Fig. 1.11 Biaxial compression tests on marble prisms (Brown 1974)

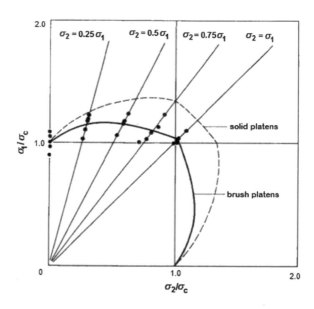

compression strength. The presence of the lateral compression σ_2 yields an increase in strength of no greater than 15 %.

The effects of the loading conditions are significant. Figure 1.12 shows the biaxial failure domain of concrete according to the test results obtained by Kupfer (1973). These results show that the behavior of concrete is similar to the marble specimen tested by Brown. These results will be reconsidered in the following sections.

1.5.2 Uniaxial Compression Tests on Mortar Specimens

Italian regulations call for measuring compression strength by testing three prismatic specimens of dimensions 40 × 40 × 160 mm. The three mortar specimens are first cast in metal molds from which they are removed after 24 h and cured in a humid environment at a constant temperature of 20 °C. A specimen is then placed on side supports and loaded with a central point load until bending failure is reached. The bending failure stress is evaluated simply as

$$f_{mf} = \frac{3}{2}\frac{PL}{b^3} \qquad (1.2)$$

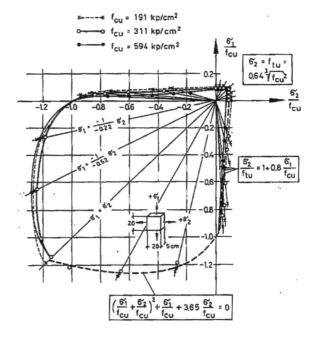

Fig. 1.12 The biaxial failure domain of concrete (Kupfer 1973)

where P is the applied collapse load, $L = 100$ mm the distance between the supports and b the length of the side of the specimen's square cross section. Six simple compression tests are then performed on the remaining half prisms and the average compression strength $f_m = Q/b^2$ is obtained from the failure force Q.

The compression strength of mortar is quite low: M_1 cement mortar has a strength of about 120 kg/cm^2; while the strength of M_4 lime mortar does not exceed 20 kg/cm^2. The strength of the different types of mortars used in historic masonries can be considered similar to that of M_4 type mortar, or lower.

1.5.3 Stress Strain Diagrams for Stone and Mortar Materials

Many rocks, when loaded in uniaxial compression, exhibit the typical load—deformation response plotted in Fig. 1.13.

An ascending branch is followed by a softening one. The peak represents the compression strength of the rock. The initial segment of the ascending branch is more or less straight up to a stress level equal to about 60 % peak stress. The slope of the diagram at the origin measures the rock's Young's modulus. A stiff testing machine is necessary to trace the full extent of the descending branch of the stress-strain curve.

Stress-strain diagrams are heavily influenced by the test conditions. In triaxial tests three conditions are of paramount importance:

(a) the cell pressure value;
(b) the temperature;
(c) the load application velocity.

Figure 1.14 shows the influence of the cell pressure $\sigma_2 = \sigma_3$ on the stress-strain diagram of a rock specimen. Increasing the cell pressure increases both the compression strength and ductility of the material.

Fig. 1.13 A typical compression stress-strain diagram for rock material

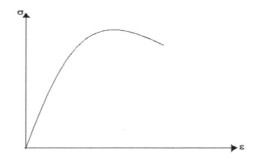

Fig. 1.14 σ–ε diagrams of a limestone in triaxial tests with various cell pressures (Brady and Brown 2004)

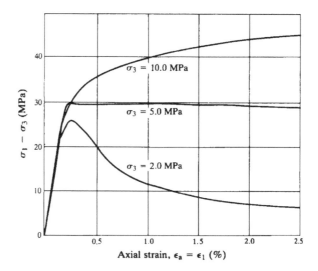

1.6 A Triaxial Failure Criterion for Stone Materials

1.6.1 Preliminary Considerations

There are many failure criteria for stone materials (Bazant and Jirasek 2002). These criteria, generally, adapt the Plasticity theory to fit, more or less, the properties of the experimentally determined failure surfaces, without to get the heart of the problem. In the next pages, on the contrary, we will present a simple criterion, (Como and Luciano 2006, 2007) founded on very different assumptions, able to describe the basic aspects of the question. Stone, together with brick, is the basic material of masonry constructions. Knowing the strength behavior of stone materials is thus essential to understanding the statics of masonry structures.

The strength of stone materials exhibits peculiar aspects, very different from those characterizing the strength of metal materials, whose behavior is ductile and is controlled by shear stresses. Consequently, the tensile and compression strengths of steel are equal.

Stone materials, on the contrary, exhibit brittle behavior and very different compression and tensile strengths. Figure 1.15 shows the fracture lines that occur in three different compressed specimens: brick, concrete and tuff. The specimens heads were treated with stearic acid, that is, more or less common wax, to eliminate friction with the platens of the test machine. Specimens are thus free to expand laterally during the test. *Vertical cracks* mark the onset of the collapse of all three specimens—a failure mode quite different from the *hourglass* mode, shown in Fig. 1.10, that occurs in standard compression tests.

What is the cause of these vertical cracks in the crushing failure of stone material?

Fig. 1.15 Compression failures of brick, concrete and tuff specimens free to expand laterally

This question is not a trivial one, because under vertical compression, the stress acting along the vertical planes of the specimen is zero.

1.6.2 Porosity Effects: Micro-Macro Stress States

A physical explanation for this apparent paradox can be given by considering the effects of the natural porosity of all stone materials, following the approach of Como–Luciano in researching a failure criterion for concrete (2006). Small cavities, ranging 0.1–10 μm in size, are in fact present in all natural or artificial rock materials, as well as in hardened concrete and mortar. By way of illustration, Fig. 1.16 shows the porous structure of tuff.

Let us consider a specimen loaded by an uniaxial state of stress σ_z, for instance, a uniform compression or tensile stress applied by the test machine to the heads of the specimen, suitably treated to avoid friction.

Fig. 1.16 The porous structure of tuff

1.6 A Triaxial Failure Criterion for Stone Materials

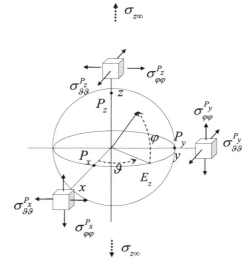

Fig. 1.17 Spherical cavity. Local stresses under the asymptotic stress σ_z

Let us now consider a generic small cavity in the material, represented to a first approximation by an ideal spherical ball: the applied stress σ_z will act at a large distance from the cavity and can be considered an *asymptotic* stress $\sigma_{z\infty}$ for the microscopic *local* stress state occurring around the cavity (Fig. 1.17).

The local stress state around the small spherical cavity can be easily evaluated (McClintock and Argon 1966). By assuming positive tensile stresses, we obtain:

- at the pole P_z:

$$\sigma_{RR}^{P_z} = 0 \quad \sigma_{\vartheta\vartheta}^{P_z} = \sigma_{\varphi\varphi}^{P_z} = A(v) \cdot \sigma_{z\infty} \quad (1.3)$$

- along the equatorial circle E_z:

$$\sigma_{RR}^{E_z} = 0; \quad \sigma_{\varphi\varphi}^{E_z} = B(v) \cdot \sigma_{z\infty}; \quad \sigma_{\vartheta\vartheta}^{E_z} = C(v) \cdot \sigma_{z\infty} \quad (1.3')$$

where the quantities $A(v)$, $B(v)$, $C(v)$, depend on the Poisson's ratio of the material and are given by

$$A(v) = -\frac{3+15v}{2(7-5v)}; \quad B(v) = \frac{27-15v}{2(7-5v)}; \quad C(v) = \frac{15v-3}{2(7-5v)} \quad (1.4)$$

The values of these coefficients, for v respectively equal to 0.20, 0.25, 0.30, are:

$A(v = 0.20) = -0.500;\quad B(v = 0.20) = 2.022;\quad C(v = 0.20) = 0$
$A(v = 0.25) = -0.587;\quad B(v = 0.25) = 2.022;\quad C(v = 0.25) = 0.065$
$A(v = 0.30) = -0.682;\quad B(v = 0.30) = 2.045;\quad C(v = 0.30) = 0.136$

Pores, on the other hand, have *irregular* shapes. The assumed spherical form of the pore cannot fully describe the strong stress concentration occurring around the cavity due to stress $\sigma_{z\infty}$ applied at a large distance from it. To take into account the effects of these irregularities it is useful to compare the local stresses occurring around a small circular or elliptical hole in a panel stretched by an uniform tensile stress, s_y, orthogonal to the major axis of the ellipse. Stress s_y can be considered asymptotic with respect to the local stress field around the small cavity.

If we compare the stress fields around the two different holes, the stress components, σ_x, at the top of the cavities will be the same. On the contrary, the stress component acting around the elliptical hole in the same direction as the applied stress, s_y, increases greatly in the neighborhood of the intersection of the ellipse with its major axis, where it becomes $k_B \sigma_y$, with $k_B \sigma_y$, with $k_B \gg 1$, where σ_y is the stress component occurring at the same point but around the circular hole. Factor k_B is an *amplification factor* that depends on the ratio between the lengths of the ellipse axes (Fig. 1.18).

These results suggest an approximate approach to describing the local stress states occurring around any of the small irregular cavities scattered throughout the specimen. Any generic stress state applied to a porous specimen has components s_x, s_y, s_z, These stresses, acting at a large distance from the cavity, are asymptotic for the local stress field around the cavity and can be indicated as $\sigma_{x\infty}$, $\sigma_{y\infty}$, $\sigma_{z\infty}$.

The corresponding local stress state can be obtained by summing up the effects of the three asymptotic stresses $\sigma_{z\infty} = s_z$. $\sigma_{y\infty} = s_y$, $\sigma_{x\infty} = s_x$. Thus, in order to account for the effects of the geometrical irregularities, we can suitably modify the expressions for the stress components around the spherical cavity in light of the foregoing comparison of the stress fields around the circular and elliptical holes.

To this end, we can assume, by way of approximation, that only the asymptotic stresses acting along the *same* direction as each of the applied asymptotic stresses

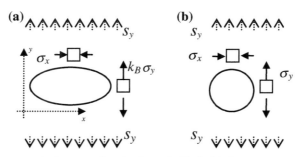

Fig. 1.18 Stress concentration around a circular and elliptical hole

1.6 A Triaxial Failure Criterion for Stone Materials

$\sigma_{x\infty}$, $\sigma_{y\infty}$, $\sigma_{z\infty}$ will be amplified by the *amplification factor* k_B with respect to the stress components occurring around the spherical cavity. Therefore, for instance, the asymptotic stress acting along the x axis will produce stress amplification of the local component directed along the same direction x. Likewise, the asymptotic stresses s_y acting along y will in turn produce stress amplification of the local stress component along the axis y. Consequently, by taking expressions (1.3) and (1.3′) into account, the local stress state occurring at the poles of an irregular pseudo-spherical cavity can be taken as:

$$\begin{aligned}
\sigma_{\phi\phi}^{P_z} &= A(v)\sigma_{z\infty} + k_B B(v)\sigma_{y\infty} + C(v)\sigma_{x\infty} \\
\sigma_{\theta\theta}^{P_z} &= A(v)\sigma_{z\infty} + C(v)\sigma_{y\infty} + k_B B(v)\sigma_{x\infty} \\
\sigma_{\phi\phi}^{P_y} &= k_B B(v)\sigma_{z\infty} + A(v)\sigma_{y\infty} + C(v)\sigma_{x\infty} \\
\sigma_{\theta\theta}^{P_y} &= C(v)\sigma_{z\infty} + A(v)\sigma_{y\infty} + k_B B(v)\sigma_{x\infty} \\
\sigma_{\phi\phi}^{P_x} &= k_B B(v)\sigma_{z\infty} + C(v)\sigma_{y\infty} + A(v)\sigma_{x\infty} \\
\sigma_{\theta\theta}^{P_x} &= C(v)\sigma_{z\infty} + k_B B(v)\sigma_{y\infty} + A(v)\sigma_{x\infty}
\end{aligned} \quad (1.5)$$

1.6.3 Micro-Macro Failure Condition: Reasons of the Different Tensile and Compression Strengths

We assume that local failure occurs when, according to the Rankine criterion, the *maximum tensile stress* around the cavity reaches the local tensile strength $f_{rt,loc}$ of the material. Thus, if $\boldsymbol{\sigma}$ indicates the local stress state around the cavity, we have:

$$\max_{\text{tensile}} \boldsymbol{\sigma} = f_{rt,loc} \quad (1.6)$$

Condition (1.6) is reached *simultaneously* around *all* the small cavities scattered throughout the material, so that material failure arises at the macroscopic level.

Let us now assume that the specimen is uniformly loaded by a *tensile stress* $s_z = t_z$ (Fig. 1.19). Consequently, the stress component

$$\sigma_{z\infty} = t_z \quad (1.7)$$

will act asymptotically at a large distance from the small cavities.

Thus from Eq. (1.5) the following stress state occurs around the pores:

$$\begin{aligned}
\sigma_{\phi\phi}^{P_z} &= A(v)t_z; & \sigma_{\theta\theta}^{P_z} &= A(v)t_z \\
\sigma_{\phi\phi}^{P_y} &= k_B B(v)t_z; & \sigma_{\theta\theta}^{P_y} &= C(v)t_z \\
\sigma_{\phi\phi}^{P_x} &= k_B B(v)t_z; & \sigma_{\theta\theta}^{P_x} &= C(v)t_z
\end{aligned} \quad (1.8)$$

Fig. 1.19 Local stresses around a spherical pore in a rock specimen under uniaxial tensile stress

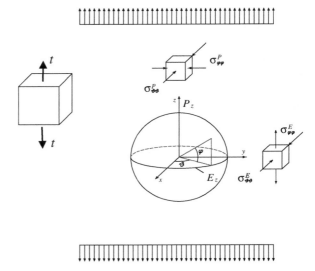

where the coefficients $A(v)$, $B(v)$, $C(v)$ are given by (1.4). The maximum tensile stresses occur at poles P_x and P_y and, in brief, around the equatorial circle E_z, taking into account the axial symmetry around the axis z. Cracks occur *orthogonally* to the directions of $\sigma^{E_z}_{\varphi\varphi}$, that is, in the direction orthogonal to the tensile stress applied to the specimen. Thus, from (1.9) the failure condition yields

$$f_{rt,loc} = \sigma^{E_z}_{\varphi\varphi} = \sigma^{P_x}_{\varphi\varphi} = \sigma^{P_y}_{\varphi\varphi} = k_B B(v) f_{rt} \tag{1.9}$$

At material failure, in fact, $t_z = f_{rt}$, where f_{rt} indicates the *macroscopic* tensile strength of the material.

Let us now consider the case of an applied macroscopic *uniform compression* (Fig. 1.20). This state of stress can be effectively applied to a specimen by treating its faces in contact with the testing machine platens with a frictionless substance. Thus, we now have

$$\sigma_z = -c_z \tag{1.10}$$

with $c_z > 0$. The local stress state at the poles of the cavity is now:

$$\begin{aligned}
\sigma^{P_z}_{\phi\phi} &= -A(v)c_z; & \sigma^{P_z}_{\theta\theta} &= -A(v)c_z \\
\sigma^{P_y}_{\phi\phi} &= -k_B B(v)c_z; & \sigma^{P_y}_{\theta\theta} &= -C(v)c_z \\
\sigma^{P_x}_{\phi\phi} &= -k_B B(v)c_z; & \sigma^{P_x}_{\theta\theta} &= -C(v)c_z
\end{aligned} \tag{1.11}$$

The coefficient $A(v)$ is negative and the stress components $\sigma^{P_z}_{\vartheta\vartheta} = \sigma^{P_z}_{\varphi\varphi}$ produce tension at pole P_z. On the other hand, compressions are acting around the equatorial

Fig. 1.20 Local stresses around a spherical pore in a rock specimen under uniaxial compressive stress

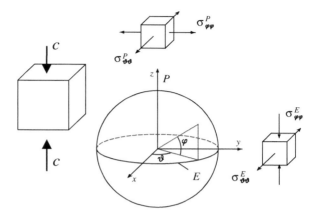

circle E_z. Material collapse is thus produced by the tensile stresses $\sigma_{\theta\theta}^{P_z} = -A(v)c_z$ at the pole P_z of the cavity, and cracks develop in the *same* direction as the applied compression. Upon compression failure the asymptotic stress c_z equals the material macroscopic compression strength f_{rc}. Thus,

$$c_z = f_{rc} \tag{1.12}$$

when the local tensile strength is reached around the pores, and we have

$$f_{rt,loc} = -A(v) \cdot f_{rc} \tag{1.13}$$

where f_{rc}, is the material's free-expansion macroscopic compression strength.

Taking into account the values attained by coefficient $A(v)$, it can be seen that the *local* tensile strength $f_{rt,loc}$ of the stone material, equal to only about *half* its macroscopic compression strength f_{rc}, is much higher than its macroscopic tensile strength f_{rt}. Figure 1.21 shows the different crack geometries predicted in compression and tensile failures.

Note that under equal intensities of the applied stress, the maximum local tensile stress occurring around the pores is much lower when the specimen is uniformly compressed rather than stretched. It follows that the intensity of the applied stress producing compression collapse of the specimen has to be much higher than the tensile stress required to produce tensile failure. Moreover, cracking will run parallel

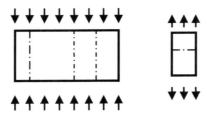

Fig. 1.21 Crack patterns in uniformly compressed or stretched specimens

to the applied compression, while it will be orthogonal to the applied tensile forces. This situation effectively occurs during specimen collapse, as illustrated in Fig. 1.15.

1.6.4 Valuation of the Pores Shape Irregularity Factor

A material's local tensile strength will be the same regardless of whether the specimen is under a compression or a tensile test. Thus, by comparing (1.12) and (1.16), the following *consistency condition* must hold

$$-A(v)f_{rc} = k_B B(v)f_{rt} \qquad (1.14)$$

from which we obtain the pore shape irregularity factor

$$k_B = -\frac{A(v)f_{rc}}{B(v)f_{rt}} \qquad (1.15)$$

The compression strength with free lateral expansion can be expressed as

$$f_{rc} = \gamma R_{rc} \qquad (1.16)$$

where R_{rc} indicates the *standard compression strength*, obtained without reducing the friction at the faces of the specimen. The compression strength of the material obtained in the compression test with free side expansion is less than R_{rc}. Thus, we can assume $\gamma \approx 0.90 \div 0.85$. Moreover, if we designate β as the ratio between the tensile and standard compression strengths, we have

$$f_{rt} = \beta R_{rc} \qquad (1.17)$$

where $\beta \approx 1/15$, and

$$f_{rt} = \frac{\beta}{\gamma} f_{rc} \qquad (1.18)$$

1.6 A Triaxial Failure Criterion for Stone Materials

and $f_{rc}/f_{rt} \approx 16 \div 17.5$. Simple calculations show that by assuming a value, $v = 0.25$, for the local Poisson's ratio, and using the foregoing values of coefficients β and γ, Eq. (1.15) yields an irregularity factor k_B of about 4.5–5.0.

1.6.5 Failure Interaction Domains

The interaction domains of stone material failure under biaxial or triaxial stress states can be obtained by applying the failure condition (1.9), obtained by Como and Luciano (2007) for concrete.

1.6.5.1 Biaxial Stress States

Compression–Compression

With refence to Figs. 1.22 and 1.23 the asymptotic stresses are

$$\sigma_{y\infty} = -c_y \quad 3\sigma_{z\infty} = -c_z \tag{1.19}$$

and the corresponding local stresses, according to (1.8), are

$$\begin{aligned}\sigma^{P_z}_{\theta\theta} &= -A(v)c_z - C(v)c_y & \sigma^{P_z}_{\phi\phi} &= -A(v)c_z - k_B B(v)c_y \\ \sigma^{P_y}_{\phi\phi} &= -k_B B(v)c_z - A(v)c_y & \sigma^{P_y}_{\theta\theta} &= -C(v)c_z - A(v)c_y\end{aligned} \tag{1.5'}$$

The two compressions c_y and c_z produce contrastino actions around the cavity. The most intense local tensiele stresses are the $\sigma^{P_z}_{\theta\theta}$ and $\sigma^{P_y}_{\theta\theta}$. The failure condition is thus reached when

$$f_{rt,loc} = -A(v)c_z - C(v)c_y \tag{1.20}$$

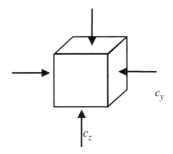

Fig. 1.22 The specimen of stone material under biaxial compression

Fig. 1.23 The local sterees for the specimen under nbiaxial compression

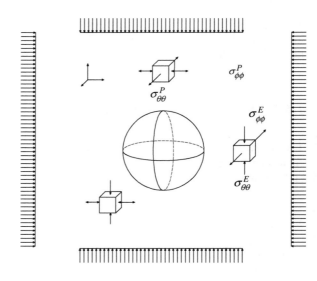

or

$$c_z = f_{rc} - \frac{C(v)}{A(v)} c_y \qquad (1.20')$$

In particularly, if the two compressions are equal to c, the corresponding biaxial strength is taking into account the (1.16),

$$f_{rc,b} = \frac{f_{rc}}{1 + \frac{C(v)}{A(v)}} \qquad (1.21)$$

The biaxial strength of the stone material is thus greater than the uniaxial compression strength.

Evaluation of the local Poisson ratio

According to (1.21) the ratio between the uniaxial and the biaxial compression strength of a stone material is

$$\frac{f_{rc}}{f_{rc,b}} = \frac{C(v)}{A(v)} + 1 \qquad (1.22)$$

The biaxial compression strength for a stone material, according to the numerous test results at disposal, can be obtained with good approximation as

$$f_{rc,b} \simeq 1.2 f_{rc} \qquad (1.23)$$

Consequently, from the (1.26) we can evaluate the corresponding value of the local Poisson ratio to satisfy (1.24). We obtain

1.6 A Triaxial Failure Criterion for Stone Materials

$$\nu \simeq 0.28 \quad (1.24)$$

The irregularity shape factor k_B, defined by the consistency condition (1.14), by taking the values of β and γ defined by (1.16) and (1.17) is about equal to 5.

Compression–Traction
In this case the asymptotic stresses are

$$\sigma_{y\infty} = t_y \quad \sigma_{z\infty} = -c_z \quad (1.25)$$

Likewise to the previous case the local stress component $\sigma_{\phi\phi}^{P_z}$ decides the failure by means of

$$\sigma_{\phi\phi}^{P_z} = -A(\nu)c_z + k_B B(\nu)t_y = f_{rt,loc} \quad (1.26)$$

from which we obtain the compression–traction failure condition

$$t_y = f_{rt} - \frac{f_{rt}}{f_{rc}} c_z \quad (1.27)$$

Traction–Traction
The high tensile stress occurring around the equatorial circle around the cavity, due to the traction is only barely weakened by the compression produced by the traction t_y. Thus the failure condtion is

$$\sigma_{\varphi\varphi}^{E_z} = A(\nu)t_y + k_B B(\nu)t_z = f_{rt,cp} \quad (1.28)$$

and we obtain the intraction failure equation

$$t_z = f_{rt} + \frac{f_{rt}}{f_{rt}} t_y \quad (1.29)$$

Particularly, if $t_y = t_z$, we have the following expression of the biaxial traction strength

$$t_{rt,b} = \frac{f_{rt}}{1 - \frac{f_{rt}}{f_{rc}}} \quad (1.30)$$

The biaxial tensile strength is a bit larger than the uniaxial tensile strength.

Biaxial Interaction Domain
Figure 1.24 shows the biaxial interaction domain in the plane σ_y, σ_z: its axis of symmetry is the bisector of the first and second quadrants f_{rt}.

The coordinates of point C define the material strength under uniform biaxial tension $f_{rt,b}$.

Fig. 1.24 The interaction biaxial failure domain (Como and Luciano 2006)

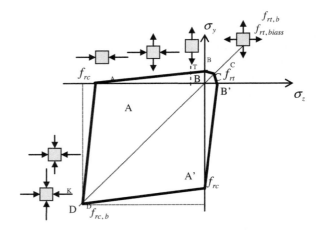

Triaxial Interaction Domain

To obtain the entire triaxial domain, we must take into account all six of the equations,

$$\begin{aligned}
\sigma^{P_z}_{\phi\phi} &= A(v)\sigma_{z\infty} + k_B B(v)\sigma_{y\infty} + C(v)\sigma_{x\infty} = f_{rt,loc} \\
\sigma^{P_z}_{\theta\theta} &= A(v)\sigma_{z\infty} + C(v)\sigma_{y\infty} + k_B B(v)\sigma_{x\infty} = f_{rt,loc} \\
\sigma^{P_y}_{\phi\phi} &= k_B B(v)\sigma_{z\infty} + A(v)\sigma_{y\infty} + C(v)\sigma_{x\infty} = f_{rt,loc} \\
\sigma^{P_y}_{\theta\theta} &= C(v)\sigma_{z\infty} + A(v)\sigma_{y\infty} + k_B B(v)\sigma_{x\infty} = f_{rt,loc} \\
\sigma^{P_x}_{\phi\phi} &= k_B B(v)\sigma_{z\infty} + C(v)\sigma_{y\infty} + A(v)\sigma_{x\infty} = f_{rt,loc} \\
\sigma^{P_x}_{\theta\theta} &= C(v)\sigma_{z\infty} + k_B B(v)\sigma_{y\infty} + A(v)\sigma_{x\infty} = f_{rt,loc}
\end{aligned} \qquad (1.31)$$

It can be obtained by equating each of the local stress components to the local material tensile strength $f_{rt,loc}$. Each of these equations represents the failure plane that separates the half-spaces of admissible stresses states from non-admissible ones in the space σ_x, σ_y, σ_z. The envelope of these planes defines a cone whose axis is concurrent with the hydrostatic axis and the vertex V with coordinates

$$V(f_{rt,triax}, f_{rt,triax}, f_{rt,triax}) \qquad (1.32)$$

where

$$f_{rt,triax} = \frac{f_{rt}}{1 - \frac{f_{rt}}{f_{rc,biax}}} \approx f_{rt}. \qquad (1.33)$$

The triaxial uniform tensile strength is only slightly larger than the uniaxial tensile strength. Figure 1.25 shows a view of the limit cone, while Fig. 1.26 shows

1.6 A Triaxial Failure Criterion for Stone Materials

Fig. 1.25 A view of the limit cone (Como and Luciano 2006)

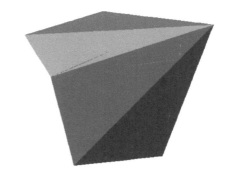

Fig. 1.26 The irregular hexagon: intersection of the limit surface with the plane orthogonal to the hydrostatic axis (Como and Luciano 2006)

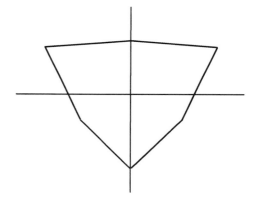

the irregular hexagon formed by the intersection of the cone with the plane orthogonal to the hydrostatic axis.

The two significant triaxial stress states for the stone materials are:

- the compression stress c_z, accompanied by two equal lateral compressions c
- the compression stress c_z, accompanied by two equal lateral tensile stresses t.

The failure conditions corresponding to these stress states, which relate the axial compression c_z to the lateral compression c, or the lateral tensile stress t (Fig. 1.27) are given by (Como and Luciano 2007):

$$c_z = f_{rc} + (1 - \frac{f_{rc}}{f_{rc,biax}} + \frac{f_{rc}}{f_{rt}})c \approx f_{rc} + \frac{f_{rc}}{f_{rt}}c \qquad (1.34)$$

$$c_z = f_{rc} - (1 - \frac{f_{rc}}{f_{rc,biax}} + \frac{f_{rc}}{f_{rt}})t \approx f_{rc} - \frac{f_{rc}}{f_{rt}}t \qquad (1.35)$$

In the following section, we will apply Eqs. (1.34) and (1.35) to evaluate masonry compression strength.

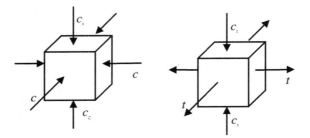

Fig. 1.27 Axial compression accompanied by side compression or tensile stresses

1.7 Masonry Compression Strength

1.7.1 Features of Compressed Masonry Failure

Masonry, with its regular patterns of blocks and mortar courses, is a complex structure whose compression strength depends on the interplay among its brick and mortar components. Tests reveal that splitting cracks in the blocks anticipate the brittle failure of uniaxially compressed masonry, which is characterized by the expulsion of brick fragments (Figs. 1.28 and 1.29).

1.7.2 Valuation of Masonry Compression Strength

Figure 1.30 shows a rough qualitative illustration of the shear stresses occurring between the mortar beds and bricks in masonry.

Let us first recall the equations of elasticity:

$$\varepsilon_x = \frac{1}{E}[\sigma_x - \nu(\sigma_y + \sigma_z)] \; \varepsilon_y = \frac{1}{E}[\sigma_y - \nu(\sigma_z + \sigma_x)] \; \varepsilon_z = \frac{1}{E}[\sigma_z - \nu(\sigma_x + \sigma_y)] \quad (1.36)$$

which describe the deformation of both bricks and mortar.

The sequence of bricks and mortar beds exhibit double symmetry, so using the reference axes from Fig. 1.31 for the stresses and strains of the bricks and mortar, we have,

$$\sigma_x = \sigma_y = \sigma_l \quad (1.37)$$

$$\varepsilon_x = \varepsilon_y = \varepsilon_l \quad (1.38)$$

where index l indicates lateral stresses and deformations. Vertical and horizontal stresses and deformations of bricks and mortar have the components σ_v, σ_l, ε_v, ε_l, hence, Eq. (1.36) become

1.7 Masonry Compression Strength

Fig. 1.28 Compression failure of a masonry pier. Specimen M0–1 (Facconi 2012)

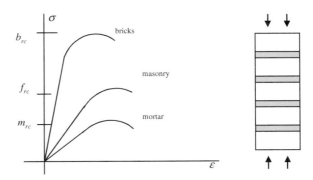

Fig. 1.29 σ–ε diagram for bricks, mortar and masonry under monoaxial compression

$$\varepsilon_l = \frac{1}{E}\left[\sigma_l(1-v) - v\sigma_v\right] \quad \varepsilon_v = \frac{1}{E}(\sigma_v - 2v\sigma_l) \qquad (1.39)$$

The mortar and brick components can be distinguished, so we have

$$\varepsilon_l^m = \frac{1}{E_m}\left[\sigma_l^m(1-v_m) - v_m\sigma_v^m\right] \quad \varepsilon_v^m = \frac{1}{E_m}(\sigma_v^m - 2v_m\sigma_l^m) \qquad (1.40)$$

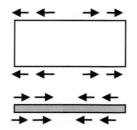

Fig. 1.30 Shear stresses between brick and joint mortar due to different lateral expansions

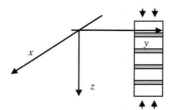

Fig. 1.31 The reference axes in compressed masonry

$$\varepsilon_l^b = \frac{1}{E_b}[\sigma_l^b(1-v_b) - v_b\sigma_v^b)] \quad \varepsilon_v^b = \frac{1}{E_b}(\sigma_v^b - 2v_b\sigma_l^b) \qquad (1.41)$$

The bricks and mortar beds are piled up along the vertical. The compressive stress on the bricks and mortar beds are equal to the applied vertical compression σ_y. Thus, we can also write

$$\sigma_z^b = \sigma_z^m = \sigma_v \qquad (1.42)$$

Now, assuming that compression is positive, by virtue of position (1.42), Eqs. (1.40) and (1.41) become

$$\varepsilon_l^m = \frac{1}{E_m}[\sigma_l^m(1-v_m) - v_m\sigma_v)] \quad \varepsilon_v^m = \frac{1}{E_m}(\sigma_v - 2v_m\sigma_l^m) \qquad (1.40')$$

$$\varepsilon_l^b = \frac{1}{E_b}[\sigma_l^b(1-v_b) - v_b\sigma_v)] \quad \varepsilon_v^b = \frac{1}{E_b}(\sigma_v - 2v_b\sigma_l^b) \qquad (1.41')$$

No sliding occurs between the mortar beds and bricks and, consequently, the corresponding lateral strains will be the same, that is,

1.7 Masonry Compression Strength

$$\varepsilon_l^b = \varepsilon_l^m. \tag{1.43}$$

Taking (1.40′) and (1.41′) into account yields

$$\frac{1}{E_m}[\sigma_l^m(1-v_m) - v_m\sigma_v)] = \frac{1}{E_b}[\sigma_l^b(1-v_b) - v_b\sigma_v)] \tag{1.44}$$

The thickness of the bricks and mortar beds is respectively h_b and h_m. Thus, equilibrium of any vertical section of the masonry pier along the horizontal direction gives

$$\sigma_l^b h_b = -\sigma_l^m h_m \tag{1.45}$$

The ratio between the thicknesses of the mortar joints and bricks is represented by the geometric factor β

$$\beta = \frac{h_m}{h_b} \tag{1.46}$$

Thus, from (1.45) we obtain

$$\sigma_l^b = -\beta \sigma_l^m \tag{1.47}$$

Substitution of (1.47) into (1.44) gives

$$\frac{1}{E_m}[\sigma_l^m(1-v_m) - v_m\sigma_v)] = \frac{1}{E_b}[-\beta\sigma_l^m(1-v_b) - v_b\sigma_v)] \tag{1.48}$$

or

$$\sigma_l^m[\frac{E_b}{E_m}(1-v_m) + \beta(1-v_b)] = \sigma_v(v_m\frac{E_b}{E_m} - v_b) \tag{1.38′}$$

We now introduce the ratio between the brick and mortar moduli

$$\phi = \frac{E_b}{E_m} \tag{1.49}$$

and the brick and mortar lateral stress components become

$$\sigma_l^m = \sigma_v \frac{(v_m\phi - v_b)}{[\phi(1-v_m) + \beta(1-v_b)]} \quad \sigma_l^b = -\beta\sigma_v \frac{(v_m\phi - v_b)}{[\phi(1-v_m) + \beta(1-v_b)]} \tag{1.50}$$

With the further position

$$\chi = \frac{v_m \phi - v_b}{\phi(1 - v_m) + \beta(1 - v_b)}, \quad (1.51)$$

we finally obtain (Haller 1947; Francis 1971; Lenczner 1972):

$$\sigma_l^m = \chi \sigma_v \quad \sigma_l^b = -\beta \chi \sigma_v. \quad (1.52)$$

Mortar is thus *compressed* along the two horizontal directions x and y, while the bricks are *stretched* horizontally. We now define σ_{zo}^m and σ_{zo}^b as the vertical compressions producing failure in the mortar and the bricks, respectively. Mortar collapse is reached when the applied vertical compression σ_{zo}^m and the corresponding lateral stress σ_l^m satisfy the failure condition (1.34). Hence, we have

$$\sigma_{zo}^m = m_{rc} + \frac{m_{rc}}{m_{rt}} \sigma_l^m \quad (1.53)$$

where m_{rc} and m_{rt} indicate the mortar uniaxial compression and tensile strengths. Hildsdorf (1965) assumed an empirical relation to describe the failure of the bricks to evaluate the compression masonry strength. Converseley we assume that failure of the bricks will be reached when the vertical compression σ_{zo}^b and the lateral tensile stresses σ_l^b satisfy previous failure condition (1.35), so we have

$$\sigma_{zo}^b = b_{rc} + \frac{b_{rc}}{b_{rt}} \sigma_l^b \quad (1.54)$$

where b_{rc} and b_{rt} indicate the brick uniaxial compression and tensile strengths.

By substituting the mortar and brick lateral stresses (1.42) into the failure Eqs. (1.43) and (1.44), we obtain the values of the vertical compressions producing mortar and brick failure, respectively

$$\sigma_{zo}^m = \frac{m_{rc}}{1 - \frac{m_{rc}}{m_{rt}} \chi} \quad \sigma_{zo}^b = \frac{b_{rc}}{1 + \frac{b_{rc}}{b_{rt}} \beta \chi} \quad (1.55)$$

Because $m_{rc}/m_{rt} \gg 1$, the denominator of σ_{zo}^m is certainly negative. The mortar can thus collapse only if the masonry is stretched instead of compressed. On the contrary, the denominator of σ_{zo}^b is > 1, so σ_{zo}^b thus represents the effective masonry compression strength. Masonry failure therefore occurs through collapse of the bricks under vertical compression and lateral stretching. The compression strength of the masonry is thus

$$f_{rc} = \frac{b_{rc}}{1 + \frac{b_{rc}}{b_{rt}} \beta \chi} \quad (1.56)$$

1.7 Masonry Compression Strength

which is lower than the brick compression strength. Masonry compression strength thus depends on both the masonry geometry, defined by the ratio β, as well as the mechanical deformability and strength of its components, defined by the ratios ϕ and χ. If the masonry is made up of standard size bricks (5.5 cm thick), the thickness of the mortar bed, h_m, will be the main factor determining masonry strength. The larger ratio β is, that is, the thicker the mortar beds are, the lower the masonry compression strength will be. Equation (1.56) gives compression failure stresses similar to the Hilsdorf (1965) result.

Examples
Brick and mortar masonry
Brick and mortar are defined by the following strength and Poisson's ratio values

$$b_{rc} = 10\,\text{MPa} \quad b_{rt} = 0.7\,\text{MPa} \quad b_{rc}^{biax} = 11\,\text{MPa} \quad v_b = 0.15$$
$$m_{rc} = 5\,\text{MPa} \quad m_{rt} = 0.5\,\text{MPa} \quad m_{rc}^{biax} = 6\,\text{MPa} \quad v_m = 0.25$$

We assume $\phi = 2$ as the ratio between the brick and mortar moduli and a mortar bed thickness of 0.5 cm. Assuming a brick thickness of 5.50 cm, we obtain $\beta = 0.0952$. Parameter χ thus takes the value

$$\chi = \frac{v_m \phi - v_b}{\phi(1 - v_m) + \beta(1 - v_b)} = \frac{0.25 \cdot 2 - 0.15}{2 \cdot 0.75 + 0.0952 \cdot 0.85} = 0.2214$$

and, according to (1.56) the corresponding masonry compression strength is

$$f_{rc} = \frac{b_{rc}}{1 + \frac{b_{rc}}{b_{rt}}\beta\chi} = \frac{10}{1 + 15 \cdot 0.0952 \cdot 0.2214} = 7.60\,\text{MPa}$$

By increasing the mortar bed thickness to 0.75 cm, we have $\beta = 0.1364$ and

$$\chi = \frac{0.25 \cdot 2 - 0.15}{2 \cdot 0.75 + 0.1364 \cdot 0.85} = 0.2166$$

The compression strength of the masonry is thereby reduced to

$$f_{rc} = \frac{10}{1 + 15 \cdot 0.1364 \cdot 0.2166} = 6.90\,\text{MPa}$$

with the bricks and mortar having the respective strengths of 10 and 5 MPa.

Tuff block and mortar masonry
We assume the following strengths for the tuff blocks and mortar

$$t_{rc} = 6\,\text{MPa} \quad t_{rt} = 0.4\,\text{MPa} \quad t_{rc}^{biax} = 7.2\,\text{MPa} \quad v_t = 0.15$$
$$m_{rc} = 5\,\text{MPa} \quad m_{rt} = 0.3\,\text{MPa} \quad m_{rc}^{biax} = 3.6\,\text{MPa} \quad v_m = 0.25$$

Table 1.5 Masonry compression strengths versus depending on block strength and mortar type

f_{rc}	M_1	M_2	M_3	M_4
1.5	1.0	1.0	1.0	1.0
3.0	2.2	2.2	2.2	2.0
5.0	3.5	3.4	3.3	3.0
7.5	5.0	4.5	4.1	3.5
10.0	6.2	5.3	4.7	4.1
15.0	8.2	6.7	6.0	5.1
20.0	9.7	8.0	7.0	6.1

and a moduli ratio $\phi = 2$. The thicknesses of the tuff blocks and mortar beds are respectively 13 and 1.5 cm. Thus $\beta = 0.1154$, and we get

$$\chi = \frac{0.25 \cdot 2 - 0.15}{2 \cdot 0.75 + 0.1154 \cdot 0.85} = 0.2190.$$

The compression strength of tuff masonry is

$$f_{rc} = \frac{6}{1 + 15 \cdot 0.1154 \cdot 0.2190} = 4.35\,\text{MPa},$$

in contrast to the assumed tuff and mortar compression strengths of 6 and 3 MPa.

Table 1.5, drawn from the Italian Building Code D.M. 20.11.87, provides indicative values for masonry strengths as functions of the compression strengths of the stone elements and mortar types (classified as M1, M2, M3 and M4; strength is expressed in Mpa). Although the strength values indicated are well-grounded, they neglect the effects of the mortar bed thickness.

1.8 Masonry Tensile Strength

Masonry tensile strength f_{rt} is generally very low—nearly negligible in comparison to its compression strength. Cracks that occur when masonry is stretched arise in the contact area between the mortar and blocks. These detachments take place due to the loss of adhesion between the mortar and blocks rather than to mortar tensile failure, as Fig. 1.32 clearly illustrates. This aspect sharply distinguishes the tensile behavior of masonry from that of concrete.

In the concrete's interior, however, the cracks are extremely small and their edges very jagged in comparison to their appearance on the surface. Thus, friction and consequent jamming between the rough crack surfaces transmit some tension, especially for crack widths of less than about 0.05 mm, and deformations concentrate across the cracks. Therefore, it is the average crack width w_c, more than the strain, that determines the degree of stretching of the concrete. Generally, Foote's law (1986) describes the relation between tension and crack opening width

1.8 Masonry Tensile Strength

Fig. 1.32 Detachment of brick and mortar joints in a section of a masonry arch

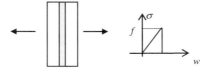

Fig. 1.33 The stress–extension law of the mortar joint

$$\frac{\sigma}{f_{rt}} = (1 - \frac{w}{w_c})^n \qquad (1.57)$$

Various softening laws can be obtained from Eq. (1.47) by assuming different values of exponent n. The case $n = 0$ corresponds to the complete absence of softening, i.e., perfectly plastic behavior. On the contrary, the case $n = \infty$ refers to perfectly brittle behavior. The behavior of masonry in tension corresponds to this latter case and the tensile force-extension relation takes the form illustrated in Fig. 1.33, where f_{rt} indicates the strength of the adhesion between brick and mortar.

1.9 Masonry Shear Strength

A shear action acting along the direction of the courses of masonry, made up of bricks or regular stones with mortar joints, can produce sliding of the courses and consequent disruption of the masonry.

Shear tests are used to check masonry's capacity to transmit shear. In such tests simple masonry panels of different shapes and mortar bed strengths are subjected to a compression force diagonal to the brick courses, as shown in Fig. 1.34. Angled metal platens positioned at the panel corners enable application of these diagonal forces. With such an experimental set-up, the height-to-width ratio of the panel also establishes the ratio between the axial and shear force components. However, using other equipment it is also possible to apply axial and shear forces at different ratios to same shaped panels.

(a) Tests show that a diagonally advancing crack causes the panel to fail by splitting the masonry into fragments of varying size. Generally, an entire half of the panel overturns with respect to the other half, as shown in Fig. 1.35. In the vertical bands of masonry walls under inplane horizontal loading, this type of failure rottura comes early the collapse by overturning.

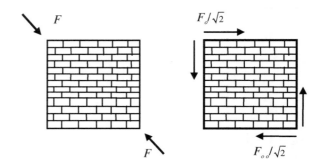

Fig. 1.34 Diagonal compression test

Fig. 1.35 Failure by diagonal cracking and overturning of one half of the panel (Carbone et al. 2005)

1.9 Masonry Shear Strength

Fig. 1.36 Failure by predominant shear sliding of courses (Borri et al. 2011)

(b) In other cases, shear failure comes about by sliding of the brick courses, as shown in Fig. 1.36. This type of failure can occur—even under low axial force values relative to the shear—when poor quality mortar has been used in the masonry.

(c) another type of failure, which exhibits features of both the previous failure modes, is shown in Fig. 1.37. Collapse occurs by the shear sliding of some courses accompanied by detachment and overturning of other masonry fragments. Such failure occurs under relatively high axial loads in the presence of poor mortar.

Fig. 1.37 Combined failure by sliding courses and overturning panel fragments (Borri et al. 2011)

Masonry shear strength is a matter of widespread discussion and even debate. The true problem is to evaluate *the masonry compression strength in presence of shear*. We will return to a consideration of the issues involved in the sections dealing with analysis of the lateral strength of masonry panels and seismic analysis of multi-plane walls. Useful information on the matter can be also obtained in the next section.

1.10 Masonry Strength and the Variable Course Inclinations

Evaluating the strength of masonry under biaxial stress whose direction of action is inclined with respect to the joints is a complex problem yet to be completely solved (Fig. 1.38). The experimental results obtained by Samarasinghe (1980) furnish useful insight into the topic. Similar results have been obtained by Page (1981, 1983). A vertical compression σ, accompanied by a horizontal tensile stress f_t, is applied to a masonry cell made with bricks courses and mortar beds laid at an angle θ with respect to the axes of the applied stresses.

Figure 1.39 gives the failure interaction curves for different values of the inclination θ. These curves connect the experimental points whose coordinates are the measured failure compression σ and the corresponding lateral tensile stress f_t.

The plots in Fig. 1.39 reveal a number of interesting characteristics of masonry that are useful to bear in mind in a wide range of applications.

Let us firstly analyze the case of $\theta = 0$ by examining the first curve in Fig. 1.39. Except for a short segment near the f_t axis, the diagram shows the *linearly decreasing* vertical failure compression with increasing horizontal tensile stress f_t. result in agreement with the general behaviour of stone materials under a biaxial stress state of compression–tension, shown in Fig. 1.22. On the contrary, when the vertical compression is small, the Samarasinghe diagram shows an opposite trend: the failure tension decreases with decreasing vertical compression. This effect can be explained by taking into account the friction strength between the horizontal joints, which decreases with decreasing vertical compression σ, or, on the contrary,

Fig. 1.38 Biaxial stresses applied to a masonry cell with inclined joints

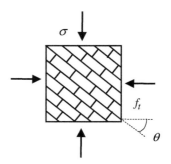

1.10 Masonry Strength and the Variable Course Inclinations

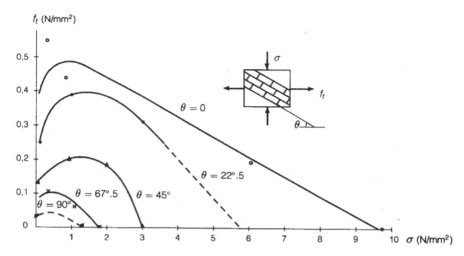

Fig. 1.39 Samarasinghe test results (1980)

increases with increasing compression σ. This last aspect will be taken up again in the study of the statics of domes.

Further, by increasing the angle θ the uniaxial compression strength reduces strongly. For $\theta = 22.5°$ and $\theta = 22.5°$ the compression strength becomes only about the 60 % or the 30 of the strength corresponding to $\theta = 0$. This result is very important because, in short, signifies the influence of the shear on the compression strength of regular brickwork or, in a similar way, the effect of the inclination of the compression with respect to the mortars beds. Doubtful is, on the contrary the monotonic strength reduction with the increasing the angle q, as far as to q = 90°.

Tests performed at the university of Brescia confirm partially these results. Five masonry specimens, with dimensions 50 cm × 71 cm × 23 cm with different joints inclination have been tested under vertical compression. The specimens are indicated as M01 and M02 ($\theta = 0°$); M22 ($\theta = 22.5°$); M45 ($\theta = 45°$); M90 ($\theta = 90°$). The angle θ is the inclination of mortar beds with horizontal, as shown in the same Fig. 1.40.

Figure 1.40 shows some of the failed specimens under vertical compression. The detected compression strengths are:

M01 ($\theta = 0°$): $\sigma_0^* = 7$ MPa; M02 ($\theta = 0°$); $\sigma_0^* = 6.1$ MPa, with a mean strength a $\sigma_0^* = 6.55$ MPa.

M22 ($\theta = 22.5°$); $\sigma_0^* = 4$ MPa; M45 ($\theta = 45°$); $\sigma_0^* = 2.6$ MPa

M90 ($\theta = 90°$); $\sigma_0^* = 5.7$ MPa.

The detected strength reductions corresponding to the inclination angle $\theta = 22.5°$ and $\theta = 45°$ are in full agreement with the above Samarasinghe results. On the contrary, when the inclination angle reaches the value $\theta = 90°$ the compression strength increases compared to the values corresponding to the less inclined joints.

Fig. 1.40 Specimesn M22 (θ = 22.5°) and M45 (θ = 45°) at failure (Facconi 2012)

Figure 1.41 shows the changes in strength by varying θ. The maximum strength reduction occurs with $\theta = 45°$. In the following the compression strength inclined with the joint inclination will be denoted

$$f_{rc<} \tag{1.58}$$

A simple descrption of the dependence of the masonry strength with the inclination angle compression/joints direction, qualitatively according to the test results shown in Fig. 1.42. is the following

Fig. 1.41 Compression strengths by varying the joint inclination (Facconi 2012)

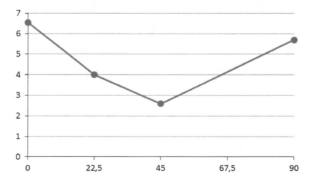

1.10 Masonry Strength and the Variable Course Inclinations

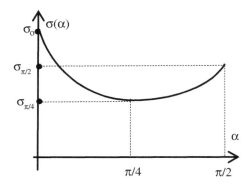

Fig. 1.42 The diagram of the function assumed to describe the variation of strength with the inclination angle compression/joint direction

$$\sigma(\alpha) = \sigma_o + \frac{8}{\pi^2}(\sigma_o + \sigma_{o\pi/2} - 2\sigma_{o\pi/4})\alpha^2 + \frac{4}{\pi}(2\sigma_{o\pi/4} - \frac{\sigma_{o\pi/2}}{2} - \frac{3}{2}\sigma_o)\alpha \quad (1.59)$$

In the particular case of $\sigma_{o\pi/4} = \sigma_o/2$, $\sigma_{o\pi/2} = 3\sigma_o/4$ the previous Eq. (1.59) becomes

$$\sigma(\alpha) = \sigma_o + \frac{8}{\pi^2}(\sigma_o + \sigma_{o\pi/2} - 2\sigma_{o\pi/4})\alpha^2 + \frac{4}{\pi}(2\sigma_{o\pi/4} - \sigma_{o\pi/2}/2 - 3\sigma_o/2)\alpha$$

Equation (1.59) can be very useful in applications.

1.11 Masonry Deformations

1.11.1 Masonry Elastic Modulus

Figure 1.43 shows a typical σ–ε diagram for compressed regular masonry. The diagram refers to a test lasting only a few minutes in order to avoid long-term effects (which will be addressed later). The corresponding instantaneous elastic modulus can be determined as the tangent to the σ–ε curve at the origin. A variety

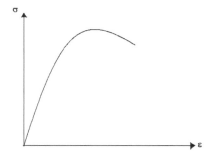

Fig. 1.43 Typical σ–ε. diagram of compressed regular masonry

of empirical formulas can yield information about the order of magnitude of the elastic modulus: the following relation being a typical example:

$$E^* = 900 f_k \qquad (1.60)$$

where f_k is the characteristic masonry compression strength. The corresponding elastic shear modulus G can be evaluated by assuming a value $v = 0.25$ for Poisson's coefficient, whence we get $G_m = 0.4 E_m$.

However, there is a simple theoretical procedure for evaluating the instantaneous elastic modulus of regular brickwork (i.e., composed of brick or regular blocks and mortar joints, as shown in Fig. 1.44).

Deformations of the masonry along the vertical direction are due to deformation of both the blocks and the mortar beds. The effects of vertical joints deformations are instead insignificant. Let us refer to the simplified scheme of a single masonry cell shown in Fig. 1.45. The cell is composed of a single brick with two horizontal mortar semi-joints, with h_m and h_b respectively the joint and brick thickness. The vertical compression σ on the bricks is the same as on the joints. The corresponding vertical deformations ε_b and ε_m of bricks and joints are thus

$$\varepsilon_v^b = \frac{\sigma}{E_b} \quad \varepsilon_v^m = \frac{\sigma}{E_m} \qquad (1.61)$$

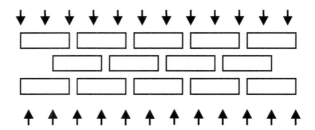

Fig. 1.44 Brickwork deformation under vertical compression

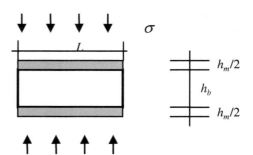

Fig. 1.45 Deformation of the masonry cell

1.11 Masonry Deformations

Recalling the expressions of the mortar and brick vertical strains ε_v^m and ε_v^b, according to (1.61') and (1.61"), we have

$$\varepsilon_v^m = \frac{1}{E_m}(\sigma_v - 2v_m \sigma_l^m) \qquad \varepsilon_v^b = \frac{1}{E_b}(\sigma_v - 2v_b \sigma_l^b) \qquad (1.61')$$

where E_m and E_b are the brick and mortar elastic moduli, respectively. Further, taking into account the (1.51) and (1.52)

$$\varepsilon_v^m = \frac{\sigma_v}{E_m}(1 - 2v_m \chi) \quad \varepsilon_v^b = \frac{\sigma_v}{E_b}(1 + 2v_b \beta \chi) \qquad (1.61'')$$

Consequently, the overall shortening of the single masonry cell is

$$\Delta = \varepsilon_v^b h_b + \varepsilon_v^m h_m = \frac{\sigma_v}{E_b} h_b[(1 + 2v_b \beta \chi) + \phi \beta(1 - 2v_m \chi)] \qquad (1.62)$$

And the overall masonry strain is

$$\varepsilon^* = \frac{\Delta}{h_b + h_m} = \frac{\sigma_v}{E_b} \frac{1}{1+\beta}[1 + \phi \beta + 2\chi \beta(v_b - v_m)] \qquad (1.63)$$

Consequently, the corresponding masonry elasticity modulus, defined as

$$E^* = \frac{\sigma}{\varepsilon^*}, \qquad (1.64)$$

takes the form

$$E^* = E_b \frac{1+\beta}{1 + \beta[\phi + 2\chi(v_b - v_m)]} \qquad (1.65)$$

or, taking into account that the quantity $2\chi(v_b - v_b)$ is negligible with respect to ϕ,

$$E^* = E_b \frac{1+\beta}{1 + \beta \cdot \phi} \qquad (1.65')$$

in full agreement with a well known formula of the International Ralways Union (Code Fiches-UIC-778.3). As can be seen from Eq. (1.74'), the masonry elastic modulus depends not only on the brick and mortar elastic moduli E_b, E_m but also on the ratio h_m/h_b between the joint and brick thicknesses. Mortars are more deformable than bricks or other stone blocks and $\phi > 1$, Hence

$$E^* \leq E_b$$

If the ratio β is negligible, from (1.65') we have $E^* \approx E_b$. The tickness of mortar beds, by means the ratio β, has thus a relevant role in defining the overall masonry elasticity modulus.

1.11.2 Masonry Deformation at the Onset of Blocks Failure

Let us evaluate the overall masonry deformation ε_o^* at the onset of the failure of blocks when it has been just reached the compression stress σ_o^*. At this stage vertical cracks crossing the blocks take place and the shear interaction between mortar beds and blocks begins to loosen: strains increase but masonry is incapable to increase its strength.

To evaluate the strain ε_o^*, let assume in (1.73) the presence of the compression strength f_{rc} given by (1.64). We have

$$\varepsilon_o^* = \frac{f_{rc}}{E^*} \tag{1.66}$$

where the modulus E^* is given by (1.65').

Numerical examples
Regular brickwork

We assume the following quantities concerning bricks and mortar:

$$b_{rc} = 10\,\text{MPa} \quad b_{rt} = 0.7\,\text{MPa} \quad v_b = 0.15 \quad E_b = 8000\,\text{MPa}$$
$$m_{rc} = 5\,\text{MPa} \quad m_{rt} = 0.5\,\text{MPa} \quad v_m = 0.25 \quad E_m = 3000\,\text{MPa}$$

Brick height: h_b = 5.5 cm. Thickness of mortar joints: h_m = 0.5 cm. Thus β = 0.0909 and ϕ = 2.67. From (1.74') we have

$$E^* = E_b \frac{1 + 0.0909}{1 + 0.0909 \cdot 2.67} = 0.878 E_b = 7022\,\text{MPa}$$

compared to an assessed compression strength equal to f_{rc} = 7.56 MPa.

$$\varepsilon_o^* = f_{rc}/E^* = 0.1077\,\%$$

Further, taking the larger thickness of mortar joint equal to 0.75 cm, we have β = 0.1364, ϕ = 2.67, and

$$E^* = E_b \frac{1 + 0.1364}{1 + 0.1364 \cdot 2.67} = 0.833 E_b = 6664\,\text{MPa}$$

compared to an assessed compression strength equal to f_{rc} = 6.77 MPa

1.11 Masonry Deformations

$$\varepsilon_o^* = f_{rc}/E^* = 0.102\,\%$$

For a masonry of worse quality, with joint thickness equal to the half of that of the brick, we have $\beta = 0.5$, $\phi = 2.67$, and

$$E^* = E_b \frac{1+0.5}{1+0.5 \cdot 2.67} = 0.642 E_b = 5136\,\text{MPa}$$

compared to an assessed compression strength equal to $f_{rc} = 3.96\,\text{MPa}$

$$\varepsilon_o^* = f_{rc}/E^* = 0.077\,\%$$

Masonry with tuff blocks and mortar beds
We assume the following quantities concerning tuff blocks and mortar:

$$t_{rc} = 5\,\text{MPa} \quad t_{rt} = 0.5\,\text{MPa} \quad E_t = 5000\,\text{MPa} \quad v_t = 0.15$$
$$m_{rc} = 2\,\text{MPa} \quad m_{rt} = 0.2\,\text{MPa} \quad E_m = 1000\,\text{MPa} \quad v_m = 0.25$$

and the ratio $\phi = 5$. The thickness of tuff blocks is equal to 13 cm and that of mortar beds is 1.5 cm. Hence $\beta = 0.1154$. We obtain $E^* = 3537$ MPa. compared to an assessed compression strength of the tuff masonry $f_{rc} = 3.76$ MPa.

$$\varepsilon_o^* = f_{rc}/E^* = 0.106\,\%.$$

With these last results we can discuss now the stress–strain law it is possible to assume for the compressed masonry.

1.11.3 Stress–Strain Diagram of the Compressed Masonry

Figure 1.46 shows the stress–strain diagram for the regular brickwork considered by many A., for instance Kaushik ed altri (2007). This diagram confirms the quasi brittle behaviour of the compreesed masonry.

Once that the peak strength has been reached, there is only a small strain increment along which masonry maintains its strength. The strain ε'_m, corresponding to the peak strength, is equal to 0.0015 while the ultimate strain, where the stress is just vanished, equals 0.003. Further, the strain to which strength begins to drop rapidly, is denoted with $\varepsilon'_{m90\%}$, to point out that the 90 % of the peak strength corresponds to this strains. We have about $\varepsilon'_{m90\%} = 0.0025$. Further, from the above, we know also that the strain where the failure strength is just reached, is about equal to 0.1 %: it can be considered as the first yielding deformation of the masonry.

Fig. 1.46 Il diagramma compressione—deformazione della muratura sotto compressione verticale e filari di mattoni orizzontali

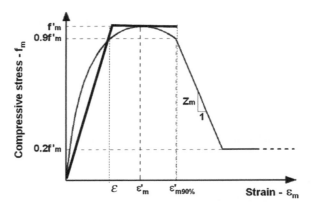

With these result it is thus possible to trace a simplified stress–strain digram for the compressed masonry that takes into account the main features of the *overall* masonry behaviour.

The first segment OA of the diagram of Fig. 1.47 describes the linear elastic relation between the applied compression $\overset{*}{\sigma}$ and the coresponding masonry strain $\overset{*}{\varepsilon}$. The end A of the linear segment is reached when the applied compression σ^* reaches the masonry strength σ_o^*, i.e. the cracking of bricks begins to occur, and the overall strain is $\varepsilon_o^* = 0.1\ \%$, i.e. equal to about the 2/3 of the peak strain $\varepsilon_m' = 0.15\ \%$, above determined. The corresponding overall elasticity modulus E^* is espressed by (1.74') and corresponds to the tgγ of Fig. 1.47.

The ultimate strain ε_u^* is obtained by test results and, with the above considerations, can be taken equal to $\varepsilon_{m90\%}' = 0.25\ \%$. It is reasonable also to assume the value of shown in Fig. 1.45. The short segment AB at constant strength is included between the strains ε_0^* and ε_u^*: its length is thus equal to 0.15 %.

Fig. 1.47 Overall simplified diagram σ–ε of compressed brickwork

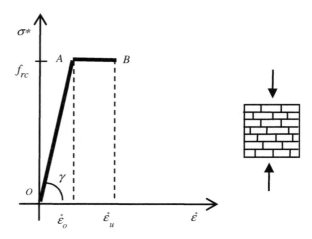

1.11 Masonry Deformations

The italian Code NTC 2008 gives the simple formula for the modulus of elasticity of masonry

$$E_m = 1000 f_k \qquad (1.60')$$

where f_k is the characteristic masonry compression strength to evaluate from the actual mean masonry strength f_m as $f'_k = 0.75 f'_m$. The actual f_m corresponds to the previous σ_0^*. Thus we have

$$E_m = 750 f_m \qquad (1.60'')$$

and the corresponding limit elastic strain ε_y^* is

$$\varepsilon_o^* = \frac{1}{750} = 0.00133$$

a bit larger than the above yield strain ε_y^* equal to 0.001. If, the mean compression strength f_m is placed in place of f_k in the (1.60'), the limit elastic strain, i.e. the yield strain ε_y^* results equal to 0.1 %.

1.11.4 Creep Deformation of Mortar

Mortar is subject to significant creep deformation that strongly influences long-term masonry deformations.

To understand the features of creep deformation it is useful to recall some aspects of viscous deformations by considering an ideal mortar specimen initially stressed by a compression σ. This stress, which we assume to be much lower than the masonry compression strength, is maintained on the specimen for a given period of time and subsequently removed. Upon application of the compression σ, an instantaneous deformation $\varepsilon(0)$ will occur, then, as consequence of mortar creep, this strain will slowly increase under the constant action of σ. When the specimen is unloaded, it will immediately recover its initial elastic deformation, but will retain a certain residual strain that will then, in turn, be recovered partially. Figure 1.48 provides a schematic illustration of this behavior.

The mortar strain occurring under the constant compression σ is thus composed of an elastic part, σ/E_m, and a viscous part $\varepsilon_v(t)$, the latter of which is slowly increasing over time,

$$\varepsilon(t) = \frac{\sigma}{E_m} + \varepsilon_v(t) \qquad (1.67)$$

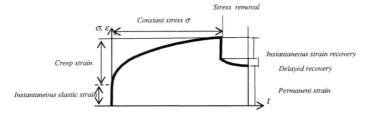

Fig. 1.48 Creep behaviour of mortar

1.11.5 The Concept of Memory in Constitutive Creep Models

Creep deformation $\varepsilon_v(t)$ at time t can be considered as the memory, at the current time t of all the stress events $\sigma(\tau)$ having occurred in the past time τ. This 'memory' can be considered proportional to the magnitude of the stress σ applied to the material in time τ. Generally, creep deformation $\varepsilon_v(t)$, will depend on both the current and past times, t and τ, as well as on the duration of action of the past stress $\sigma(\tau)$. Thus, we can write

$$\varepsilon_v(t,\tau) = \frac{\Phi(t,\tau)}{E_m} \qquad (1.68)$$

where the $\Phi(t,\tau)$ is a function suitably defined so as to represent the memory at current time t of a unit stress $\sigma(\tau) = 1$ applied during a unit time interval $\Delta\tau = 1$ at past time τ. Thus, the expression

$$d\varepsilon_v(t) = \Phi(t,\tau)\frac{\sigma(\tau)}{E_m}d\tau \qquad (1.69)$$

represents the differential of the viscous strain. The viscous deformation $\varepsilon_v(t)$ is the sum of the memories at time t of all stresses $\sigma(\tau)$ applied in the past. Thus, we obtain

$$\varepsilon_v(t) = \int_{t_i}^{t} \Phi(t,\tau)\frac{\sigma(\tau)}{E_m}d\tau \qquad (1.70)$$

where t_i indicates the start time of the initial loading. In general, the memory function depends on both the current and the past times, t and τ. More precisely, it depends on the difference

1.11 Masonry Deformations

$$(t - \tau). \tag{1.71}$$

In fact, the memory of stress $\sigma(\tau) = 1$ applied at time τ during the unit interval $\Delta \tau = 1$ may fade with the passing of time and the function $\Phi(t,\tau)$ is thus decreasing with increasing $(t-\tau)$. However, tests have revealed that mortar, like concrete, has almost permanent memory, in the sense that the effects of a unitary stress event $\sigma(\tau) = 1$ that occurred at past time τ fade to a minimal degree over time. We can therefore assume with good approximation that

$$\Phi(t, \tau) \approx \Phi(\tau). \tag{1.72}$$

However, the degree of memory at current time t of an event $\sigma(\tau) = 1$ will also depend on whether the mortar is 'young' or 'old' at time t, that is, whether t follows closely or long after the mortar's time of curing t_o. According to Krall and Whitney (1951), we can assume

$$\Phi(t, \tau) = \alpha \beta e^{-\beta(\tau-t_o)}, \tag{1.73}$$

where α is a factor, ranging from 1 to 4, representing the magnitude of the creep, $\beta = 1$ year^{-1} is a scale factor and t_o the mortar curing time. Consequently, if the mortar is stressed under *constant* compression σ applied at initial time $t_i > t_o$, the total strain (the sum of elastic and viscous parts) will be

$$\varepsilon(t) = \varepsilon_e(t) + \varepsilon_v(t) = \frac{\sigma}{E_m} + \alpha\beta \frac{\sigma}{E_m} \int_{t_i}^{t} e^{-\beta(\tau-t_o)} d\tau \tag{1.74}$$

By integrating (1.60), we get

$$\begin{aligned}\varepsilon(t) = \varepsilon_e(t) + \varepsilon_v(t) &= \frac{\sigma}{E_m} - \frac{\sigma}{E_m} \alpha \left[e^{-\beta(\tau-t_o)} \right]\Big|_{t_i}^{t} \\ &= \frac{\sigma}{E_m} \left\{ 1 - \alpha [e^{-\beta(t-t_o)} - e^{-\beta(t_i-t_o)}] \right\} \end{aligned} \tag{1.75}$$

In this simplified model the residual creep strain is permanent, as shown in Fig. 1.49. Other formulations can more accurately describe the fading of memory over time, particularly for concrete (Chiorino 2005). On the other hand, such formulations are more complex and in practice the simplifying assumption (1.73) can be considered acceptable. If we first consider the simpler case, with $t_i = t_o = 0$ we get

$$\varepsilon(t) = \frac{\sigma}{E_m}[1 + \alpha(1 - e^{-\beta t})] \tag{1.76}$$

and the long-term strain, for $t \to \infty$, is

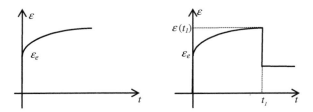

Fig. 1.49 Simplified creep behavior of a viscous material under constant compression (*left*) and after the stress removal (*right*)

$$\varepsilon_\infty = \frac{\sigma}{E_m}(1+\alpha) \qquad (1.77)$$

The ratio between the long-term and the initial strain ε_e defines the long-term deformation factor, expressed by

$$\varphi_\infty = \frac{\varepsilon_\infty}{\varepsilon_e} = 1 + \alpha \qquad (1.77')$$

If, on the contrary, the initial loading time t_i is not simultaneous with the curing time t_o, i.e., $t_i \neq t_o$, in place of (1.63'), we have

$$\varphi_\infty = \frac{\varepsilon_\infty}{\varepsilon_e} = (1 + \alpha e^{-\beta(t_i-t_o)}). \qquad (1.77'')$$

This formulation can be applied to study the long-term behavior of masonry walls. In this context, the use of the delayed elastic masonry modulus can be very useful, as will be shown in later chapters.

1.11.6 Mortar Shrinkage

A mortar specimen in the open air undergoes a contraction of its volume that gradually ceases over time. This contraction is independent of stress. If the mortar specimen is a long prism, the prevailing deformation is shortening strain, which is defined through the uniaxial shrinkage strain function represented by the expression

$$\varepsilon_s(t) = \varepsilon_S(1 - e^{-\beta(t-t_o)}), \qquad (1.78)$$

where ε_S is the long-term shrinkage strain, equal to about 0.3–0.35×10^{-3}.

Mortar shrinkage of the horizontal joints of a masonry wall causes only a small reduction in the joint thickness and, consequently, leads to only a slight overall shortening of the wall. A brick wall undergoes long-term shortening of the order of magnitude of 0.1–0.2 mm per meter of height.

References

Adam, J. P. (1988). *L'arte di costruire presso i Romani*. Longanesi, Milano: Materiali e tecniche.
Bazant, Z. P., & Jirasek, M. (2002). *Inelastic Behavior of Structures*. Chichester, England: J. Wiley.
Brady, B. H. G., & Brown, E. T. (2004). *Rock mechanics for underground mining* (3rd ed.). The Netherlands: Springer.
Brown, E. T. (1974). *Fracture of Rock under uniform biaxial compression*, Proceedings 3rd ISRM Congress (Vol. 2).
Brooks, J. J., & Abu Bakar, B. H. (2004). Shrinkage and creep of masonry mortar. *Materials and Structures, 37*(3).
Chiorino, M. A. (2005). Effetti statici dei fenomeni differiti del calcestruzzo...., in *Moderni Orientamenti di Ingegneria Strutturale e Geotecnica*, Franco Angeli, Torino.
Choisy, A.*L'art de bâtir chez les Romains*, Ducher et C. Éditeurs, Paris 1873 (rist. Arnaldo Forni Ed., Sala Bolognese 1984).
Como, M., & Luciano, R. (2007). A New multi–axial failure criterion for the concrete, 2006–2007. In: *FraCmos Catania, Giugno and International Conference FIB*, Napoli 2006.
Fieni, S. (2006–2007). *Ricerca del dominio triassiale di rottura per i materiali rocciosi*, supervisor M. Como, a.y.
Forino, G. (2006–2007). *Sulla resistenza a compressione della muratura*, supervisor M. Como, a.y.
Francis, A. J. M., Horman, C.B., & Jerems, L.E. (1971). The effect of joint thickness and Other factors on Compressive Strength of Brickwork. In: *Proceedings 2nd International BrickMasonry* Conference, Stoke-on-Trent.
Giuffrè, A. (1988). *Monumenti e terremoti*. Multigrafica Editrice, Roma: Aspetti statici del restauro.
Giuffrè, A. (1990). *Letture sulla meccanica delle murature storiche*. Roma: Dipartimento di Ingegneria Strutturale e Geotecnica.
Giuliani, C. F. (1995). *L'edilizia nell'Antichità*. Roma: La Nuova Italia Scientifica.
GNDT/SNN, *Manuale per la compilazione della scheda di 1° livello di rilevamento danno per edifici ordinari nell'emergenza post–sismica*, Dipartimento della Protezione Civile, Editrice Adel Grafica, Roma 2002.
Haller, P. (1947). *Physik des backsteins*. Zurch: Festigkeitseigenshaften.
Hildsdorf, H. K. (1965). Untersuchunggenuber die Grundlagen des Mauerwerks-Festigkeit, berich n. 40, *Materialprufungsamt fur des Bauwesender Technische Hochschule*, Munchen.
Krall, G. (1947). Statica dei mezzi elastici cosiddetti viscosi e sue applicazioni, *Accademia Nazionale dei Lincei*, fasc. 3–4, Roma.
Kupfer, H. (1973). Behavior of concrete under multiaxial short term loading, with emphasis on biaxial loading. In: *Deutscher Ausschuss fur Stahlbeton*, Vol. 254, Berlin.
Lenczner, D. (1970). Creep in brickwork. In: *Proceedings of the 2nd International Conference on Brick Masonry*, SIBMAC.
Lenczner, D. (1972). *Elements of Loadbearing Brickwork*. Oxford: Pergamon press.
Lenczner, D. (1981). Brickwork: A Guide to Creep. *International Journal of Masonry Constructions, 2*(4).
Page, A.W. (1981). The biaxial compressive strength of brick masonry. In: *Proceedings of Institution of Civil Engineering*, (Part 2, Vol. 71).
Page, A.W. (1983). The strength of Brick Masonry under Biaxial Compression – Tension. *International Journal of Masonry Constructions, 3(1)*.
Samarasinghe, S.W., & Hendry, A.W. (1980) The strength of brickwork under biaxial tensile and compressive stress. In: *Proceedings of the 7th International Symposium on Load Bearing Brickwork*, London.

Shrive, N. G., & England, G. L. P. (1981). Elastic creep and shrinkage behavior of masonry. *International Journal of Masonry Constructions, 2*, 3.

Sparacio, R. (1999). *La, Scienza e i Tempi del Costruire*. Torino: UTET Università.

Theses (University of Rome, Tor Vergata, Rome, Italy).

Venturini, V. W. (2003–2004). *Caratteristiche della resistenza del calcestruzzo*, supervisor M. Como, a.y.

Warren, D., & Lenczner, D. (1981). A creep-time function for single leaf brickwork walls. *International Journal of Masonry Constructions, 2*, 1.

# Chapter 2
Fundamentals of Statics of Masonry Solids and Structures

Abstract Groundings of statics of masonry solids and structures are the subject of the chapter. Masonry behavior is strongly influenced by the dramatically lower strength in tension than in compression. Masonry structures can thus suffer cracks generating displacement fields, called *mechanisms*, which develop without any internal opposition of the material. Collapse can occur without any material failure. The Heyman masonry model, the idealized rigid in compression no tension material, is fruitfully assumed as basis of the approach followed in the chapter. The extension of this model to the masonry continuum is then developed. Strains and detachments occurring in a no tension masonry solid can thus obtain a suitable mathematical definition together with the admissible equilibrium. Both a proper virtual work equation, that considers the boundary of the body including the crack surfaces, as a condition on the loads, necessary and sufficient to the existence of the masonry equilibrium, can be formulated. This last condition governs the collapse strength of masonry structures. The notion of the minimum thrust, from both static and kinematical approaches, is then introduced, widening the field of application of the Limit Analysis also to the study of the actual stress states. A critical analysis of the recent failure of the cathedral of Noto, in Sicily (Italy), useful to a better understanding of the above discussed mechanical concepts, ends the chapter.

2.1 Introduction

Under a given loading path, a masonry structure can reach a collapse condition solely due to loss of equilibrium, that is to say, in the absence of any material failure. Such a condition can therefore arise even in masonry with infinite compression strength. As discussed in the previous chapter, the tensile strength of masonry is, in fact, very low, near zero. Consequently, masonry structures can suffer cracks or detachments that may in turn generate displacement fields, often called *mechanisms*, which develop without any internal opposition from the material.

As soon as the pushing loads begin to exceed the action of the resistant loads along one of these mechanisms, the structure fails. It is thus easy to understand how the presence of negligible tensile strength can disrupt the behavior of structures as compared to the common elastic ones.

Clearly, other failure modes can occur, such as those depending on the compression strength of the material, described in Chap. 1, or those involving the destabilizing effects of axial loads, covered in Chap. 6. However, this first collapse mode affects a wide variety of structures and is, in practice, the most relevant. It stems from the essential aspects of the behavior of masonry structures, aspects which were fully understood by ancient builders, and which have therefore shaped the course of architecture since the origins up to the 19th century.

The aim of this chapter is to analyze these issues involved in such failure mechanisms. The choice of the most convenient model to use for masonry materials will be addressed first. Any model for describing masonry behavior must be as simple as possible, but at the same time be able to represent its most salient aspects. In this regard, the ingenious Heyman model of masonry as a no-tension material that is rigid under compression (1966) is the most satisfactory for our purposes and will be discussed in the following and constantly referred to throughout this book.

2.2 No-Tension Masonry Models

2.2.1 No-Tension Assumption

There are sound reasons for adopting the assumption of no tensile strength in masonry. First of all, as evidenced in the previous chapter, most masonry materials exhibit very low tensile strength. This is due, rather than to the mortar's low tensile strength, to the very low adhesion between mortar and bricks, which thus represents the weakest link. Moreover, the mortar in historic constructions may be very poor. Masonry may, in some exceptional cases, exhibit non-negligible tensile strength and its behavior could, at first sight, be modeled as a traditional elastic material. However, random dynamic actions, which can produce cracks in the masonry mass, will eventually cause the material to revert to no-tension behavior.

The effects of subsequent slow penetration of humidity into the cracks can then make things even worse. In such cases, it is possible that a masonry structure which in its pristine state is able to sustain the action of given loads by virtue of its initial non-negligible tensile strength, will not be able to sustain the same loads later, when this strength is fading. In such cases the long-term behavior of masonry can be conservatively assumed to follow the no-tension model. A number of examples, some quite striking, will be discussed in the following.

To sum up then, it is clear that the no-tension assumption is well-grounded. Indeed, it is widely adopted in nearly all the mechanical models proposed for historic masonry structures.

2.2.2 The Problem of Elastic Compressive Strains

In order for any model to adequately describe masonry structures, it must account not only for elastic compressive strains, but for the continuous or discontinuous extensional strains associated with cracking as well.

In-depth research studies into the behavior of *elastic no-tension* bodies have been conducted by many authors, among which the works of Di Pasquale (1984), Del Piero (1989), Lucchesi and co-workers (2003), Romano and Romano (1985), Romano and Sacco (1986), Baratta (1982), Angelillo (2010), Trovalusci (1993), Bacigalupo and Gambarotta (2011) etc. All have addressed the general problem of the elastic equilibrium of no-tension bodies and, although numerous, noteworthy stress solutions have been provided (Lucchesi et al. 2008), the much more complex goal of solutions expressed in terms of displacement and strain fields remains incompletely solved to date. The difficulties encountered in this latter research field stem from the fact that the no-tension elastic model cannot easily account for the presence of shear strains. One example in this regard is the masonry panel subjected to bending and axial loads illustrated in Fig. 2.1. The panel, of thickness s, is loaded at its top and base by the pressure p which remains constant along the band a and then varies linearly from p to zero on the remaining band of width b. The borders of the panel are free to deform. The band of width a, under uniform compression p, shortens with respect to the horizontal axis of symmetry, $c-c$. The top sections of the panel band, of width a, move vertically by the amount

$$\Delta = \frac{pas}{Eas}\frac{L}{2} = \frac{p}{E}\frac{L}{2} \qquad (2.1)$$

The side band of width b is axially loaded eccentrically. The left-hand corner of the top section of this band moves vertically and remains in contact with the right-hand corner of the top section of the band of width a. The right-hand band is loaded by the axial load, N, and bending moment, M

$$N = \frac{pbs}{2} \qquad M = \frac{pb^2s}{12} \qquad (2.2)$$

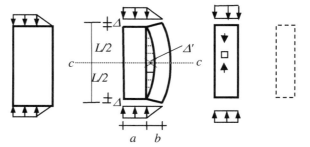

Fig. 2.1 The behavior of an elastic no-tension panel under variable compressions applied at its *base* and *top* sections

The top section of this band sinks under the action of N and rotates counter clockwise under the bending moment M. The total vertical displacement of left-hand corner of this section is

$$\Delta = \frac{pbs}{2}\frac{L}{2Ebs} + M\frac{12L}{2Esb^3}\frac{b}{2} = \frac{p}{E}\frac{L}{4} + \frac{pb^2s}{12}\frac{12L}{2Esb^3}\frac{b}{2} = \frac{p}{E}\frac{L}{4} + \frac{p}{E}\frac{L}{4} = \frac{p}{E}\frac{L}{2} \quad (2.3)$$

which is equal to displacement (2.1). Cracks arise along the vertical line of connection between the two bands. The maximum width Δ' of these discontinuities or detachments is

$$\Delta' = M\frac{12}{2Esb^3}\left(\frac{L}{2}\right)^2 = \frac{pL}{8E}\frac{L}{b} \quad (2.4)$$

Now, as the width b of the band becomes smaller and smaller, at the limit $b \to 0$, we have

$$\lim_{b \to 0} \Delta' = \infty \quad (2.5)$$

In particular, if only one band of the panel is uniformly compressed, while the other band is unloaded, the stress state is defined unequivocally. The loaded side is uniformly compressed and the side band remains unloaded. Strains, on the contrary, behave in a singular way. The unloaded side of the panel detaches from the loaded one and tends towards infinity because a no-tension elastic material is unable to accept the presence of shear strains along the contact zone between the loaded and unloaded sides.

None of the linear segments, all contained within the unloaded masonry solid, can in fact be shortened during deformation. This result emphasizes the singular behavior of such panels and raises doubts about the ability of the elastic no-tension model to realistically represent the response of masonry structures partially loaded over their boundaries. By using the rigid in compression no-tension model, which neglects the elastic strains, the solution is more regular. In this case, in place of the diverging solution, the unloaded part of the panel can detach from the loaded part by an arbitrary, but finite quantity (Fig. 2.2).

Neglecting the elastic compression strains has *no influence* on evaluation of the limit loads. Indeed, from the perspective of Limit Analysis, we know that the presence of elastic strains has no effect on the collapse load, except in the event that the changes in geometry become relevant. This question has been thoroughly studied in the general analysis of elastic-plastic bodies. During the onset of the failure mechanism all stresses remain constant and new elastic strains do not develop (Prager 1959). The same occurs for both elastic and rigid no-tension structures, as will be shown in the following (Fig. 2.3).

Currently, only advanced, computationally demanding programs can provide information on the effects of the elastic compressive strains within masonry model of structures by assuming a small finite tensile strength.

2.2 No-Tension Masonry Models

Fig. 2.2 The rigid no-tension panel partially compressed at its *base* and *top* sections

Fig. 2.3 Collapse of the arch

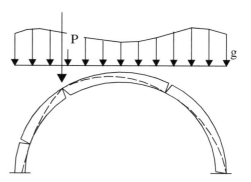

The problem is similar if we want to check the actual equilibrium state. At a first sight the problem can be solved by assuming the *elastic* in compression no tension model that permits to obtain the lacking equations: they express the compatibility of the deformation of the structure with its external environment. On the contrary, the research of equilibrium configuration of the *rigid no tension* masonry structures under a given load distribution becomes *statically undetermined*. The external constraints, representing the structure environment, will suffer a certain amount of deformation, that we will call *settlement*. These deformations can be represented by the settlement of the soil, as in the example of the arch of Fig. 2.4, or by the same deformation of supporting substructures: for instance the deformation of the drum sustaining a dome.

The elastic no tension solution, on the other hand, is dependent on the magnitude of settlement only if this displacement is very small because rapidly it matches the solution corresponding to the rigid no tension model.

Fig. 2.4 Settlement deformation of the masonry arch

These considerations can be illustrated by the following example regarding the behavior of an arch, studied by assuming the elastic in compression no tension model. The example was studied by N. Zani of the university of Florence. The arch, loaded by its own weight, suffers an horizontal settlement at its springing. It is required to evaluate the variation of the thrust of the arch with the magnitude of the imposed settlement. Figure 2.5 gives the plot of the thrust of the arch versus the magnitude of the imposed settlement.

Starting from the fixed springings condition, the thrust of the arch drops immediately as soon as the settlement occurs and approaches the value of the minimum thrust, about equal to 874 kg, corresponding to the reaching of the mechanism state. Further, Fig. 2.6 shows the pressure lines developing in the arch respectively at the initial state, with fixed springings, and just after the thrust drop.

The assuming the rigid in compression no tension model prevents to describe, during the loading progress, the *gradual* development of cracking in the structure before the reaching the settlement mechanism state. But this phase of working of masonry structures should be *meaningless*, owing to the strong sensitivity of masonry structures to the magnitude of the settlement, as shown by the arch behavior of Fig. 2.5.

Fig. 2.5 Plot of the thrust of the elastic no tension arch versus the horizontal settlement (private communication by N. Zani, Departm. of Constructions, Univ. of Florence, Italy, July 2011)

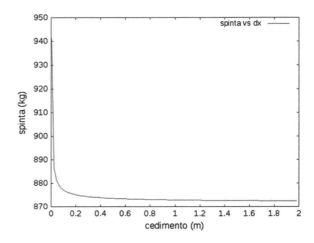

2.2 No-Tension Masonry Models

Fig. 2.6 Pressure lines in the elastic no tension arch at its initial state, with fixed springings, and near the mechanism state, just after the thrust drop. (private communication by N. Zani, Departm. of Constructions, Univ. of Florence, Italy, July 2011)

After the settling displacements have reached a certain level, the structure effectively becomes a mechanism and can adjust by maintaining its internal stresses constant. With reference to the example masonry arch in Fig. 2.4, which has undergone a slight increase in span due to settling of its springers, according to Heyman (1995), it can be seen that the arch is able to adapt itself to the settlement by maintaining the stresses constant regardless of the degree of settling.

In studying the real equilibrium states of masonry structures, equivalent results can thus be obtained by applying the rigid no-tension model. As will be shown in the following, the minimum thrust states can provide the additional equations needed to solve the problem. On the wake of this Heyman suggestion, the so called equilibrium approach (Huerta 2001) can permit to tackle with great simplicity the research of the actual set up of masonry structures. In the end, summing up all the previous discussions, we will make reference in our analysis to the rigid no tension Heyman model.

2.3 The Rigid No-Tension Model

2.3.1 The Heyman's Assumptions

The constitutive assumptions originally formulated by Heyman (1966) are as follows:

(i) *masonry is incapable of withstanding tensions;*
(ii) *masonry has infinite compressive strength;*
(iii) *elastic strains are negligible;*
(iv) *slidings cannot occur because masonry has infinite shear strength;*

The corresponding uniaxial stress–strain relation is shown in Fig. 2.7. The first two of Heyman's assumptions above involve stresses; the latter two, strains.

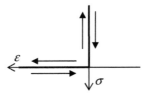

Fig. 2.7 Uniaxial stress–strain relation of rigid no-tension masonry material

2.3.2 The Unit Resistant Masonry Cell

The foregoing assumptions turn out to be very clear if we refer to the elementary resistant cell of the masonry structure, represented by two idealized rigid masonry bricks compressed one against the other by a more or less eccentric axial load and possibly loaded by a shear force (Fig. 2.8).

The two rigid bricks of the unit resistant cell, of height h, cannot deform internally, but they can detach from each other (Fig. 2.9). A *crack* occurs in the cell.

The overall stress state is determined by an axial force N applied at the section's centre, a moment M and a shear force T. Thus, the stress state acting on the unit cell can be represented by the vector

$$\Sigma = \begin{bmatrix} M \\ N \\ T \end{bmatrix} \qquad (2.6)$$

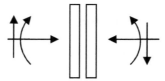

Fig. 2.8 The unit resistant masonry cell

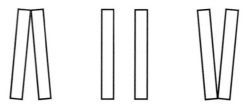

Fig. 2.9 Admissible displacements of the unit cell blocks

2.3 The Rigid No-Tension Model

According to assumption (i), only compressive stresses are consistent. Thus, the eccentricity of the axial load N

$$e = M/N \tag{2.7}$$

must satisfy the inequalities

$$-h/2 \leq e \leq h/2 \tag{2.8}$$

Assumption (iv) prohibits sliding. Consequently, the shear force T will not be bound by any restrictions. Using the reference system N, M, T, any point in this space thus defines a possible loading condition. Since shear T is uninvolved in defining the limit equilibrium between the two ideal bricks, we can consider the projection Σ' of Σ on the coordinate plane N, M, as shown in Fig. 2.10.

According to (2.8), the eccentric axial loading state, defined by the values of the axial force N and moment M (and which we will continue to indicate as Σ for the sake of simplicity), cannot extend beyond the two limit lines represented in the plane (N, M) by the two straight lines OA and OB in Fig. 2.10.

$$M = N\frac{h}{2} \quad M = -N\frac{h}{2} \tag{2.9}$$

The set of *all consistent stress states* in the space M, N, T is thus the region between the two π planes orthogonal to plane $T = 0$ having intersections with the two limit lines $M = Nh/2$ and $M = -Nh/2$. In particular, region Y of the consistent stress states is the region in plane M, N delimited by angle OAB. Specifically, a vector Σ placed along either line OA or line OB, represents an axial force with eccentricity respectively equal to $h/2$ or $-h/2$, as shown in Fig. 2.11a, b. These peculiar stress states are denoted as Σ_o^+ and Σ_o^-, and are expressed as

$$\Sigma_o^+ = \begin{bmatrix} Nh/2 \\ N \\ T \end{bmatrix} \quad \Sigma_o^- = \begin{bmatrix} -Nh/2 \\ N \\ T \end{bmatrix} \tag{2.10}$$

Fig. 2.10 The region Y of the admissible stresses

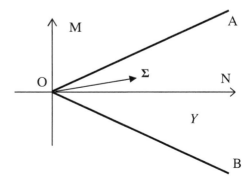

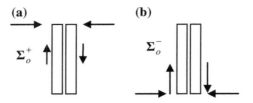

Fig. 2.11 Limit stress states. a, b Limit state

Any deformation of the unit resistant cell will be either zero or a detachment strain and will be represented by the strain vector

$$\mathbf{E} = \begin{bmatrix} \phi \\ \Delta \\ \gamma \end{bmatrix} \qquad (2.11)$$

whose components ϕ, Δ and γ are the elementary strains along which the force components M, N and T respectively do work. No strains can occur until the eccentric axial force reaches the upper or the lower edge of the brick cell section (Fig. 2.11a or b).

In the case of Fig. 2.11a, the stress is Σ_o^+, and the corresponding detachment strain, the vector $\mathbf{E}^+$, is produced by the opening of the hinge situated at the upper edge of the section; in the case of Fig. 2.11b, the strain $\mathbf{E}^-$ corresponds to the stress Σ_o^- These strains are thus defined as

$$\mathbf{E}^+ = \begin{bmatrix} \phi \\ -\phi h/2 \\ 0 \end{bmatrix} \quad \mathbf{E}^- = \begin{bmatrix} -\phi \\ -\phi h/2 \\ 0 \end{bmatrix}. \qquad (2.12)$$

Strains (2.12) are kinematically consistent. Stresses at the limit state (+), represented by the vector Σ_o^+, do not perform work on the corresponding detachment strain $\mathbf{E}^+$. Likewise, at the limit state (−), Σ_o^- does no work on the strain $\mathbf{E}^-$ (Fig. 2.12).

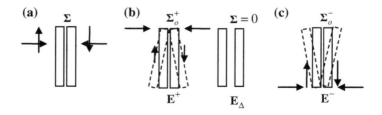

Fig. 2.12 Stress states not producing or producing detachment strains in the cell

2.3 The Rigid No-Tension Model

In fact

$$\Sigma_o^+ \cdot \mathbf{E}^+ = \begin{bmatrix} Nh/2 & N & T \end{bmatrix} \begin{bmatrix} \phi \\ -\phi h/2 \\ 0 \end{bmatrix} = 0$$

$$\Sigma_o^- \cdot \mathbf{E}^- = \begin{bmatrix} -Nh/2 & N & T \end{bmatrix} \begin{bmatrix} -\phi \\ -\phi h/2 \\ 0 \end{bmatrix} = 0$$
(2.13)

The detachment strain $\mathbf{E}^+$ is thus *orthogonal* to the limit line $M = Nh/2$, while vector ε^- is orthogonal to the other limit line $M = -Nh/2$, as shown in Fig. 2.13. Let us now consider a generic consistent state of stress Σ, that is, within the angular region Y: the resistant cell cannot thus be opened, so the stress Σ cannot do any positive work on any detachment strain $\mathbf{E}$ (Fig. 2.14a). Thus the following inequality holds

$$\Sigma \cdot \mathbf{E} \leq 0, \quad \forall \Sigma \in Y, \tag{2.14}$$

where

$$\Sigma \cdot \mathbf{E} = 0, \tag{2.15}$$

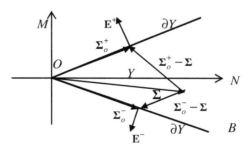

Fig. 2.13 Admissible stress states and detachment strains

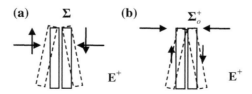

Fig. 2.14 The non-positive work of stresses on the cell strains

iff $\Sigma = \Sigma_o^+$ and $E = E^+$, or $\Sigma = \Sigma_o^-$ and $E = E^-$, (Fig. 2.14b). In short, we have

$$\Sigma_o^\pm \cdot E^\pm = 0 \qquad (2.16)$$

This last inequality represents the *normality rule* connecting the limit stress with the associated strain vectors. Taking both conditions (2.14) and (2.16) into account, we also have

$$(\Sigma_o^\pm - \Sigma) \cdot E^\pm \geq 0, \quad \forall \Sigma \in Y \qquad (2.17)$$

which shows that the vector

$$(\Sigma_o^\pm - \Sigma)$$

is directed outside region Y, as shown in Fig. 2.13. Inequality (2.17) means that the angle between vectors $(\Sigma_o^\pm - \Sigma)$ and $E^\pm$ cannot be larger than $\pi/2$.

It is worthwhile examining the case in which all the stresses acting on the cell are equal to zero. In this case, the cell can deform with all its degrees of freedom, as shown in Fig. 2.15. Any consistent deformation of the cell can be expressed by a linear combination of the basic strain components shown in Fig. 2.15: the two strains E^+ and E^- and the uniaxial extension E_Δ. Vector E^+, originating at the vertex O of the angular region Y in Fig. 2.11, is orthogonal to the limit line $\Sigma = \Sigma_o^+$; likewise vector E^- is orthogonal to the limit line $\Sigma = \Sigma_o^-$.

Finally, the uniaxial detachment strain, E_Δ, is represented by a vector originating at the vertex O and acting along the positive direction of the axis $N(\Delta)$ in Fig. 2.16. The overall strain, obtained by the linear combination of all the three basic vectors E^+, E^-, and E_Δ, lies within the angular region OAB having its vertex at the origin O and bounded by the lines OA and OB, respectively orthogonal to the limit lines $\Sigma = \Sigma_o^+$ and $\Sigma = \Sigma_o^-$.

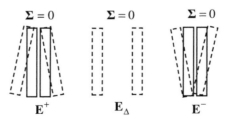

Fig. 2.15 Possible basic strains occurring on the unloaded cell

Fig. 2.16 Overall strain of the unloaded cell as the sum of all the basic strains $\mathbf{E}^+$, $\mathbf{E}^-$ and $\mathbf{E}_\Delta$ of the cell

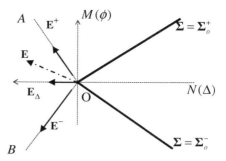

2.3.3 Properties of the Rigid No Tension Material

2.3.3.1 Stability

Let us imagine that a consistent stress state $\mathbf{\Sigma}_A$ is applied to the cell by the agent (A) and that another agent, defined as (B), applies additional stresses $\mathbf{\Sigma}_B$ to the cell, so that the overall stress reaches the limit state $\mathbf{\Sigma}_o^\pm$ and the detachment strain $\varepsilon^\pm$ can occur. The additional stress state $\mathbf{\Sigma}_B$ applied by agent (B) can be expressed as (Fig. 2.17)

$$\mathbf{\Sigma}_B = \mathbf{\Sigma}_o^\pm - \mathbf{\Sigma}_A \tag{2.18}$$

When the $\mathbf{E}^\pm$ occurs, agent (B) does the work $(\mathbf{\Sigma}_o^\pm - \mathbf{\Sigma}_A) \cdot \mathbf{E}^\pm$ (Fig. 2.17), which according to (2.17) cannot be negative.

The unit masonry cell thus requires that agent (B) expends energy to produce detachment strains. According to Drucker (1959), given the assumed constitutive equation, the masonry material may be defined *stable*. However, the behavior of the material would be quite different if, on the contrary, its constitutive equation were based on friction.

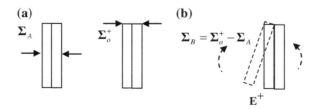

Fig. 2.17 Stresses applied by agent **a** and subsequent agent **b** to reach the limit state

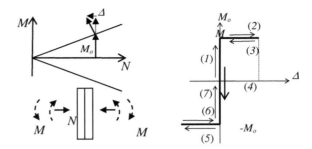

Fig. 2.18 Loading and unloading cycles

2.3.3.2 Elasticity

Despite the foregoing results, the difference between masonry behavior and plastic behavior is significant. Even according to the rigid in compression no-tension model, masonry exhibits a behavior that, due to the lack of internal dissipation, can be considered nonlinear elastic. To illustrate, let us examine the unit resistant cell in Fig. 2.18 under a constant axial load N applied by agent (A) and a sequence of loading–unloading cycles of the additional moment M produced by agent (B).

During the first additional loading, due to an increase in moment M, the eccentricity of the axial load N increases gradually and the moment reaches the limit moment M_o, at which point a small detachment strain increment, $\varDelta\phi$, ensues. Agent (B) expends energy to produce this strain $\varDelta\phi$. Then, in the return cycle, the expended energy is once again *restored* to agent (B) and the diagram of moment M —rotation $\varDelta\phi$ takes the form illustrated in Fig. 2.18. Though in many aspects similar to plastic deformations, the occurrence of detachment strains does not involve energy expense. This result highlights a difference between the response of classical elastic-plastic, or rigid-plastic materials, and masonry materials. This question will be taken up again in Chap. 8 by examining some aspects of the dynamic behavior of some elementary masonry structures.

2.3.3.3 The Coulomb Definition of the Masonry Material

The first three previously cited Heyman assumptions, (i), (ii) and (iii), which define the assumed masonry model, can be easily understood as soon as masonry's low tensile strength is taken into account. It is noteworthy, however, that assumption (iv), at first sight very different from the first three, can also be considered as following directly from them (Como and Grimaldi 1985). This can be corroborated in the context of so-called Mohr-Coulomb materials. In fact, with reference to a plane stress state, according to the Coulomb criterion (1773), the shear strength along the plane under the compressive stress σ is (Fig. 2.19)

2.3 The Rigid No-Tension Model

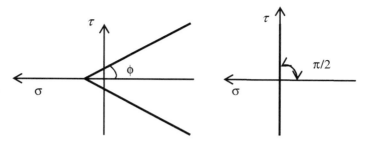

Fig. 2.19 The Coulomb failure criterion, which for $\phi \rightarrow \pi/2$ overlaps the failure criterion of zero maximum tensile stress

$$\tau = c + \sigma \cdot tg\phi \qquad (2.19)$$

where ϕ is the angle of the internal friction. According to the Coulomb criterion, the ratio between uniaxial compressive and tensile stresses takes the form

$$\frac{\sigma_{rc}}{\sigma_{rt}} = \frac{1 + \sin\phi}{1 - \sin\phi} \qquad (2.20)$$

By gradually reducing the ratio σ_{rt}/σ_{rc}, at the limit we obtain

$$\frac{\sigma_{rc}}{\sigma_{rt}} \rightarrow \infty \quad \Rightarrow \quad \frac{1 + \sin\phi}{1 - \sin\phi} \rightarrow \infty \quad \Rightarrow \quad \phi \rightarrow \frac{\pi}{2}. \qquad (2.21)$$

The two limit lines of the Coulomb criterion, which form angle ϕ with the horizontal axis, become vertical when $\phi \rightarrow \pi/2$ and the Coulomb criterion overlaps the criterion of maximum tensile stress (Fig. 2.19). The internal friction strength becomes unbounded and the condition $\sigma \leq 0$ satisfies assumptions (i) and (ii) above.

2.4 The Masonry Continuum

It can seem strange to use the model of "*continuum*" to describe the mechanical behavior of a material having a discrete structure, composed by bricks or stones and eventually by mortar beds.

This is indeed possible because essentials of the masonry behavior don't require to specify the internal composition of the material but only its unilateral response, so conditioned by the dramatically lower strength in tension than in compression. This can be made assuming the above discussed basic features of the masonry behavior and the simple model of the material rigid in compression but without tension strength.

According to this no-tension model, a masonry body can be considered an assemblage of rigid particles held together by the compressive stresses produced by loads. The small size of the stones compared to the dimensions of the body enables it to be considered a continuous body instead of a discrete system of many individual particles. When the compression stresses that held stones together cancel out in some regions of the masonry body, it can get deformed. Cracks can thus occur in the masonry mass: they represent discontinuities or detachments of the displacement fields describing the deformation of the body.

The above assumptions of the rigid in compression no-tension model can thus be adopted for the masonry continuum in a more general form by means of suitable conditions that we will go to look for (Como 1992). At the same time, the use of the continuous medium immediately makes the powerful methods of calculus available and able to describe the discontinuous mechanism displacements, due to cracks formation. The study of the failure of masonry bodies due to the cracking development can thus be made in a general form.

In the next sections the compatibility conditions on loads and stresses will be established and then the compatibility conditions for displacements and strain fields.

Various kinds of mechanism displacements can be in fact defined for rigid in compression no tension bodies. Figure 2.20 shows the mechanism displacement due to relative rotations of the three parts in which an arch has been subdivided. In this case no extension strains occur in the arch segments. Figure 2.21 shows the case of the panel that suffers continuous extension strains producing an inner crack.

Fig. 2.20 Mechanism induced by rigid rotations of parts of the arch

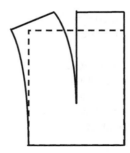

Fig. 2.21 Mechanism induced by extension strains distributed over the upper zone of a panel

2.4 The Masonry Continuum

2.4.1 Compatibility Conditions on Loads

Let us consider a masonry body occupying the region Ω, whose boundary is denoted as $\partial\Omega$, which we assume to be sufficiently regular (Fig. 2.22). The body is loaded by mass and surface loadings $\rho(\Omega)$ and **p**. The loaded part of the body surface $\partial\Omega$ is $\partial\Omega_p$. The surface region $\partial\Omega_r$ is subjected to appropriate boundary conditions. Furthermore, the line f indicates a *crack*. Unlike linear elastic bodies, for masonry structures external loads and internal stresses must necessarily satisfy some compatibility conditions, which are what we are now seeking to define.

For example, tensile forces cannot be applied on the boundary of a no-tension masonry body. Indeed, the surface loads, **p**, must be exerted on the surface $\partial\Omega_p$, so that at any point on surface $\partial\Omega_p$ the following condition holds

$$\mathbf{p}(P) \cdot \mathbf{n} \leq 0, \quad \forall P \in \partial\Omega_p \tag{2.22}$$

where **n** is the unit vector of the outward normal to $\partial\Omega_p$ at point P.

Likewise, the reactions $\mathbf{r}(P)$ that take place along the boundary $\partial\Omega_r$ will also act on $\partial\Omega_r$, and we thus have

$$\mathbf{r}(P) \cdot \mathbf{n} \leq 0, \quad P \in \partial\Omega_r. \tag{2.23}$$

Inequalities (2.22) and (2.23) represent the compatibility conditions on the surface loads **p** and reactions **r**. As consequence of condition (2.22), *self-equilibrated* load distributions cannot be applied to the masonry body.

2.4.2 Compatibility Conditions on Stresses

Tensile stresses can never develop inside the masonry mass, hence

$$\sigma(P) \leq 0. \tag{2.24}$$

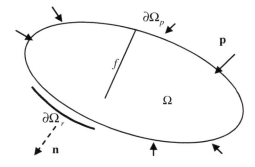

Fig. 2.22 The masonry body occupying the region Ω with boundary $\partial\Omega$

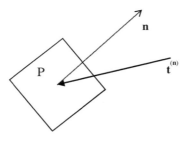

Fig. 2.23 Vector stress on a generic plane without tensile components

Condition (2.24) establishes that at each point P of the body the largest of the principal stresses must be either a compression or equal to zero. Condition (2.24), which was originally formulated by Di Pasquale (1984), defines the domain Y of the statically admissible stress tensors σ.

Let P be an arbitrary point inside the region Ω of the body, dS an oriented surface element passing through P and having the unit vector $\hat{\mathbf{n}}(P)$ on the outward normal originating at P, and $\mathbf{t}^{(\mathbf{n})}(P)$ the associated tension vector representing the force transmitted across the oriented surface element dS. Tensile interactions are not admissible: for any normal $\hat{\mathbf{n}}$ of surface element dS (Fig. 2.23), the following inequality thus holds

$$\mathbf{t}^{(\mathbf{n})}(P) \cdot \hat{\mathbf{n}}(P) \leq 0 \qquad (2.25)$$

Apart from restating the material's incapacity to sustain tensile stresses, condition (2.25) also determines several important properties of the stress, which is the topic of the next section.

2.4.3 First Consequence of the No-Tension Assumption

The main consequences of the assumption that masonry materials cannot withstand any tension at all are:

(1) *Masonry is incompatible with load scattering*
(2) *The internal resistant structures arising in the body depend on the geometry of the applied loads*

Let us consider a masonry wall loaded only on its inner band, as illustrated in Fig. 2.24. It is immediately evident that *no load scattering*, or dispersion, occurs inside the wall (Di Pasquale 1984). Let us section the wall along the line a–a and consider the equilibrium of the corresponding side band of the wall bounded by the line a–a and the corresponding external edge. By considering the vector stress

2.4 The Masonry Continuum

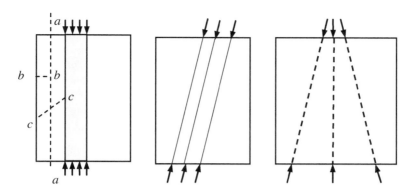

Fig. 2.24 Resistant masonry band without load scattering

$\mathbf{t}^{n^{a-a}}(P)$ and its component along the normal $\mathbf{n}^{a-a}$ to line a–a at each point on line a–a, we get

$$\int_{a-a} \mathbf{t}^{n^{a-a}}(P) \cdot n^{a-a} dS = 0. \qquad (2.26)$$

On the other hand, by accounting for the fact that along line a–a, the stress vectors $\mathbf{t}^{n^{a-a}}(P)$ must satisfy condition (2.25), from condition (2.26) we obtain

$$\mathbf{t}^{(\mathbf{n}^{a-a})}(P) \cdot \hat{\mathbf{n}}^{a-a}(P) = 0, \qquad (2.26')$$

whence, if

$$\mathbf{t}^{(\mathbf{n}^{a-a})}(P) \neq 0, \qquad (2.27)$$

at each point P on the line a–a, a distribution of shear stresses could develop along the edge a–a. This is however inadmissible, because a positive principal stress could occur along the line a–a, which contrasts with (2.24). Consequently, the stress states along a–a is *null*, that is

$$\mathbf{t}^{(\mathbf{n}^{a-a})}(P) = 0. \qquad (2.27')$$

This also occurs along any section b–b transverse to the considered band, as well as along any inclined section c–c in Fig. 2.20. The side bands of the considered masonry are thus unloaded.

In Chap. 6 it will be shown that the lack of loads dispersion actually occurs in masonry walls. This result does not imply that there cannot exist stress distributions radiating from a point of a solid, related to the peculiar load distribution on the surface of the body. The rightmost scheme in Fig. 2.24 shows such a case.

Masonry *channels* the applied loads within its interior to its boundaries along well-defined compression bands determined by the loads' geometry. *The loads determine the resistant masonry structure within the actual masonry body:If the loads change, the resistant masonry structure will consequently change.* Viollet–Le Duc (1854–1868) clearly grasped this peculiar behavior of masonry constructions.

The behavior of linear elastic bodies is completely different, in that such bodies, to the contrary, spread out the action of point loads and, according to the so-called St. Venant principle, 'soften' the actions of self-equilibrated load distributions.

2.4.4 Impenetrability Condition on the Displacement Fields

The strains and detachments that can occur in a masonry body are defined by displacement fields

$$\mathbf{u}(P), \ P \in \Omega, \tag{2.28}$$

called *mechanisms*, where Ω is the region occupied by the body of boundary $\partial\Omega$, which we assume to be sufficiently regular. These displacement functions must satisfy suitable *kinematic compatibility conditions,* which we shall now examine. *Impenetrability* between the rigid stones requires that the displacement function **u** (P) cannot produce any contraction between points connected by segments *entirely contained within the body*. Thus, if (P_1, P_2) is such a pair of points in Ω, and (Q_1, Q_2) is the corresponding pair after the transformation

$$d(Q_1, Q_2) \geq d(P_1, P_2) \tag{2.29}$$

where $d(Q_1, Q_2)$ denotes the distance of the segment connecting the points (Como 1992). According to these assumptions, *no internal sliding can occur.* Consequently during the development of body deformation, cracks or detachments, representing point discontinuities of the displacement function $\mathbf{u}(P)$, must represent *openings*. In short, masonry material can only *expand or be opened*. Thus, the relative displacement between a pair of points located across the line of the crack will occur along the direction *normal* to the crack.

2.4.5 Compatibility Conditions on Strains and Detachments

Let us consider the line f of the crack and its two edges f^- and f^+ (Fig. 2.25). We choose a point P on f and the corresponding points P^- on edge f^- and P^+ on the other edge f^+, obtained by intersecting f^- and f^+ with the line of the unit vector, for instance $\mathbf{n}^-$, located along the outward normal to f^- and passing through P^-

2.4 The Masonry Continuum

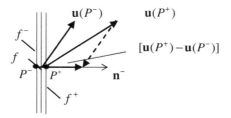

Fig. 2.25 The opening of a crack

(Fig. 2.25). Cracks can *open* only along the direction of $\mathbf{n}^-$. We can thus define the *crack opening vector* or the *detachment vector* as follows:

$$\Delta^{(n^-)}\mathbf{u}(P) = [u(P^+) - u(P^-)]\mathbf{n}^-, \tag{2.30}$$

whence we obtain

$$[\mathbf{u}(P^+) - \mathbf{u}(P^-)] \cdot \mathbf{n}^- = u(P^+) - u(P^-) > 0, \tag{2.31}$$

where $u(P^+)$ and $u(P^-)$ are the scalar values of $\mathbf{u}(P^+)$ and $\mathbf{u}(P^-)$. Consequently, we can define the scalar *crack opening* by means of the *positive* quantity

$$\Delta^{(n^-)}u(P) = u(P^+) - u(P^-) > 0. \tag{2.32}$$

Inequality (2.32) is the first kinematic compatibility condition to be satisfied by the mechanism displacement $\mathbf{u}(P)$ along the cracks (Como 1992).

Strains, expressed by mean derivatives of function $\mathbf{u}(P)$, can conversely develop within those regions of the masonry body where the displacement function $\mathbf{u}(P)$ is sufficiently regular. These strains can only involve expansions, and so at each point of these regions we have the condition

$$\varepsilon(P) \geq 0, \tag{2.33}$$

which signifies simply that the smallest of the principal strains cannot be negative. The set of all strain tensors satisfying the inequality (2.33) is denoted by Y'. Lastly, the presence of external constraints along boundary $\partial\Omega$ of the body requires further restrictions to the displacement fields.

Let $\partial\Omega_r$ be the part of the surface of the body where restraints are applied and $\mathbf{n}$ be the outward normal at the any given point P of $\partial\Omega_r$. During deformation parts of the body's boundary can *detach* from the surface $\partial\Omega_r$, initially in contact with the body. Thus, if $\mathbf{n}$ is the outward normal to $\partial\Omega_r$, the following condition will hold (Fig. 2.26)

Fig. 2.26 Unilateral boundary constraint of the masonry body

$$\mathbf{u}(P) \cdot \mathbf{n} \leq 0, \quad \forall P \in \partial\Omega_r. \tag{2.34}$$

Inequalities (2.32–2.34) represent the kinematic compatibility conditions imposed on displacement functions $\mathbf{u}(P)$. In particular, inequalities (2.32) and (2.33) must be satisfied on the sets of points defined by the displacement field $\mathbf{u}(P)$ itself. Redefinition of the boundary and interior of the cracked masonry body is required and will be addressed in the following.

2.4.6 The Boundary of the Cracked Body

Developing a general analysis of the equilibrium of masonry bodies is generally a very difficult task due to the discontinuities present in the corresponding displacement functions. Volpert and Hujiadev's idea (1985) of including the set of all discontinuity points of the function $\mathbf{u}(P)$ within the body's boundary turns out to be quite fruitful. First of all, the previous conditions on stress and strain and on stress vectors and detachments. Following this suggestion and Como's formulations (1992), let us now consider, for any displacement field $\mathbf{u}(P)$ of the masonry body satisfying all previous compatibility conditions, the set

$$\Gamma(\mathbf{u}) \tag{2.35}$$

of all points of discontinuities, that is, the set of all cracks, *each with its two edges*. This set is a new part of the boundary of the body, generated by the cracks associated to mechanism $\mathbf{u}(P)$. Consequently, we can define, the region $\Omega(\mathbf{u})$ *lacking* cracks associated to mechanism $\mathbf{u}(P)$, that is, the region

$$\Omega(\mathbf{u}) = \Omega/\Gamma(\mathbf{u}). \tag{2.36}$$

As per customary representations, the left-hand scheme in Fig. 2.27 shows the boundary of the masonry body crossed by crack f; the right-hand scheme instead shows the boundary $\partial\Omega(\mathbf{u})$ that includes the two edges of crack f. This latter shows that we can cover the entire boundary $\partial\Omega(\mathbf{u})$, for instance, by circling region $\Omega(\mathbf{u})$ in the counter clockwise direction, that is, having region $\Omega(\mathbf{u})$ always on the left. The kinematic compatibility conditions (2.32–2.34) can now be more thoroughly specified. Thus, following (2.32) and (2.33), the following conditions hold

2.4 The Masonry Continuum

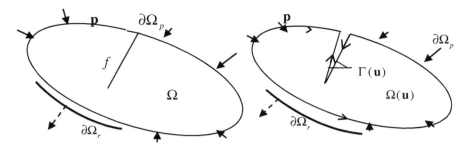

Fig. 2.27 The boundary of the masonry body and the new boundary of the cracked body corresponding to mechanism **u**

$$\Delta^{(n^-)}u(P) = u(P^+) - u(P^-) > 0 \quad P \in \Gamma(\mathbf{u}) \tag{2.37}$$

$$\varepsilon(P) \geq 0 \quad P \in \Omega(\mathbf{u}). \tag{2.37'}$$

The first is defined along the cracks, i.e. on the region $\Gamma(\mathbf{u})$ and the second in the region $\Omega(\mathbf{u})$, free from cracks.

Let us consider now the stress vector along a crack, i.e. on the region $\Gamma(\mathbf{u})$ at the actual stress state of the cracked body.

Recalling the previous definition of crack orientation, let us reconsider line f of the crack having two edges f^- and f^+ and the two points P^- and, P^+ located respectively along the edges f^+ and f^-. The orientation of the crack is defined by the unit vector $\mathbf{n}^-$ of the outward normal to f^- passing through P^- (Fig. 2.28). The *actual* tension vector

$$\mathbf{t}_a^{(n^-)}(P^-) \tag{2.38}$$

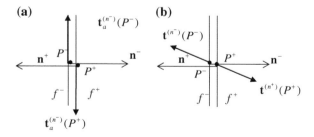

Fig. 2.28 a Actual stress vector, not admissible unless null. **b** Generic stress vector acting on the crack edge

cannot thrust on the *actual* crack edge. Thus, we have

$$\mathbf{t}_a^{(n^-)}(P^-) \cdot \mathbf{n}^- = 0, \quad P^- \in f^-. \tag{2.39}$$

Likewise, considering the opposite tension vector,

$$\mathbf{t}_a^{(n^+)}(P^+) \tag{2.38'}$$

applied at P^+ on the opposite edge f^+ of the crack, equal and opposite to $\mathbf{t}_a^{(n^-)}(P^-)$, such that

$$\mathbf{t}_a^{(n^+)}(P^+) = -\mathbf{t}_a^{(n^-)}(P^-), \tag{2.40}$$

will satisfy the condition

$$\mathbf{t}_a^{(n^+)}(P^+) \cdot \mathbf{n}^+ = 0, \quad P \in f^+ \tag{2.39'}$$

Consequently, if not equal to zero, $\mathbf{t}_a^{(n^-)}(P^-)$, will be orthogonal to $\mathbf{n}^-$ to edge f^- and $\mathbf{t}_a^{(n^+)}(P^-)$ will be orthogonal to the normal $\mathbf{n}^+$ to edge f^+. But shearing stresses cannot be exist along the crack edges f^- and f^+, as shown at Sect. 2.3.3. Consequently, the stress vector is *null* along the crack edges and *no stress transmission* occurs across the cracks at the actual state of the masonry body. Thus we have

$$\mathbf{t}_a^{(n^-)}(P^-) = 0, \quad P^- \in f^- \quad \mathbf{t}_a^{(n^+)}(P^+) = 0, \quad P \in f^+ \tag{2.41}$$

Likewise, if we consider, in place of the actual tension vector acting along the crack, *any admissible* tension vector $\mathbf{t}_a^{(n^-)}(P^-)$ applied at P^-, or $\mathbf{t}_a^{(n^+)}(P^+)$ applied at P^+, we can write

$$\mathbf{t}_a^{(n^-)}(P^-) \cdot \mathbf{n}^- \leq 0, \quad P^- \in f^- \mathbf{t}_a^{(n^+)}(P^+) \cdot \mathbf{n}^+ \leq 0, \quad P^+ \in f^+. \tag{2.42}$$

2.4.7 Coupled Conditions on Stresses and Strains and on Stress Vectors and Detachments

Within the region lacking cracks $\Omega(\mathbf{u})$ only purely stretching strains can occur. These can effectively develop at each point $P \in \Omega(\mathbf{u})$ and along a given direction only if the actual stress component at P and along the same direction vanishes. In brief, extensions can occur only in the directions along which the compressive stress is zero. Thus, the following normality condition holds (Fig. 2.29)

2.4 The Masonry Continuum

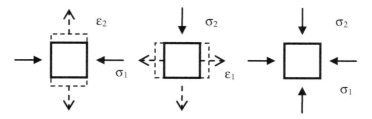

Fig. 2.29 Admissible stresses and strains in masonry material

$$\boldsymbol{\sigma}_a(P) \cdot \boldsymbol{\varepsilon}(P) = 0 \quad P \in \Omega(\mathbf{u}), \tag{2.43}$$

where $\boldsymbol{\sigma}_a(P)$ is the *actual* stress and $\boldsymbol{\varepsilon}(P)$ the *actual* strain at P. It is worth pointing out that condition (2.43) extends to the continuum the simple, intuitive condition (2.15) discussed for the unit resistant masonry cell.

Likewise, as consequence of (2.24) and (2.33), the *generic* admissible stress and the generic admissible strain at a point P of region $\Omega(\mathbf{u})$ will be linked by the following inequality

$$\boldsymbol{\sigma}(P) \cdot \boldsymbol{\varepsilon}(P) \leq 0 \quad P \in \Omega(\mathbf{u}) \quad \forall \boldsymbol{\sigma} \in Y, \ \forall \boldsymbol{\varepsilon} \in Y'. \tag{2.44}$$

Equivalent relations hold along the cracks. By coupling the conditions involving generic admissible stress vectors and detachments, we can immediately establish the following inequalities

$$\mathbf{t}_a^{(n^-)}(P^-) \cdot \Delta^{(n^-)}\mathbf{u}(P) \leq 0, \quad \mathbf{u} \in M, \ P \in f^- \tag{2.45}$$

or,

$$\mathbf{t}_a^{(n^+)}(P^+) \cdot \Delta^{(n^-)}\mathbf{u}(P) \geq 0, \quad \mathbf{u} \in M, \ P \in f^+, \tag{2.45'}$$

which must be satisfied by *any* admissible stress vector $\mathbf{t}^{(n)}(P)$ and admissible detachment $\Delta^{(n^-)}(P)\mathbf{u}$. Conditions (2.45) and (2.45') show that the admissible stress vectors, cannot never *pull* within the interior of the body.

2.4.8 Specifications for One-Dimensional Masonry Systems

The previous general definitions can be specified for the simple case of one-dimensional structures, as, for instane, for a masonry arch.

Let us consider, for instance, the arch illustrated in Fig. 2.30, whose pressure line, wholly within the arch, meets the arch extrados at points A and C and its intrados at B and D. Hinges are thus formed at A, B, C and D. A mechanism displacement ensues. The corresponding vertical displacements of the arch are

Fig. 2.30 The pressure line of the arch

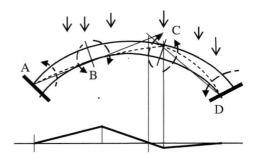

Fig. 2.31 Symmetric five-hinge mechanism in an arch

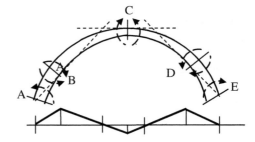

shown in Fig. 2.30. The mechanism is kinematically compatible. In fact, the counter clockwise rotation at hinge A of segment AB detaches section A from the left springing of the arch; likewise, the rotation of CD at D, also counter clockwise, detaches section D from the right springing. Relative rotations at B and C occur with the formation of opening hinges at B and C. Figure 2.31 shows a symmetric mechanism composed of the four rigid segments AB, BC, CD and DE connected by the five hinges A, B, C, D and E. The relative rotations occurring between contiguous rigid segments give rise to compatible deformations.

Lastly, Fig. 2.32 shows two mechanisms, the left compatible, the right incompatible.

2.4.9 Indeformable Masonry Structures

Due to their geometries and constraints some masonry structures cannot be deformed to give rise to mechanisms: interpenetration of the material arises for any hinge position. Examples of structures of this type are the flying buttress or the platbands inserted between fixed springers (Fig. 2.33, left and middle). For such structural systems, in short, we have

$$M = \varnothing. \tag{2.46}$$

This condition is a consequence of the assumption of the compressionally rigid material. Another example is the stair ramp shown at the left in Fig. 2.33.

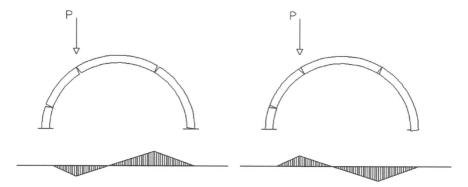

Fig. 2.32 Compatible mech. Incompatible mech

Fig. 2.33 Examples of no mechanisms masonry structures

2.5 Equilibrium and Compatibility

2.5.1 Principle of Virtual Work

Let us consider a masonry body under the action of the loads **p** at an admissible equilibrium state. Let $\delta\mathbf{u}(P) \in M$ be a mechanism field, representing a kinematically admissible virtual displacement of the body. Cracks will arise during the development of the virtual mechanism $\delta\mathbf{u}(P)$ and $\Gamma(\delta\mathbf{u})$ will be the region representing the cracks' boundaries. Consequently,

$$\Omega(\delta\mathbf{u}) = \Omega/\Gamma(\delta\mathbf{u}) \tag{2.47}$$

is the corresponding region occupied by the *crack-free* body. The new boundary of the body associated to the virtual displacement $\delta\mathbf{u}$ is then obtained by adding to the initial boundary $\partial\Omega$ the crack boundaries, that is,

$$\partial\Omega(\delta\mathbf{u}) = \partial\Omega \cup \Gamma(\delta\mathbf{u}). \tag{2.48}$$

The virtual displacement $\delta\mathbf{u}$ satisfies the above kinematic compatibility conditions along the discontinuities' surfaces $\Gamma(\delta\mathbf{u})$ and we thus have

$$\Delta^{(n^-)}\delta u(P) = \delta u(P^+) - \delta u(P^-) > 0 \quad P \in \Gamma(\delta\mathbf{u}). \tag{2.49}$$

Moreover, in the inner crack-free region $\Omega(\delta\mathbf{u})$, the corresponding strains, in conformity with (2.33), will satisfy the following inequality

$$\delta\varepsilon(P) \geq 0, \quad \delta\varepsilon = D\delta\mathbf{u}(P), \quad P \in \Omega(\delta\mathbf{u}), \tag{2.50}$$

where D is the known operator that associates the strains $\delta\varepsilon$ to the displacement $\delta\mathbf{u}$ in $\Omega(\delta\mathbf{u})$. Finally, according to conditions (2.34), on the boundary $\partial\Omega_r$ we have

$$\delta\mathbf{u}(P) \cdot \mathbf{n} \leq 0, \quad P \in \partial\Omega_r \tag{2.51}$$

The virtual mechanism $\delta\mathbf{u}$ will satisfy the *kinematic compatibility* conditions (2.49–2.51). Likewise, we can define the *static compatibility* conditions for the admissible stresses in equilibrium with loads $\mathbf{p}$. Thus, from (2.44) we have

$$\boldsymbol{\sigma}(P) \cdot \delta\boldsymbol{\varepsilon}(P) \leq 0, \quad \delta\varepsilon(P) = D\delta\mathbf{u}(P), \quad P \in \Omega(\delta\mathbf{u}), \quad \delta\mathbf{u} \in M. \tag{2.52}$$

At the same time, considering the two points P^- and P^+ located along the direction of the outward normal to a virtual crack, where the jump $\Delta^{(n^-)}\mathbf{u}(P^-)$ of $\delta\mathbf{u}(P)$ occurs, from (2.45) we have

$$\mathbf{t}^{(n^+)}(P^+) \cdot \Delta^{(n^-)}(P^-)\delta\mathbf{u} \geq 0, \quad \delta\mathbf{u} \in M, \, P \in f^+. \tag{2.53}$$

Lastly, the reaction $\mathbf{r}$, acting along $\partial\Omega_r$ will satisfy the condition

$$\mathbf{r}(P) \cdot \delta\mathbf{u}(P) \geq 0, \quad P \in \partial\Omega_r. \tag{2.54}$$

Inequalities (2.52–2.54) are the coupled compatibility conditions associated to the virtual mechanism $\delta\mathbf{u}(P)$. These conditions, together with the internal equilibrium equations, define the admissible equilibrium state of the masonry solid, which, for the sake of simplicity, we will indicate as AE.

The equilibrium of the body is governed by the principle of the so called *virtual works* or of the *virtual displacements*. This principle will take a *particular form* that is representative of the compressionally rigid no-tension bodies that will be analyzed along the lines previously set forth by Como (1992, 2012). At any point P within the region $\Omega(\delta\mathbf{u})$, the stress field $\boldsymbol{\sigma}$ will satisfy inequality (2.52) together with the following internal equilibrium equation

$$\sigma_{ij,j} + \rho_i = 0. \tag{2.55}$$

Now let dV be a generic volume element around P in $\Omega(\delta\mathbf{u})$. The virtual work done to displace this element is

2.5 Equilibrium and Compatibility

$$(\sigma_{ij,j} + \rho_i)\delta u_i dV. \quad (2.56)$$

According to the equilibrium Eq. (2.55), this work vanishes. Integration of (2.56) over the volume $\Omega(\delta\mathbf{u})$ thus yields

$$\int_{\Omega(\delta u)} (\sigma_{ij,j} + \rho_i)\delta u_i dV = 0. \quad (2.57)$$

From (2.57), the Gauss–Green theorem, together with some tensor calculations and previous specifications, enables us to obtain

$$\int_{\Omega(\delta u)} \sigma_{ij,j}\delta\varepsilon_{ij}dV = \int_{\partial\Omega(\delta u)} t_i^{(\mathbf{n})}\delta u_i dS + \int_{\Omega(\delta u)} \rho_i \delta u_i dV, \quad (2.58)$$

where **n** is the unit vector along the outward normal to the crack surface.

Figure 2.34a shows a masonry arch in an admissible equilibrium state under the action of loads **p** and internal stress **σ**. Figure 2.34b also shows the displacement field $\delta\mathbf{u}$ with hinges A, B, C and D, together with the corresponding internal cracks BB' and CC'. Figure 2.34a, b also show:

- the cracks' boundaries $\Gamma(\delta\mathbf{u})$;
- the region $\Omega(\delta\mathbf{u}) = \Omega/\Gamma(\delta\mathbf{u})$ lacking cracks;
- the overall boundary of the body, including the crack boundaries $\partial\Omega(\delta\mathbf{u}) = \partial\Omega \cup \Gamma(\delta\mathbf{u})$.

The entire boundary can also be specified by the union of the boundaries $\Gamma(\delta\mathbf{u})$, $\partial\Omega_r$ and $\partial\Omega_p$

$$\partial\Omega(\delta\mathbf{u}) = \Gamma(\delta\mathbf{u}) \cup \partial\Omega_r \cup \partial\Omega_p. \quad (2.59)$$

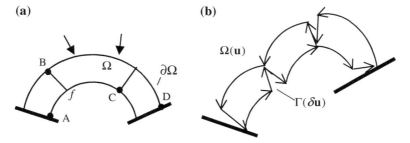

Fig. 2.34 The boundary of the crack-free arch and the new boundary including the cracks associated to virtual mechanism $\delta\mathbf{u}$

The internal work (2.58) can now be written in a more explicit form. In fact, according to (2.59), we have

$$\int_{\Omega(\delta u)} \sigma_{ij,j}\delta\varepsilon_{ij}dV = \int_{\Gamma(\delta u)} t_i^{(\mathbf{n})}\delta u_i dS + \int_{\partial\Omega_r} r_i^{(\mathbf{n})}\delta u_i dS + \int_{\partial\Omega_p} p_i^{(\mathbf{n})}\delta u_i dS + \int_{\Omega(\delta u)} \rho_i\delta u_i dV. \tag{2.60}$$

To work out the first integral in the second member of (2.60), by moving around the whole contour of the body, the virtual work of the interactions $t_i^{(\mathbf{n})}$ can be evaluated along each of the two edges of the cracks (Fig. 2.34b). For the sake of simplicity, we can refer to a single crack alone and write

$$\Gamma(\delta\mathbf{u}) = \Gamma_1(\delta\mathbf{u}) \cup \Gamma_2(\delta\mathbf{u}), \tag{2.61}$$

where $\Gamma_1(\delta\mathbf{u})$ and $\Gamma_2(\delta\mathbf{u})$ are the two equal surfaces representing the two edges of the crack. Evaluating the first integral in the second member of (2.60) thus gives

$$\int_{\Gamma(\delta u)} t_i^{(\mathbf{n})}\delta u_i dS = \int_{\Gamma_1(\delta u)} t_i^{(\mathbf{n}^-)}\delta u_i(P^-)dS + \int_{\Gamma_2(\delta u)} t_i^{(\mathbf{n}^+)}\delta u_i(P^+)dS. \tag{2.62}$$

On the other hand, using expression (2.32) for the crack opening $\Delta^{(n^-)}u(P)$, we have

$$\delta u_i(P^-) = \delta u_i(P^+) - \Delta^{(n^-)}\delta u_i(P); \tag{2.63}$$

Substituting (2.63) into (2.62) gives

$$\int_{\Gamma(\delta u)} t_i^{(\mathbf{n})}\delta u_i dS = \int_{\Gamma_1(\delta u)} t_i^{(\mathbf{n}^-)}\delta u_i(P^+)dS$$
$$- \int_{\Gamma_1(\delta u)} t_i^{(\mathbf{n}^-)}\Delta^{(n^-)}\delta u_i(P)dS + \int_{\Gamma_2(\delta u)} t_i^{(\mathbf{n}^+)}\delta u_i(P^+)dS. \tag{2.62'}$$

Furthermore, by taking into account that

$$t_i^{(\mathbf{n}^-)} = -t_i^{(\mathbf{n}^+)}, \tag{2.64}$$

2.5 Equilibrium and Compatibility

we get

$$\int_{\Gamma(\delta u)} t_i^{(\mathbf{n})} \delta u_i dS = - \int_{\Gamma_1(\delta u)} t_i^{(\mathbf{n}^+)} \delta u_i(P^+) dS$$
$$- \int_{\Gamma_2(\delta u)} t_i^{(\mathbf{n}^-)} \Delta^{(n^-)} \delta u_i(P) dS + \int_{\Gamma_2(\delta u)} t_i^{(\mathbf{n}^+)} \delta u_i(P^+) dS. \quad (2.62')$$

On the other hand,

$$\int_{\Gamma_1(\delta u)} t_i^{(\mathbf{n}^+)} \delta u_i(P^+) dS = \int_{\Gamma_2(\delta u)} t_i^{(\mathbf{n}^+)} \delta u_i(P^+) dS. \quad (2.65)$$

In fact, the integral is evaluated on the same surface because $\Gamma_1(\delta \mathbf{u})$ and $\Gamma_2(\delta \mathbf{u})$ are equal, hence

$$\int_{\Gamma(\delta u)} t_i^{(\mathbf{n})} \delta u_i dS = - \int_{\Gamma_2(\delta u)} t_i^{(\mathbf{n}^-)} \Delta^{(n^-)} \delta u_i \quad (2.66)$$

or

$$\int_{\Gamma(\delta u)} t_i^{(\mathbf{n})} \delta u_i dS = \int_{\Gamma_1(\delta u)} t_i^{(\mathbf{n}^+)} \Delta^{(n^-)} \delta u_i dS \quad (2.66')$$

Finally, summing up the work along all the crack surfaces, we get

$$\int_{\Omega(\delta u)} \sigma_{ij,j} \delta \varepsilon_{ij} dV = \sum_k \int_{\Gamma_k^1(\delta u)} t_i^{(\mathbf{n}^+)} \Delta^{(n^-)} \delta u_i dS + \int_{\partial \Omega_r} r_i^{(\mathbf{n})} \delta u_i dS + \int_{\partial \Omega_p} p_i^{(\mathbf{n})} \delta u_i dS + \int_{\Omega(\delta u)} \rho_i \delta u_i dV.$$
$$(2.67)$$

With the following definitions

$$\{\mathbf{t}^{(\mathbf{n}^+)}, \Delta^{(\mathbf{n}^-)} \delta \mathbf{u}\} = \sum_k \int_{\Gamma_k^1(\delta u)} t_i^{(\mathbf{n}^+)} \Delta^{(n^-)} \delta u_i dS \quad \langle \mathbf{r}, \delta \mathbf{u} \rangle = \int_{\partial \Omega_r} r_i^{(\mathbf{n})} \delta u_i dS; \quad (2.68)$$

$$\langle \mathbf{p}, \delta \mathbf{u} \rangle = \int_{\partial \Omega_p} p_i^{(\mathbf{n})} \delta u_i dS + \int_{\Omega(\delta u)} \rho_i \delta u_i dV; \quad \langle \mathbf{\sigma}, \delta \mathbf{\varepsilon} \rangle = \int_{\Omega(\delta u)} \sigma_{ij,j} \delta \varepsilon_{ij} dV, \quad (2.68')$$

condition (2.67) becomes

$$\langle \boldsymbol{\sigma}, \delta\boldsymbol{\varepsilon} \rangle = \{\mathbf{t}^{(\mathbf{n}^+)}, \Delta^{(\mathbf{n}^-)} \delta\mathbf{u}\} + \langle \mathbf{r}, \delta\mathbf{u} \rangle + \langle \mathbf{p}, \delta\mathbf{u} \rangle \quad \forall \delta\mathbf{u} \in M. \tag{2.69}$$

Moreover, inequalities (2.52), (2.53) and (2.54) can be expressed as

$$\langle \boldsymbol{\sigma}, \delta\boldsymbol{\varepsilon} \rangle \leq 0 \quad \{\mathbf{t}^{(\mathbf{n}^+)}, \Delta^{(\mathbf{n}^-)} \delta\mathbf{u}\} \geq 0 \quad \langle \mathbf{r}, \delta\mathbf{u} \rangle \geq 0. \tag{2.70}$$

Vice versa, working back step by step from Eq. (2.69), we arrive at Eq. (2.55), of course, in obedience of all compatibility conditions.

Thus, conditions (2.69) and (2.70) are *necessary and sufficient* for the admissible equilibrium and provide a suitable representation for the principle of virtual work for rigid no-tension masonry bodies (Como 1992). Comparing the current formulation of the same principle for linear elastic solids with this one concerning no-tension bodies, the difference is that here the work of the tension vectors of the virtual detachments $\Delta\delta\mathbf{u}$ must be added, as must also the associated compatibility conditions (2.70). Figure 2.41 shows the two systems, of forces and deformations, respectively statically and kinematically compatible, connected together by condition (2.69), representing the principle of virtual works.

2.5.2 Variational Inequality for the Existence of the Admissible Equilibrium State

A correctly constrained linear elastic structure is always able to reach a consistent, equilibrated configuration. In brief, the problem of the linear elastic equilibrium admits a solution for any loads distribution. However, for no-tension masonry structures this no longer holds true. Masonry structures can, on the contrary, collapse under loading **p**. It is therefore useful to seek conditions, involving only *known* quantities, that enable predicting whether any given body made of a rigid no-tension material can withstand the action of the assigned loads. Although conditions (2.69) and (2.70) are necessary and sufficient to guarantee the existence of admissible equilibrium, they must be satisfied by both the loads and the internal stresses. However, these latter may be a priori unknown.

In this section we shall prove that the variational inequality on loads **p** alone

$$\langle \mathbf{p}, \delta\mathbf{u} \rangle \leq 0, \quad \forall \, \delta\mathbf{u} \in M \tag{2.71}$$

is necessary and sufficient to guarantee the existence of the AE state.

It should be noted that the mechanisms $\delta\mathbf{u}$ represent the various deformation modes of the body. Inequality (2.71) thus simply means that the body is in an AE state under loads **p** *iff* the work of these loads **p** *is not positive* along any possible deformation of the body.

2.5 Equilibrium and Compatibility

Necessity follows immediately from (2.69) and (2.70). In the context of elastic no-tension models, proofs of the sufficiency of condition (2.71) have been furnished by Romano and Romano (1985) and Romano and Sacco (1986). Another simple proof, in the framework of the rigid no-tension model, has been given by Como (1992). The main lines of this latter proof are the following.

If the variational inequality (2.71) was only necessary, but insufficient, it could be also satisfied by loads **p** unsustainable by the body in the AE state. Such a situation is however impossible. It will in fact be shown that any load **p** that is unsustainable by the body in an AE state and that consequently sets the body in motion (Fig. 2.35), does *positive* work on displacement **v** along which the body begins to move. This contradiction with the assumption proves the statement.

Let us therefore assume, *ad absurdum*, that, in spite of condition (2.71), the body is *not* in an AE state under loads **p**, and let us then consider the motion defined by the velocity field $\mathbf{v}(P, t)$ initiated just after application of the loads. A simple example is represented by the collapse of an arch loaded by its weight and a central point load, as in Fig. 2.36. The body will begin to move. At any instant during the motion, the stress field $\boldsymbol{\sigma}$ will satisfy the internal constraints, i.e., condition (2.24), and the normality rule (2.43). Thus,

$$\boldsymbol{\sigma}(P,t) \leq 0, \quad \boldsymbol{\sigma}(P,t) \cdot \underline{\boldsymbol{\varepsilon}}(P,t) = 0, \; P \in \Omega(\mathbf{v}), \quad \forall t \geq 0. \tag{2.72}$$

By applying the virtual work equation in which we take, as virtual displacement $\delta \mathbf{u}$, the actual displacement occurring during movement of the body over time interval dt, we have

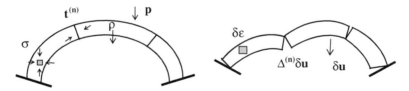

Fig. 2.35 The two systems, forces and deformations, respectively statically and kinematically compatible, connected by the virtual work equation

Fig. 2.36 The masonry arch at incipient failure

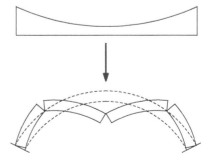

$$\delta \mathbf{u} = \mathbf{v}(P,t)dt. \tag{2.73}$$

Thus, with

$$\delta \varepsilon = \underline{\dot{\varepsilon}}(P,t)dt, \quad \Delta^{(\mathbf{n})}\delta \mathbf{u} = \Delta^{(\mathbf{n})}\mathbf{v}(P,t)dt \tag{2.73'}$$

where $\dot{\varepsilon} = d\varepsilon/dt$ and taking into account the inertial forces produced in the body due to acceleration $\dot{\mathbf{v}}$, we obtain

$$\langle \boldsymbol{\sigma}, \dot{\varepsilon} \rangle = \{\mathbf{t}^{(\mathbf{n}^+)}, \Delta^{(\mathbf{n}^-)}\mathbf{v}\} + \langle \mathbf{r}, \mathbf{v} \rangle + \langle \mathbf{p}, \mathbf{v} \rangle - \langle \rho \dot{\mathbf{v}}, \mathbf{v} \rangle, \ t > 0. \tag{2.74}$$

Thus, during the motion we have, with the previous notation,

$$\langle \boldsymbol{\sigma}(P), \dot{\varepsilon} \rangle = 0, \ \{\mathbf{t}^{(\mathbf{n}^+)}, \Delta^{(\mathbf{n}^-)}\mathbf{v}\} = 0 \quad \langle \mathbf{r}, \mathbf{v} \rangle = 0. \tag{2.75}$$

The first of conditions (2.75) follows from the second of (2.72). Regarding the second condition, note that when cracks begin to open, the stress interaction $\mathbf{t}^{(\mathbf{n}^+)}$ vanishes since $\Delta^{(\mathbf{n}^-)}\mathbf{v} \neq 0$ along them.

Likewise, if during the motion, the body is displaced away from the constraint boundary $\partial \Omega_r$, there $\mathbf{v} \neq 0$ and consequently $\mathbf{r} = 0$, and the last condition in (2.75) also holds. Thus, condition (2.74) becomes

$$\langle \mathbf{p}, \mathbf{v} \rangle - \langle \rho \dot{\mathbf{v}}, \mathbf{v} \rangle = 0, \quad t > 0. \tag{2.76}$$

The kinetic energy of the body is

$$T = 1/2 \langle \rho \mathbf{v}, \mathbf{v} \rangle \tag{2.77}$$

and its rate of change is

$$\frac{dT}{dt} = \langle \rho \dot{\mathbf{v}}, \mathbf{v} \rangle. \tag{2.78}$$

Equation (2.76) thus becomes

$$\langle \mathbf{p}, \mathbf{v} \rangle = \frac{dT}{dt}, \quad t > 0. \tag{2.79}$$

However, when the body begins to move the derivative dT/dt of the kinetic energy can only be positive. Thus, if loads $\mathbf{p}$ are applied and cannot be sustained, the body begins to move and the work done by loads $\mathbf{p}$ along the displacement of this motion is positive. Such a result contradicts assumption (2.71), whence we can conclude that if $\langle \mathbf{p}, \delta \mathbf{u} \rangle \leq 0, \forall \delta \mathbf{u} \in M$, the body is in an AE state.

2.5.3 No-Existence of Self-Equilibrated Stress Fields in Deformable Structures

Within the framework of no-tension models, another typical aspect of the behavior of masonry bodies that can be deformed with mechanisms, is their inability to sustain self-equilibrated stresses. Likewise, the reactions of constraints vanish in the absence of external loads.

The proof of this property follows immediately from application of the principle of virtual work (2.69), together with compatibility conditions (2.70). In fact, with vanishing loads, the following conditions hold

$$\langle \sigma, \delta\varepsilon \rangle = \{ \mathbf{t}^{(\mathbf{n}^+)}, \Delta^{(\mathbf{n}^-)} \delta\mathbf{u} \} + \langle \mathbf{r}, \delta\mathbf{u} \rangle \quad \forall \delta\mathbf{u} \in M; \tag{2.69'}$$

$$\langle \sigma, \delta\varepsilon \rangle \leq 0 \quad \{ \mathbf{t}^{(\mathbf{n}^+)}, \Delta^{(\mathbf{n}^-)} \delta\mathbf{u} \} \geq 0 \quad \langle \mathbf{r}, \delta\mathbf{u} \rangle \geq 0 \quad \forall \delta\mathbf{u} \in M. \tag{2.70'}$$

Both conditions (2.69') and (2.70') can be satisfied, $\forall \delta\mathbf{u} \in M$, only if

$$\langle \sigma, \delta\varepsilon \rangle = 0 \{ \mathbf{t}^{(\mathbf{n}^+)}, \Delta^{(\mathbf{n}^-)} \delta\mathbf{u} \} = 0 \quad \langle \mathbf{r}, \delta\mathbf{u} \rangle = 0. \tag{2.80}$$

We can now assume that any straight segment S leaving from any point of the *constrained* boundary $\partial\Omega_r$ intersects the opposite side of the body only at points of its *free* boundary, as certainly it turns out for all the structures rising from the ground (Fig. 2.37). Extensions can thus develop along the segments S.

For any point P along S. It will be thus possible to choose a mechanism $\delta\mathbf{u}$ such that in a neighborhood I(P) of P the strains $\varepsilon(\delta\mathbf{u})$ along S will be positive. Consequently, by virtue of (2.80), $\sigma \equiv 0$ in I(P). We can repeat the argument for any other point on S so that $\sigma \equiv 0$ along all the segment S. All the points in the body can be intercepted by segments S and $\sigma \equiv 0$ in the whole body. The same results is obtained considering the occurrence of detachments.

In conclusion, taking into account that conditions (2.80) must be satisfied for each mechanism $\delta\mathbf{u}$, we get

$$\sigma \equiv 0; \quad \mathbf{t}^{(\mathbf{n})} \equiv 0; \quad \mathbf{r} \equiv 0. \tag{2.81}$$

in the *unloaded* masonry structure. Thus, self-stresses and reactions vanish in unloaded masonry structures that can be deformed through mechanisms. Deformable masonry structures can thus be considered *statically determinate systems*.

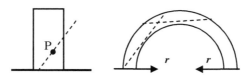

Fig. 2.37 Deformable systems: no self-equilibrated stresses and constraint reactions

2.5.4 Indeformable Structures: Statically Indeterminate Behaviour

For indeformable masonry systems the set M of the kinematically admissible mechanisms is empty. Self-equilibrated stresses do exist in indeformable masonry systems. In such cases, conditions (2.80) are in fact satisfied by

$$\delta\boldsymbol{\varepsilon} \equiv 0 \quad \Delta^{(\mathbf{n}^-)}\delta\mathbf{u} \equiv 0 \quad \delta\mathbf{u} \equiv 0, \tag{2.82}$$

which, by virtue of (2.80), yields

$$\boldsymbol{\sigma} \neq 0 \quad \mathbf{t}^{(\mathbf{n})} \neq 0 \quad \mathbf{r} \neq 0. \tag{2.83}$$

Figure 2.37 shows some examples of self-stresses acting in indeformable masonry systems. Note that such a state requires the presence of fixed constraints.

The existence of constraint reactions in absence of external loads allows to better define the indeformable structures. For these systems it is in fact possible, starting from a point of the laterally constrained sections, to trace at least one straight line wholly contained within the structure (Fig. 2.38). The flying buttress, sketched in left of Fig. 2.35, is one example of such an indeformable system.

2.5.5 Admissible Equilibrium in One-Dimensional Systems

All the foregoing conditions governing the admissible equilibrium of masonry bodies take simpler forms when referred to a one-dimensional structure.

Let us consider the masonry arch shown in Fig. 2.39. It is in an AE state under the action of loads **p**. Figure 2.38 shows the pressure line in the arch as the curve joining all points traversed in each section of the arch by the resultant of all the forces preceding or following the section.

The internal stresses, σ, are all compressive in each section of the arch. In one-dimensional systems potential stretching strains of the voussoirs lead to displacements negligible with respect to those produced by the relative rotations at the

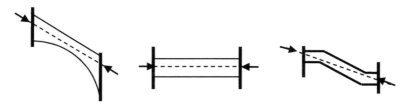

Fig. 2.38 Indeformable systems: existence of self-equilibrated stresses and reactions

2.5 Equilibrium and Compatibility

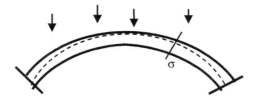

Fig. 2.39 The arch in a statically admissible equilibrium state: the pressure line is wholly contained within the arch

hinges. In defining the corresponding mechanisms, it is thus possible to consider only detachments $\Delta^{(n^-)} \mathbf{u}$ arising in the voussoirs, where hinges develop, and consequently neglect any strains, ε, that may spread into the voussoirs, occupying the region, $\Omega(\mathbf{u})$, defined above.

We can also assume that the external constraints are fixed. Hence, neither the work of reactions, $\mathbf{r}$, nor the work of stresses, $\boldsymbol{\sigma}$ on the strains, ε, will appear in the virtual work equation. With these restrictions, this equation takes the simpler form

$$\{\mathbf{t}^{(\mathbf{n}^+)}, \Delta^{(\mathbf{n}^-)} \delta \mathbf{u}\} + \langle \mathbf{p}, \delta \mathbf{u} \rangle = 0 \quad \forall \delta \mathbf{u} \in M, \tag{2.69'}$$

associated to the admissibility condition

$$\{\mathbf{t}^{(\mathbf{n}^+)}, \Delta^{(\mathbf{n}^-)} \delta \mathbf{u}\} \geq 0. \tag{2.70'}$$

Recalling previous definitions, note that the symbol in parentheses is the sum of the product of the stress vectors by the corresponding virtual detachments.

The forces acting on the side sections of a small voussoir of the arch are equal and opposite to the resultant of the stress vectors, $\mathbf{t}^{(\mathbf{n}^+)}$ and $\mathbf{t}^{(\mathbf{n}^-)}$, acting on the sections of the arch facing the side sections of the voussoir.

They can be expressed in terms of the components N, M and T of the resultant vector, Σ (Fig. 2.40). Consequently, if, according to (2.45'), the work of $\mathbf{t}^{(\mathbf{n}^+)}(P^+)$ on the detachment $\Delta^{(\mathbf{n}^-)} \delta \mathbf{u}$ is non-negative, the work of the equal and opposite actions on the detachments themselves will be non-positive (Fig. 2.41).

Fig. 2.40 Internal actions and reactions inside the masonry arch

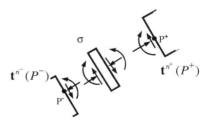

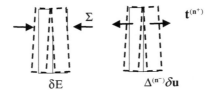

Fig. 2.41 The opposite in sign works of stresses σ and of the stress vectors $\mathbf{t}^{(n+)}$

The resultant of forces $\mathbf{t}^{(n^+)}$ or $\mathbf{t}^{(n^-)}$ acting on the transverse sections delimiting the crack, where a hinge is formed, can be decomposed into the components of the axial force, N, the bending moment, M, and the shear, T, of the resultant vector $\mathbf{\Sigma}$, as defined by (2.6). At the same time, the detachments, $\Delta^{(n^-)}\delta\mathbf{u}$, can, in turn, be expressed in terms of the virtual deformation vector, $\delta\mathbf{E}$, whose components are the axial displacement, $\delta\Delta$, and the relative rotation, $\delta\phi$, defined according to (2.11). In brief, for the sake of simplicity, we can write

$$\{\mathbf{t}^{(n^+)}, \Delta^{(n^-)}\delta\mathbf{u}\} = -\langle \mathbf{\Sigma}, \delta\mathbf{E}\rangle, \qquad (2.84)$$

and the equation of virtual work (2.69′) becomes

$$\langle \mathbf{p}, \delta\mathbf{u}\rangle = \langle \mathbf{\Sigma}, \delta\mathbf{E}\rangle \quad \forall \delta\mathbf{u} \in M, \qquad (2.69')$$

and the admissibility condition on the stresses, finding (2.14), is

$$\langle \mathbf{\Sigma}, \delta\mathbf{E}\rangle \leq 0. \qquad (2.14')$$

The virtual work Eq. (2.80) thus takes the typical form expressing the equality between internal and external virtual work (Fig. 2.42).

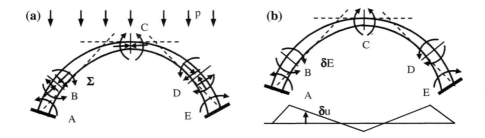

Fig. 2.42 **a** System of external and internal forces. **b** System of virtual displacements and strains

2.5.6 Admissible Equilibrium of Elastic No Tension One-Dimensional Systems

Condition (2.88), with its associated compatibility conditions, can be generalized to the case of structures with *elastic* ashlars free to detach each other (Fig. 2.43).

The virtual strain field can be decomposed into the compressive elastic and the extensive anelastic share

$$E(\delta\mathbf{u}) = E_e(\delta\mathbf{u}) + E_f(\delta\mathbf{u}) \quad \forall \delta\mathbf{u} \in M \qquad (2.85)$$

where,

$$E_e(\delta\mathbf{u}) \leq 0; \quad E_f(\delta\mathbf{u}) \geq 0 \quad \forall \delta\mathbf{u} \in M \qquad (2.86)$$

Condition (2.69′) thus becomes

$$\langle \mathbf{p}, \delta\mathbf{u} \rangle = \langle \mathbf{\Sigma}, \mathbf{E}_f(\delta\mathbf{u}) \rangle + \langle \mathbf{\Sigma}, \mathbf{E}_e(\delta\mathbf{u}) \rangle \quad \forall \delta\mathbf{u} \in M \qquad (2.69′)$$

with

$$\langle \mathbf{\Sigma}, \mathbf{E}_e(\delta\mathbf{u}) \rangle \geq 0 \quad \langle \mathbf{\Sigma}, \mathbf{E}_f(\delta\mathbf{u}) \rangle \leq 0 \quad \forall \delta\mathbf{u} \in M \qquad (2.86′)$$

At the same section at least one of strains $\varepsilon_e(\delta\mathbf{u})$ and $\varepsilon_f(\delta\mathbf{u})$ is zero. Thus

$$\mathbf{E}_f(\delta\mathbf{u}(P)) \cdot \mathbf{E}_e(\delta\mathbf{u}(P)) = 0 \quad \forall \delta\mathbf{u} \in M \qquad (2.87)$$

Using the same argument of Sect. 2.5.3 we can affirm that self stresses cannot develop in the elastic no tension systems, as for the rigid no tension ones.

2.5.7 Weight and Live Loads

Loads acting on a structure can be subdivided into two broad categories having different characteristics. On the one hand, there are the so-called *dead* loads or weight loads **g**, which are permanent and generally quite large, and on the other,

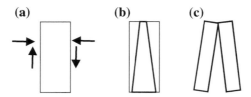

Fig. 2.43 **a** Admissible Σ. **b** admissible $\mathbf{E}_g$. **c** admissible $\mathbf{E}_f$

live loads, **p**, which can be considered to be determined by the loading parameter λ. Thus, we can write

$$\mathbf{p} = \mathbf{g} + \lambda \mathbf{q}. \tag{2.88}$$

As a rule, the weight, **g**, represents the resistant load for a masonry structure. Consequently, recalling condition (2.71), the structure will certainly be *safe* under the action of its own weight **g**, and the following condition will be satisfied

$$\langle \mathbf{g}, \mathbf{v} \rangle < 0, \quad \forall \mathbf{v} \in M, \tag{2.71'}$$

or

$$\langle \mathbf{g}, \mathbf{v} \rangle < -k, k > 0, \quad \forall \mathbf{v} \in M. \tag{2.71''}$$

The weight will always *oppose* any deformation of the masonry structure. For instance, with reference to Fig. 2.43, conditions (2.71') or (2.71'') imply that the pressure line, corresponding to the weight **g**, will always be contained within the arch. In particular, condition (2.71'') dictates that the pressure curve can never touch the arch extrados or intrados, at any section, as shown in Fig. 2.44.

In the case of the arch illustrated in Fig. 2.44, the weight **g**, evaluated per unit length on the horizontal projection, is symmetrical and increases from the key to the springers. All kinematically admissible mechanisms develop vertical displacements in which lifting is dominant. It is thereby clear why arch equilibrium is as a rule strictly admissible under their own weight alone, unless, of course, the arch is too slender. Live loads, **q**, can exert a pushing action along mechanisms. Thus, for any assigned distribution of live loads **q**, it is admissible that at least one mechanism exist along which load **q** will do positive work.

The contribution to resistance of the weight **g** come by virtue of the structure's *geometry*. Masonry structures must be designed so that the mechanisms produce vertical displacements in which lifting is always dominant, thereby satisfying condition (2.87) for any mechanism. It is the *geometry* alone that ensures that the structure's weight counters the emergence of any mechanisms.

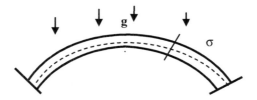

Fig. 2.44 Pressure line strictly contained within the arch

2.5.8 Mechanism States

The previous Fig. 2.41 shows the sketch of a masonry arch in equilibrium under loads **p** and having its pressure line wholly contained within its thickness. In this state internal stresses oppose any deformation of the arch. It is however possible that a structure in admissible equilibrium under loads **p** be freely deformed by a given mechanism displacement $\mathbf{u}_m$, defined by an arbitrary constant. In such a state, instead, the internal stresses and constraint reactions do not counter the emergence of the mechanism. Such a masonry structure is said to be in a *mechanism state* defined by the displacement $\mathbf{u}_m$. The mechanism, which implies the occurrence of a small movement of the structure while the admissible equilibrium is maintained, has arbitrary magnitude.

More precisely, the mechanism state, defined by the displacement field $\mathbf{u}_m$, is considered to be effectively activated in any body in admissible equilibrium under the loads **p** when the following conditions are simultaneously satisfied:

- equilibrium between loads **p** and internal stresses **σ**

$$\langle \boldsymbol{\sigma}, \delta\boldsymbol{\varepsilon} \rangle = \{\mathbf{t}^{(\mathbf{n}^+)}, \Delta^{(\mathbf{n}^-)}\delta\mathbf{u}\} + \langle \mathbf{r}, \delta\mathbf{u} \rangle + \langle \mathbf{p}, \delta\mathbf{u} \rangle \quad \forall \delta\mathbf{u} \in M; \quad (2.89)$$

- admissibility of the internal stress state

$$\langle \boldsymbol{\sigma}, \delta\boldsymbol{\varepsilon} \rangle \leq 0 \quad \{\mathbf{t}^{(\mathbf{n}^+)}, \Delta^{(\mathbf{n}^-)}\delta\mathbf{u}\} \geq 0 \quad \langle \mathbf{r}, \delta\mathbf{u} \rangle \geq 0; \quad (2.90)$$

- lack of opposition by the internal stresses to activation of the mechanism displacement $\mathbf{u}_m$

$$\begin{aligned} \{\mathbf{t}^{(\mathbf{n}^+)}(P^+), \Delta^{(\mathbf{n}^-)}\mathbf{u}_m(P^-)\} &= 0, \quad P^+, P^- \in \Gamma(\mathbf{u}_m) \\ \langle \boldsymbol{\sigma}(P), \boldsymbol{\varepsilon}(\mathbf{u}_m(P)) \rangle &= 0, \quad P \in \Omega(\mathbf{u}_m); \\ \langle \mathbf{r}(P), \mathbf{u}_m(P) \rangle &= 0, \quad P \in \partial\Omega_r. \end{aligned} \quad (2.91)$$

Conditions (2.89), (2.90) and (2.91) are not altered if the mechanism displacement $\mathbf{u}_m$ is affected by a constant factor: the displacement $\mathbf{u}_m$, which is small with respect to the structure's dimensions, thus has indefinite amplitude.

One consequence of conditions (2.91) is that the external loads **p** also offer no opposition to the development of the mechanism displacement $\mathbf{u}_m$. In fact, setting $\delta\mathbf{u} = \mathbf{u}_m$, conditions (2.89) and (2.91) yield

$$\langle \mathbf{p}, \mathbf{u}_m \rangle = 0 \quad (2.92)$$

Fig. 2.45 Pier failure under force S

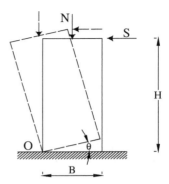

A mechanism state occurs in masonry structures under various peculiar equilibrium conditions, for example, just at the limit equilibrium attained upon collapse, or due to settling, as shown schematically in the two sketches in Figs. 2.43 and 2.44. Figure 2.45 shows the state of a masonry pier loaded by a central vertical force N and a horizontal one,

$$S = NB/2H. \qquad (2.93)$$

The value of S has been chosen so that the resultant of N and S passes precisely through the toe O. The internal forces are due to axial force N and shear S: at the base section their resultant passes through the toe O. The mechanism $\mathbf{v}_c$ is represented by the counterclockwise rotation of the pier around the hinge at O. No detachment strain occurs at any section of the panel except at its base: here the resultant of the stresses passes exactly through hinge O. The mechanism condition (2.92) is thus satisfied.. The pier is at a mechanism state under loads N and S and can rotate in the counterclockwise direction around hinge O.

The case shown in Fig. 2.46 shows another example of a mechanism state differing from the previous condition of limit equilibrium. A slight increase in the span of an arch has been caused by settling of its foundation.

Fig. 2.46 The settled arch

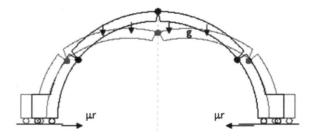

2.5 Equilibrium and Compatibility

The arch, now free at its springings, is loaded by the forces **g** and the thrust μ**r**. The pressure line of the arch, without ever leaving the interior of the arch thickness, passes through the hinges indicated in the figure. The arch is thus in a mechanism state. The extension of the span of the arch can increase arbitrarily under the action of the constant loads **q** and the constant thrust μ**r**.

2.6 Collapse State

2.6.1 *Definitions*

Let us now consider a masonry structure under a loading path **p**(λ), with λ the loading parameter. According to (2.86), we assume that the loads **p**(λ) are made up of the resistant component **g**, i.e. the dead loads, and the pushing forces λ**q**. We moreover assume that by increasing λ, the structure, initially in a *safe* AE state, will pass through a sequence of safe admissible equilibrium states, in the sense that condition (2.71) will always be satisfied in the strict form:

$$\langle \mathbf{g} + \lambda \mathbf{q}, \delta \mathbf{u} \rangle < 0, \quad \lambda > 0, \quad \forall \delta \mathbf{u} \in M. \tag{2.94}$$

Lastly, we assume that at some point during the loading process, when λ attains a critical value λ_c, the structure will *reach a mechanism state* defined by the mechanism $\mathbf{u}_c$. It is moreover admissible for the live loads, **q**, to *push* along $\mathbf{u}_c$, or in other terms that

$$\langle \mathbf{q}, \mathbf{u}_c \rangle > 0. \tag{2.95}$$

Consequently, at $\lambda = \lambda_c$, condition (2.94) continues to be satisfied along all mechanisms other than $\mathbf{u}_c$, so that

$$\langle \mathbf{g} + \lambda_c \mathbf{q}, \delta \mathbf{u} \rangle < 0, \quad \lambda_c > 0, \quad \forall \delta \mathbf{u} \neq \mathbf{u}_c \in M, \tag{2.96}$$

while, on the contrary, the work done by the forces $\mathbf{p} = \mathbf{g} + \lambda_c \mathbf{q}$ vanishes along the mechanism $\mathbf{u}_c$, which is to say

$$\langle \mathbf{g} + \lambda_c \mathbf{q}, \mathbf{u}_c \rangle = 0, \quad \lambda_c > 0, \quad \mathbf{u}_c \in M. \tag{2.97}$$

Thus, as soon as the loading parameter λ is further increased beyond λ_c, by accounting for (2.95), we have

$$(\frac{d}{d\lambda} \langle \mathbf{p}(\lambda), \mathbf{u}_c \rangle)_{\lambda_c} = \langle \mathbf{q}, \mathbf{u}_c \rangle > 0. \tag{2.98}$$

Accordingly, condition (2.71), necessary and sufficient for the *existence of an admissible equilibrium state*, is violated and the structure *collapses*. At this collapse state an exchange occurs from conditions of existence to those of non-existence of an admissible equilibrium state. Condition (2.95) evidences the presence of a pushing action by live loads, **q** along displacement $\mathbf{u}_c$, the *failure mechanism*.

The *development* of this displacement can be represented by a sequence $k\mathbf{u}_c$ of mechanisms of increasing amplitude. Collapse thus occurs under *constant loads*, because by gradually increasing the constant $k > 0$, we consistently have

$$\langle \mathbf{g}, k\mathbf{u}_c \rangle + \lambda_c \langle \mathbf{q}, k\mathbf{u}_c \rangle = 0, \quad k\mathbf{u}_c \in M, k > 0 \tag{2.97'}$$

for any amplitude of mechanism $k\mathbf{u}_c$. Constant loads also imply constant stresses. The failure mechanism thus develops under frozen loads and stresses. There is no energy dissipation at collapse. Nevertheless, the masonry structure is able to maintain its limit strength during the development of the failure mechanism, as occurs for a steel bar upon yielding.

Despite the lack of dissipation, the behavior at collapse of masonry structures is similar to that of ductile steel structures, as predicted by Limit Analysis.

The possibility of reaching collapse during a loading process represents the *most relevant aspect* of the behavior of masonry structures: for such structures, weight and geometry are the only strength resources countering failure.

Summing up all the foregoing general assumptions, it can be stated that a collapse state under loads $\mathbf{g} + \lambda_c \mathbf{q}$, is attained when the following four conditions are satisfied simultaneously:

(1) *equilibrium under loads* $\mathbf{g} + \lambda_c \mathbf{q}$ *and stresses* σ_c.
Consequently, the virtual work equation holds

$$\langle \sigma_c(P), \varepsilon(\delta\mathbf{u}) \rangle = \langle \mathbf{g}, \delta\mathbf{u} \rangle + \lambda_c \langle \mathbf{q}, \delta\mathbf{u} \rangle, \quad \forall \delta\mathbf{u} \in M; \tag{2.99}$$

(2) *compatibility of internal stresses* σ_c

$$\langle \sigma_c(P), \varepsilon(\delta\mathbf{u}) \rangle \leq 0; \tag{2.100}$$

(3) *existence of a mechanism state* $\mathbf{u}_c$ *under loadings* $\mathbf{g} + \lambda_c \mathbf{q}$:

$$\exists \mathbf{u}_c \in M : \langle \sigma_c, \varepsilon(\mathbf{u}_c) \rangle = 0; \tag{2.101}$$

(4) *positive work performed by live loads* **q** *along mechanism* $\mathbf{u}_c$:

$$\langle \mathbf{q}, \mathbf{u}_c \rangle > 0. \tag{2.102}$$

In the following, we will prove the *static* and the *kinematic* theorems of Limit Analysis in their specific form for masonry structures. The earliest proofs of the validity of Limit Analysis to masonry structures dates back to Kooharian (1954), Prager (1959) and Heyman (1966).

The proofs described in the next sections follow the lines of reasoning formulated by Como (1992, 1996a, b). A number of different checks of the theorems have been provided within the no-tension framework by others, such as for example, Sinopoli et al. (1998).

2.6.2 The Static Theorem

Let us consider a masonry structure loaded by fixed dead loads **g** and the additional live load $\lambda^-\mathbf{q}$, with λ^- the multiplier of loads **q**. A given known stress distribution $\boldsymbol{\sigma}^-$ is in an admissible equilibrium with the assigned loads. The statement of the theorem is thus the following:

The loads $\mathbf{g} + \lambda^-\mathbf{q}$ *are not greater that the collapse loads if admissible equilibrium exists between the loads* $\mathbf{g} + \lambda^-\mathbf{q}$ *and the internal stresses* $\boldsymbol{\sigma}^-(P)$

The assumptions underlying the theorem are:

- equilibrium between $\mathbf{g} + \lambda^-\mathbf{q}$ and $\boldsymbol{\sigma}^-(P)$:

$$\langle \boldsymbol{\sigma}^-(P), \varepsilon(\delta\mathbf{u}) \rangle = \langle \mathbf{g}, \delta\mathbf{u} \rangle + \lambda^- \langle \mathbf{q}, \delta\mathbf{u} \rangle \quad \forall \delta\mathbf{u} \in M \qquad (2.103)$$

- static compatibility of stresses $\boldsymbol{\sigma}^-(P)$:

$$\langle \boldsymbol{\sigma}^-(P), \varepsilon(\delta\mathbf{u}) \rangle \leq 0, \quad \forall \delta\mathbf{u} \in M. \qquad (2.104)$$

Together, assumptions (2.103) and (2.104) yield:

$$\lambda^- \leq \lambda_c. \qquad (2.105)$$

The proof of the theorem starts by considering that at collapse, under loads

$$\mathbf{g} + \lambda_c \mathbf{q},$$

the following condition will be satisfied

$$\langle \boldsymbol{\sigma}_c(P), \varepsilon(\mathbf{u}_c) \rangle = \langle \mathbf{g}, \mathbf{u}_c \rangle + \lambda_c \langle \mathbf{q}, \mathbf{u}_c \rangle. \qquad (2.106)$$

Such condition, in which $\boldsymbol{\sigma}_c$ denotes the corresponding stresses, is obtained by applying the virtual work equation to the collapse state with $\delta\mathbf{u} = \mathbf{u}_c$.

At collapse, however, the mechanism condition (2.101) also holds, because the internal stresses $\boldsymbol{\sigma}_c$ do not oppose the development of the mechanism $\mathbf{u}_c$. Consequently, from (2.106) we get

$$0 = \langle \mathbf{g}, \mathbf{u}_c \rangle + \lambda_c \langle \mathbf{q}, \mathbf{u}_c \rangle. \qquad (2.99')$$

Furthermore, taking into account the assumed existence of admissible equilibrium between the loads $\mathbf{g} + \lambda^-\mathbf{q}$ and the internal stresses $\boldsymbol{\sigma}^-$, from (2.103) with $\delta\mathbf{u} = \mathbf{u}_c$, we obtain

$$\langle \boldsymbol{\sigma}^-(P), \boldsymbol{\varepsilon}(\mathbf{u}_c) \rangle = \langle \mathbf{g}, \mathbf{u}_c \rangle + \lambda^- \langle \mathbf{q}, \mathbf{u}_c \rangle. \tag{2.107}$$

By subtracting the previous equality (2.99′) from this, we get

$$\langle \boldsymbol{\sigma}^-(P), \boldsymbol{\varepsilon}(\mathbf{u}_c) \rangle = (\lambda^- - \lambda_c) \langle \mathbf{q}, \mathbf{u}_c \rangle. \tag{2.108}$$

On the other hand, the static compatibility of stresses $\boldsymbol{\sigma}^-(P)$, expressed by inequality (2.104) with $\delta \mathbf{u} = \mathbf{u}_c$, gives

$$\langle \boldsymbol{\sigma}^-(P), \boldsymbol{\varepsilon}(\mathbf{u}_c) \rangle \leq 0 \tag{2.109}$$

whence, from (2.108)

$$(\lambda^- - \lambda_c) \langle \mathbf{q}, \mathbf{u}_c \rangle \leq 0. \tag{2.110}$$

At collapse, according to (2.102), the work of the live loads $\mathbf{q}$ along the failure mechanism $\mathbf{u}_c$ is positive; thus, from (2.110) we have $(\lambda^- - \lambda_c) \leq 0$, or

$$\lambda^- \leq \lambda_c. \tag{2.111}$$

The static multiplier λ^- is thus *not greater* than the collapse multiplier λ_c. The static theorem is one of most important theorems in Structural Engineering. Frequently, when checking the behavior of complex structures under the action of given load distributions, we make conjectures about different internal resistant systems, many of which are unable to sustain the loads (with the exception of at least one).

The static theorem, perhaps surprisingly, tells us that the structure will not fail; that it is certainly able to produce at least one resistant system able to support the loads (though not necessarily the one identified).

The search for states of admissible equilibrium by applying the static theorem begins with construction of a preliminary equilibrium configuration of the structure, for instance, by using funicular polygons, and then verifying its static admissibility. For example, in the case of an arch under given loads, a funicular curve of the load distribution is first drawn and then checked to see whether it is wholly contained within the arch. If it is, the loads are not greater than the collapse ones and the structure is in equilibrium.

2.6.3 The Kinematic Theorem

Assuming that a structure is in *a mechanism state* under loads $\mathbf{g} + \lambda^+ \mathbf{q}$ along a given mechanism $\mathbf{u}^+$, the multiplier λ^+ is initially unknown. However, according to (2.92), we have

2.6 Collapse State

$$0 = \langle \mathbf{g}, \mathbf{u}^+ \rangle + \lambda^+ \langle \mathbf{q}, \mathbf{u}^+ \rangle \tag{2.112}$$

and the kinematic multiplier λ^+, satisfying (2.112), is thus defined as

$$\lambda^+ = -\frac{\langle \mathbf{g}, \mathbf{u}^+ \rangle}{\langle \mathbf{q}, \mathbf{u}^+ \rangle} \tag{2.113}$$

Condition (2.112) implies equilibrium along the mechanism $\mathbf{u}^+$ between the pushing load $\lambda^+ \mathbf{q}$ and the resistant weights $\mathbf{g}$. We also assume that the loads $\mathbf{q}$ push along $\mathbf{u}^+$, hence

$$\langle \mathbf{q}, \mathbf{u}^+ \rangle > 0. \tag{2.114}$$

Under such conditions, the theorem states that the kinematic multiplier λ^+ cannot be lower than the collapse multiplier λ_c, that is,

$$\lambda^+ \geq \lambda_c. \tag{2.115}$$

Once again in this case the proof depends on conditions defining the collapse state, particularly the condition of limit equilibrium. Thus, from condition (2.99) with $\delta \mathbf{u} = \mathbf{u}^+$, we obtain

$$\langle \boldsymbol{\sigma}_c(P), \boldsymbol{\varepsilon}(\mathbf{u}^+) \rangle = \langle \mathbf{g}, \mathbf{u}^+ \rangle + \lambda_c \langle \mathbf{q}, \mathbf{u}^+ \rangle, \tag{2.99'}$$

which by subtracting equality (2.112) yields

$$\langle \boldsymbol{\sigma}_c(P), \boldsymbol{\varepsilon}(\mathbf{u}^+) \rangle = (\lambda_c - \lambda^+) \langle \mathbf{q}, \mathbf{u}^+ \rangle. \tag{2.116}$$

On the other hand, simply because an admissible equilibrium still exists at collapse, from (2.100), with $\delta \mathbf{u} = \mathbf{u}^+$, we have

$$\langle \boldsymbol{\sigma}_c(P), \boldsymbol{\varepsilon}(\mathbf{u}^+) \rangle \leq 0, \tag{2.100'}$$

and from (2.116) we obtain

$$(\lambda_c - \lambda^+) \langle \mathbf{q}, \mathbf{u}^+ \rangle \leq 0. \tag{2.116'}$$

Hence, taking into account to our initial assumptions (2.114)

$$\lambda^+ \geq \lambda_c. \tag{2.117}$$

Multiplier λ^+, i.e., the kinematically admissible multiplier of loads $\mathbf{q}$, represents the *upper bound* of the collapse multiplier λ_c.

2.6.4 Uniqueness of the Collapse Multiplier

Let us assume, *ad absurdum*, that two different values λ_{c1} and λ_{c2} of the collapse load multiplier exist. Both λ_{c1} and λ_{c2} satisfy the equilibrium conditions (2.99) between loads and stresses. Moreover, the stress compatibility conditions (2.100) are satisfied, as are conditions (2.101) required for the existence of the mechanism state, as well as (2.102). Let us now assume, for instance, that

$$\lambda_{c1} \leq \lambda_{c2}. \tag{2.118}$$

However, the failure multiplier λ_{c2} is also a statically admissible multiplier, since it satisfies the equilibrium conditions with the corresponding internal compatible stresses $\boldsymbol{\sigma}_{c1}$. Thus, if λ_{c1} is a collapse multiplier, from (2.111) we also have

$$\lambda_{c2} \leq \lambda_{c1}. \tag{2.119}$$

Comparison of (2.118) with (2.119) yields

$$\lambda_{c1} = \lambda_{c2}.$$

The result should be the same if, on the contrary, it were assumed that $\lambda_{c2} \leq \lambda_{c1}$. The collapse multiplier is thus unique, though, in general, the failure mechanism is not.

The two theorems, static and kinematic, set a bounding interval for the collapse multiplier, because

$$\lambda^{-} \leq \lambda_c \leq \lambda^{+}. \tag{2.120}$$

It should be recognized that the collapse load does not depend on the material properties, but only on the geometry of the structure and the magnitude of the dead loads. Subsequent chapters will provide in-depth examples of the numerous applications of the theorems presented here.

2.6.5 Indeformable Systems: Lack of Collapse

Under the action of external loads, an *indeformable* system tends to become deformed, resulting in dilatation and cracking. On the other hand the interpenetration strength of the material, together with the presence of fixed constraints, prevents any deformation. Thus, only compressions can take place within the body. The indeformable structure will be thus always able to sustain the applied loads. Collapse mechanisms cannot therefore exist and such systems never fail, unless the material undergoes crushing or the constraints are displaced through settling.

2.6 Collapse State

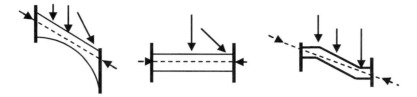

Fig. 2.47 Lack of collapse for indeformable systems

The same conclusion can be reached from another perspective. As noted previously, self-equilibrated stresses do exist in these systems. An arbitrary self-equilibrated stress distribution can be added to any stress field in equilibrium with the loads in such a way that the overall stress state turns out to be solely compressive. Referring to any of the three illustrations in Fig. 2.47, once a funicular polygon of the applied loads is traced, for instance, passing through the intrados at the springers and the extrados at the key section, we can apply two equal and opposite forces able to modify the polygon so that it remains entirely contained within the structure. These self-equilibrated stresses are produced by the same external loads that tend to deform the structure and force the external constraints. Such structures are thus able to sustain any load.

The real problem, on the other hand, is evaluating the most probable thrust transmitted by the structure at its supports. The indeterminacy of the static solutions can be overcome by seeking solutions within the framework of the principles of minimum thrust. The thrust also slightly deforms the supports and activates *minimum thrust states*.

It should lastly be noted that for such systems the usual assumption of infinitely strong masonry can be opportunely removed with in order to obtain more realistic evaluations of a structure's maximum capacity, as we will show in the next chapter.

2.6.6 Collapse State for the Elastic No Tension Systems

The definition of collapse state given at Sect. 2.6.1 for the rigid in compression no tension bodies holds also for the elastic no tension systems. During the loading failure is met when a mechanism $\mathbf{u}_c$ turns up along which the structure deforms under constant loads $\mathbf{g} + \lambda_c \mathbf{q}$ and, consequently, under constant stresses. *No elastic strain increments $\Delta \sigma$ take place at the collapse.* To prove this statement, let us admit, by contradiction, that the failure displacement is composed both by cracking and elastic strain increments $\varepsilon_f (\mathbf{u}_c)$ and $\varepsilon_e (\mathbf{u}_c)$. Thus, if the stress increment $\Delta \boldsymbol{\sigma}$ takes place, the equilibrium condition under constant loads and along the failure displacement $\mathbf{u}_c$ gives

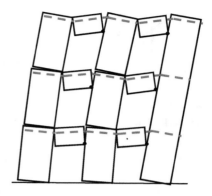

Fig. 2.48 Failure of a masonry wall reinforced by steel ties. During the mechanism, plartuc stretching occurs in the steel ties

$$\langle \Delta\boldsymbol{\sigma}, \varepsilon_f(\mathbf{u}_c)\rangle + \langle \Delta\boldsymbol{\sigma}, \varepsilon_e(\mathbf{u}_c)\rangle = 0 \qquad (2.121)$$

But along the failure displacement we have

$$\langle \Delta\boldsymbol{\sigma}, \varepsilon_f(\mathbf{u}_c)\rangle = 0 \qquad (2.122)$$

because the stress increments $\Delta\sigma$ cannot oppose the opening deformations of the structure. Hence, condition (2.123) becomes

$$\langle \Delta\boldsymbol{\sigma}, \varepsilon_e(\mathbf{u}_c)\rangle = 0 \qquad (2.123)$$

But this result cannot occur because, when elastic strains take place, the work $\langle \Delta\boldsymbol{\sigma}, \varepsilon_e(\mathbf{u}_c)\rangle$ represents an positive elastic strain energy. Thus

$$\Delta\sigma = 0 \qquad (2.124)$$

The failure of the elastic no tension systems thus occurs with a mechanism displacement, in absence of elastic strain. Internal stresses maintain frozen during the collapse. This result is useful for a better inderstandig of the behaviour at the collapse of masonry structures reinforced by elasto-plastic systems, as for instance, steel ties (Fig. 2.48). At the collapse these devices, at the plastic state, follow at constant stress the occurrence of the failure mechanism.

2.7 Incipient Settled States

2.7.1 *Definitions*

Let us consider a masonry structure at a *safe admissible equilibrium* state at configuration C_i under the actions of loads g and the corresponding internal stresses σ_i. For deformable systems inequality (2.71') holds, so we have

2.7 Incipient Settled States

$$\langle g, \delta \mathbf{u} \rangle < 0, \quad \forall \delta \mathbf{u} \in M. \tag{2.125}$$

Evidently, according to Sect. 2.6.5, the assumption of a safe admissible equilibrium state at the initial configuration C_i undr the loads g is still valid also if the structure, for its initial constraint conditions, is indeformable. Recall that the equilibrium at C_i is *safe* because no mechanism exists along which the work of loads **g** vanishes.

Let us now assume that the structure becomes slightly deformed as a consequence of the incipient *settlement mechanism,* $\mathbf{v}_s$, occurring at one of its external constraints. By way of example, consider the arch in Fig. 2.49, which undergoes a slight increase in span due to settling.

Let C_s be the configuration taken on by the structure once settlement has occurred. By assuming that C_s is *very near* C_i, we can refer to the geometry of the initial configuration C_i when expressing the equilibrium equations. The settlement mechanism, $\mathbf{v}_s$, is the displacement field that moves the structure from C_i to C_s.

As the settlement occurs, the structure's internal equilibrium shifts from initial configuration C_i to the displaced one C_s. Changes in the internal stresses and constraint reactions will occur during the transition from C_i to C_s, so that the initial stress state, $\boldsymbol{\sigma}_i$, admissible and in equilibrium with loads **g**, is altered and becomes

$$\boldsymbol{\sigma}_s.$$

This internal stress state, $\boldsymbol{\sigma}_s$, which accounts for settlement $\mathbf{v}_s$, is statically admissible and thus satisfies the inequality

$$\langle \boldsymbol{\sigma}_s, \varepsilon(\delta \mathbf{u}) \rangle \leq 0 \tag{2.126}$$

Likewise, the settled constraint, which *before* the settling, produced the reaction $\mu_i \mathbf{r}$, *after* the settling produces the new reaction

$$\mu_s \mathbf{r}, \tag{2.127}$$

where **r** is a given force having the direction of reaction of the settled constraint and μ is the corresponding multiplier. In brief, during the development of the settlement mechanism, $\mathbf{v}_s$ the structure will remain in a state of admissible equilibrium while

Fig. 2.49 Arch with increased span due to settling at its springers

the stresses vary from $\boldsymbol{\sigma}_i$ to $\boldsymbol{\sigma}_s$, the corresponding pressure line shifts from π_i to π_s and the reaction of the settled constraint changes from $\mu_i\mathbf{r}$ to $\mu_s\mathbf{r}$.

In the case of the masonry arch that has undergone a slight increase in span, its pressure line π_s will pass through the hinges corresponding to mechanism $\mathbf{v}_s$. Consequently, no work will be done by the internal stresses $\boldsymbol{\sigma}_s$ on the deformations corresponding to $\mathbf{v}_s$. The same occurs for any structure that is slightly deformed by a mechanism and adapts itself to settling. Thus, at the settlement state the following *mechanism state* holds

$$\langle \boldsymbol{\sigma}_s, \boldsymbol{\varepsilon}(\mathbf{v}_s) \rangle = 0. \tag{2.128}$$

We can *release* the structure by removing the settled configuration constraint by applying the reaction $\mu_s\mathbf{r}$ to the eliminated constraint. The set of all mechanisms of the released structure is denoted by $\bar{M}$. In the released structure in the settled state, the applied loads are represented by both the weights $\mathbf{g}$ and the reactions $\mu_s\mathbf{r}$. Thus, considering the released structure in the admissible settled equilibrium state, from the virtual work equation we get

$$\langle \mathbf{g}, \delta\mathbf{u} \rangle + \mu_s \langle \mathbf{r}, \delta\mathbf{u} \rangle = \langle \boldsymbol{\sigma}_s, \boldsymbol{\varepsilon}(\delta\mathbf{u}) \rangle, \quad \forall \delta\mathbf{u} \in \bar{M}, \tag{2.129}$$

which for $\delta\mathbf{u} = \mathbf{v}_s$, according to (2.128), yields

$$\langle \mathbf{g}, \mathbf{v}_s \rangle + \mu_s \langle \mathbf{r}, \mathbf{v}_s \rangle = 0. \tag{2.130}$$

Loads $\mathbf{g}$ perform positive work along the mechanism displacements $\mathbf{v}_s$, while the reaction $\mu_s\mathbf{r}$ of the released constraint opposes settling, so that

$$\langle \mathbf{g}, \mathbf{v}_s \rangle > 0 \tag{2.131}$$

and

$$\mu_s \langle \mathbf{r}, \mathbf{v}_s \rangle < 0. \tag{2.132}$$

An *admissible settlement equilibrium state* is therefore defined once the mechanism settlement $\mathbf{v}_s$ has taken place. Conditions (2.126–2.132), define the *admissible settlement equilibrium state*.

2.7.2 Features of Incipient Settled States

Firstly, once the slight settlement, $\mathbf{v}_s$, has occurred and the structure has shifted into configuration C_s, the settled constraint can once again be made active.

In spite of the settling that occurred, the work of the loads due to any mechanism, $\delta\mathbf{u}$ is still the same as the work evaluated at the initial configuration C_i, assuming, of course, that displacements, $\mathbf{v}_s$, are *very small* (as stated above) and that

2.7 Incipient Settled States

the changes in geometry are consequently *negligible*. Thus, if at the initial state C_i, the admissibility condition $\langle \mathbf{g}, \delta \mathbf{u} \rangle < 0, \forall \delta \mathbf{u} \in M$ is satisfied, the same condition will still be satisfied by the new configuration C_s. In this regard, recall Heyman's statement (1966): *"if the foundations of a stone structure are liable to small movements, such movements will never, of themselves, promote the collapse of the structure"*. Moreover, if settlement $\mathbf{v}_s$ returns, increases and becomes

$$k\mathbf{v}_s, \quad k > 1, \qquad (2.133)$$

the static arrangement of the structure will not change, and the internal stresses will remain fixed at σ_s. Indeed, as the stresses σ_s are forced to satisfy condition (2.128), they will continue to satisfy the same condition when the structure deforms by displacements k times larger than the prior ones. Thus, if the pressure curve skims the extrados and the intrados of the arch at the hinges corresponding to mechanism $\mathbf{v}_s$, and condition (2.128) is consequently satisfied, the same condition (2.128) will continue to be satisfied by assuming the mechanisms $k\mathbf{v}_s$, $k > 1$. Consequently the work of the same stresses, σ_s, on the deformations associated to displacements $k\mathbf{v}_s$ continues to be zero. Likewise, the thrust $\mu_s \mathbf{r}$, which according to (2.129), satisfies the equilibrium with the loads $\mathbf{g}$ along the displacements $\mathbf{v}_s$, will continue to satisfy the same equilibrium condition (2.129) along the displacement field $k\mathbf{v}_s$, $k > 1$. In short, the structure freely follows any increase in the settlement, maintaining its configuration in admissible equilibrium: settling develops with frozen internal stresses, σ_s, and constraint reactions, $\mu_s \mathbf{r}$. The actual degree of settling is difficult to quantify. Despite this uncertainty, the internal stress state of the structure is, to the contrary, well-defined. No equilibrium loss will occur during the settling. This is a peculiar aspect of masonry structures that can explain the great durability and longevity of so many historic buildings. *How do we evaluate this stress state and the corresponding reaction of the settled restraint?* The answer to this question will be provided in the following sections.

2.7.3 Statically Admissible Thrusts: The Static Theorem of Minimum Thrust

Let us now look at the static equilibrium of a structure that has previously undergone settling. Two relevant examples are represented by a masonry arch whose span is lengthened by settling at the springings and the vertical settlement of the central pier of a masonry bridge. The settled structure is certainly at AE equilibrium under the loads $\mathbf{g}$ and internal stresses σ. We know, a priori, nothing about the internal stresses σ occurring in the settled state, except that they are statically admissible. Let S be the set of all statically admissible internal stresses, σ, with $\boldsymbol{\sigma} \in S$. We choose any one such distribution of statically admissible stresses σ. A thrust $\mu \mathbf{r}$ of the released constraint will correspond unequivocally to this distribution, hence let

$$\mu \mathbf{r}(\boldsymbol{\sigma} \in S) \tag{2.134}$$

be the reaction of the settled constraint associated to the statically admissible stresses $\boldsymbol{\sigma}$. For any $\boldsymbol{\sigma} \in S$, we can associate the reaction $\mu \mathbf{r}(\boldsymbol{\sigma} \in S)$, which defines the thrust of the settled constraint.

According to our assumptions, in the released structure the following equation between loads $\mathbf{g}$, internal stresses $\boldsymbol{\sigma}$ and the corresponding thrust $\mu \mathbf{r}(\boldsymbol{\sigma} \in S)$ for the AE holds

$$\langle \mathbf{g}, \delta \mathbf{u} \rangle + \mu \mathbf{r}(\boldsymbol{\sigma}) \langle \mathbf{r}, \delta \mathbf{u} \rangle = \langle \boldsymbol{\sigma}, \varepsilon(\delta \mathbf{u}) \rangle, \quad \boldsymbol{\sigma} \in S, \ \forall \delta \mathbf{u} \in \bar{M}. \tag{2.135}$$

Moreover, the following inequality

$$\langle \boldsymbol{\sigma}, \varepsilon(\delta \mathbf{u}) \rangle \leq 0 \quad \forall \delta \mathbf{u} \in \bar{M} \tag{2.136}$$

will also be satisfied for the admissibility of the stresses $\boldsymbol{\sigma} \in S$. Thrust $\mu \mathbf{r}(\boldsymbol{\sigma} \in S)$, in equilibrium with loads $\mathbf{g}$ and internal stresses $\boldsymbol{\sigma}$, that is to say, satisfying condition (2.135), represents *any state of statically admissible thrust*. Now, of all the statically admissible thrusts, which one corresponds to the settled state?

We know that the actual thrust in the settled state corresponds to an admissible stress state that *does no work* on the deformations occurring during the settlement mechanism, that is, it satisfies (2.128). If we define $\mathbf{v}_s$ as the effective settlement mechanism, we specify condition (2.135) using $\delta \mathbf{u} = \mathbf{v}_s$ to obtain

$$\langle \mathbf{g}, \mathbf{v}_s \rangle + \mu(\boldsymbol{\sigma}) \langle \mathbf{r}, \mathbf{v}_s \rangle = \langle \boldsymbol{\sigma}, \varepsilon(\mathbf{v}_s) \rangle \quad \boldsymbol{\sigma} \in S. \tag{2.137}$$

We now subtract equality (2.130), regarding the actual settlement state, from equality (2.137) to obtain

$$(\mu(\boldsymbol{\sigma}) - \mu_s) \langle \mathbf{r}, \mathbf{v}_s \rangle = \langle \boldsymbol{\sigma}, \varepsilon(\mathbf{v}_s) \rangle \quad \boldsymbol{\sigma} \in S. \tag{2.138}$$

By taking (2.136) into account, from (2.138), with $\delta \mathbf{u} = \mathbf{v}_s$, we get

$$(\mu(\boldsymbol{\sigma}) - \mu_s) \langle \mathbf{r}, \mathbf{v}_s \rangle \leq 0 \quad \boldsymbol{\sigma} \in S, \tag{2.139}$$

and by virtue of (2.132), we obtain

$$\mu_s \leq \mu(\boldsymbol{\sigma}) \quad \boldsymbol{\sigma} \in S \tag{2.140}$$

The multiplier, μ_s, of the settled thrust $\mathbf{r}$ is thus *lower* than all the statically admissible multipliers μ. The thrust in the settled state is consequently the lowest of all the statically admissible thrusts.

This finding (Como 1996a, b, 1998) relates to a previously described, well-known property of masonry arches that undergo an increase in span due to settling at the springings: such an arch is in the state of minimum thrust because its pressure line corresponds to the minimum span and the maximum sag, as shown by Heyman (1966).

2.7.4 Kinematically Admissible Thrusts: The Kinematic Theorem of the Minimum Thrust

Le us now examine the settlement equilibrium state from a *kinematic* point of view. The actual settlement mechanism is unknown: for instance, for the case of the arch, we cannot know the position of the internal hinge. We only know that, during the development of the mechanism, loads **g** will do positive work, while the work of the reaction of the settled constraint is, to the contrary, negative.

Let $\bar{M}$ be the set of all kinematically admissible settlement mechanisms of the released structure: they do not allow for any internal interpenetration of masonry and respect all the restrictions for the other unsettled constraints. Let us consider some settlement mechanism

$$\mathbf{v} \in \bar{M} \tag{2.141}$$

of the released structure. The loads **g** will push along **v** and consequently

$$\langle \mathbf{g}, \mathbf{v} \rangle > 0. \tag{2.142}$$

We define the *kinematic multiplier*, λ, of the reaction **r** of the settled constraint as that multiplier able to ensure equilibrium of the structure along the assumed settlement mechanism, **v**, or, in other terms, that the following condition holds

$$\langle \mathbf{g}, \mathbf{v} \rangle + \lambda \langle \mathbf{r}, \mathbf{v} \rangle = 0. \tag{2.143}$$

Reaction $\lambda \mathbf{r}(\mathbf{v})$ opposes the development of settling, **v**, given that, by taking (2.142) into account, we get

$$\lambda \langle \mathbf{r}, \mathbf{v} \rangle < 0. \tag{2.144}$$

where **r** is any resistant reaction of the settled constraint. The kinematic multiplier $\lambda(\mathbf{v})$ of reaction **r** is thus defined as

$$\lambda(\mathbf{v}) = -\frac{\langle \mathbf{g}, \mathbf{v} \rangle}{\langle \mathbf{r}, \mathbf{v} \rangle}, \quad \mathbf{v} \in \bar{M}. \tag{2.145}$$

Let us now search for the conditions under which kinematic reaction $\lambda \mathbf{r}(\mathbf{v})$ may represent the actual settled state. This latter is represented by the reaction, $\mu_s \mathbf{r}$, that satisfies the foregoing settlement conditions (2.126), (2.128), (2.129), (2.130), (2.131) and (2.132). Thus, let us assume $\delta \mathbf{u} = \mathbf{v}$ in (2.139) to get

$$\langle \mathbf{g}, \mathbf{v} \rangle + \mu_s \langle \mathbf{r}, \mathbf{v} \rangle = \langle \boldsymbol{\sigma}_s, \boldsymbol{\varepsilon}(\mathbf{v}) \rangle, \quad \mathbf{v} \in \bar{M}, \tag{2.146}$$

where, according to (2.126)

$$\langle \boldsymbol{\sigma}_s, \boldsymbol{\varepsilon}(\mathbf{v}) \rangle \leq 0 \quad \mathbf{v} \in \bar{M}. \tag{2.147}$$

Now, subtracting (2.143) from equality (2.146) yields

$$(\mu_s - \lambda(\mathbf{v})) \langle \mathbf{r}, \mathbf{v} \rangle = \langle \boldsymbol{\sigma}_s, \boldsymbol{\varepsilon}(\mathbf{v}) \rangle \quad \mathbf{v} \in \bar{M}. \tag{2.148}$$

Moreover, from (2.126), with $\delta \mathbf{u} = \mathbf{v}$, we get $(\mu_s - \lambda(\mathbf{v})) \langle \mathbf{r}, \mathbf{v} \rangle \leq 0$ and, consequently, from (2.144) (Como 1996a, b, 1998)

$$\mu_s \geq \lambda(\mathbf{v}) \quad \mathbf{v} \in \bar{M}. \tag{2.149}$$

For any settlement mechanism, $\mathbf{v} \in \bar{M}$, the corresponding kinematic multiplier, $\lambda(\mathbf{v} \in \bar{M})$, can never be greater than the actual settlement multiplier, μ_s. Thus μ_s is *the maximum* of all kinematic multipliers, $\lambda(\mathbf{v} \in \bar{M})$, for varying $\mathbf{v}$ in the set of all settlement mechanisms $\bar{M}$, or in other terms

$$\mu_s = MAX(-\frac{\langle \mathbf{g}, \mathbf{v} \rangle}{\langle \mathbf{r}, \mathbf{v} \rangle}) \quad \mathbf{v} \in \bar{M}. \tag{2.150}$$

2.7.5 Uniqueness of the Settlement Multiplier

The proof of the uniqueness of the settlement multiplier follows the same path as that for the collapse multiplier.

2.7.6 The Class of the Statical and Kinematical Settlement Multipliers

$\lambda(\mathbf{v} \in \bar{M})$ is any admissible *kinematic* multiplier, defined according to (2.145). The corresponding thrust $\lambda(\mathbf{v} \in \bar{M}) \mathbf{r}$ is not, as a rule, statically admissible. Since

$$\lambda(\mathbf{v} \in \bar{M}) \leq MAX \lambda(\mathbf{v} \in \bar{M}) = \mu_s, \tag{2.151}$$

the thrust $\lambda(\mathbf{v} \in \bar{M}) \mathbf{r}$ is weaker than the minimum thrust $\mu_s \mathbf{r}$. Moreover, $\mu(\boldsymbol{\sigma} \in S)$ is any *statically* admissible multiplier that, according to previous definitions, is not

2.7 Incipient Settled States

kinematically admissible, in the sense that the internal stresses are not generally associated to a mechanism. In the case of an arch, for instance, the pressure line corresponding to the stress distribution, $\sigma \in S$, does not skim the extrados and intrados of the arch to form hinges in numbers sufficient to produce a kinematically admissible mechanism. The thrust $\mu(\sigma)\mathbf{r}$ is greater than the minimum $\mu_s \mathbf{r}$ and

$$\mu_s = \underset{\sigma}{MIN}\,\mu(\sigma) \leq \mu(\sigma). \tag{2.152}$$

In conclusion, we obtain

$$\lambda(\mathbf{v}) \leq MAX\lambda(\mathbf{v}) = \mu_s = MIN\mu(\sigma) \leq \mu(\sigma). \tag{2.153}$$

The actual settlement mechanism, $\mathbf{v}_s$, together with the reaction of the settled constraint, may be determined via (2.153). These results will be applied in the following to analyze the statics of a number of masonry structures.

The problem of the evaluation of the minimum thrust for the rounded arch was analyzed by Coulomb (1773). Coulomb guessed that such a state was reached by the settled arch that endured a light widening at its springers. The pressure line touches the intrados and the extrados of the arch to form the hinges of the settlement mechanism. Figure 2.57 shows the rounded arch with the three hinges symmetricaly placed. The position of the hinge C, having distance d from the horizontal straight line passing through the key hinge A, is defined by the angle β.

Figure 2.57 shows the force V, resultant of the weights of the segment AC of the arch.

Position of V is given by its distance L_A from the internal hinge C. Coulomb, esamining the equilibrium of the segment AC of the arch and for a given position of the hinge C, i.e. for a given β, valuated the thrust of the arch as

$$H = \frac{VL_A}{d} \tag{2.154}$$

Thus thrust H depends on the angle β. Coulomb noticed, as pointed out by Ochsendorf (2006), that the minimum thrust of the arch was attained by the maximum of thrust (2.154) by varying the angle β, as shown in Fig. 2.57. This remark can be explained taking into account that the thrust (2.154) is just the kinematical thrust corresponding to definition (2.150). According to the kinematical theorem the minimum thrust has to be searched as the maximum of the kinematical ones

$$H_{Min} = \underset{\beta}{Max}\,H = \underset{\beta}{Max}\,\frac{\langle g,v \rangle}{\Delta} \tag{2.155}$$

Expression (2.155) corresponds to (2.154). In fact the work of the weight $g(x)$ of the arch along the vertical displacements $v(x)$ produced by the settlment mechanism is

$$\langle g, v \rangle = \theta \int_{AC} g(x) \cdot x \cdot dx \qquad (2.156)$$

The force V is on the other hand defined as

$$V = \int_{AC} g(x) \cdot dx \qquad (2.157)$$

The distance L_A

$$V \cdot L_A = \int_{AC} g(x) \cdot x \cdot dx \qquad (2.158)$$

gives the position of the force V. Consequently we get, with (2.158) or

$$H = \frac{\theta \int_{AC} g(x) \cdot x \cdot dx}{d \cdot \vartheta} \qquad (2.159)$$

Finally, taking into account that $\Delta = d\,\theta$ (Fig. 2.50) we can write

$$H = \frac{V \cdot L_A}{d} = \frac{\langle g, v \rangle}{\Delta} \qquad (2.160)$$

The thrust (2.146) is properly the kinematical thrust defined by (2.140). On the other hand the minimumthrust of the arch is the maximum of all the kinemathical thrusts H by varying the angle β, i.e. the position of the internal hinges C. Thus we have

$$H_{Min} = \underset{\beta}{Max}\, H = \underset{\beta}{Max}\, \frac{\langle g, v \rangle}{\Delta} \qquad (2.161)$$

The kinematical thrust of the rounded arch of minimum thickness, i.e. with t/R = 0.1075, attains its maximum at the angle $\beta = 54.5°$, (Ochsendorf 2006).

Fig. 2.50 The research of the minimum thrust of the arch

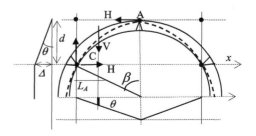

2.7.7 An Application of the Kinematic Approach. A Minimum Thrust Assessment

Let us consider the masonry bridge sketched out in Fig. 2.51. The bridge's central pier has undergone vertical settling. The settlement mechanism is traced out in the figure, where $\mu_s \mathbf{r}$ indicates the vertical reaction of the central pier.

The figure also shows the critical sections where hinges are located: they are the two points O at the base corners of the right and left abutments, point A at the connection section between the abutments and girders, and point B at an intermediate section along the girder extrados. The distance between hinges B and C is denoted by x. The vertical settlement of the central pier defines the position of hinges O and A of the abutments, which will rotate outwardly. Hinges B, whose positions are instead unknown, are each at a distance $a + x$ from the internal edge of the abutments. The same figure shows the deformation of the bridge corresponding to the assumed mechanism. Our aim is to evaluate the reaction R_c of the central pier. Let us consider any settlement mechanism, v, and the corresponding kinematic reaction of the pier, denoted by μr defined by the equilibrium equation along v

$$\langle g, v \rangle + \mu \langle r, v \rangle = 0. \tag{a}$$

The corresponding kinematic multiplier, which depends on the position of hinge B, is thus given by

$$\mu(x) = -\frac{\langle g, v \rangle}{\langle r, v \rangle}. \tag{b}$$

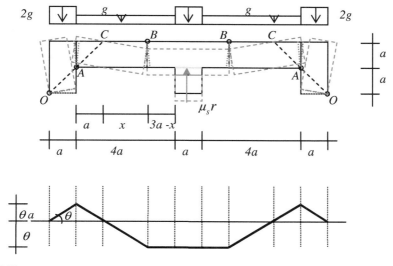

Fig. 2.51 A masonry bridge whose central pier has settled vertically

The actual pier reaction R_c can thus be obtained as the *maximum* of all kinematic reactions (b) by varying v in the set of settlement mechanisms. We now evaluate the work, $\langle g, v \rangle$ and $\langle r, v \rangle$. The work of weight g along the mechanism is given by

$$\langle g, v \rangle = 2[-2g\frac{a^2\theta}{2} - g\frac{a^2\theta}{2} + g\frac{x^2\theta}{2} + g(3a - x)x\theta + 2g\frac{ax\theta}{2}]$$
$$= g(-3a^2 - x^2 + 8ax)\theta \tag{c}$$

while the work $\langle r, v \rangle$ of the pier's reaction is $-x\theta$. According to (b), the kinematic reaction is thus given by

$$\mu r(x) = \frac{g(-3a^2 - x^2 + 8ax)}{x} \quad 0 \leq x \leq 3a, \tag{d}$$

depending on the position x of hinge B, which lies in the range $0 \leq x \leq 3a$. When $x \to 0$, $\mu r \to -\infty$, and for $x = 3a$, $\mu r = 4ga$ (Fig. 2.52).

The function $\mu(x)$ effectively reaches *a maximum* for $0 \leq x \leq 3a$. The value $\bar{x}$ at which $\mu r(x)$ attains its maximum is thus obtained by solving the equation

$$\frac{d(\mu r)}{dx} = g(\frac{3a^2}{x^2} - 1) = 0, \tag{e}$$

which yields

$$x = \bar{x} = a\sqrt{3} \approx 1.73a. \tag{f}$$

The reaction of the central pier is evaluated by substituting (f) into (d), which yields

$$R_c = \mu r_{max} = \mu r(x = a\sqrt{3}) = 2ga(4 - \sqrt{3}) \approx 4.54ga. \tag{g}$$

We will now show that the internal stress corresponding to the evaluated reaction is statically admissible. Figure 2.53 illustrates the equilibrium of the central part of the bridge, including the girder segments from hinges B to the pier.

Fig. 2.52 Finding the maximum of function $\mu r(x)$

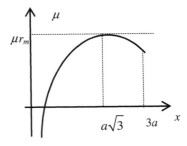

2.7 Incipient Settled States

Fig. 2.53 Stress state around the central pier

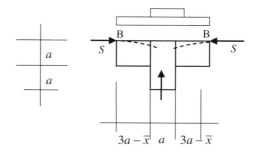

Fig. 2.54 Stresses in the span

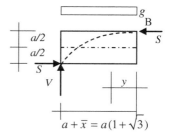

The pressure line passes through hinge B and has a horizontal tangent at B. With the aim of evaluating the thrust, S, shown in the figure, let us examine the equilibrium of the portion of the girder between B and the abutment, as shown in Fig. 2.53. The length of this portion is $a + \bar{x} = a(1 + \sqrt{3})$ (Fig. 2.54). The shear force V transmitted to the abutment can be obtained via the equilibrium equation along the vertical direction:

$$V = ga(\sqrt{3} + 1).$$

We can also evaluate the thrust S. At $y = a + \bar{x} = a(1 + \sqrt{3})$, we have $Sa - g(a + \bar{x})^2 / 2 = 0$ and

$$S = \frac{ga}{2}(\sqrt{3} + 1)^2 = ga(2 + \sqrt{3}).$$

To obtain the corresponding pressure line, we can evaluate the moment acting at the section center located at distance y from B, which gives us

$$M(y) = S\frac{a}{2} - g\frac{y^2}{2} = g\frac{a^2}{2}(2 + \sqrt{3}) - g\frac{y^2}{2}.$$

The axial force N equals the thrust S, while the eccentricity of N is $M(y)/S$. Moreover, from the equilibrium along the vertical direction of the abutment, we get (Fig. 2.55)

Fig. 2.55 Stresses in the abutments

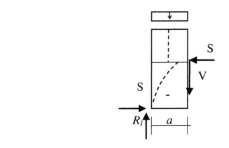

Fig. 2.56 The pressure line in the bridge

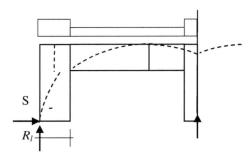

$$R_l = 2ga + V = 2ga + ga(\sqrt{3}+1) = ga\sqrt{3}(1+\sqrt{3}).$$

The abutment is loaded by a central axial force, N, from the head to midway along its height. Beneath this section the axial load varies and becomes $ga\sqrt{3}(2+\sqrt{3})$, and its eccentricity ranges from 0 to $a/2$.

It can thus be concluded that the structure of the bridge, whose central pier has settled by $\sqrt{3}a\theta$ and is subject to the assumed loads and reaction $\mu r(\bar{x})$ at the base of the central pier, is in admissible equilibrium.

passing through the hinges of the settlement mechanism. The stress state is also statically admissible, thus proving that the thrusts, determined as the maximum of all kinematically admissible ones, is the sought for minimum thrust.

Figure 2.56 shows the overall pressure line on the entire structure of the bridge.

The minimum thrust can also be obtained by searching among the admissible funicular curves of the loads: the curve exhibiting the minimum slope at the sections where the girders join the central pier corresponds to this minimum.

2.7.8 Masonry Structures at Their Actual State

Limit Analysis, whose bases have been discussed above, offers the possibility of performing useful checks of the bearing capacity of masonry structures within the framework of the rigid no-tension model. However, Limit Analysis alone does not provide the means to analyze the *actual* equilibrium state of a masonry structure. The admissible equilibrium equations alone are not enough to evaluate the internal stresses, as such problem is statically indeterminate.

To this end, other additional equations are required: the compatibility equations, as in the case of elastic structures. However, according to the rigid no-tension model, masonry structures cannot be deformed unless a mechanism is activated. Consequently, these supplementary equations cannot be formulated. Heyman (1995) however found that useful information can instead be obtained in the same context of the simple rigid no-tension model, that is, providing that the deformability of the constraints connecting the structure to the surrounding environment are taken into account. In brief, the assumption that settling of some constraints will inevitably occur may provide the required supplementary equations. Such an assumption generally reflects the *real behavior* of masonry structures, which frequently push against their supports that are in turn deformed by these thrusting actions. An arch or a dome, for instance, is inserted into a more complex structural system that must sustain their thrusts. These auxiliary structures undergo lateral deformations and can displace the springings of the arch or dome to follow their deformation. A minimal thrust state takes place in the arch or dome. The same occurs, for example, in the settling of the foundation of a masonry bridge pier. In many cases even a very thin crack can signal the occurrence of deformation.

The degree of settling can be predicted only with difficulty. Thankfully, the compatibility equations expressing the occurrence of settling do not require defining the magnitude of the settlement, but only indication of the constraints to the settled state. In more complex structures, as for instance a multi-span bridge or a masonry wall with openings, inspection of cracking patterns can provide useful information regarding the settlement mechanism effectively produced. According to this approach, determination of the actual stress state in masonry structures becomes statically determinate and Limit Analysis can once again be fruitfully applied, as will be shown. A number of useful applications of this approach will be described in later chapters.

2.8 Geometry and Strength: The Theory of Proportions of the Past Architecture

Masonry constructions have a long history. They have been built, studied, tested for about 7000 years but throughout all this time the material masonry, in spite of the large variety of its typologies, has maintained the same mechanical features. It is an unilateral material which can resist compression, but not tension.

This aspect has influenced the history of constructions and has marked out a forced path in the long research of the various structural solutions.

Construction experiences condensed during the time in the form of structural rules. The essence of these rules is that *proportion* controls the overall form of the structure of the building. A Theory of Proportions developed slowly since Vitruvius up to Leon Battista Alberti and Palladio.

According to this theory Statics of masonry structures has to be ruled solely by geometry. Knowledge of the most suitable proportions amongst the various components of a structure and of their basic measure, *the modulus,* irrespective of its absolute magnitude, represented the essence of the art of past constructions. Since antiquity master builders have always used simple geometrical rules for designing arches or buttresses for a cross vault.

An 'ideal city' was conceived according these rules, as shown by the famous painting of an unknown artist of the fifteenth century (Fig. 2.57). Then, later, in the nineteenth century, masonry architecture felt into decay due the appearance of new materials and new structural forms.

Galileo confuted the rules of proportional design in his *Dialogues* (Fig. 2.58). Galileo observed that, given any structure which supports its own weight, if we multiply its size by a certain factor k maintaining its geometrical form, it becomes weaker.

To illustrate this statement by way of example, let us examine two similar beams a and A built with the same material and loaded by their own weights. The beam

Fig. 2.57 The ideal city conceived according to the "proportionality theory"

Fig. 2.58 The Galileo example of beam

2.8 Geometry and Strength: The Theory of Proportions of the Past Architecture

A is k times larger than the beam a. In the transverse direction the beams have the same width s. Stresses σ and σ' in the beams a and A are respectively

$$\sigma = \frac{M}{W} \quad \sigma' = \frac{M'}{W'} = k\sigma \tag{2.162}$$

As a consequence, the beam A, as it grows in size, becomes weaker than the beam a. If we want maintain the same strength, the cross sections of the beam A must become thicker, as shown in the classical sketch of Fig. 2.59, taken from Galileo *Dialogues*.

Galileo realized that his discovery contradicted the rules of the proportional design of his days.

Some scholars, as Parson (1970), Benvenuto (1981), Mark (1990), identified in the Galileo judgement the irrefutable proof of the error rooted in the theory of proportions. On the contrary, for other scholars, as Dorn (1970), Heyman (1995), Baratta (1982) and Huerta (2001), the Galileo conclusion was not applicable to masonry constructions because for them the material strength plays no role.

We will show now, by means a direct proof and in the strict context of the no tension masonry model, that the Theory of Proportions is correct.

In previous sections we have seen that weight and geometry represent the essential elements in the strength of masonry structures. More precisely, it is *the proportions* among a structure's various constituent parts that define a its geometry, irrespective of the actual absolute dimensions.

To illustrate this by way of example, let us examine the two similar arches a and A in Fig. 2.60. Arch A, on the right, is k times larger than the arch a, on the left, or in other words, arch A is a k times *magnified copy* of arch a. In the transverse direction, i.e. in the direction orthogonal to their plane, the structures have the same width s. Each segment in structure A is thus k times longer than the corresponding segment in structure a. We can moreover consider other similar structures, such as s and S shown in Fig. 2.47, and *refer to the same system of coordinate axes* with origin O defined at the same position (Fig. 2.61). By definition, two points $\mathbf{p_i}$ and $\mathbf{P_i}$, having respective coordinates (x_i, y_i) and (X_i, Y_i), are *conjugated* if

Fig. 2.59 Larger thickness of the great bone in order to have the same strength of the small

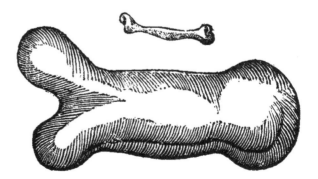

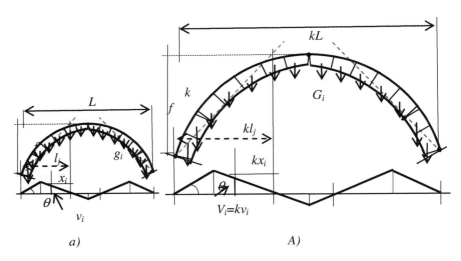

Fig. 2.60 Geometries of two similar arches and two corresponding mechanisms governed by dimension ratio k

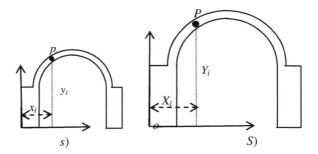

Fig. 2.61 Another example of two similar structures, s and S, in which the latter is the k times magnification of the former

$$X_i = kx_i \quad Y_i = ky_i. \qquad (2.163)$$

So, the two arches a and A, for instance, are subdivided into an equal number of *conjugated* voussoirs v_i and V_i (1, 2,...i,...N), i.e. having centers having coordinates

$$(b_{xi}, b_{yi}) \ (B_{xi}, B_{yi}) \qquad (2.164)$$

with

$$B_{xi} = kb_{xi} \quad B_{yi} = kb_{yi}. \qquad (2.163')$$

At the same time the dimensions of the voussoirs c_i of arch a are length d_i and height h_i, while those of voussoirs C_i of arch A are D_i and H_i (i = 1, 2,...N), with

2.8 Geometry and Strength: The Theory of Proportions of the Past Architecture

$D_i = k\,d_i$ and $H_i = kh_i$. Consequently, if g_i and G_i are the weights of the voussoirs, we have

$$G_i = k^2 g_i. \tag{2.165}$$

Let us now consider the *conjugated mechanisms* m and M, respectively for structures s and S: the latter is the k magnification of the former. The two mechanisms present the same rotation parameter θ and their hinge points c_i and C_i are conjugated. Hence, if l_J is the distance of the hinges of the arch a mechanism from the left springing, the corresponding distance of the hinges of arch A will be kl_j, and the lines connecting the two corresponding hinges are parallel to each other, as shown in Fig. 2.60. The centre, b_i, of voussoir i of the arch a, being at distance x_i from the corresponding point of rotation, moves vertically by v_i.

Likewise,

$$V_i = kv_i. \tag{2.166}$$

and kx_i are the vertical displacement and analogous distance of centre B_i of the corresponding voussoir i of arch A. Let us now assume that structure a is *stable* under its own weight g, as defined according to (2.71′),

$$\langle g, v \rangle = \sum_1^N g_i v_i < 0 \tag{2.167}$$

for any mechanism v. The work $\langle g,v \rangle$ is evaluated considering the work of the weight forces g_i on the corresponding vertical displacements v_i of the mechanism.

We will now show that the k magnified structure A is thus also *stable* under its own weight, in the sense that, analogously for any mechanism V, we will have

$$\langle G, V \rangle = \sum_1^N G_i V_i < 0. \tag{2.167′}$$

In fact, according to the foregoing assumptions and definitions, due to the similarity between s and S and associated mechanisms m and M from (2.127) and (2.128)

$$\langle G, V \rangle = \sum_1^N G_i V_i = k^3 \sum_1^N g_i v_i < 0. \tag{2.168}$$

Thus, to conclude, *if a structure under its own weight is stable, a k times magnified copy will also be stable.*

The *same* outcome holds in a more general sense. Let us consider the two similar structure s and S of Fig. 2.62 where now *only* the weights, g_a and G_a, of their central spans increase via loading parameters λ and Λ. According to the kinematic approach, the maximum load $\lambda_0 g_a$ that structure s can sustain (see Sect. 2.6.3) can be obtained as

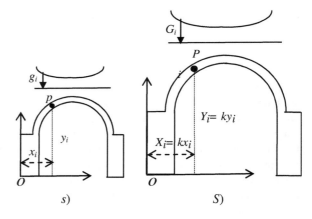

Fig. 2.62 The k similar structures loaded by increasing weights along their spans

$$\lambda_o = Min\left(-\frac{\langle g, v\rangle}{\langle g_a, v\rangle}\right) = Min\left(-\frac{\sum g_i v_i}{\sum g_{ai} v_i}\right) \quad v \in M \tag{2.169}$$

where g is the pier weight and g_{ai} and g_i the corresponding weights of the voussoirs into which the structure has been divided. Likewise, the maximum load $\Lambda_o G$ that the k magnified structure S can sustain is

$$\Lambda_o = Min\left(-\frac{\langle G, V\rangle}{\langle G_a, V\rangle}\right) = Min\left(-\frac{\sum G_i V_i}{\sum g_{ai} V_i}\right) \quad V \in M. \tag{2.170}$$

However, according to previous assumptions

$$G_i = k^2 g_i. \quad G_{ai} = k^2 g_{ai} \quad V_i = k v_i, \tag{2.171}$$

and the two structures s and S exhibit the same strength under loads λg and ΛG. In fact. we have

$$\Lambda_o = Min\left(-\frac{k^3 \sum g_i v_i}{k^3 \sum g_{ai} v_i}\right) = \lambda_o. \tag{2.172}$$

This result holds even if we consider that, instead of *vertical* loads λg, there are *horizontal* forces λg, still proportional to weight g, acting on the structure. Such a loading condition is frequently considered representative of *seismic* actions. The two similar structures s and S in Fig. 2.63 thus exhibit the same horizontal strength under the action of horizontal forces λg and ΛG.

These results, here directly proved, were well known to architects of the past and formed the basis for their fundamental rules of construction. As set down in the theory of proportions by Andrea Palladio and Leo Battista Alberti, the statics of masonry structures is governed solely by their geometry and, consequently, by their basic measurement, the modulus, irrespective of their absolute measurements

2.8 Geometry and Strength: The Theory of Proportions of the Past Architecture

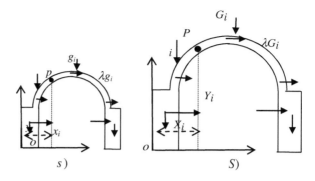

Fig. 2.63 Similar structures under horizontal load λg and λG

Knowledge of the most suitable proportions amongst the various components of a masonry structure (knowledge often jealously guarded by past masters) represented the essence of the art of construction.

As discussed throughout this book, these results, arrived at through centuries' long experience, is a direct consequence of the unique, fundamental behavior of masonry. By way of example, Iori (2009) has recalled two Romans dome constructions: the Pantheon and the Temple of Romulus, similar but different in scale (Fig. 2.64). Both the constructions have the maximum height equal to the dome diameter and a drum thickness equal to 0.3 of the radius.

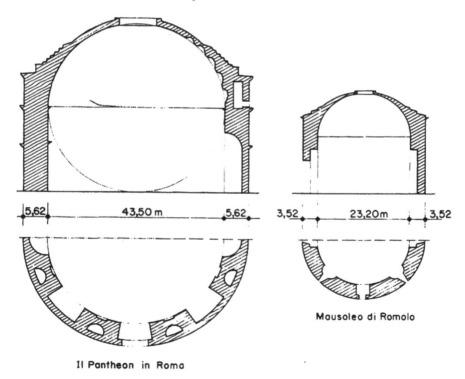

Fig. 2.64 The Pantheon in Rome and the Temple of Romulus (Iori 2009)

Appendix

A.1 Lessons from the Failure of Cathedral of Noto

The Church

In order to better illustrate the foregoing basic assumptions regarding the behavior of masonry structures, it is useful to analyze the causes of the 1996 collapse of the Cathedral of Noto (Sicily). The Noto cathedral has two side aisles and a central nave. The beautiful 40 m-wide, 32 m-high baroque facade is flanked by two bell towers, as shown in Fig. 2.65.

The bases of the towers and facade are 3.40 m below the level of the Church floor (assumed to be at 0.00 m). The tower's foundations are partially exposed near the side streets due to some excavation work done in the late 19th century. The lateral walls of the Church are faced with regular blocks and mortar beds, while rubble masonry with poor mortar makes up the walls' inner cores. The five piers that run alongside the central nave, are similar in structure to the lateral walls: facings made of regular square blocks of local travertine with inner rubble masonry (Fig. 2.66). Only their bases were built with the more substantial Noto limestone.

The piers have a roughly rectangular cross section, 3.25 m × 1.60 m, with their main lengths laid along the church's longitudinal direction. The piers sustaining the dome at the transept are larger in cross section: 6.50 m × 1.65 m.

The side aisles are roofed by small domes set on small drums, in turn, sustained by the side walls and longitudinal and transverse arches spanning from walls to piers (Figs. 2.66 and 2.67). All the domes and arches of the aisles have a thick inner rubble structure and regular facings with squared stone blocks.

A long wall running along the top of the interior piers sustains a reinforced concrete floor that was built to replace the original wooden roof during past restructuring work. Fortunately, the operations spared the high transverse masonry arches spanning the nave. The spherical dome, with an inner diameter of 11.20 m

Fig. 2.65 Cathedral façade

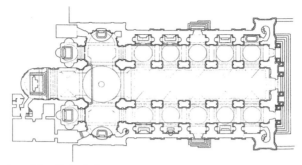

Fig. 2.66 The plan of the Cathedral

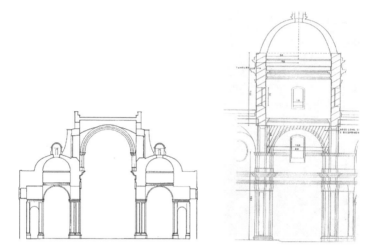

Fig. 2.67 Cross section of the church at the nave and transept

was built using strong blocks of Modica limestone on top of a drum containing several large windows. The dome and drum were 1.30 m thick. The foundations of the church's lateral walls are made up of continuous masonry walls set on a thick formation of sandstone and clays. The more superficial foundations of the piers rest on an arenaceous formation 1 m beneath the level of the church floor.

Construction of the Cathedral was begun in 1753 and completed in 1769. The dome has collapsed a number of times: first during its construction, a second time during an 1848 earthquake, after which it was rebuilt for the third time between 1860 and 1862 and remained up to the most recent collapse in 1996. It has since been reconstructed. In 1990 an earthquake measuring 4.7 on the Richter scale struck the entire region of Syracuse and seriously damaged many structures of the church. In the area of Noto the damage was rated at level VI on the Modified Mercalli Intensity scale (MMI). The Church was seriously damaged in the earthquake: the transverse arches together with the aisle roofs suffered deep cracking, particularly on the right aisle. Afterwards, some damaged piers were buttressed with simple scaffolding and the church remained open while its condition continued to worsen, until its near complete collapse in 1996.

The 1996 Collapse

The failure occurred suddenly, without any storm or earthquake, and affected the entire interior of the church. All the piers and coverings of the right nave collapsed together with the dome. Only the beautiful façade was spared. Figure 2.68a, b show the interior of the cathedral with the remnants of the dome after the collapse.

Investigations of the Causes of the Failure

A committee appointed by the Court of Syracuse, composed of M. Como, G. Croci, M.T. Lo Balbo, A. Migliacci, and F. Selleri, carried out a survey to ascertain the cause of the collapse (1998, 2001). Compression strengths of the various stones: local travertine: $\sigma_{rm} = 61$ kg/cm^2; Noto limestone, $\sigma_{rm} = 195$ kg/cm^2; giuggiulena stone, $\sigma_{rm} = 95$ kg/cm^2; Modica limestone, $\sigma_{rm} = 220$ kg/cm^2.

Fig. 2.68 a The interior of the church after collapse; **b** The collapsed dome

Appendix

Fig. 2.69 Upper fragment of pier 4

Fig. 2.70 The base of pier 4, weakened by the passageway to the pulpit

After all the various investigations, geotechnical drillings, material testing and ground-penetrating radar explorations, it became possible to examine the bases of the failed piers as soon as the rubble was completely cleared (Figs. 2.69 and 2.70).

The Structural Collapse: How Can We Explain the Failure?

Despite the presence of many cavities in the soil beneath the church floor, accurate numerical and geotechnical investigations excluded the possibility that settling of the foundation had occurred, a conclusion that was also confirmed by visual inspection of the pier bases. The elements that were fundamental to enabling definition of the failure kinematics were the counter rotation of the upper parts of the failed piers near the transept and the sequence of cracks that signalled failure of the upper reinforced concrete ring beam. Figure 2.71 shows the strong wrenching action exerted on the reinforced concrete ring beam running along the top of the upper wall lining the nave, revealing the point where the collapse started—precisely at the position of pier n. 4. This pier was weaker than the others due to the presence of an old interior passageway providing access to the pulpit (Fig. 2.70). The presence of the overhead transverse arches at the level of the roof over the nave determined a two-hinge mechanism, one at the pier base and the other high up at the springing of the transverse arch. This mechanism caused counter-rotation of the upper part of the piers, which was in effect confirmed by examination of the position of the parts of the failed piers (Fig. 2.69). The structure of the cathedral was thick and resistant. In particular, the transverse arches, drums and small domes roofing the aisles, constructed with heavy concrete (Fig. 2.74), give the impression of a solid monolithic structure. The roofs also gave the appearance of behaving as a single unified mass, able to transmit only axial loads to their supports.

However, contrary to appearances, these thick roofing structures were actually particularly vulnerable. The concrete had very little tensile strength, so these structures were able to sustain their own weight and transmit it to the underlying piers only as long as the concrete was intact. However, if its strength were to decay, for instance due to damage from sudden seismic actions or slow masonry decay by the action of rainwater seeping into the masonry, the behavior of the structure would change drastically (Fig. 2.72).

Fig. 2.71 Cracking of the ring beam showing the strong pulling action due to the collapse of underlying pier 4

Appendix   129

Fig. 2.72 The massively thick concrete of the arches and small domes

In fact, the 1990 earthquake caused extensive damage to many of the cathedral's structures. Alarming cracks appeared in its support structures, especially on the right side. When the thick structures covering the right aisle cracked, their monolithic behavior was lost and strong thrusts were activated. Static checks of the piers lining the nave revealed that the compression stresses acting upon them were admissible only for the heavy vertical loads acting alone; if instead the action of the

Fig. 2.73 The failure mechanism

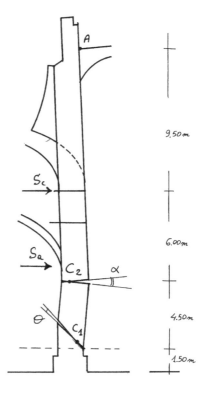

Fig. 2.74 The limit equilibrium of the pier

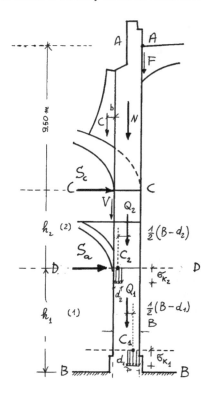

thrusts of the domes and arches were taken into account, the check showed that the piers were at the very limit for failure. Figures 2.73 and 2.74 show the collapse mechanism and the distribution of thrusts and internal stresses of pier n.4.

Lesson to Be Learned from the Failure

The lesson imparted by the failure of the Noto Cathedral serves as confirmation that, in the interplay between weight and geometry, only the no-tension model can provide reliable indications as to the true strength of masonry structures. Thus, although a certain amount of tensile strength may be allowed for, in the end, it is the no-tension model that best predicts the true behavior of masonry structures.

We must therefore distinguish between structures that are *intrinsically stable*—that is, those that, given their geometry and load distribution, can sustain the loads within the assumptions of no-tension behavior—and those that are instead *intrinsically unstable*—unable to sustain such thrusts under the no-tension model.

In this sense, the structure of the cathedral was intrinsically unstable. Unfortunately, the apparent solidity of the vaulted concrete systems was misleading. In fact, once the thrusts of the arches and domes covering the right aisle were activated by the cracking, the piers were unable to sustain these thrusts and consequently collapsed.

Ironically, the church's collapse could have been avoided by simply fixing chains under the transverse arches and right aisle domes just after the damage caused by the 1990 earthquake.

The cathedral of Noto has since been rebuilt and currently represents one of the most extraordinary examples of baroque architecture in Italy, particularly in its magnificent façade.

References

Angelillo, M., Cardamone, L., & Fortunato, A. (2010). A new numerical model for masonry like structures. Journal of Engineering and Material and Structures, 5.
Baratta, A., & Toscano, R. (1982). Stati tensionali in pannelli di materiale non resistente a trazione, AIMETA, National Conference, Genova.
Benvenuto, E. (1981). *La Scienza delle Costruzioni e il suo sviluppo storico*. Roma: Firenze, Sansoni, Ristampa, Edizioni di Storia e Letteratura.
Benvenuto, E. (1991). *An Introduction to the history of structural mechanics, part ii, vaulted structures and elastic systems*. NY: Springer-Verlag.
Briccoli, B.S., Paradiso, M., & Tempesta, G. (1988). Analisi Statica, Cinematica ed Equilibrio Limite di Strutture ad arco a Vincoli unilateri, IX Congr. Naz.le AIMETA, University of Bari, 4–7 Ott., Tipolito Il Globo, Bari.
Como, M., & Grimaldi, A. (1985). An unilateral model for the limit analysis of masonry walls. In *International Congress on Unilateral Problems in Structural Analysis, Ravello, 1983, CISM Udine*. Springer Verlag.
Como, M. (1992) Equilibrium and collapse of masonry bodies. Meccanica, 27, 185–194. Kluwer Academic Publications, London.
Como, M. (1996a). *On the role played by Settlements in the Statics of Masonry Structures in the Conference Geotechnical Engineering for the Preservation of Monuments and Historic sites*, Napoli, Italy, 3–4, October. Rotterdam: A.A. Balkema.
Como, M. (1996b). In Pitagora (Ed.), *Multiparameter Loading nd Settlements in Statics of Masonry Structures, Proceedings Conference Meccanica delle Murature tra teoria e progetto*, Messina, 18–20 September 1996, Bologna.
Como, M. (1998) Minimum and maximum thrust states in Statics of ancient masonry buildings. In *Proceedings 2nd International Arch Bridge Conference*, Venice, Italy, 6–9 October. Sinopoli, Rotterdam: A.A. Balkema.
Como, M. (2012). On the statics of bodies made of compressionally rigid no tension materials. In M. Fremond & F. Maceri (Eds.), *Mechanics, models and methods in civil engineering*. Berlin Heidelberg: Springer Verlag.
Como, M., Croci, G., Lo Balbo, M.T., Migliacci, A., & Selleri, F. (1998). sintesi a cura di, tratta da *Consulenza Tecnica sulla Chiesa Madre San Nicolò di Noto*, Proc. della Repubblica presso il Tribunale di Siracusa.
Como, M., Croci, G., Lo Balbo, M.T., Migliacci, A., & Selleri, F. (2001). Problematiche ed indagini sul crollo nella Cattedrale di Noto del 13-03-1996, Atti del Convegno Nazionale *Crolli e Affidabilità delle Strutture Civili*, IUAV, Venezia.
Coulomb, C.A. (1821). Essai sur une application des règles..., *Mémoires de Mathématique et de Physique*, présentés à l'Académie Royale des Sciences, Vol. 17, 1773, pp. 343–382. Paris 1776, reprinted in *Theorie des machines simples*, Paris.

Del Piero, G. (1989). Constitutive equation and compatibility of the external loads for linear elastic masonry–like materials. *Meccanica, 24*, 150–162.
Di Pasquale, S. (1984). *Statica dei solidi murari: teorie ed esperienze*, Atti del Dipartimento di Costruzioni, Università di Firenze.
Di Pasquale, S. (1996). *L'Arte del Costruire, Tra conoscenza e scienza*, Ed. Marsilio, Venezia.
Drucker, C. K. (1959). A definition of stable inelastic material. *Journal of Applications Mechanics, 26*, 1.
Fontana, C. (1694). *Il tempio vaticano e la sua origine*, Roma.
Bacigalupo, A. & Gambarotta, L. (2011). Non-local computational homogenization of periodic masonry. International Journal Multiscale Computational Engineering, *9*.
Giuffrè, A. (1990). *Letture sulla Meccanica delle Murature Storiche*. Roma: Facoltà di Architettura dell'Università di Roma La Sapienza.
Giuffrè, A. (1991). *La meccanica nell'Architettura, La statica*. Roma: La nuova Italia Scientifica.
Heyman, J. (1966). The stone skeleton. *Intern, Journ. Solids Structures, 2*, 249.
Heyman, J. (1982). *The masonry arch*. Chichester: Ellis Horwood.
Heyman, J. (1995). *The stone skeleton, structural engineering of masonry architecture*. Cambridge: Cambridge University Press.
Horne, M.R. (1971). Plastic Theory of structures, T. Nelson and Sons Ltd, London.
Huerta, S. (2001). Mechanics of masonry vaults: The equilibrium approach. In P.B. Lourenco & P. Roca (Eds.), Proceedings 3rd International Semester on Historic Constructions, Guimaraes, Portugal, University of Minho.
Kooharian, A. (1952). Limit analysis of voussoir (Segmental) and concrete arches. ACI Journal (December, Proc. V, 49).
Iori, I. (2009). Il quadro storico-scientifico in *Santa Maria del Quartiere in Parma, Storia, rilievo e stabilità di una fabbrica farnesiana* a cura di Giandebiaggi P., Mambriani C., Ottoni F., University of Parma, Fondazione Cariparma, ed. Grafiche Step, Parma.
Lucchesi, M., & Zani, N. (2003). Some explicit solutions to plane equilibrium problem for no–tension bodies. *Structural engineering and Mechanics, 16*(3), 295–316.
Lucchesi M., Padovani C., Pasquinelli G., & Zani N (2008). Masonry Constructions: Mechanical Models and Numerical Applications, Lecture Notes in Applied and Computational Mechanics (Vol. 39). Berlin Heidelberg: Springer.
Ochsendorf, J. (2006). The masonry arch on spreading supports. *The Structural Engineer, 84*(2).
Prager, W. (1959). *An Introduction to Plasticity*. U.S.A: Addison Wesley Publications, Reading, Mass.
Romano G., & Romano M. (1985). Elastostatics of sstructures with unilateral conditions on strains and displacements. In *Intern. Congr. On Unilateral Problems in Struct. Analysis'*, Ravello 1983, CISM, Springer.
Romano G., & Sacco E. (1986) Sul calcolo di strutture murarie non resistenti a trazion*e, Atti Istit. Scienza delle costruzioni*. Università di Napoli.
Sinopoli, A., Corradi, M., Foce, F. (1998). Lower and upper bond therems for masonry arches as rigid systems with unilateral contacts. In *Proceedings 2nd International Arch Bridge Conference, Venice, Italy, 6–9 October*. Rotterdam: A.A. Balkema.
Trovalusci, P. (1993). Sulla modellazione meccanica dei solidi murari. In *Sicurezza e conservazione dei centri storici. Il caso di Ortigia*, a cura di A.Giuffré, Laterza, Bari.
Trovalusci, P., & Masiani, R. (2005) A multifield model for blocky materials base on multiscale description. Int.mJ. Solids and Structures, *42*.
Viollet–le–Duc, E.E. (1854–1868). *Dictionnaire raisonné de l'architecture*, Paris.
Vol'pert, A. I., Hudjaev, S. I. (1985). *Analysis in Classes of Discontinuous Functions and equations of Mathematical Physics*, Nijoff.

Chapter 3
Masonry Arches

Abstract Aim of this chapter is studying statics of masonry arch. Limit Analysis of the arch, under vertical and horizontal forces, together with the research of its minimum thrust, is pursued by direct application of the general theory presented in the foregoing chapter. The minimum thrust is determined for the round and for the depressed arch, examining the strict connection existing between the geometry of the pressure line and the hinge position of the corresponding settlement mechanisms. The chapter ends by examining and discussing the tests performed in laboratories to study the masonry arch behavior.

3.1 Definitions and History

A masonry arch is a curved structural element that spans an opening. It is able to sustain loads solely *by virtue of compressions*. It is a structure whose geometry, perhaps better than any other epitomizes the ingenuity of masonry constructions.

In order to erect an arch with bricks or stone voussoirs, the individual masonry elements must be placed with radial joints to prevent shear slippage. Before achieving the skill necessary to construct true stone arches, with radial joints, ancient builders first learned to erect so-called *false arches*. These are the corbelled structures formed by offsetting successive stone courses so that they project towards the archway's center from each supporting side (Fig. 3.1).

Ancient examples of these constructions are the walls built above the Lion Gate at Mycenae (Fig. 3.2) and other Minoan structures.

The safety of a false arch depends on positioning the vertical joints in such a way as to prevent the overturning of any of the blocks, which are set and therefore act in groups (Fig. 3.3).

In the false arch the horizontal stones at the key section are unable to transfer thrust. In fact, only friction opposes their sliding, which is difficult to exploit at the top of the arch, as there is no vertical compression. Each of the two halves of the arch must therefore be freestanding, that is, able to remain upright on its own.

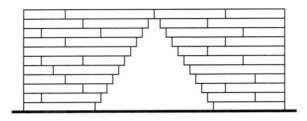

Fig. 3.1 False arch

Fig. 3.2 Lion gate at mycenae

Fig. 3.3 False arch safety—preventing the overturning of various block groups

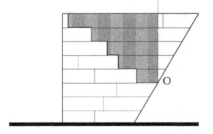

These more rudimentary constructions are conceptually very different from true arches. In fact, true arches were developed much later, though some ancient Sumerian buildings contain examples of simple small arches built with bricks joined through radial joints. It was however the Etruscans who finally made systematic use of true masonry arches.

Fig. 3.4 An Etruscan arch at Volterra

One outstanding example is the arch of the Volterra gateway (Fig. 3.4) from the 4th century BCE., in which the radial arrangement of the joints between the stone voussoirs are clearly visible.

Another example is the round arch over the so-called *Porta Rosa* (Pink Gate, Fig. 3.5) of the city wall of Velia, the ancient Greek settlement of Elea, also built in the 4th century BCE. It is a rare example of Greek arch construction in Magna Græcia. As can be seen in the photo, the voussoirs are wedge-shaped, a geometry that increased the masonry's strength and enabled reducing the thickness of the joints.

The Romans later made widespread use of the arch, particularly the round arch, in the construction of aqueducts, bridges, etc. Figure 3.6 shows a view of the bridge of Fabricius, Rome's best-preserved Roman bridge, from the 1st century BCE. Figure 3.7, from a drawing by the architect Andrea Palladio, shows an example of a typical Roman bridge, with its round arches and constant thickness.

One beautiful example of a more modern masonry bridge is the Turin 'Mosca' bridge erected in 1827 to span the Dora Riparia River. Each arch is made up of 93 granite voussoirs: it has a span of 45 m, a rise at the intrados of 5.5 m and a thickness varying from 2 m at the springers to 1.5 m at the keystone (Fig. 3.8).

Mortar was used to cement only the first 11 joints at the springers and the 22 joints near the key. The bridge was studied by Castigliano in (1879) to verify the

Fig. 3.5 The double-arched *Porta Rosa* at Velia

Fig. 3.6 The Fabricius bridge in Rome

advantages of using mortar joints on the pressure line, a question which will be taken up again later.

Nowadays most bridges are constructed of reinforced concrete and steel. The last large masonry bridges were built in Europe and the USA with spans ranging

3.1 Definitions and History

Fig. 3.7 A Roman bridge as drawn by Palladio (from D'Agostino et al. 2001)

Fig. 3.8 The Mosca bridge in Turin

between 80 and 100 m. The Plauen bridge (1903–1906) with a span length of 90 m (Fig. 3.9) is a good example of these longer masonry bridges.

Though new ones are rarely constructed, still today there are myriad stone bridges in service along European motor and railway networks. For small-span bridges, from 3 to 6 m in length, the arch thickness is generally kept constant roughly according to the formula

$$s = a(1 + \sqrt{L}), \tag{3.1}$$

where a varies between 0.15 and 0.18 for roadway bridges and s and L are expressed in m. For larger spans, the thickness at the key is generally equal to

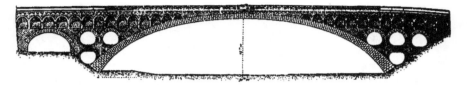

Fig. 3.9 The Plauen bridge (1903–1905) with 90 m span (from D'Agostino et al. 2001)

$$s_c = 0.10 + 0.20\sqrt{L}, \tag{3.1'}$$

while at the springers it is 1.1 ÷ 1.6 times the thickness at the key (from D'Agostino and Bellomo 2001). The variation in the cross sections' moment of inertia for arches of variable thickness is frequently taken to be

$$I = \frac{I_c}{\cos \alpha} \quad I = \frac{I_c}{\cos^3 \alpha}, \tag{3.2}$$

where I_c is the key section moment of inertia and α is the angle between the tangent to the arch axis and the horizontal. The left-hand formula furnishes the variation law from springers to key of the moment of inertia of arches having transverse sections with constant height and linearly increasing width, while the right-hand expression is for arches with constant width and linearly increasing section height.

3.2 The Birth of the Statics of the Arch and Its Evolution

The first mechanical definition of the arch (Fig. 3.10) was formulated by Leonardo Da Vinci (1451–1519).

According to his definition, reported by Marcolongo (1937): "[... *arch is no more than a strength caused by two weaknesses, in that in its construction the arch is composed of two quarter circles, each of which, being in and of itself very weak, tends to fall, but as each opposes this tendency in the other, the two weaknesses are transformed into one strength*]".

Leonardo not only gave us this insightful interpretation of the arch, but he also designed equipment to measure arch thrusts, as shown in the lower sketch in Fig. 3.10, taken from the Forster Codex and reported by Benvenuto (1991).

The research conducted by Robert Hooke at Cambridge led to formulation of the *resistant model* of the arch, using the funicular curve of the loads. The arch, whose axis is upside down with respect to a hanging chain (catenary), can sustain the same loads determining the equilibrium curve of the chain.

In 1675 Hooke realized the importance of studying upside down, or inverted, systems to analyze the equilibrium of masonry arches. He announced his discovery in a now famous Latin anagram to secure his authorship: "...*ut pendet continuum*

3.2 the Birth of the Statics of the Arch and Its Evolution 139

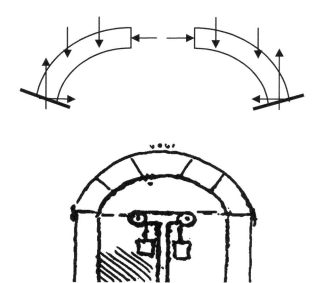

Fig. 3.10 The first definition of the arch, by Leonardo da Vinci, with the equipment designed to measure thrust

flexile, sic stabit contiguum rigidum inversum,...", meaning "... as hangs a flexible cable so, inverted, stands the rigid arch" (Fig. 3.11). His finding marked the beginning of the discipline of Statics of masonry arches, thanks to which the study of the behavior of masonry arches made great progress by using the concept of the inverted funicular curve. The configuration assumed by a heavy chain suspended between two fixed points, B and C, is represented by the catenary, also called the "chainette". Leibniz, Huygens and Johann Bernoulli, together with his brother, Jakob, discovered this curve (Timoshenko 1953).

An important property of the catenary is that, at any point G along the curve, the resultant of all forces applied to the chain to the left of point G—that is, the resultant of the weight of the catenary segment between B and G, together with the reaction of the suspension point B—passes precisely through G and is tangent to it. It is worthwhile deriving the equation of the catenary, also known as Hooke's curve, BAC, as illustrated in Fig. 3.11. Figure 3.12 shows an element *ds* of the chain in equilibrium under its own weight and the axial tensile forces acting along the tangents to the end sections of the element.

The problem is to determine the equilibrium curve, $z = z(x)$, of the chain, where x indicates the horizontal abscissa. The function $z = z(x)$, which defining the horizontal projections of the ends of the chain, varies between the values $x = 0$ and $x = L$. We moreover assume that $z = 0$ at $x = 0$.

Let g be the weight per unit length of the chain with constant cross section. Each segment *ds* weighs *gds*. The horizontal component of the tensile force N in the chain is constant, while the variation along x of its vertical component equals the weight of the element *ds* of the chain. Thus, we have

Fig. 3.11 Hooke's Law
(from Heyman 1997)

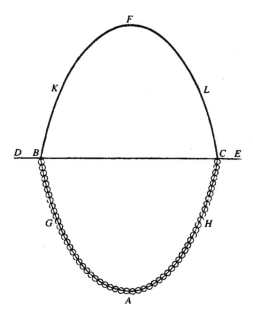

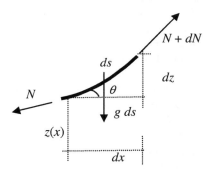

Fig. 3.12 Internal equilibrium of a heavy chain

$$\frac{d}{dx}(N\cos\theta) = 0 \quad \frac{d}{dx}(N\sin\theta)dx + gds = 0, \qquad (3.3)$$

where θ indicates the angle between the tangent to the catenary and the horizontal axis x. The horizontal component of the pull N in the chain is denoted by H and is given by

$$H = N\cos\theta = \text{cost.} \qquad (3.4)$$

Therefore, taking the first of conditions (3.3) into account, from the second we obtain

3.2 the Birth of the Statics of the Arch and Its Evolution

$$H\frac{d}{dx}(tg\theta) = g\frac{ds}{dx} \tag{3.5}$$

and, because

$$ds\cos\theta = dx. \tag{3.6}$$

we also have

$$H\frac{d}{d\theta}(tg\theta)\frac{d\theta}{dx} = \frac{g}{\cos\theta} \tag{3.7}$$

and

$$H\frac{1}{\cos\theta}\frac{d\theta}{dx} = g. \tag{3.7'}$$

Splitting the variables and integrating gives

$$\int_0^\theta \frac{d\theta}{\cos\theta} = \frac{g}{H}\int_0^x dx. \tag{3.8}$$

Under the assumption that, by symmetry, the origin $x = 0$ is found at the middle of the chain, so that for $x = 0$, we have $\theta = 0$, integration of (3.8) gives

$$\ln tg\left(\frac{\pi}{4}+\frac{\theta}{2}\right) = \frac{g}{H}x, \quad tg\left(\frac{\pi}{4}+\frac{\theta}{2}\right) = e^{\frac{g}{H}x}. \tag{3.9}$$

However,

$$tg\left(\frac{\pi}{4}+\frac{\theta}{2}\right) - ctg\left(\frac{\pi}{4}+\frac{\theta}{2}\right) = 2tg\theta \tag{3.10}$$

and

$$tg\theta = \frac{1}{2}\left(e^{\frac{g}{H}x} - e^{-\frac{g}{H}x}\right). \tag{3.11}$$

Recalling that the sought-for equilibrium configuration must be defined by the equation

$$z = z(x), \tag{3.12}$$

by virtue of (3.11), we also have

$$\frac{dz}{dx} = \frac{1}{2}(e^{-\frac{g}{H}x} - e^{\frac{g}{H}x}) \tag{3.13}$$

and we get

$$z(x) = \frac{1}{2}\frac{H}{g}(e^{\frac{g}{H}x} + e^{-\frac{g}{H}x}) + \text{cost} \tag{3.14}$$

Using the hyperbolic function cosh$x$ and taking into account that $z = 0$ at $x = 0$, we finally obtain

$$z(x) = \frac{H}{g}(\cosh\frac{g}{H}x - 1). \tag{3.15}$$

Equation (3.15) defines the segment *AHC* of the chain drawn in Fig. 3.11. Once function $z = z(x)$ has been determined, the pull H can be evaluated.

Let us consider now the case of the chain with suspension points A and F, loaded by a sequence of point loads F_{12}, F_{23}, ... The equilibrium configuration of the chain is thus defined by the polygon ABCDEF drawn in Fig. 3.13. This polygon can be obtained by using the so-called funicular polygons. The definition of a funicular polygon is the same as that of the catenary. The resultant of the forces acting to the left, for instance, at each node A, B, ... is directed along the stretched sides AB, BC, etc. of the polygon. The same inverted catenary, or inverted polygon, will define the funicular polygon with compressed sides (Fig. 3.14).

De la Hire (1712) continued the studies of Hooke and pointed out that the arch could be able to sustain a given sequence of loads if its pressure curve, obtained as the funicular of all loads applied at the centers of the various voussoirs, was contained entirely within the arch's thickness (Fig. 3.15). In 1729, Belidor, in his book *La Science des Ingénieurs*, applied the procedure proposed by De la Hire to the problem of dimensioning the arches, in particular, their abutments. The consequences were enormous. Using this approach, for instance, Poleni (1748) was able to check the stability of the dome of St. Peter's Basilica in Rome, which at the

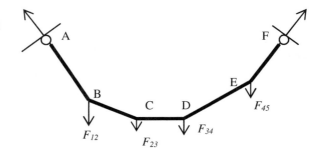

Fig. 3.13 Equilibrium polygonal of the chain loaded by vertical point forces

3.2 the Birth of the Statics of the Arch and Its Evolution

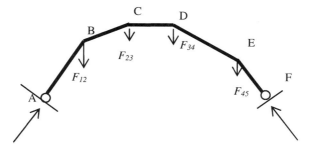

Fig. 3.14 The inverted equilibrium polygon

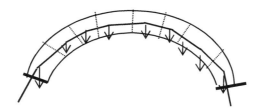

Fig. 3.15 Masonry arch in admissible equilibrium: the funicular of the loads is wholly contained within the arch

time aroused serious concerns about its cracking. Based on these concepts the statics of arches continued to develop gradually throughout the 18th and part of the 19th century. Important contributions to the advancement of the statics of arches came from French engineers during the planning and construction of French road networks, for which many masonry bridge were built. A particularly significant contribution was made by Coulomb in his famous essay, *Sur une Application des Régles de maximis et minimis à quelques problèmes de statique relatifs à l'àrchitecture,* presented to the French Academy of Sciences in 1773 and brilliantly commented on two centuries later by Heyman (1972).

In his work Coulomb showed some results he obtained while working as a military engineer in Martinique directing the reinforcement operations on walls and bastions. His milestone essay also established modern soil mechanics and addressed many other problems, such as the bending of beams, column fracture and the statics of abutments under the action of the thrust of masonry arches.

Coulomb thoroughly examined the problem of arch safety by considering both shear and axial failures. He observed that the presence of a pressure curve entirely contained within the arch did not actually furnish any indication about the level of arch collapse safety. Coulomb analyzed a symmetrical arch loaded symmetrically as in Fig. 3.16, which shows half of the arch between the key and springer sections. Only the thrust H acts on the middle section, AB. The shear $T(\alpha)$, acting on any given section, defined by the angle α shown in Fig. 3.16, is

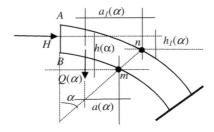

Fig. 3.16 The search for critical arch sections according to Coulomb (1773)

$$T(\alpha) = Q(\alpha)\cos\alpha - H\sin\alpha$$

and is opposed by the friction strength, which is in turn proportional to the compressive force $N(\alpha)$, which is given by

$$N(\alpha) = Q(\alpha)\sin\alpha + H\cos\alpha.$$

In his analysis, Coulomb evaluates the smallest value of thrust able to oppose sliding of the arch segment, *ABmn*, on the plane *mn* by writing

$$Q(\alpha)\cos\alpha - H\sin\alpha = \mu[Q(\alpha)\sin\alpha + H\cos\alpha], \tag{3.16}$$

where μ is the coefficient of friction. Actually, besides friction strength, Coulomb added another term, independent of the compression force and equal to τA, where A is the area of the section *mn*. Thus, in place of (3.16), we have

$$Q(\alpha)\cos\alpha - H\sin\alpha = \mu[Q(\alpha)\sin\alpha + H\cos\alpha] + \tau A, \tag{3.16'}$$

whence Coulomb obtained

$$H(\alpha) = \frac{Q(\alpha)(\cos\alpha - \mu\sin\alpha) - \tau A}{\sin\alpha + \mu\cos\alpha}. \tag{3.17}$$

It is now necessary to identify the smallest value of thrust H able to oppose slipping between all arch sections by varying angle α. Alternatively, the same result can be achieved by determining the position of the critical section with regards to slippage in the opposite direction. Coulomb also analyzed the possibility of hinge formation at the arch extrados or intrados and defined the two other conditions

$$H'h(\alpha) = Q(\alpha)a(\alpha) \quad H''h_1(\alpha) = Q(\alpha)a_1(\alpha), \tag{3.18}$$

which depend on the position, α, of the critical section.

Two limit values, H' and H'', bound the arch thrust able to avoid such hinge formation. They can be determined by means of suitable conditions of *maximum* and *minimum* of appropriate functions of α. Numerical applications have shown that hinging conditions are more critical than those of shear slipping.

3.2 the Birth of the Statics of the Arch and Its Evolution

Coulomb's static analysis showed for the first time the need to verify masonry arches with regard to both slipping and hinging. His findings were not well understood by engineers of the time because finding the maximum and minimum of the functions was a demanding task that only few scientists were able to manage.

However, in the 19th century, thanks to the work of Poncelet (1852), which lead to the development of Graphic Statics, the Coulomb approach was rediscovered and applied anew. Foce (2002) has published a thorough commentary on Coulomb's research. The most hotly debated problems, however, remained the determination of the thrust and tracing of the pressure line. The English engineer Moseley, in his book, *The Mechanical Principles of Engineering and Architecture* (1843), introduced French methods of static analysis of constructions into English schools. In his paper, *On the Theory of the Arch* (1839), Moseley had once again taken up the problem of finding the pressure line of an arch by searching for the optimal position of this curve via the so-called principle of *minimum pressure*. His results are closely related to the considerations that originally gave rise to the concept of minimum thrust and will be addressed further on. Moseley also defined the so-called *strength curve*, represented by the polygon connecting all the intersection points of the funicular polygon with the sections between the arch voussoirs.

Moseley's admissibility condition was based on the strength curve: the presence of a strength curve completely contained within the arch ensured the static admissibility of the arch. It can easily be appreciated that the strength curve does not coincide with the funicular polygon corresponding to the assigned load distribution, even though, as the voussoirs become thinner and thinner, the strength curve tends to overlap the pressure curve.

The 1840 paper, *Sur l'équilibre des voutes en berceau,* by Mery, then set forth a new practical procedure for evaluating the pressure curve of arches and established that the pressure curve should be contained within the *core* sections of the arch so that all arch sections are entirely compressed. Mery did not however clarify whether the true pressure line of the arch should actually correspond to the proposed curve. Since the Mery procedure used a simple graphical construction, it spread rapidly throughout Europe.

Meanwhile, in the latter half of the 19th century, with the development of steel and, later, reinforced concrete constructions, the new theory of elasticity received great impetus. The study of arches was absorbed into the field of elastic structures. Castigliano (1879), working on some technical problems involving the Turin Mosca bridge, defined masonry arches as *imperfect arches,* given their susceptibility to cracking. Bresse (1859) determined the expression for the thrust of elastic arches with different profiles under various loading conditions. The introduction of the elasticity solved, at least apparently, the difficulties involved in the search for the pressure line. Winkler (1858) studied the flexural energy of an elastic arch

$$V = \frac{1}{2} \int_0^L \frac{M^2(s)}{EI} ds, \qquad (3.19)$$

Fig. 3.17 The moment acting on the arch sections according to Winkler

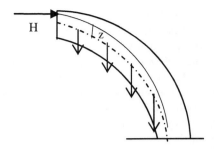

where M indicates the bending moment in the arch and s is the curvilinear abscissa along the arch axis.

Let us consider the funicular of the loads, which represents the pressure line of the arch. Figure 3.17 shows half a masonry arch: the axis of the arch and the pressure line are represented by a dotted and solid line, respectively. Only thrust, H, acts at the key section. If z is the vertical distance of any given point on the pressure line from the arch's curved axis, according to Winkler the bending moment $M(s)$ can be expressed as

$$M(s) = H \cdot z. \tag{3.20}$$

In the late 19th century the principle of minimum strain energy was first set forth by Menabrea (1858) and subsequently proved by Castigliano (1879).

According to this principle, in a statically undetermined structure the actual reactions of the redundant constraints make their strain energy a minimum. By applying this principle to the elastic arch, Winkler was able to establish the position of the pressure line. Of all the possible funicular curves of the loads that can be traced within the arch, Winkler showed that the actual pressure line is that curve that deviates *as little as possible* from the arch axis. By substituting (3.20) into (3.19), the actual pressure line turns out to be the line that *minimizes* the integral

$$V = \frac{1}{2} \int_0^L \frac{H^2 z^2}{EI} ds. \tag{3.21}$$

For an arch under solely vertical loads, the thrust H is constant along the arch's length, and the condition of minimum becomes

$$\int_0^L z^2 ds. \tag{3.22}$$

The search for the actual pressure line of the arch thus hinges on finding the minimum of integral (3.22). Clearly, the minimum is attained when this integral is

3.2 the Birth of the Statics of the Arch and Its Evolution

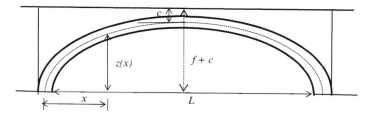

Fig. 3.18 The arch with its superstructure

zero, i.e., when the pressure line coincides with the arch axis. This explains why many late 19th century road and railway bridges were designed with their axes coinciding with the funicular curve of the dead loads.

Winkler's solution, however, seemed inappropriate for masonry arches, which do not exhibit linear elastic behavior, so research aimed at defining the most appropriate profile for bridge axes thus continued to represent an ongoing issue in bridge design. In addition to the weight of the arch, there are also the weights of the filling material, the spandrels, the roadbed and any ballast, which as dead loads should be taken into account in the arch design. This problem was studied by Inglis (1863) and later by Irvine (1981).

According to Inglis' formulation (Fig. 3.18), the axis of the arch, defined as the equation of the funicular curve of the loads, is given by

$$H\frac{d^2z}{dx} = -\gamma[f + c - z(x)], \tag{3.23}$$

where γ is the mean specific weight of the masonry and superstructure, f the rise of the arch, and c the height of the filling at the key of the arch. Integration of (3.23) thus gives

$$z = f + c\left\{1 - \cosh\left[(1 - \frac{2x}{L})\cosh^{-1}(\frac{c+f}{c})\right]\right\}. \tag{3.24}$$

However, this equation was seldom applied in the construction of road and railway bridges during the 19th century. Instead, many arch bridges were built with an *elliptical* profile.

Figure 3.19 shows a photograph of an elliptical profile bridge that spans the River Calore in Benevento, in the south of Italy. A thorough research study on construction of bridges of this type in southern Italy was conducted by Belli (2008).

Note that the problem of finding the most suitable arch profile becomes much simpler if the weight of the deck is greater than that of the arch itself: the weight of the deck can thus be considered uniformly distributed (Fig. 3.20).

The equation of the axis curve of the arch, a funicular of the uniform load q, with H the thrust of the arch, takes the form

Fig. 3.19 An elliptical profile bridge on the River Calore in Benevento

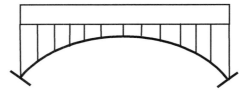

Fig. 3.20 Arch with predominating deck weight

$$H\frac{d^2z}{dx^2} = -q. \quad (3.25)$$

The solution to this equation, satisfying the boundary conditions $z = 0$ at $x = 0$ and at $x = L$, is

$$z = \frac{q}{2H}x(L-x), \quad (3.26)$$

and the corresponding thrust of the arch is thus

$$H = \frac{qL^2}{8f}, \quad (3.27)$$

where f denotes the rise of the arch. This simple solution for calculating the thrust of a parabolic arch was actually already known well before Inglis' time. The Russian engineer Fuss, Euler's son-in-law, had already evaluated it in 1794 when designing a bridge spanning the Neva River in St. Petersburg (Timoshenko 1953).

On the other hand, the ability to define the failure loads for an arch, involving the emergence of mechanisms, was not achieved until well after Winkler and Inglis' times. Indeed, it was only in the late 20th century that it was discovered that Limit Analysis, initially formulated for steel structures, was valid for masonry arches as well (Kooharian 1952). This discovery, which inextricably linked the statics and kinematics of arches, also led to a revival of earlier, 18th century studies and the consequent development of a new line of research on the statics of masonry structures, in which Heyman (1966, 1982a, b) figured prominently (Kurrer 2008). In this context, the old Mery procedure, so widespread among engineers, received a valid mechanical interpretation: the arch pressure line sustaining a given load distribution and plotted using Mery's procedure represents a statically admissible stress state. Thus, according to the static theorem, the loads acting on the arch cannot be greater than the loads producing arch collapse.

The history of the masonry arch shows that the efforts of ancient builders were concentrated above all on building arches able to sustain their own weight, with their axis near to the funicular of the load distribution. To this end, their aim was to erect arches that satisfied the condition

$$\langle \mathbf{g}, \delta \mathbf{v} \rangle < 0, \quad \forall \delta \mathbf{v} \in M, \tag{3.28}$$

which has been covered in depth in Chap. 2. The arch geometry determines whether or not the weight of the arch will consistently counter the development of any mechanism.

Figure 3.21 shows a generic mechanism displacement, v, satisfying condition (3.28), which must hold for any and all mechanisms belonging to the set M of kinematically admissible arch mechanisms.

The action of loads having distributions different from that of the weight, for instance, a point load at any position along its length, modifies the initial profile of the pressure line and can lead to arch failure. Figure 3.22a, b show the failure

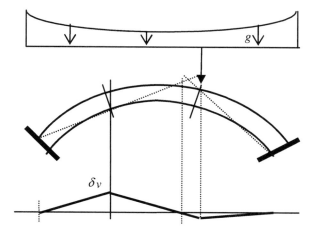

Fig. 3.21 A generic mechanism of the arch opposed by weight **g**

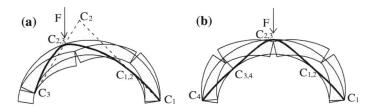

Fig. 3.22 Asymmetrical and symmetrical failure mechanisms of an arch loaded by its own weight and a point load

mechanisms of an arch under its own weight and a point load applied at any given section along its length and its middle section, respectively. In the first case, the collapse mechanism is asymmetrical, in the second case symmetrical.

At the limit equilibrium state, the pressure line of the arch passes through the hinges of the corresponding mechanism. The point load applied on the arch has thus modified the initial pressure line, forcing it to pass precisely through the aforementioned hinges.

Such behavior can explain the difficulties often encountered in building masonry bridges spanning rivers or streams. The centering structures required to build an arch could only be kept in place for short times, when the river was dry. As a rule, only the stone arch was built on the centering: completion of the bridge was instead left until later, by working on the already erected arch without the support of the centering. This was the most hazardous stage of construction, because the progressive addition of the deck could seriously alter the arch pressure line and shift it near a mechanism state. Construction of the bridge deck had to proceed with great care by placing successive loads in a symmetrical arrangement. Many bridges were completed only after much trial and error.

Frequently, for structural reasons, the axis curve of the arch shifted from the funicular curve of its own weight, something which occurred most commonly in circular arches. For these arches under their own weights the pressure line cannot coincide with the circular axis: the pressure line threads through the arch, just barely within its thickness. Thus, arches whose thickness is less than a certain *minimum* cannot completely contain the pressure line and consequently cannot sustain their own weight. It was Couplet (1731), who, following Belidor, first managed to approximate the minimum admissible thickness of a circular arch with given thickness t. For a round arch the ratio t/r_i between t and its internal radius r_i, cannot be less than

$$(t/r_i)_{\min} = 0.108. \qquad (3.29)$$

This value (3.29) was evaluated by Heyman (1966) and later checked by Ochsendorf (2006). When the arch has the bare minimum thickness, the pressure line takes on the course shown in Fig. 3.23: in this state the arch will attain a mechanism condition. For a masonry arch bridge it is clearly important to know the

Fig. 3.23 Minimum thickness of a round arch (from Heyman 1997)

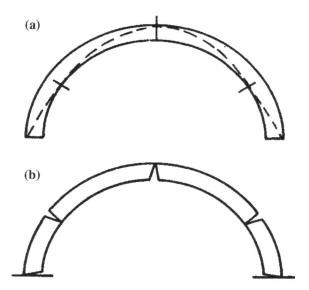

minimum thickness taking into account the weight not only of the arch itself, but its accessory structures as well (i.e., deck, filling, etc.).

Table 3.1 gives the values of the critical ratio t/R between thickness t and radius R for a round arch sustaining its own weight and the weight of the deck. Table 3.1 also includes the values of the angle θ, measured from the springing line, defining the position of the corresponding internal hinges, as determined by Irvine (1981) under conditions of minimum thickness.

As can be seen from the values, the presence of filling only at the haunches, that is when $c = 0$, has a stabilizing effect. Instead, when the filling is also laid midway along the arch, it has the opposite effect: the filling destabilizes the structure.

Table 3.1 Values of the minimum ratio t/R and the angle θ defining the position of internal hinges for various values of the ratio c/R between the height c of the filling at the arch key and radius R for round masonry arches

c/R	θ (°)	t/R	c/R	θ (°)	t/R	c/R	θ (°)	t/R
0.00	21.6	0.047	0.2	32.0	0.094	2.0	36.8	0.133
0.001	21.7	0.047	0.4	33.6	0.104	4.0	35.1	0.138
0.005	22.3	0.049	0.6	34.2	0.117	6,0	35.2	0.140
0.01	22.0	0.051	0.8	34.5	0.122	10.0	35.2	0.141
0.5	26.0	0.063	1.0	34.7	0.125	1000	35.3	0.144
0.1	29.5	0.078	1.5	34.9	0.130	∞	35.3	0.144

3.3 Internal Equilibrium in the Arch

It is useful to examine the internal equilibrium of the arch in terms of the axial force N, the shear T and the bending moment M. In the following analysis it will be assumed that the joints between the voussoirs are orthogonal to the tangent of the arch axis. Along this curved axis, where we define the curvilinear abscissa s, both the tangent $p_t(s)$ and normal loads $p_n(s)$ are applied and act within the plane of the arch (Fig. 3.24). These loads are, as a rule, components of the vertical weight $g(s)$ per unit length of arch. The equations for the translational equilibrium of a unit element ds of the arch are

$$N + dN - N\cos d\phi + T\sin d\phi + p_t ds = 0 \tag{3.30}$$

$$T + dT - T\cos d\phi - N\sin d\phi + p_n ds = 0, \tag{3.31}$$

which give

$$\frac{dN}{ds} + \frac{T}{\rho} + p_t = 0 \tag{3.30'}$$

and

$$\frac{dT}{ds} - \frac{N}{\rho} + p_n = 0. \tag{3.31'}$$

The rotational equilibrium equation around the centroid of the section at the abscissa $s + ds$ is expressed as

$$M + dM - M - Tds = 0. \tag{3.32}$$

which yields

$$T = \frac{dM}{ds}. \tag{3.32'}$$

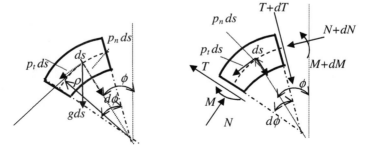

Fig. 3.24 Free-body diagrams for the arch element

3.3 Internal Equilibrium in the Arch

Equations (3.30′–3.32′) define the equilibrium equations of rectilinear beams for $\rho \to \infty$.

3.3.1 Shear Force in Arches

The shear acting in the arch is influenced by its curvature. The bending moment M can be evaluated as the product of the axial force N by its eccentricity e with respect to the section's center. Thus, with position

$$M = Ne, \tag{3.33}$$

from (3.32′) and using (3.33), we get

$$T = \frac{\frac{de}{ds}N - p_t e}{1 + e/\rho}. \tag{3.34}$$

Under symmetrical loading conditions, such as those holding in a symmetrical arch under its own weight, at the middle section defined by $\phi = 0$, we have $T = 0$. In this case, in fact, $p_t = 0$ and $de/ds = 0$.

The springing section is a critical point for the shear in a round arch of radius R under vertical loads, where $p_t = 0$, and hence the shear is

$$T = \frac{\frac{de}{ds}N}{1 + e/R}. \tag{3.34′}$$

The shear is non-zero at the internal hinge sections, where the pressure line runs tangentially to the arch intrados or extrados. In these sections we have $e = t/2$, where t is the thickness and

$$\frac{de}{ds} = 0. \tag{3.35}$$

The shear in these sections is

$$T_{\text{inth}} = -\frac{p_t t}{2(1 + t/2\rho)}. \tag{3.36}$$

When only the weight g acts on the arch, the tangent and the normal components of these loads are

$$p_t = g \sin \varphi \quad p_n = g \cos \varphi, \tag{3.37}$$

in which case the shear at the internal hinge, defined by angle ϕ, is

$$T_{\text{int}}h = -\frac{gt}{2(1+t/2\rho)}\sin\phi. \qquad (3.36')$$

Taking into account that $t/2\rho \ll 1$, we have $T_{\text{int}}h \approx (gt \sin\phi)/2$.

Note that this evaluation of the shear refers to the direction orthogonal to the arch axis. Thus, when the joints of voussoirs are arranged precisely along the direction orthogonal to the arch axis, the evaluated shear must be compared to the shear strength at the voussoirs. The radial disposition of the voussoirs in a semi-circular arch, for instance, conforms to this hypothesis. Safety against slipping requires that for each section of the arch

$$T \leq fN. \qquad (3.38)$$

where f is the friction coefficient between mortar and brick or stone. According to (3.34), we thus have

$$\frac{\frac{de}{ds}N - p_t e}{1 + e/\rho} \leq fN. \qquad (3.38')$$

3.4 Limit Analysis

Figure 3.25 is taken from the paper, *Limit Analysis of Voussoir (Segmental) and Concrete Arches,* published in the ACI Journal (December, Kooharian 1952, Proc. Vol. 49, p. 317) by A. Kooharian, a Ph.D. student at Brown University, where he prepared his thesis under the guidance of D.C. Drucker.

Kooharian discovered that Limit Analysis, initially formulated for ductile steel structures, could also be applied to structures composed of concrete voussoirs. Heyman would later recognize that this result was also valid for masonry structures satisfying the no-tension assumptions previously discussed in Sect. 3.3 of Chap. 2.

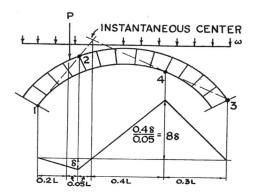

Fig. 3.25 The arch with concrete voussoirs studied by Kooharian (1952)

3.4 Limit Analysis

Fig. 3.26 A symmetrical arch mechanism

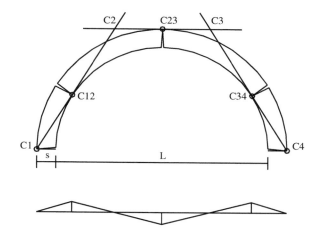

Application of Limit Analysis to masonry arches provides a great deal of useful information. The failure mechanisms of an arch, as seen earlier, may be symmetrical or asymmetrical. Figure 3.26, for instance, shows a symmetrical mechanism. Figure 3.27 refers to the case of the arch loaded by both its own weight g and the point load P having increasing magnitude and variable position.

The plot on the right of Fig. 3.27 shows how the value of the failure load P varies with its position x between the key ($x = 0$) and the springer section.

The plot has been obtained with reference to a depressed masonry arch with a depression angle of 10°; the arch has a span length of 1.50 m, thickness t of 0.35 m and a width B of 0.35 m.

This simple example clearly reveals how the resistance of an arch to point loads is greatly reduced when the load is applied at the haunches.

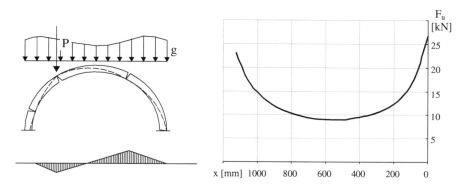

Fig. 3.27 The arch at collapse under its own weight and a generic point load

3.5 Minimum and Maximum Thrust

Let us consider an arch in an admissible equilibrium state. Thus, at least one pressure line will exist that is a funicular of the applied loads contained within the arch's thickness. This holds, for example, for a round arch whose thickness is greater than the minimum (3.29) under its own weight. An unique thrust will correspond to each pressure line in the arch and an unique pressure line will pass through three different generic points, hence, there are ∞^3 possible pressure lines. Of all these curves, one curve will exist that corresponds to the minimum thrust and another one to the maximum. Figure 3.28 shows an example of such pressure lines: one corresponding to the minimum thrust and the other to the maximum. The dotted line in Fig. 3.28 is the curve furnishing the minimum thrust, the continuous line, the maximum. The curve corresponding to the minimum thrust traces the line of the arch with the minimum span and highest rise, while the other curve, corresponding to the maximum thrust, instead has the maximum span and shortest rise.

The *actual* thrust value of the arch falls somewhere between all these possible thrusts. However, determining this value has proved to be quite a difficult task; it is not a simple matter to resolve according to the methods briefly recalled above in the history of the statics of arches. On the other hand, in most cases the *minimum* thrust provides a better approximation the actual thrust of the arch, as will be shown in the following.

3.5.1 Effects of the Elastic Deformation on the Thrust of the Arch

Knowing the evolution of the stresses within an arch during the various stages of construction can be useful for identifying the various factors that influence the thrust in a masonry arch. Arch construction begins on some centering structures. During the initial stages of construction, the arch is unloaded because all its weight is sustained by the centering. However, after stripping, that is, removal of the centering structures, the arch sustains its own weight and pushes against the springers. Let us assume that the axis of the arch is a *funicular* of the loads acting upon dismantling of the centering. Such assumption is generally made in the design of large-span arches. Immediately after stripping, the arch, with fixed springers, is

Fig. 3.28 Maximum and minimum pressure lines in an arch

3.5 Minimum and Maximum Thrust

subjected to a uniform compression and the corresponding thrust can be evaluated quite simply. In fact, if L is the arch span, f its rise, both measured along its axis, and

$$M_c \tag{3.39}$$

denotes the bending moment at the midsection of the equivalent beam with the same span and loads supported at its ends, the thrust of the arch is

$$H = \frac{M_c}{f}. \tag{3.40}$$

More specifically, if the load p is constant and the arch is parabolic, we have

$$H = \frac{pL^2}{8f}. \tag{3.40'}$$

Elastic shortening of all voussoirs will take place due to the uniform compression acting on the arch. An instantaneous *thrust drop* ΔH thus occurs upon stripping, and the pressure line rises at the key section. The thrust drop ΔH can be determined by accounting for the fact that it will produce a detachment between the springers that is equal and opposite to that determined by the elastic shortening (Fig. 3.29).

Let ds be the length of a single voussoir and Δds the shortening of the voussoir due to elastic deformation. The shortening between the springers produced by the elastic deformation of a single symmetrical pair of voussoirs, is

$$2\Delta ds \cdot \cos\alpha = 2\frac{Nds}{EA}\cos\alpha = 2\frac{H\cos\alpha}{EA\cos\alpha}\frac{dx}{\cos\alpha} = 2\frac{H}{EA}ds. \tag{3.41}$$

The total shortening ΔL at the springers consequent to elastic shortening of all the voussoirs is thus

$$\Delta L = 2H \int_0^{L/2} \frac{ds}{EA} = H \int_0^L \frac{ds}{EA}. \tag{3.41'}$$

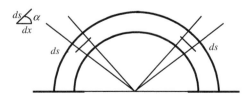

Fig. 3.29 Elastic shortening of symmetrical voussoirs

Fig. 3.30 Extension of the springers due to the thrust drop ΔH

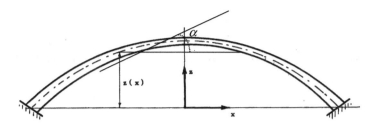

Fig. 3.31 The arch with variable sections

This shortening is however incompatible with the constraints at the springers. With firm springers the thrust ΔH will thus produce an extension equal and opposite to (3.41′). In order to evaluate this extension, we can refer to the scheme of an arch with hinges at its ends, as shown in Fig. 3.30.

Let Δ' be the displacement at the springers produced by two equal and opposite unit forces applied to the arch, as in Fig. 3.30. Thus, we have

$$\Delta H \cdot \Delta' = -\Delta L. \tag{3.41'}$$

From (3.41″) we can obtain the thrust drop for any specific case. For instance, with the two-hinge, parabolic profile arch under constant load p (Fig. 3.31), with a variable section corresponding to the variation in inertial moment given by the first equation in (3.2), the thrust drop ΔH is (Santarella 1974)

$$\Delta H = H \frac{v}{1+v}, \tag{3.42}$$

where H is the thrust given by (3.40′), that is, the thrust corresponding to an axially indeformable arch. The thrust, H', actually acting at the springers, is then

$$H' = \frac{pL^2}{8f} \frac{1}{1+v}, \tag{3.42'}$$

where the quantity v is defined by

$$v = \frac{15}{8} \frac{J_c}{A_c f^2}. \tag{3.42'}$$

Due to creep deformation, the vertical mortar joints can produce further shortening of the arch axis.

The presence of mortar joints thus increases the thrust drop and leads to lower thrust at the arch springers. Such effects were probably taken into account by Castigliano in his (1879) study of the statics of the Mosca bridge in Turin.

3.5.2 Cracking

The presence of cracks near the intrados of the midsection of masonry arches is very common. Attentive inspection may sometimes reveal the presence of similar cracking even at the extrados, near the haunches. The origin of these cracks might be traced back to the thrust drop occurring upon stripping. However, attributing the emergence of cracks wholly to the thrust drop seems dubious, considering the limited magnitude of this drop. By way of illustration, considering a parabolic arch with unit width, a span $L = 40$ m, a rise $f = 4.5$ m, and a key section height $h = 1.20$ m, under uniform loads of 4 t/m^2, and assuming the following parameter values, $J_c/A_c = h^2/12 = 0.12$ m^2, $v = 0.011$, $pL^2/8f = 177.8$ t, the thrust drop ΔH would be only 1.95 t. The effective thrust H' equals 175.8 t/m, and the bending moment at the key section is $M_c = 8.7$ tm/m. Thus, the eccentricity e of the axial load at the key section is only about 5 cm, and the key section remains completely compressed. Even if we add the effects of creeping of the mortar joints, which is about the same order of magnitude as the elastic deformation, the eccentricity e of the axial load would be about 15 cm, and cracking would still not take place. Analogous results can be obtained for other types of arches. Thus, if the arch profile is designed as a funicular of the dead loads present at stripping, the pressure line of the arch will not deviate much from its axis. Even if the dead loads change during completion of the arch, any further deviations of the pressure line will still be quite small. All things considered, if the arch profile has been designed to match the arch axis, upon its completion, the pressure line will substantially match this axis. Under such conditions, the occurrence of cracking can be excluded. The frequent presence of cracking in masonry arches, particularly at the intrados of the key section, can in many cases be explained by settling at the springers. In other cases, specifically when the masonry arch is embedded between piers or side walls, slight lateral rotation of the supporting structures can lead to an increase in the arch span and consequent cracking. Thus, all told, the occurrence of cracking at the intrados of the key section of an historic arch can generally be attributed to settling of the foundations.

3.5.3 Minimum Thrust State

An arch that has settled through a mechanism deformation adapts itself to the slight increase in span. Hinges form at the extrados near the key and at the intrados of the haunches. The initial pressure line of the arch changes its configuration and will now pass through the hinges of the settlement mechanism (Fig. 3.32).

This pressure line will be neither that calculated by Mery, wholly contained within the thickness of the arch sections, nor that determined by Winkler's minimum deviation principle. Instead, it follows the hinge distribution of the settlement mechanism and matches the line obtained by applying the Moseley principle (1839) of minimum pressure. The minimum thrust state attained in the arch can be determined by applying the static or kinematic approach discussed in Chap. 2. The value of the minimum thrust is thus obtained graphically by tracing the funicular loads curve that is completely contained within the arch thickness and has the greatest vertical inclination at the springers. The pressure line will pass through the extrados of the key section and will skim the intrados at the haunches. The kinematic procedure is used to determine the location of the hinges in the minimum thrust mechanism by looking for the settlement mechanism for which the kinematic multiplier is at a maximum. For the parabolic arch shown in Fig. 3.32, the settlement mechanism has lower hinges located at the intrados of the haunches sections. The minimum thrust is $H_{min} = pL^2/8f$, where L is the span along the intrados profile, and the rise, f, measures the distance between the key hinge and the horizontal axis passing through these lower hinges. For a semicircular arch, instead, the pressure line corresponding to the minimum thrust exhibits various geometries depending on the geometry of the arch, a result which is the topic of the next sections.

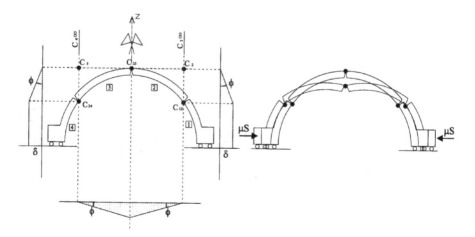

Fig. 3.32 The settlement mechanism producing arch span widening

3.5.4 Minimum Thrust in the Round Arch

Figure 3.33 shows the pressure line for a round arch in the minimum thrust state under its own weight: the dotted line indicates the intrados circle.

The pressure line is traced in the right half of the arch. The angle $\bar{\alpha}$ marks the position of the inner hinge of the settlement mechanism. The mean radius of the arch and radius of the intrados circle are respectively denoted by r and r_i, and t is the thickness.

A generic mechanism corresponding to horizontal settling δ at the arch springers is shown in Fig. 3.32. In this mechanism the position of the inner hinge is unknown. According to the kinematic approach, the minimum thrust can be determined through condition (2.170′) in Chap. 2 (Como 1996), as

$$\mu = \underset{\bar{M}}{Max}\left(-\frac{\langle \mathbf{g}, \mathbf{v} \rangle}{\langle S, \delta \rangle}\right) \quad (3.43)$$

where:

- **v** is a generic settlement mechanism corresponding to an assigned position of the inner hinge;
- $\bar{M}$ is the set of all settlement mechanisms considering all possible inner hinge positions;
- $\langle \mathbf{g}, \mathbf{v} \rangle$ is the work of the weight **g** on the vertical displacements induced by mechanism **v**;
- $\langle S, \delta \rangle$ is the work of the thrust S on the horizontal displacement δ of the springers induced by mechanism **v**.

The problem of finding the maximum, (3.43), has only one unknown: the position $\bar{\alpha}$ of the hinge at the arch intrados (Fig. 3.33). This position has been determined (Fabiani 2007) via a suitable parametric search by varying the ratio t/r_i between the arch thickness t and the radius r_i of the intrados circle. The results reveal that angle $\bar{\alpha}$, defining the position of the inner hinge, decreases gradually as the thickness of the arch increases. The position of the inner hinge reaches its lowest point in the arch, defined by the angle $\bar{\alpha}'$, for the limit value of the ratio t/r,

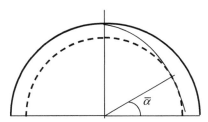

Fig. 3.33 The pressure line for a round arch at the minimum thrust state with the location of the inner hinge

denoted by $(t/r_i)^*$. Thus, with any further increases in t/r, angle $\bar{\alpha}$ remains fixed at the value $\bar{\alpha}'$. The limit value of t/r turns out to be

$$(t/r_i)^* = 0.534. \qquad (3.44)$$

The minimum thrust can thus be expressed as (Fabiani 2007)

$$(\mu S)_{\min} = \gamma b r_i^2 \bar{f}(k). \qquad (3.45)$$

where γ is the specific weight of the masonry, b the transverse width of the arch and $\bar{f}(k)$ a suitable function of the ratio $k = t/r_i$.

Consequently, the search for the minimum thrust is conducted for $0.108 \leq t/r_i \leq 0.534$, where the value of $t/r_i = 0.108$ corresponds to the minimum admissible arch thickness. Table 3.2 shows the values of $\bar{\alpha}$ and $\bar{f}(k)$ with varying values of ratio k.

By way of example, for a round arch made of stone voussoirs, with $\gamma = 2.6$ t/mc, a mean radius $r = 6.5$ m, thickness $t = 1.00$ m and consequently an intrados circle of radius $r_i = 6.0$ m and transverse width $b = 1.00$ m, we obtain $t/r_i = 0.167$. Interpolating the values in Table 3.2 yields $\bar{\alpha} = 0.559$ rad and $\bar{f}(k) = 0.099$. The minimum thrust is thus

$$(\mu S)_{\min} = \gamma b r_i^2 \bar{f}(k) = 2.6 \cdot 1.0 \cdot 6.0^2 \cdot 0.099 = 9.26 \, t.$$

3.5.5 Minimum Thrust in the Depressed Arch

Determining the settlement mechanism for a depressed round arch is rather more complicated. Figure 3.34 illustrates the arrangement of such an arch with springer angle θ. In this case, we must consider both angles θ and $\bar{\alpha}$, which defines the position of the inner hinge for the round arch.

Table 3.2 Position $\bar{\alpha}$ of the inner hinge and the function $\bar{f}(k)$ of $k = t/r_i$

$k = t/r_i$	$\bar{\alpha}$(rad.)	$\bar{f}(k)$
0.108	0.619	0.075
0.15	0.574	0.092
0.20	0.530	0.112
0.25	0.499	0.129
0.30	0.478	0.143
0.35	0.462	0.155
0.40	0.452	0.165
0.45	0.466	0.172
0.534	0.442	0.181

3.5 Minimum and Maximum Thrust

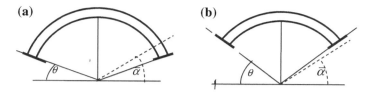

Fig. 3.34 Depressed round arches with: **a** $\theta < \bar{\alpha}$ or **b** $\theta > \bar{\alpha}$

Fig. 3.35 Equilibrium of the depressed arch with $\theta > \bar{\alpha}$

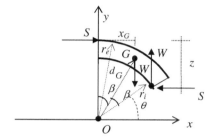

Two different situations can occur, depending on whether the springer angle θ is smaller or larger than $\bar{\alpha}$ (Fig. 3.34a, b). Let us indicate with S' the thrust of the corresponding round arch. When the springer angle θ of a depressed arch is the less than $\bar{\alpha}$, as in Fig. 3.35a, the pressure line touches the arch intrados at the same position as the round arch: the thrust is S'. If, on the contrary, $\theta > \bar{\alpha}$, the pressure line passes through the *intrados* of the arch springer.

The minimum thrust, denoted by S, is less than S'. The minimum thrust S for a depressed arch with $\theta > \bar{\alpha}$ is evaluated by considering the equilibrium of one half the arch.

Referring now to Fig. 3.35, if G indicates the position of the center of the half-arch, d_G the radial distance of G from O, and x_G the distance of this center G from the vertical axis y, we have

$$S \cdot z - W(r_i \cos \theta - x_G) = 0, \tag{3.46}$$

where

$$z = (r_e - r_i \cos 2\beta), \tag{3.47}$$

in which z denotes the distance of the springer section intrados from the top of the midsection, and r_i and r_e the radii of the arch intrados and extrados circles, respectively. The weight W of the half-arch is

$$W = \gamma b A = \gamma b \iint_A dA = \gamma b \int_0^{2\beta} d\theta \int_{r_i}^{r_e} \rho d\rho = \gamma b \beta (r_e^2 - r_i^2) \qquad (3.48)$$

where A is the area of the half-arch, γ the specific weight of the masonry, b the transverse width of the arch and ρ the radial distance of any given point within the arch from the origin O.

The distance x_G of the center G of the half-arch from axis y is linked to the radial distance d_G of G from point O through the relation

$$x_G = d_G \sin \beta, \qquad (3.49)$$

where 2β is the amplitude of the arch. Given the position

$$d_G A = M_O, \qquad (3.49')$$

we thus have

$$A = \beta(r_e^2 - r_i^2); M_O = \iint_A \rho dA = \int_0^{2\beta} d\theta \int_{r_i}^{r_e} \rho^2 d\rho = \frac{2}{3} \beta (r_e^3 - r_i^3) \qquad (3.50)$$

and the radial distance d_G is

$$d_G = \frac{M_O}{A} = \frac{2(r_e^3 - r_i^3)}{3(r_e^2 - r_i^2)}. \qquad (3.51)$$

Finally, from (3.46), the minimum thrust S of a depressed arch (Fabiani 2007) is thus given by

$$S = \gamma b \beta \frac{(r_e^2 - r_i^2)}{(r_e - r_i \cos 2\beta)} [r_i \cos \theta - \frac{2(r_e^3 - r_i^3)}{3(r_e^2 - r_i^2)} \sin \beta]. \qquad (3.52)$$

An analogous procedure can be followed for pointed arches.

3.6 Coupled Systems of Arches of Different Spans

An interesting case for analysis is the behavior of a number of arches with different spans connected in series, for instance, the system of coupled arches illustrated in Fig. 3.36. The arches are loaded by their own weight and the weight of the filling. The two external arches are the same, but their spans are greater than the central arch. These two can be considered to be in a minimum thrust state because their lateral abutments are not constrained.

3.6 Coupled Systems of Arches of Different Spans

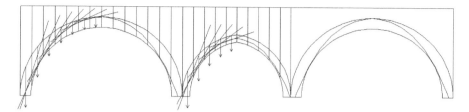

Fig. 3.36 System of arches of different spans

The inner piers, not constrained by buttresses, can only sustain vertical forces, so the thrusts transmitted to their heads by the lateral and central arches balance out. The thrust of the external arches will thus be equal and opposite to that transmitted by the shorter-span central arch. The middle arch will thus be in an intermediate state, between minimum and maximum thrust.

Starting with such considerations, the analysis of coupled arch systems involves the following steps:

- evaluation of the vertical loads on the various voussoirs, taking into account the weight of both the stone and the section's share of filling;
- tracing the funicular curve of the loads on the external arches in the minimum thrust state and estimating the corresponding thrust;
- tracing the pressure line of the central arch, so that it transmits, at its springers, a thrust equal and opposite to that of the external arches.

Figure 3.36 shows the plots of the various pressure lines for the three arches. In many other arch systems, assuming the existence of minimum thrust states can help obtain a description of their stress states.

3.7 Masonry Arches Loaded by Horizontal Forces

A particularly interesting, and important, case is that of an arch subjected to horizontal forces (Fig. 3.37). The assumed loading condition is that the weight of the arch remains constant while the horizontal forces increase according to the loading parameter λ (Abruzzese et al. 1985).

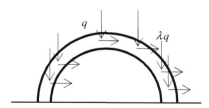

Fig. 3.37 The arch under constant weight and increasing horizontal forces

Fig. 3.38 Vertical and horizontal displacements of the failure mechanism for an arch under horizontal forces

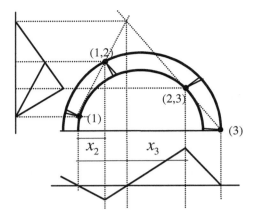

The failure mechanism involves the formation of four hinges and the development of vertical and horizontal displacements, as shown in Fig. 3.38. The failure multiplier can be determined through the following steps (Como and Lanni 1988).

(a) finding the location of the four hinges compatible with an admissible failure mechanism u', as shown in Fig. 3.38. The horizontal distances of the hinges (1), (1, 2), (2, 3) and (3) from the intrados of the left springer are respectively denoted by x_1, x_2, x_3 and x_4;
(b) evaluating the kinematic multiplier $\lambda(u')$;
(c) tracing the funicular polygon corresponding to both the assumed hinge positions and the resulting horizontal loads, multiplied by the factor $\lambda(u')$;
(d) checking whether the funicular polygon is wholly contained within the arch. If it is, the kinematical multiplier can also be assumed to be statically admissible, as can the sought-for failure multiplier; if not, the procedure must be repeated.

The following details a practical method for simple evaluation of the failure multiplier. Firstly, let us define:

$y_i(x)$ intrados profile of the arch
$y_e(x)$ extrados profile of the arch;
$M_v(x)$ moment at section x of the entire vertical load distribution $p(x)$ situated to the left of section x;
M_{oi} moment of all the horizontal loads $p(x)$ situated to the left of section x with respect to the *intrados* of section x;
M_{oe} moment of all the horizontal loads $p(x)$ situated to the left of section x with respect to the *extrados* of section x;
R_V and R_H the vertical and horizontal reactions of hinge (1).

3.7 Masonry Arches Loaded by Horizontal Forces

Numerical trials have enabled establishing that at arch failure:

- hinge (1) is near the intrados of the left springer section
- hinge (3) is located at the extrados of the right springer section
- the distance between hinges (1, 2) and (2, 3) is approximately equal to L/2, where L is the internal span of the arch.

These results allowing reducing the four unknowns, x_1, x_2, x_3 and x_4, to only two: x_1 and x_2. Therefore, for any choice of x_1 and x_2, the corresponding mechanism is defined. The corresponding kinematic multiplier $\lambda(u')$ will depend on the chosen values of x_1 and x_2. The failure mechanism will be obtained by finding the minimum value of the kinematic multiplier from amongst all possibilities by varying the positions of hinges (1) and (1, 2), that is, by varying the abscissas x_1 and x_2. Function $\lambda(x_1, x_2)$ can be easily defined by rewriting the following equilibrium equations in the unknowns λ, R_V and R_H (Fig. 3.39). We can now analyze the equilibrium of the arch along the assumed mechanism by imposing the vanishing of the moment of all forces around the corresponding hinges. Thus, with some small approximations and assuming positive moment in the counter clockwise direction, we have

(a) equilibrium around hinge (1, 2)

$$-R_V x_2 + R_H[y_e(x_2) - y_i(x_1)] + M_V(x_2) + \lambda M_{oe}(x_2) = 0; \qquad (3.54)$$

(b) equilibrium around hinge (2, 3), where $x_3 = x_2 + L/2$

$$-R_V(x_2 + L/2) + R_H[y_i(x_2 + L/2) - y_i(x_1)] + M_V(x_2 + L/2) + \lambda M_{oi}(x_2 + L/2) = 0; \qquad (3.55)$$

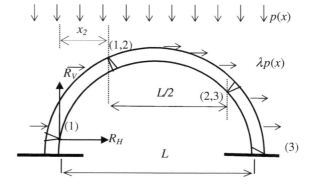

Fig. 3.39 Valuation of the function $\lambda_c(x_1, x_2)$ by imposing zero moment at hinges (1, 2), (2, 3) and (3)

(c) equilibrium around hinge (3)

$$-R_V L + R_H [y_e(L) - y_i(x_1)] + M_V(L) + \lambda M_{oe}(L) = 0. \quad (3.56)$$

The solution to Eqs. (3.54–3.56) furnishes the value of the kinematic multiplier, $\lambda(x_1, x_2)$, together with the values of the reactions R_V and R_H for the assigned values of the abscissas, x_1 and x_2. Finding the minimum of function $\lambda(x_1, x_2)$ is thus easily pursued by suitably varying x_1 and x_2.

A numerical example

Consider the round arch illustrated in Fig. 3.40 and defined by:

- span length, L = 15 m;
- thickness, s = 1.20 m;
- transversal width, B = 4 m;
- tuff block and mortar masonry: γ = 1600 kg/mc.

Accordingly, simple calculations furnish the radii of the intrados r_i, extrados r_e, and mean r_g circles (Fig. 3.41): r_i = 7.50 m; r_e = 8.70 m; r_g = 8.10 m. The corresponding equations for the intrados and extrados profiles are (Fig. 3.41):

$$y_i(x) = \sqrt{r_i^2 - (x - r_i)^2} \quad y_e(x) = \sqrt{r_e^2 - (x - r_i)^2}.$$

The positions of the hinges are defined as in Fig. 3.42. The arch has been subdivided into 11 equal voussoirs of amplitude $\Delta\alpha$ = 15°;

Given the assumed hinge distribution, a first segment of the arch remains fixed (Fig. 3.41). The area of the outer face of the single voussoir measures 07854 m². The weight Pi of the single voussoir, accounting for a transverse width of 4.00 m, is, Pi = 0.7854 × 4.00 × 1600 = 5027 kg.

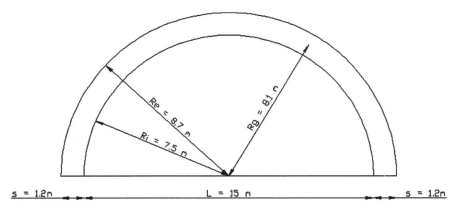

Fig. 3.40 Intrados, extrados and mean radii of the example round arch

3.7 Masonry Arches Loaded by Horizontal Forces

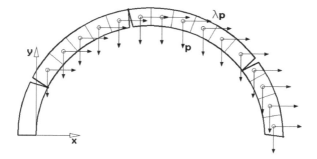

Fig. 3.41 The arch subdivided into 11 equal voussoirs with the vertical and horizontal load distributions and hinge positions at the first step

Fig. 3.42 Position of the hinges for the collapse mechanism

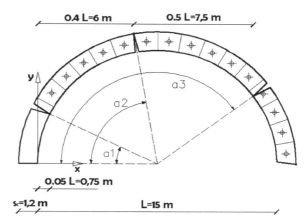

The reference system has its origin O at the intrados of the springer section. Accordingly, the coordinates of the voussoirs' centers are:

x_{gi}	y_{gi}
0.017	3.100
1.074	4.931
2.569	6.426
4.400	7.483
6.443	8.031
8.557	8.031
10.600	7.483
12.431	6.426
13.926	4.931
14.983	3.100
15.530	1.057

Numerical calculation of the kinematic multipliers λ is performed by varying the hinge coordinates, x_1 and x_2. From these calculations the smallest value of λ, which represents the failure multiplier of the horizontal load distributions is

$$\lambda_c = 0.1417 \quad R_H = 25810 \text{ kg} \quad R_V = 65750 \text{ kg}.$$

Figure 3.42 shows the corresponding hinge positions of the collapse mechanism, with the following hinges coordinates:

(1) $x_1 = 0.75$ m, $y_1 = 3.07$ m;
(2) $x_2 = 5.81$ m, $y_1 = 8.53$ m;
(3) $x_3 = 13.31$ m, $y_3 = 4.74$ m;
(4) $x_4 = 16.20$ m, $y_4 = 0.00$ m.

3.8 Some Experimental Results and Comments on Failure Tests of Masonry Arches

3.8.1 Test Description

We shall now look at some of the experimental results obtained at the Laboratory of the Department of Civil Engineering of the University of Tor Vergata in Rome.

These tests were conducted by U. Ianniruberto, Z. Rinaldi and, more recently, A. Carratelli, within the framework of a research program on the behavior of fiber-reinforced masonry arches. Although the trials involved unreinforced masonry arches only marginally, the results are nonetheless of interest to the current topic. The tests refer to two round masonry arches made of bricks 5 cm × 11 cm × 24 cm in size and cement mortar (Fig. 3.43). The characteristic compression strength of the bricks and mortar, tested 28 days after casting, were 250 kg/cm^2 and equal to or greater than 35 kg/cm^2, respectively. The round arches have an internal radius of 130 cm and a thickness of 24 cm. The bricks were laid radially with the support of a suitable metallic scaffolding and the thickness of the joints varied from 0.8 cm to 1.9 cm. Table 3.3 shows the compression strengths of the mortar, bricks and overall masonry, while Table 3.4 reports the tensile and compression strengths of the mortar.

The testing equipment, represented mainly by a hydraulic jack, applied a point load to the key section of the arch extrados. The head of the jack was forced vertically downwards at a constant speed of 0.3 mm/s, and the value of the acting load for each value of the displacement was measured by a pressure cell. The arch was loaded by its own weight and the point load applied by the test equipment. The recorded value of the ultimate failure point load was about 390 kg, much larger than the 100 kg estimated via Limit Analysis.

3.8 Some Experimental Results and Comments on Failure Tests of Masonry Arches

Fig. 3.43 The masonry arch during a test

Table 3.3 Strengths of the various materials

Materials	Mortar (sika)	Bricks (rdb)	Masonry
R_{bk} (kg/cm^2)	35	250	65

Table 3.4 Tensile and compression strengths of mortar specimens

Specimens	Tension (kg/cm^3)	Compression (kg/cm^3)
1	3.4	38
2	3.6	38
3	3.5	41
Mean versus (kg/cm^3)	3.5	39

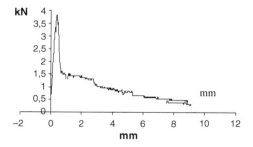

Fig. 3.44 Point load versus vertical displacement at the key section of the arch during testing

Fig. 3.45 Asymmetrical arch collapse mechanism with four hinges

Upon reaching the maximum load of 390 kg, the arch begun to crack and deform considerably, after which any further displacement caused an initial rapid drop in the resistant force of the arch, followed by a stage of very slow decline (Fig. 3.44).

Under the maximum load, the vertical displacement of the arch key section was measured to be about 0.4 mm. Collapse turned out to be largely asymmetrical, through the development of four hinges. Figure 3.45 shows the arch at collapse. The instant the resistance dropped suddenly, cracks opened on the extrados of the haunches, as well as on the intrados at both the key and right springer, as can been seen in the figure. As the deformation of the arch increased, the cracks gradually enlarged to the point of forming hinges. As is evident from Fig. 3.46, the cracks opened at the interface between the mortar joint and the bricks, indicating that the strength of the mortar-to-brick adhesion was lower than the tensile strength of the mortar itself. Such tests thus demonstrate that:

- arch strength is significantly greater than that predicted by Limit Analysis;
- arch strength falls sharply upon the formation of cracks;
- arch strength approaches the Limit Analysis value after cracking.

3.8.2 Comments on Test Results

The considerable overstrength of the arch revealed by the tests is due to the high adhesion strength at the interface between the mortar and bricks. The arch initially presents brittle behavior and then, as it transforms into a mechanism, fully develops strongly ductile behavior upon hinge formation.

3.8 Some Experimental Results and Comments on Failure Tests of Masonry Arches

Crack at the right haunch Crack at the left haunch

Hinge at the key Hinge at the right springer

Fig. 3.46 .

Regarding the test results, it is foremost worth underscoring that the behavior of actual historic arches is quite different from the behavior of arches built in laboratories using cement mortars. The mortar-to-stone or mortar-to-brick adhesion strength of historic masonry is considerably lower than that of laboratory-built masonry.

An even more relevant point regards a further difference between actual historic arches and the arches tested in the trials described. This refers to the fact that historic arches were often used to support masonry bridges or walls built over them. For such structures, the dead loads acting on the arch are far greater than the loads due only to the arch self-weight. The compression forces acting between the stones, which hold the stones or bricks together, are thus more efficient than the weak tensile strengths at the joints. Such effects can explain the generally large discrepancies found between actual strength and the results of Limit Analysis.

One condition that can be used checks the import of the compressive forces maintaining the voussoirs firmly fixed one to the other, relative to the adhesion between mortar and stone, is the following

$$\chi = \frac{qL}{\sigma_o A}\frac{L}{8f} \gg 1. \tag{3.57}$$

Only if such condition is verified will the weight have the predominating effect on arch strength. The ratio between the mean compression acting at the arch key section, which represents on average the forces holding the voussoirs together, and the adhesion strength between mortar and brick, may represent the term of comparison between the actual arch and the model.

In brief, arch models formulated from laboratory tests can describe the behavior of actual arches only for large values of parameter χ. For instance, with reference to an actual masonry arch, defined by the following parameters:

arch + filling weights: q = 5 t/m;
mortar-brick adhesion strength: σ_o = 0.5 kg/cm^2;
arch span = 10 m;
arch rise = 4 m;
arch cross-sectional area: A = 0.8 × 0.4 = 0.32 m^2;

The resulting value of the χ ratio is $\chi \approx 10$.

If instead we make reference to a laboratory model of arch, with:

arch weight: $q \approx 50$ kg/m^2;
mortar brick adhesion strength: σ_o = 3.5 kg/cm^2;
arch span = 2.0 m;
arch rise = 0.5 m
arch cross-sectional area: A = 10 × 5 = 50 cm^2

we obtain

$$\chi = \frac{qL}{\sigma_o A}\frac{L}{8f} = \frac{50 \cdot 2.0}{3.5 \cdot 50}\frac{2.0}{8 \cdot 0.5} = 0.286.$$

and the similitude condition (3.57) is not verified.

References

Abruzzese, D., Como, M., & Lanni, G. (1985). Some results on the strength evaluation of vaulted masonry structures. In C. A. Brebbia & B. Leftheris, (Eds.), *Architectural studies, materials and analysis*. Southampton, Boston: Computational Mechanics Publ.
Belidor, B. F. (1729). *La science des ingénieurs dans la conduit des travaux de fortification et d'architecture civile*, Paris.
Belli, P. (2008). Ponti in muratura di fine 800 nell'Italia Meridionale, Atti del 2° *Convegno Naz.le di Storia dell'Ingegneria,* Naples, 7–9 Aprile 2008.
Benvenuto, E. (1991). *An introduction to the history of structural mechanics, Part II, Vaulted structures and elastic Systems*. NY: Springer.
Bresse, J. A. C. (1859). *Cours de mécanique appliquée*, Paris.

Castigliano, C. A. P. (1879). *Theorie de l'equilibre des systemes élastiques et ses applications.* Turin: Augusto Federico Negro.
Como, M. (1996). Multiparameter loading nd settlements in statics of masonry structures. In Pitagora (Ed.), *Proceedings of the Conferences Meccanica delle Murature tra teoria e progetto*, September 18–20 1996, Bologna, Messina.
Como, M. (1998). Minimum and maximum thrust states in Statics of ancient masonry buildings. In A. A. Balkema (Ed.), *Proceedings 2nd International Arch Bridge Conference*, Venice, October 6–9, Rotterdam, Italy 1998.
Como, M., & Lanni, G. (1988). Sulla verifica alle azioni sismiche di complessi monumentali in muratura, *Cinquantenario della Facoltà di Architettura di Napoli.* Naples: Istituto di Costruzioni, Giannini Ed,.
Coulomb, C. A. (1773). Sur une Application des Régles de maximis et minimis à quelques problems de statique relatives à l'architecture, *Mém. Acad. Sci. Savants Etrangers*, (Vol. VII). Paris.
Couplet C. A. (1731). De la poussée des voutes, *Memoires de l'Academie Royale des Sciences*, Paris 1731.
D'Agostino, S., & Bellomo, M. (2001). *Ponti in muratura e in calcestruzzo armato, Manuale di ingegneria Civile* (Vol. II). Buologne: Zanichelli/ESAC.
De la Hire, P. (1712). *Sur la Construction des voutes dans les édifices.* Paris: Mémoires de l'Académie Royale des Sciences.
Di Carlo, F., Ciampa, D., & Vaccaro, G. (2007–2008). *Analisi di un sistema di archi accoppiati e diversa luce*, a.y., *Exercises in the Course Statics of masonry historic constructions.* M. Como, Faculty of Engineering, Un. of Roma Tor Vergata, Rome (Italy).
Fabiani, F. M. (2007). *Analisi statica delle volte a crociera in muratura, Tesi di dottorato.* Rome: Dep. of Civil Eng., University of Rome Tor Vergata.
Foce, F. (2002). Sulla teoria dell'arco murario: una rilettura storico-critica, In di A. Becchi e F. Foce (Ed.), *Degli Archi e delle Volte.* Venice: Saggi Marsilio.
Heyman, J. (1966). The stone skeleton. *International Journal of Solids Structures, 2*, 249.
Heyman, J. (1972). *Coulomb's memoir on statics.* Cambridge: Cambridge Univ. Press.
Heyman, J. (1982a). *The masonry arch.* Chichester: Ellis Horwood.
Heyman, J. (1982b). *The masonry arch.* Cambridge: Cambridge Press.
Heyman, J. (1997) *The stone skeleton.* Cambridge: Cambridge Univ. Press.
Hooke, R. (1675). *A description of helioscopes and some other instruments*, London.
Inglis Sir, C. (1863). *Applied mechanics for engineers.* Dover, New York.
Irvine, M. (1981). *Cable structures.* NY: MIT Press and Dover Publ. (1992).
Kooharian, A. (1952). Limit analysis of voussoir (segmental) and concrete arches. *ACI Journal, Proceedings V, 49*, 317.
Kurrer, K. E. (2008). *The history of the theory of structures.* Ernst & Sohn: Verlag, Berlin.
Luffarelli, D. (2005–2006). *Modellazioni e Sperimentazione di Archi in Muratura rinforzati con F. R.P.*, supervisors U. Ianniruberto e Z. Rinaldi, a.y., Theses. (University of Rome, Tor Vergata, Rome, Italy).
Marcolongo, R. (1937). *Studi vinciani*, (Vol. 7). Naples.
Menabrea, L. F. (1858) Nouveau principe sur la distribution des tensions dans les systems elastiques. *Comptes rendus, 46.*
Mery, E. (1840). Sur l'equilibre des voutes en berceau, *Annales des Ponts et Chaussées.*
Moseley, H. (1839). On the theory of the Arch. In J. Hann, (Ed.), *The theory, practice and architecture of bridges.* London.
Moseley, H. (1843). *The mechanical principles of engineering and architecture.* London.
Nerilli, F., & Cortesini, F. (2008–2009). *Valutazione dell'intensità delle forze orizzontali di collasso per un arco ad asse circolare*, a.y., *Exercises in the Course Statics of masonry historic constructions.* M. Como, Faculty of Engineering, Un. of Roma Tor Vergata, Rome (Italy).
Ochsendorf, J. (2006). The masonry arch on spreading supports. *The Structural Engineer, 84*(2).

Poleni, G. (1988) *Memorie istoriche della gran cupola del tempio vaticano e de' danni di essa, e de' ristoramenti loro*, Rome, Padova 1748: Ristampa della Biblioteca Facoltà di Arch. Univ. di Roma La Sapienza.

Poncelet, J. V. (1852). Examen critique et historique des principales théories ou solutions concernant l'équilibre des voutes. *Comptes Rendus, 35*.

Santarella, L. (1974). *Il Cemento Armato* (Vol. 1). Milan: U. Hoepli.

Timoshenko, S. P. (1953). *History of strength of materials*. New York: Mc Graw-Hill Book Comp. Inc.

Winkler, E. (1858). Formanderung und Festigkeit gekrummter Korper, insbesondere der Ringe, *Civilingieur, 4*.

Chapter 4
Masonry Vaults: General Introduction

Abstract Statics of masonry vaults is the subject of this chapter, divided into four sections. The construction of these vaults has a long history because they were the first curved structures built to cover an underlying space. Masonry vaults have signed the history of architecture since the origins up to the 19th century. Many of them have been destroyed during time but many survived and frequently represent outstanding examples of our historic heritage. The study of their statics is therefore mainly addressed to their conservation and maintenance. Masonry vaults present different static behavior according to their geometry, masonry strength, constructional techniques and so forth. Stability of a broad class of masonry vaults depends upon the stone elements being held together by gravity and friction. Most vaults fall within this category and their static behavior is strongly conditioned by the low tensile masonry strength. After some historical notes, of introductory character, the implemented approach to analyze statics of these masonry vaults is described in the first section of the chapter. Two different stress states can be detected: the initial state, without cracking, and the final cracked stress state, governed by the vanishing tensile strength. In the early stages of their lifetime vaults, in fact, behave as solid structures, able to sustain tensile stresses: for them the membrane solution provides a reasonably static accurate description. The final state, occurring after cracking, will be conversely modeled within the framework of the no tension masonry materials.

4.1 Brief Historical Notes

Historically, a vault, rather than an arch, was probably the first curved structure built to cover an underlying space. The first traces of these curved structures are ruins of small barrel vaults with inclined courses discovered in the Mesopotamian region and lower Egypt, which date back to between the 4th and 5th millennium BCE. (Choisy 1883). A thick backing wall was built first, and then inclined brick courses were erected leaning against it, thereby avoiding the need for centering structures. This construction technique allowed for covering long expanses, though the spans possible were rather limited (Fig. 4.1a).

Domed structures also have ancient origins, as far back as the 4th millennium BCE The ruins of a number of primitive dwellings were topped by clay domes resting on circular walls made of rubble and clay, such as the houses of Khirokitia in Cyprus. The first domes built with proper stone masonry date back to the late Bronze Age, about 1500 BCE. These were the so–called *tholoi*, tombs built by corbelling circular courses of stones, a technique developed by the Mycenaean civilization. The Treasure of Atreus and the Tomb of Clytemnestra in Mycenae are the best conserved examples of *tholoi* (Fig. 4.1b).

Domes covering large areas were built by the Ancient Romans using the technique called *opus caementicium*. One of the earliest existing examples is the vault of the temple of Jupiter Anxur in Terracina, built in the early 1st century BCE. The technique of *opus caementicium* attained it most widespread use during the Age of the Roman Empire, the Pantheon being its uncontested highest expression.

Roman builders often inserted brick 'ribs' in the concrete mass. The concrete was then fractioned and cast between the ribs. The static requirement of building vaults as light as possible led to the use of *opus caementicium,* with the *caementa* (or cement) growing gradually lighter from the springings towards the center, for example by laying in sequence first travertine blocks, then bricks, tuff, pumice and lastly slag. Later, vaults were also built by inserting *"olle"*, that is amphorae, at the haunches or above the openings of the dome, where stresses are particularly low, one example being the temple of Minerva Medica in Rome from the 4th century (Lancaster 2005).

Later, more sophisticated techniques were used to lighten the masonry mass, such as the insertion of *"tubi fittili"*—hollow terracotta tubes placed vertically, one beside the other, in concentric circles into the concrete, as in the dome of San Vitale in Ravenna from the 5th century.

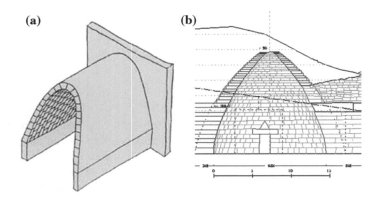

Fig. 4.1 a Mesopotamian vault **b** Mycenaean Tholos

4.1 Brief Historical Notes

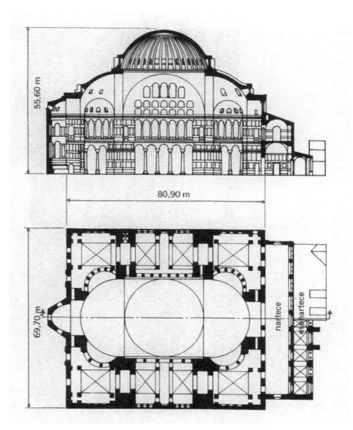

Fig. 4.2 Plan and section of the *Hagia Sophia* Cathedral

In the Byzantine era hemispherical vaults evolved further: they were sustained by piers through the insertion of pendentives in place of the drum. This enabled covering much wider areas. Moreover, the vaults were constructed of brickwork rather than concrete. An extraordinary example of these techniques is the dome of the Constantinople Cathedral, *Hagia Sophia*, from the 6th century (Fig. 4.2).

The large central dome of the cathedral is sustained by two smaller semi–domes, in turn supported by arches, piers and walls. The central dome has an inner diameter of 32 m and is connected to two semi–cupolas on its eastern and western sides and two large arches on the south and north. The striking view from the interior of the church is shown in Fig. 4.3.

The earliest church on the site of the current Cathedral was built by order of Emperor Constantine and consecrated in the year 360 CE. The original church was however destroyed by fire and rebuilt in 415, only to be destroyed once again in 532. Finally, the emperor Justinian commissioned Anthemius of Tralles and Isidore

Fig. 4.3 Interior of the *Hagia Sophia* Cathedral

of Miletus to rebuild the church for the third time, resulting in the church we can still admire today, though over the centuries it has suffered many accidents and consequently undergone many changes.

Despite its complexity and size—60 m high with each side over 70 m wide—*Hagia Sophia* was built in a short time and completed by 537. The church plan is that of a traditional basilica with three aisles reached through two narthexes. The interior of the church is a wide expanse with a central plan. The vastness of the central nave, spanning over 30 m, is emphasized by the dome, with the two lateral semi—cupolas leaning on the four piers through pendentives. The presence of forty ribs in the dome enabled forty windows to be opened around its base, whence light filters with a striking effect.

During its construction the foundations settled, causing many structural changes. The dome failed in 558 and was rebuilt, higher by several meters, in 562, with a consequent reduction in the thrust. The tangent line to the dome's meridian section at its base makes an angle of about 20° with the vertical. Thrust was thus acting, even in the original stress state, primarily on the dome, before any cracking emerged. Thus, while thrust would be activated on the hemispherical domes only later, when meridian cracks had already appeared, the dome of Hagia Sophia, on the contrary, began to push immediately, as soon as the centering was dismantled.

In spite of its static problems and its bold design, Hagia Sophia, was to have a great influence on all subsequent Byzantine and Ottoman architecture. Indeed, even the 15th-century architect Sinan, who was responsible for many of Turkey's most magnificent mosques, took inspiration from the innovative architecture of Hagia Sophia.

In the Middle Ages financial constraints compelled builders to construct lighter, less expensive vaults, leading to the use of smaller bricks or stone ashlars. Thus, the vaults built at the time did not follow the solid, heavy Roman design, with thick, rigid foundations, but were instead overall much lighter, with flexible foundations able to adapt to frequent soil settlements.

Cross vaults, also known as a double barrel or groin vaults, are typical of the Middle Ages, though some the most magnificent examples date back to Roman times. During the Middle Ages, however, the cross vault, with pointed arch and cross ribs, developed into an architectonic element of imposing aesthetic quality. The great vaults covering cathedral naves were linked externally by sequences of flying buttresses, which conveyed the thrusts to the large external counterforts.

Then, in the 15th century Roman architecture was rediscovered. Domes, pavilions, as well as segmental and ribbed vaults, which had long been abandoned, were built once again in both religious and civil structures. The construction techniques used, however, were quite different from the Romans': the vaults were made with brickwork. They were also lightened by replacing the heavy filling of crushed stone or gravel with a sequence of ribs and lunettes, the latter built on the vault haunches to provide much improved interior lightening. The dome was thus to become the most important architectural element in nearly all places of worship. In this regard, an exceptional example is the renowned ogival dome engineered by Brunelleschi on the Florence Cathedral of Santa Maria del Fiore.

Michelangelo carried on and further developed this tradition with the dome of St Peter's Basilica in Rome, with its circular profile and double ribbed cupola. Later, in the Baroque era, domes with elliptic plans and elaborate lanterns were also built. Guarini's vaults with crossed arches, the London Cathedral of St Paul's, with its double–capped dome designed by Wren, the triple cap on Soufflot's dome of Sainte–Geneviève in Paris, and finally the daring vaults by Antonelli represent some of the latest, most spectacular achievements in the construction of masonry domes, which subsequently came to an end with the advent of the new materials steel and reinforced concrete.

Of the most recent achievements in masonry vault construction, the work of the Valencian engineer Guastavino deserves special mention. Towards the end of the 19th century, Guastavino, working in the U.S., studied, developed and eventually patented techniques based on age–old Catalan methods for designing and building vaults. He termed his system "Cohesive Construction".

The system enabled constructing robust, self–supporting arches and vaults using interlocking terracotta tiles and layers of mortar to form a thin skin, whose strength depends upon the strong cohesion between its small components. The technique is known as timbrel vaulting because of its likeness to the skin of a timbrel or a tambourine. Tiles are usually set in three herringbone–pattern courses with a sandwich of thin layers of Portland cement. Unlike heavier stone constructions, these tile vaults could be built without centering. Each tile was in fact cantilevered out over the open space.

By virtue of its lightness and high tensile strength, this tile system provided solutions that were impossible with traditional masonry arches and vaults. The static behavior of this system is mainly elastic and so the no–tension model is not applicable to studying the statics of this class of vaults, which though limited in extension, offer some extraordinary features (Ochsendorf 2009). Unfortunately, at the time, the ever–worsening lack of skilled craftsmen and easy availability of inexpensive steel made Guastavino's work too cost–inefficient, two factors that combined to make also the architecturally extraordinary and efficient system of 'Guastavino tiles' a thing of the past.

4.2 The Implemented Static Approach

Masonry vaults present different static behavior depending on their geometry, the strength of the masonry, the techniques used in their constructional and so forth. A broad class of masonry vaults, which has been built since ancient times, is one whose stability depends upon the elements being held together by gravity and friction. Most vaults belonging to the world's historic architectural heritage fall into this category. For such vaults the constituent elements, either bricks or stone blocks, are held together by compressive stresses: as in nearly all traditional masonry constructions, the influence of masonry's weak tensile strength on their statics is negligible. Cracks mainly determine the deformation of these vaults. Thus, analysis of their cracking deserves further attention, because cracking, if it occurs, is generally delayed, particularly in domes.

In a masonry arch cracks occur as soon as stresses exceed the weak adhesion strength between the stones and the mortar. Compressive forces and shears continue to spread across the hinges that have formed and friction does not oppose the formation of cracks. Instead, in masonry vaults the friction increases their tensile strength.

In a dome of revolution when the hoop stresses reach the masonry tensile strength, cracks develop and run along the meridians. Upon cracking the dome widens and the circular courses slip one over the other (Fig. 4.4). The friction

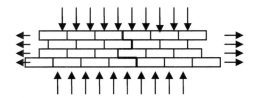

Fig. 4.4 Meridian compression opposes cracking through friction between courses

4.2 The Implemented Static Approach

strength produced by meridian compression opposes this slipping. Thus, the masonry's tensile strength is actually increased by the meridian compression. Humidity, which generally penetrates slowly into masonry, weakens the mortar and reduces this friction strength. Other causes, such as the dynamic actions due to an earthquake or settling of the foundation, can hasten cracking. These considerations explain why cracks generally occur only many years after a vault's construction, particularly in domes. The history of cracking of many famous vaults and domes bears this out. For example, both the domes of St Peter's in Rome and Santa Maria del Fiore in Florence, began to crack 50 years or more after their completion. This delay in cracking implies that in the early stages of their lifetime vaults behaves as solid structures able to sustain tensile stresses. The so–called membrane solution can provide reasonably accurate descriptions of the stress states involved.

The stress state of a vault may remain constant over time if its structure is very light and the masonry offers sufficient tensile resistance. This is true of Guastavino vaults, but not of *most* masonry vaults. For these latter, cracking generally occurs. Once a dome has cracked, a completely different stress state ensues. Upon cracking, tensile hoop stresses in the lower rings of the dome disappear and meridian cracks lengthen upwards, towards the center, until a new internal equilibrium, if any, is reached. If such equilibrium configuration does indeed exist, it will be considerably different from the original one. This new stress state, which can be reasonably modeled within the framework of no–tension masonry materials, represents the *actual* or *final* stress state occurring in the vault.

Cracking brings about a new resisting system in the vault. Now the question must be asked: how do we find this new resisting structure? It can be identified by determining the new system of force transmission that develops inside the vault when the primary tensile forces have been canceled out. In such determination, the influence of the elastic compressive stresses can be neglected and the behavior of a cracked vault can be explained via the *no–tension, rigid in compression masonry model*, described previously in detail. According to Heyman (1966), after cracking, the resisting vault can be defined as a so–called *sliced vault*.

The internal forces in the early stage of a vault's life, which we will call the *initial* stresses, can be represented with relatively good approximation by *membrane stresses*. According to the membrane solution, initially no thrust acts at the springing of an hemispherical dome. Instead, thrust does generally arise when cracking begins. The vault support structures will thus settle and become deformed under the action of this thrust and further deformation of the dome will occur. The response of the cracked vault to such deformation can be evaluated using the sliced vault model. Within this framework, recalling the considerations put forth in Chap. 2, the cracked vault can be said to adapt to the settling of its supporting structures through a mechanism that activates *the minimum thrust* on its supports. The main lines of the approach followed can be outlined as follows. The initial stresses are first determined via membrane stress analysis. Then, by examining where the tensile forces act, it is possible to obtain a sketch of the cracking pattern occurring in the vault. The cracking pattern is very useful in formulating *simple no–tension models* that can describe the real behavior of vaults governed by the vanishing tensile strength.

References

Heyman J. (1966). The stone skeleton, *International Journal of Solids Structures*, 2.
Ochsendorf, J. (2009). *Guastavino vaulting: The art of structural tile*. New York: Princeton Architectural Press.

Chapter 5
Masonry Vaults: Domes

Abstract Beautiful domes have been built firstly by Mycenaeans and later by Romans and Byzantines; domes were practically neglected in the Middle Ages, later rediscovered by the Renaissance and Baroque architecture. Following the approach outlined in the previous section, this chapter firstly deals with the analysis of membrane stresses occurring in rotational shells that describe, with sufficient accuracy, the initial un-cracked stress state of the masonry dome. Later, the masonry dome probably cracks when the tension stresses in the hoop rings near the springing will reach the masonry's weak tensile strength, usually fading in the course of time. The initial membrane equilibrium is thereby lost and meridian cracks will arise and spread along the dome. The emergence of thrust is the main consequence of cracking of the hemispherical domes. The occurrence of this thrust yields a subsequent deformation of the supporting structures and the dome, as a rule, activates its minimum thrust state. The research of the dome thrust is the main subject of the chapter and it will be searched both by static and kinematic approaches: in this last case as maximum thrust among all the kinematical ones. In addition to the hemispherical dome with constant thickness, four outstanding examples of actual masonry domes are then analyzed in detail: the ancient Mycenaean tholoi, the Pantheon, the dome of St. Maria del Fiore in Florence and the St. Peter dome in Vatican.

5.1 Some Recalls on Membrane Equilibrium of Rotational Shells

The primary stresses that are activated when a masonry vault is first loaded (and cracking has not arisen) can be represented quite realistically by membrane solutions. As discussed before, studying this state, through determination of the tensile stress field, can be very useful to formulate a model for the cracked vault.

The aim of this section is to recall the features of dome membrane stresses. In doing so, we will refer to shells with a positive double Gaussian curvature: they have the shape of a so-called *dome of revolution*. The shell will possess a high

Fig. 5.1 Dome composed of meridian and ring bands

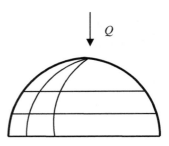

degree of both stiffness and strength, that is, the so-called *shape strength,* if the constituent material of the shell is linearly elastic. The rotational shell can, in fact, be represented as a sequence of elastic slices connected by circular elastic rings. These rings develop hoop stresses that strongly counter any widening or constricting of the dome. These rings stretch or shrink uniformly and, consequently, attain their full stiffness and strength. Any bending of the meridian bands is restrained by deformation of the rings under the action of axial symmetric loads. Meridian bending is consequently very limited and the dominating stresses can be considered constant throughout the shell's thickness.

The following recalls the most relevant aspects of this particular stress state and refers to the rotational shell whose middle surface is sketched as in Fig. 5.1. The surface element is bounded by curved meridians and parallel lines. The position of any given point P on this surface is defined by two spherical coordinates, θ and ϕ. The angle θ is the *longitude*: it measures the angular distance between the reference meridian and the meridian passing through P. The angle ϕ, between the normal to the spherical surface at P and the axis of rotation, is the complement of the latitude, the so-called co-latitude.

At each point P on the shell's middle surface we can identify the meridian plane, passing through the axis of the shell, and the corresponding orthogonal plane passing through the normal to the surface at P: these are the planes of the principal curvatures (Fig. 5.2), whose radii are denoted by

$$r_1, r_2. \tag{5.1}$$

Fig. 5.2 Sections of the principal curvatures

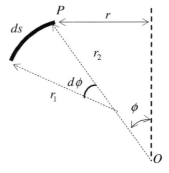

5.1 Some Recalls on Membrane Equilibrium of Rotational Shells

where r_1 is the radius of curvature of the meridian at P and r_2 the radius of curvature of the curve on the surface obtained by its intersection with the plane orthogonal to the meridian plane. The radii (5.1) are the minimum and maximum of all the curvature radii of curves obtained by sectioning the middle surface with the planes of the bundle whose axis is the normal to the surface at P. The radius r_2 is the length of the segment OP connecting the point P with the point O, the intersection of the normal to the surface in P with axis of revolution. With regard to Fig. 5.2, which shows a small arc of a meridian of the dome, let r_1 be the radius of curvature of the meridian at P. For the line element ds of the meridian, we thus have

$$ds = r_1 d\phi \tag{5.2}$$

Moreover, let r be the distance of point P on this meridian from the axis of rotation; it is the radius of the parallel obtained by sectioning the dome with a plane orthogonal to its axis passing through P. Thus, from Fig. 5.2, we have

$$r = r_2 \sin \phi. \tag{5.3}$$

which represent, for a line of unit length, the resultant along the thickness of the stresses σ_ϕ and σ_θ acting along the coordinate lines ϕ and θ (Fig. 5.3).

Likewise, the small surface element dA, with side lengths $r_1 d\phi$ and $rd\theta$, respectively, has the area

$$dA = r_1 r d\phi d\theta \tag{5.3'}$$

The stress state acting on the middle surface of the vault, loaded axial-symmetrically, is defined by the components

$$N_\phi, N_\theta, \tag{5.4}$$

Owing to the axial symmetry, forces N_ϕ and N_θ, vary only with the co-latitude ϕ. The unknown internal forces and the available equilibrium equations are both two in number. The stress state in the axial-symmetric shell is thus statically

Fig. 5.3 Membrane stresses in the rotational shell under axial symmetrical loading

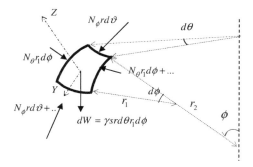

determinate. The upper side of the element dA, running along the parallel circle, has length $r(\phi)d\theta$; while the lower has length

$$r(\phi+d\phi)d\theta = \left[r\phi + \frac{dr}{d\phi}d\phi\right]d\theta.$$

Figure 5.4 shows a sketch of the projection of the surface element on the vertical and horizontal planes, together with the corresponding forces. Those acting on the sides of the element lying along the parallel circles are respectively

$$N_\phi r d\theta \quad \left(N_\phi + \frac{dN_\phi}{d\phi}d\phi\right)\left(r + \frac{dr}{d\phi}d\phi\right)d\theta$$

and are directed along the tangent to the meridian at the coordinates ϕ and $\phi + d\phi$. These two forces act in opposite directions and include the angle $d\phi$. Consequently, they produce the resultant along the normal Z

$$N_\phi r d\vartheta d\phi$$

Likewise, the two hoop forces

$$N_\theta r_1 d\phi$$

on either side of the element, lie in the plane of a parallel circle and include the angle $d\theta$. They produce the resultant force situated in the same horizontal plane,

$$N_\theta r_1 d\phi d\theta,$$

pointing outside the shell element, as sketched out in Fig. 5.4. We assume the shell has constant thickness s, unit weight g, and that only the dead load acts on it. Thus, the weight of element dA is

$$dW = gdA = grr_1 d\theta d\varphi$$

Fig. 5.4 Forces acting on the unit element of the revolution shell

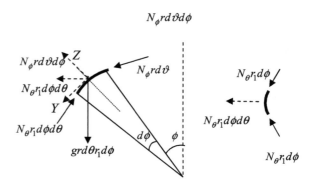

5.1 Some Recalls on Membrane Equilibrium of Rotational Shells

The equilibrium condition of the element dA along the normal Z is thus

$$N_\phi r d\theta d\phi + N_\theta r_1 d\theta d\phi \sin\phi - g r_1 r d\theta d\phi \cos\phi = 0, \tag{5.5}$$

or, by simple manipulations,

$$\frac{N_\phi}{r_1} + \frac{N_\theta}{r_2} - g \cos\phi = 0, \tag{5.6}$$

Likewise, the other equilibrium equation along the tangential direction Y is

$$\frac{d}{d\phi}(N_\phi r) d\theta d\phi + N_\theta r_1 d\phi d\theta \cos\phi + g r_1 r d\phi \sin\phi = 0 \tag{5.7}$$

or

$$\frac{d}{d\phi}(N_\phi r) + N_\theta r_1 \cos\phi + g r_1 r \sin\phi = 0. \tag{5.8}$$

The unknowns N_ϕ, N_θ are obtained by solving the two Eqs. (5.6) and (5.8). It is however simpler to obtain the forces N_ϕ, that directly solve the equilibrium equation along the vertical direction of the generic cap, as sketched out in Fig. 5.5.

The vertical force $Q(\phi)$, representing the resultant of all forces acting on the cap defined by angle ϕ, is sustained by the vertical components of the meridian forces N_ϕ uniformly distributed along the parallel circle at the co-latitude ϕ: hence we have

$$Q(\phi) - N_\phi 2\pi r \sin\phi = 0 \tag{5.9}$$

and

$$N_\phi = \frac{Q(\phi)}{2\pi r \sin\phi}. \tag{5.10}$$

Force N_θ can thus be obtained from Eq. (5.6) by using expression (5.10) for N_ϕ. The case of the hemispherical shell is quite meaningful. Both principal curvature radii are equal to the radius R of the middle spherical surface and Eq. (5.6) can be simplified to

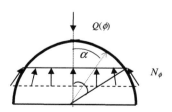

Fig. 5.5 Equilibrium along the vertical direction of a spherical cap

$$N_\phi + N_\theta - gR\cos\phi = 0. \qquad (5.11)$$

The spherical cap lies between the co-latitude values $\alpha = 0$ and $\alpha = \phi$. The weight $Q(\phi)$ of the cap, with constant thickness, is

$$Q(\phi) = 2\pi gR^2 \int_0^\phi \sin\alpha\, d\alpha = 2\pi gR^2(1 - \cos\phi). \qquad (5.12)$$

Substitution of (5.12) into (5.10) gives

$$N_\phi = g\frac{R}{1 + \cos\phi}, \qquad (5.13)$$

and evaluation of forces N_θ follows immediately from Eq. (5.11), whence we get

$$N_\theta = gR\left(\cos\phi - \frac{1}{1 + \cos\phi}\right). \qquad (5.14)$$

At the top of the hemispherical shell, that is at $\phi = 0$, we have $(N_\theta)_{\phi=0} = \gamma sR/2$, while, at the springing, at $\phi = \pi/2$, $(N_\theta)_{\phi=\pi/2} = -\gamma sR$. The distribution of the forces along the parallel circles has the contour shown in Fig. 5.6. Forces N_θ exert compression on the upper shell and tension further down.

A *graphical* procedure can be applied to evaluate the membrane stresses. This approach is particularly useful in cases of domes with more complex profiles and varying thickness.

A slice of the dome (Fig. 5.7) is subdivided into N voussoirs, each of whose weight and center position are to be determined. The weight forces are applied at the voussoirs centers and the procedure starts from the first voussoir, at the crown. The ring sections at the top of the slice transmit a compressive force.

According to the membrane stress state, the magnitude of the resultant S_1 of these compressive forces is such that, when summed with the vertical weight W_1 of the first voussoir, the direction of the resultant vector N_1 coincides with the

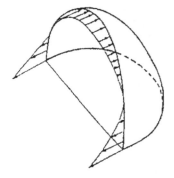

Fig. 5.6 Variation of stresses N_θ along the meridian section (from Heyman 1995)

Fig. 5.7 The slice of the dome

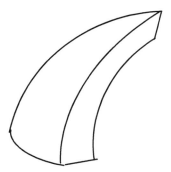

direction of the line connecting the centers of the first and second voussoirs. The axial action N_1 transmitted by the first voussoir can thus be determined.

Beginning with the second voussoir, we consider the resultant R_2 of axial force N_1 with the weight W_2 of the second voussoir. We then evaluate the resultant S_2 of the compressive forces transmitted by the second ring, so that, summing this resultant S_2 with the previous force R_2, the line of action of resultant vector N_2 coincides with the line connecting the center of the second voussoir with the center of the third (Fig. 5.8). The procedure is then applied iteratively.

The rings in the upper part of the dome are compressed. However, descending along the meridian, the horizontal force S transmitted by the rings could *change sign*, i.e. become a tensile force, to ensure that compressive force N is directed along the center line of the slice. This change in sign requires that the rings at that height be stretched, which is however impossible given masonry's inability to withstand tensile stresses. Thus, in searching for an admissible equilibrium, axial force N acting along the meridian will diverge from the curved shell axis and be displaced to the slice's interior, even in the upper sections where the sign inversion of the ring stress was not occurred.

Fig. 5.8 Membrane stresses obtained by evaluating successive resultants

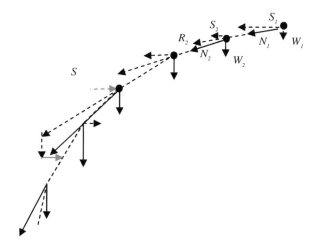

5.2 Meridian Cracking: Definitive Stress State in Masonry Domes

5.2.1 From the Membrane to the Cracked State

The dome will crack when the tension stress in the hoop rings near the springing reach masonry's weak tensile strength. The initial membrane equilibrium is thereby lost and meridian cracks will arise and spread along a band on the dome much higher than that subjected to the tensile stresses in the membrane equilibrium (Fig. 5.9). Although the number of these meridians cracks is indeterminate, only one is sufficient to disrupt the hoop actions of the rings. By way of example, only four major cracks have been be detected on the intrados of the Brunelleschi dome in Florence, while fourteen are present in the Pantheon dome and about the same number of meridian cracks were detected in the dome of St. Peter's before the restoration works performed by Poleni and Vanvitelli.

The cracked dome tends to open wide along a large band and break up into slices that behave as independent pairs of semi-arches leaning on each other.

Predictably, cracking brings about a profound change in the dome's statics. The forces, N_θ, in the cracked zone are eliminated and the forces N_ϕ, if acting along the slices centerline, are no longer able to ensure equilibrium. The pressure curve thus tilts towards the horizontal and deviates away from the central surface of the dome. A small cap at the top of each slice will be subjected to the thrusting action transmitted by the other slices, which will be transmitted all the way to the springings. Figure 5.10 shows a rough sketch of the pressure curve of a cracked hemi-spherical dome. The dotted line shows the position of this curve, which inclines towards the horizontal at the springing. The horizontal component of the reaction of the supports represents the thrust S per unit length of the dome's base circumference.

The emergence of thrust in the dome represents the most consequential outcome of meridian cracking in typical masonry round domes. Loaded by the dome's thrust, the sustaining structures (e.g., the drum and underlying piers) deform and splay. The slices, no longer restrained from deforming by the rings, bend under the loads and can form mechanisms.

The weight of a particularly heavy lantern, for example, could even cause the dome to fail on cracking.

Fig. 5.9 Typical meridian cracks in a masonry dome (from Heyman 1995)

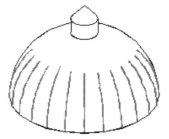

5.2 Meridian Cracking: Definitive Stress State in Masonry Domes

Fig. 5.10 Rising thrust due to meridian cracking

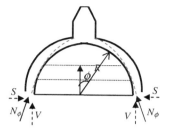

Figure 5.11 shows two different collapse mechanisms of the sliced dome: the first involves uniform lowering of the central zone of the dome, while the second is subject to rotation, as well as a lowering of the central portion of the slices. These mechanisms entail the presence of both full-thickness meridian cracks, as well as two ring cracks due to the presence of two hinges: one high up on the extrados, and the other further down at the intrados. Generally, cracks on the extrados are not detectable because dome extradoses are usually covered, for instance, by sheets of lead. The circumferential cracks at the intrados sometimes develop very high, near the connection of the dome with the lantern locking ring.

Figure 5.12 shows a settlement mechanism by which the dome adapts itself to a small widening of its springings. Such a mechanism, however, may also describe dome collapse involving deformation of the drum as well.

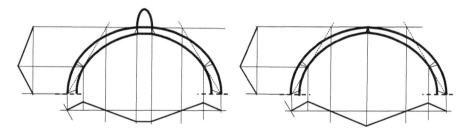

Fig. 5.11 Symmetric collapse mechanism in a dome

Fig. 5.12 Symmetrical mechanism involving widening of the dome springing

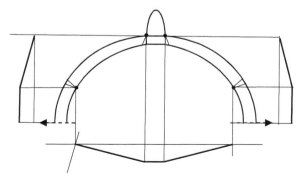

Analyzing the cracking pattern in a dome can provide useful information regarding its safety. In the presence of thin, nearly hairline cracks, elastic unloading of the lower rings due to meridian cracking offsets the widening of the rings due to cracking. More specifically, from Eq. (5.14), the tension force N_θ near the springings equals $\gamma\, sR$ and the corresponding tensile stress σ_θ is γR. Therefore, with $R = 15$ m, $\gamma = 1.6$ t/m^3, for instance, we have $\sigma_\theta = 2.4$ kg/cm^2. Upon the unloading of σ_θ, the contraction strain $\varepsilon_\theta = \sigma_\theta/E_m$ occurs, where E_m is the masonry elastic modulus. By assuming $E_m = 50{,}000$ kg/cm^2, $\varepsilon_\theta = 2.4/50{,}000$, which is about 0.5×10^{-4}. The circumference of the dome at the springing shrinks by $0.5 \times 10^{-4} \times 2\pi \times 15$ m $= 4.7$ mm. If four meridian cracks arise, the width of each is thus equal to $4.7/4 = 1.1$ mm.

When cracks are barely perceptible, only elastic unloading of the dome rings has occurred. In the presence of numerous broader cracks, on the other hand, the widening of the drum will be the main factor responsible for the dome's deformation. Knowing the dome thrust and, consequently, the safety level afforded by the stresses present in its retaining structures, allows for verifying whether the dome arrangement is sufficiently stable.

5.2.2 Safety Check of Domes: Static and Kinematic Approaches

Dome failure occurs when the pushing work due to the lowering of the dead loads distributed around the dome's center equals the resisting work of the vertical loads near the springing. A safety check can be carried out, for instance, by evaluating the lantern weight that would cause the dome to collapse. If the mechanism involves the drum, the work done by the weight of the drum slice will also be included in the resisting work.

Figure 5.13 shows a failure mechanism of this kind. In this case the quantity Q_L is the share of the weight of the lantern corresponding to the slice considered. The analysis can be conducted by applying the static or kinematic theorem of Limit Analysis. The safety check considers all possible failure mechanisms, such as those including or excluding drum deformation, as sketched out in Figs. 5.12 and 5.13. If the dome is safe it could be useful to check the safety level of its sustaining structures. The widening of the drum at its top is due to thrust activation in the dome. Consequently, the dome deforms according to a settlement mechanism. As discussed in Chap. 2, the thrust of the *settled* dome is the *minimum* S_{Min} from among all the thrusts S transmitted by statically admissible pressure curves.

The minimum thrust S_{Min} can be obtained via the *static,* as well as the *kinematic* approach. The *static* approach calls for tracing the statically admissible funicular curves of the loads. In this case we can neglect the small hooped cap situated at the

5.2 Meridian Cracking: Definitive Stress State in Masonry Domes

Fig. 5.13 Failure mechanism of the dome involving the drum

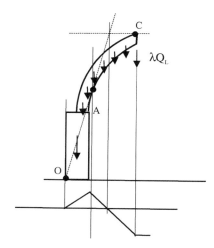

top, near the dome zenith. Any pair of facing slices of the settled dome behaves like an arch that has undergone a small increase in span. The arch deforms through a mechanism and thereby adapts to the sideways settlement. A sliced dome behaves in the same manner. In the settled state the pressure curve passes through the extrados at the key section of the slices and then runs within their interior, *skimming* over the intrados of the dome. Domes with lanterns have a top ring to sustain it. Thus, instead of a single hinge, two symmetric hinges will form at the extrados of the section connecting the slice with the top ring. The thrust S of the settled dome is transmitted along the pressure curve passing through these hinges and corresponds to the minimum S_{Min} of all thrusts of statically admissible pressure lines wholly contained within the slice. Thus, following the *static* approach, we must identify, from among all the statically admissible pressure lines, the one that releases the minimum thrust at the dome springing.

The *kinematic* approach is dual with respect to the static one. Let us consider any kinematically admissible settlement mechanism, v, describing the adjustment of the dome to the side deformation of its sustaining structures and define the kinematic thrust multiplier $\lambda(v)$ as

$$\lambda(v) = \frac{\langle g, v \rangle}{\langle r, v \rangle}. \tag{5.15}$$

In (5.15) $\langle g, v \rangle$ represents the work, undoubtedly positive, of the dead loads on the vertical displacements of mechanism v, and $\langle r, v \rangle$ the work, undoubtedly negative, performed by the thrust on the corresponding horizontal displacement. The kinematic thrust $\lambda(v)r$ can be directly expressed as

Fig. 5.14 The settlement mechanism of the slice following widening of the drum *top*

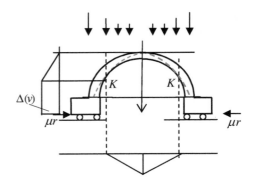

$$\lambda(v)r = S_{kin}, \qquad (5.16)$$

where, according to definition (5.15),

$$S_{kin}(v) = \frac{\langle g, v \rangle}{\Delta(v)}, \qquad (5.16)$$

where $\Delta(v)$ is the radial widening of the dome at its base due to settlement mechanism v. According to this approach the minimum thrust can be evaluated as the *maximum* of all kinematic thrusts $S_{kin}(v)$ (Como 1996, 1998)

$$S_{Min} = Max\, S_{kin}(v) = Max \frac{\langle g, v \rangle}{\Delta(v)} \qquad (5.17)$$

by varying v in the set of all kinematically admissible settlement mechanisms.

Figure 5.14 shows a generic dome mechanism produced by base widening. In this mechanism the position of the internal hinge K is unknown. The set of all these kinematically admissible mechanisms is described by varying the position of the hinge K between the springing and the key section of the slice. Identifying the maximum of function $\lambda(v)$ by varying the position of hinge K enables us to obtain the sought-for thrust. Many applications of this approach will be described in the following.

5.2.3 Minimum Thrust for the Hemispherical Dome with Constant Thickness

5.2.3.1 Analytical Search for the Unknown Pressure Surface

For some simple dome arrangements, it is a relatively simple matter to obtain the analytical expression for the minimum thrust. By way of illustration, we will now analyze the case of a hemispherical dome with constant thickness without lantern.

5.2 Meridian Cracking: Definitive Stress State in Masonry Domes

Fig. 5.15 Equilibrium of a small element of the unknown pressure surface

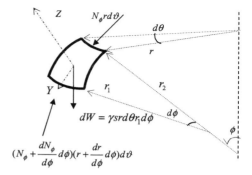

The following formulation is based on the analysis performed by Bossut (1778) and reported by Benvenuto (1990). It has been modified here in order to search for the minimum thrust of the dome. An interesting historical note is that Bossut originally applied this analysis to check the stability of the dome of Sainte Geneviève in Paris, which was engineered by Soufflot and in 1791 was declared a secular shrine under the name of the Pantheon.

Bossut's approach starts out by considering the equilibrium of a small element of the *unknown* pressure surface to be determined. This surface has a regular rotational shape and runs continuously between the dome extrados and intrados. We define the unit load g per mean surface of the dome; approximately the same load g will act on the unknown surface. Figure 5.15 shows an element, dA, of this pressure surface: the forces N_ϕ are the only ones applied on this element. Equilibrium of the element along the Y direction gives

$$N_\phi r d\theta - \left(N_\phi + \frac{dN_\phi}{d\phi}d\phi\right)\left(r + \frac{dr}{d\phi}d\phi\right)d\theta + gr_1 r d\phi d\theta \sin\phi = 0, \quad (5.18)$$

where all the quantities, including the ring radius r and the curvature radius r_1 of the meridian, refer to the unknown pressure surface. From Eq. (5.18) we get

$$\frac{d}{d\phi}(N_\phi r) = g r r_1 \sin\phi. \quad (5.18')$$

Equilibrium of the element along the Z direction, normal to the pressure surface, is instead expressed by

$$-N_\phi r d\theta d\phi + g r r_1 d\theta d\phi \cos\phi = 0 \quad (5.19)$$

and we have

$$N_\phi = g r_1 \cos\phi. \quad (5.20)$$

Substituting (5.20) into (5.18′) thus gives

$$-\frac{d}{d\phi}(grr_1 \cos\phi) + grr_1 \sin\phi = 0. \tag{5.21}$$

Moreover,

$$\frac{d}{d\phi}(rr_1 \cos^2\phi) = \cos\phi\left[-\frac{d}{d\phi}(rr_1 \cos\phi) + rr_1 \sin\phi\right], \tag{5.22}$$

which, accounting for (5.21) yields

$$rr_1 \cos^2\phi = C = \text{cost}. \tag{5.23}$$

By looking at the equilibrium condition along the horizontal projection of a slice element of magnitude $d\theta$, we can explain the significance of condition (5.23) (Fig. 5.16). We have

$$N_\phi rd\theta \cos\phi - \left(N_\phi + \frac{dN_\phi}{d\phi}d\phi\right)\left(r + \frac{dr}{d\phi}d\phi\right)d\theta \cos(\phi + d\phi) = 0 \tag{5.24}$$

or

$$-\cos\phi\frac{d}{d\phi}(N_\phi r) + N_\phi r \sin\phi = -\frac{d}{d\phi}(N_\phi r \cos\phi) = 0, \tag{5.25}$$

whence we obtain

$$\frac{d}{d\phi}(N_\phi r \cos\phi) = 0 \tag{5.26}$$

and taking Eq. (5.20) into account

$$\frac{d}{d\phi}(grr_1 \cos^2\phi) = 0. \tag{5.27}$$

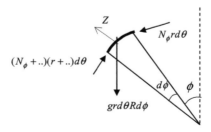

Fig. 5.16 Projection on the meridian plane of the forces acting on the surface element

5.2 Meridian Cracking: Definitive Stress State in Masonry Domes

Fig. 5.17 Coordinates r, z

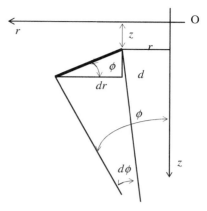

Thus, we derive condition (5.23) on the slice of unit magnitude $\theta = 1$ rad, (Fig. 5.17)

$$grr_1 \cos^2 \phi = H = \text{cost.} \tag{5.28}$$

From Eq. (5.25) we can write

$$grr_1 \cos^2 \phi = gr \frac{ds}{d\phi} \frac{(dr)^2}{(ds)^2} = Cg = H, \tag{5.29}$$

where H is the thrust acting on the slice of unit magnitude $\theta = 1$. From (5.29) we also have

$$rdr = C \frac{ds}{dr} d\phi. \tag{5.30}$$

and, by integrating,

$$\frac{r^2}{2} = C \ln \frac{1 + \sin \phi}{\cos \phi} + C_1, \tag{5.31}$$

that is,

$$\left(\frac{r^2}{2C} - \frac{C_1}{C} \right) = \ln \frac{1 + \sin \phi}{\cos \phi}. \tag{5.32}$$

For the sake of simplicity, we can set

$$\xi = \frac{r^2}{2C} - \frac{C_1}{C} \tag{5.33}$$

and, by virtue of (5.32)

$$\xi = \ln \frac{1 + \sin \phi}{\cos \phi}, \tag{5.34}$$

hence,

$$e^{\xi} = \frac{1 + \sin \phi}{\cos \phi}, \quad e^{-\xi} = \frac{\cos \phi}{1 + \sin \phi}. \tag{5.35}$$

Now by taking into account the definition of sin hξ,

$$\frac{1}{2}(e^{\xi} - e^{-\xi}) = \sinh \xi = \frac{1}{2}\left(\frac{1 + \sin \phi}{\cos \phi} - \frac{\cos \phi}{1 + \sin \phi}\right) = \text{tg}\,\phi, \tag{5.36}$$

from (5.28) we get

$$\frac{dz}{dr} = tg\phi = \sinh\left(\frac{r^2}{2C} - \frac{C_1}{C}\right). \tag{5.37}$$

By integrating (5.37) we arrive at the equation for the sought-after pressure surface in terms of the coordinates r, z

$$z(r) = \int_0^r \sinh\left(\frac{x^2}{2C} - \frac{C_1}{C}\right) dx + A. \tag{5.38}$$

The pressure surface is symmetrical and the symmetry condition at the zenith requires

$$\left(\frac{dz}{dr}\right)_{r=0} = 0 \tag{5.39}$$

Consequently, taking in account of (5.38) we obtain

$$\sinh\left(\frac{C_1}{C}\right) = 0 \tag{5.39'}$$

Hence

$$C_1 = 0 \tag{5.40}$$

Further, by using the variable change

$$\frac{x^2}{2C} = y^2 \tag{5.41}$$

5.2 Meridian Cracking: Definitive Stress State in Masonry Domes

we have

$$z(r) = \sqrt{2C} \int_0^{r/\sqrt{2C}} \sinh y^2 dy + A \qquad (5.42)$$

Equation (5.42) defines a family of pressure surfaces, each one corresponding to a particular choice of the constant C. Among these we have to choose that corresponding to the minimum thrust.

5.2.3.2 The Minimum Thrust Pressure Surface

At the meridian cracking of the hemispherical dome, thrust takes place. The supporting structures, as the drum, will thus deform and a small widening of the dome springing occur. The dome adapts itself to the settlement and a minimum thrust state develops in it. This state is so characterized:

(a) the thrust surface is all contained inside the dome
(b) the extrados and the intrados of the dome will be touched by the thrust surface along two circles corresponding to the two internal hingings of the settlement mechanism.

Compared with the arch with constant width, in the dome slices narrow towards the crown. This fact highly influences the geometry of the pressure surface that remains almost horizontal in the neighbourhood of the zenith. Thus, to maintain the pressure surface within the slice, the settlement mechanism cannot admit an hinge at the dome zenith. Here we have

$$z(0) = A \qquad (5.43)$$

where, if s denotes the dome thickness,

$$0 < A \leq s \qquad (5.44)$$

The constant A, that dimensionally is a length, defines the position of the pressure curve at the zenith. Going down along the meridian, slices gradually expand and the pressure curve bends down. The two hinges of the mechanism form at points M and N of Fig. 5.18. Suitable conditions will define the position of the pressure curve in the slice. The pressure curve will be *tangent* at the points M and N, respectively at the extrados and at the intrados of the slice. We have (Fig. 5.18)

$$\left(\frac{dz}{dr}\right)_{r_1} = \sinh \frac{r_1^2}{2C} = tg\phi_1 \quad \left(\frac{dz}{dr}\right)_{r_2} = \sinh \frac{r_2^2}{2C} = tg\phi_2 \qquad (5.45)$$

where, if R is the mean radius of the dome and s its thickness

Fig. 5.18 The pressure line in a slice at e minimum thrust state

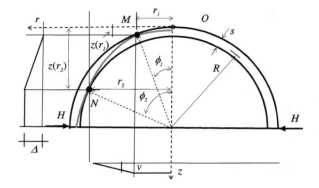

$$r_1 = \left(R + \frac{s}{2}\right) \sin \phi_1 \quad r_2 = \left(R - \frac{s}{2}\right) \sin \phi_2 \tag{5.46}$$

Thus conditions (5.45) become

$$tg\, \phi_1 = \sinh\left(R + \frac{s}{2}\right)^2 \frac{\sin^2 \phi_1}{2C} \quad tg\, \phi_2 = \sinh\left(R - \frac{s}{2}\right)^2 \frac{\sin^2 \phi_2}{2C} \tag{5.45'}$$

or

$$\left(R + \frac{s}{2}\right)^2 \frac{\sin^2 \phi_1}{2C} = \sinh^{-1}(tg\phi_1) \quad \left(R + \frac{s}{2}\right)^2 \frac{\sin^2 \phi_1}{2C} = \sin h^{-1}(tg\, \phi_1) \tag{5.45'}$$

Taking into account that

$$\sinh^{-1} tg\, \phi_1 = \ln(tg\, \phi_1 + \sqrt{tg^2 \phi_1 + 1}); \quad \sin h^{-1} tg\phi_2 = \ln(tg\phi_2 + \sqrt{tg^2 \phi_2 + 1}); \tag{5.47}$$

finally we obtain the following expressions of r_1 and r_2

$$\frac{r_1^2}{2C} = f(\phi_1) \quad \frac{r_2^2}{2C} = f(\phi_2) \tag{5.48}$$

where

$$f(\phi_1) = \ln(tg\phi_1 + \sqrt{tg^2\phi_1 + 1}) \quad f(\phi_2) = \ln(tg\phi_2 + \sqrt{tg^2\phi_2 + 1}) \tag{5.49}$$

The same constant $2C$ is contained in the two conditions (5.47) and (5.47').

5.2 Meridian Cracking: Definitive Stress State in Masonry Domes

Hence

$$2C = R^2 \frac{(1+\chi)^2 \sin^2 \phi_1}{f(\phi_1)} = R^2 \frac{(1-\chi)^2 \sin^2 \phi_2}{f(\phi_2)} \quad (5.50)$$

where

$$\chi = \frac{s}{2R} \quad (5.51)$$

Condition (5.50) is one of the two equations required to obtain the angles ϕ_1 and ϕ_2.

We consider now the other two conditions, those prescribing *the passage* of the pressure curve through the points M and N. We have

$$z(r_1) = R(1+\chi)(1-\cos \varphi_1) \quad z(r_2) = R(1+\chi)(1-\cos \varphi_2) \quad (5.52)$$

or, in a more explicit form by using (5.40),

$$\int_0^{r_1} \sinh \frac{x^2}{2C} dx + A = \left(R + \frac{s}{2}\right)(1 - \cos \varphi_1);$$

$$\int_0^{r_2} \sinh \frac{x^2}{2C} dx + A = \left(R - \frac{s}{2}\right)(1 - \cos \varphi_2) \quad (5.53)$$

These last conditions, with the change of variable (5.41), give

$$\sqrt{2C} \int_0^{\frac{r_1}{\sqrt{2C}}} \sinh y^2 dy + A = R(1+\chi)(1 - \cos \varphi_1) \quad (5.54)$$

$$\sqrt{2C} \int_0^{\frac{r_2}{\sqrt{2C}}} \sinh y^2 \, dy + A = R(1-\chi)(1 - \cos \varphi_2); \quad (5.55)$$

The same constant A is contained in the two conditions (5.54 and 5.55). Thus

$$A = R(1+\chi)(1 - \cos \varphi_1) - \sqrt{2C} \int_0^{\frac{r_1}{\sqrt{2C}}} \sinh y^2 \, dy$$
$$= R((1+\chi) - (1-\chi) \cos \varphi_2) - \sqrt{2C} \int_0^{\frac{r_2}{\sqrt{2C}}} \sinh y^2 \, dy \quad (5.56)$$

that, together with (5.50), is the other equativo required to obtain the two unknowns ϕ_1 e ϕ_2. Solving the two Eqs. (5.50) and (5.56) gives the angles ϕ_1 and ϕ_2 defining the positions of the internal hinges of the slices at their minimum thrust state. Figure 5.19 gives the plot of the solution angles ϕ_1 and ϕ_2 versus the factor χ, the ratio thickness/mean diameter of the dome.

With the increasing of the ratio χ. the hinge at the intrados, defined by the angle ϕ_2, moves towards the springing while the estrados hinge, defined by the angle ϕ_1, moves towards the crown.

Fig. 5.19 The solution angles ϕ_1 dan ϕ_2 versus the ratio χ (Coccia, Como, Di Carlo 2015)

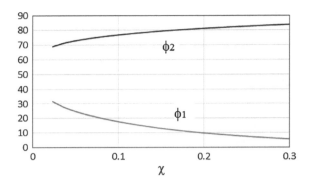

The solution angles ϕ_1 and ϕ_2, represent the real positions of the internal hinges of the pressure curve at its state of minimum thrust, *provided* the compatibility conditions at zenith and at springings are satisfied

$$0 < A \leq s \tag{5.57}$$

$$R(1 - \chi) \leq z^{-1}[R(1 + \chi)] \leq R(1 + \chi) \tag{5.58}$$

The minimum thrust is thus given by

$$h_{min} = \frac{H_{min}}{gR^2} = \frac{(1+\chi)^2 \sin^2 \phi_1}{2f(\phi_1)} = \frac{(1+\chi)^2 \sin^2 \phi_2}{2f(\phi_2)} \tag{5.59}$$

Figure 5.20 gives the plot of the non dimensional ratio $h_{min} = H_{min}/gR^2$ versus the ratio $\chi = s/2R$.

The following expression obtained by interpolation the curve of Fig. 5.20 is (Coccia, Como and Di Carlo 2015).

$$h_{min} = -2.99\left(\frac{s}{2R}\right)^3 + 2.65\left(\frac{s}{2R}\right)^2 - 1.15\left(\frac{s}{2R}\right) + 0.269 \tag{5.60}$$

The minimum thrust given by (5.61) is in good agreement with the curve obtained by Lau (2006) analysing the dome model without the presence of hoop stresses.

Fig. 5.20 The plot $h_{min} = H_{min}/gR^2$ versus the ratio $\chi = s/2R$

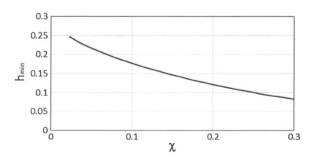

5.2.3.3 Hemispherical Dome of Minimum Thickness

Let us consider the equation of the pressure surface in which its dependance on the angles ϕ_1 and ϕ_2 is espresse in a more explicit form. Thus, according (5.42) with (5.50) and (5.56) we can write:

$$\frac{z(r)}{R} = \frac{(1+\chi)\sin\phi_1}{\sqrt{f(\phi_1)}} \left[\int_0^{\frac{r\sqrt{f(\phi_1)}}{R(1+\chi)\sin\phi_1}} \sinh y^2 dy - \int_0^{\sqrt{f(\phi_1)}} \sinh y^2 dy \right] \quad (5.61)$$
$$+ (1+\chi)(1-\cos\phi_1)$$

Likewise to the case of the arch, let us assume that the pressure surface *touches at the base the extrados of the dome*, i.e. let us admit the condition

$$z[R(1+\chi_{min})] = R(1+\chi_{min}) \quad (5.62)$$

from which we can evaluate the corresponding χ_{min}. The solution of Eq. (5.62) gives:

$$\chi_{min} = 0.0215 \quad (5.63)$$

But, according to this value of χ_{min} the compatibility inequality (5.58) is not satisfied. By means (5.62) in fact we have (Fig. 5.21)

$$\frac{z(0)}{R} = 0.046 > 2\chi_{min} = 0.043 \quad (5.64)$$

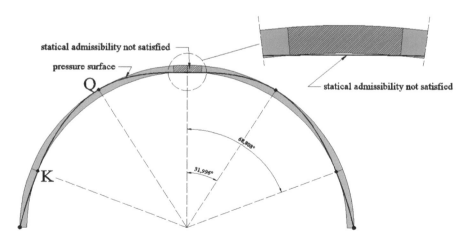

Fig. 5.21 Statical inadmissibility at the crown: when reducing the thickness, the pressure surface touches the dome extrados at the springer

Searching for the minimum thickness of the dome, the thickness can be reduced as far as the pressure surface, that goes up from the springing and passes through the above defined points M and N, touches the intrados at the zenith. The minimum thickness ratio can be thus obtained under the condition that at the zenith we have (Fig. 5.22):

$$z(0) = t_{min} = 2\chi_{min}R \tag{5.65}$$

and, according to (5.62), we have

$$\chi_{min} = 0.0226 \tag{5.66}$$

and the minimum thickness ratio t/R should be equal to 0.0452, i.e. *a bit larger* than the value 0.042 given by Heyman (1977). Only if the ratio t/R is not lower than 0.0452, the compatibility conditions (5.44) and (5.45) are fulfilled everywhere. Figure 5.22 shows the pressure curve in the minimum thickness dome, i.e. with the ratio χ given by (5.66). We can notice that at the spingings the pressure curve passes barely within the dome. Any further thickness reduction pushes the pressure curve downwards, outside the crown. The ratio $t/R = 0.0452$ thus represents the minimum thickness of the dome. It is worth to point out that, in spite of the presence of a sufficient number of hinges, the mechanism cannot develop because not kinematically admissible: it yields ring contraction towards the crown.

Actually, real domes are generally much more complex than the simplified case analyzed here, so that graphical or numerical approaches are frequently adopted as useful aids.

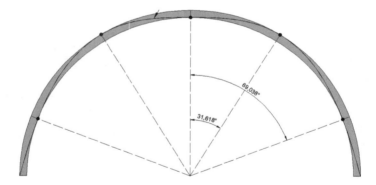

Fig. 5.22 The pressure surface in the state of minimum thickness. The surface touches the intrados of the crown, then skims the extrados and then the intrados at the haunches and is only barely inside the dome at the springings (Coccia, Como, Di Carlo 2015)

5.2.4 Domes of More Complex Shape: The Kinematic Approach

It is a relatively simple matter to apply the kinematic approach to evaluate the minimum thrust of masonry domes. The settlement mechanisms are obtained releasing the slices by positioning hinges to allow horizontal sliding of the dome at its springings. The hinges must thus be positioned:

- at the *extrados*, on the section linking the central closing ring with the slice;

at the *intrados*, at the haunches, as shown in Fig. 5.23. The position of this hinge is unknown and is indicated by σ in the figure.
- Thus, the minimum thrust $\mu_{min}S$ is evaluated by seeking the maximum of the function

$$\mu_{min}S = \mathrm{Max}\frac{\langle g, v(\sigma)\rangle}{\delta(\sigma)} \tag{5.67}$$

by varying angle σ along the intrados and where

$$\delta(\sigma) = (h - R\sin\sigma)\theta. \tag{5.68}$$

is the horizontal displacement of the slice at springing. The search for the minimum thrust thus translates into searching for the maximum of the function

$$\mu_{min}S = \mathrm{Max}\frac{\langle g, v(\sigma)\rangle}{(h - R\sin\sigma)\theta}. \tag{5.69}$$

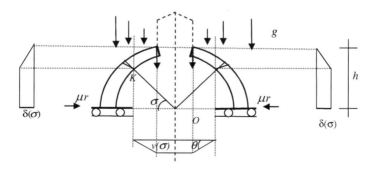

Fig. 5.23 Finding the minimum thrust via the kinematic approach

5.3 Mycenaean Tholos

5.3.1 Description and Historical Notes

The Mycenaean domed tomb, usually called *tholos*, is an admirable architectural achievement of late Bronze Age civilizations. The first *tholoi* discovered were in Messenia, in the Peloponnesian region of Greece, and date back to 1500 BCE. Later, the construction of *tholos* also spread to other regions of Greece, reaching its maximum expression in the so-called Tomb of Clytemnestra, the Orchomenos tholos and the Treasure of Atreus, in around 1300 BCE. Tholoi are usually set in hills: a chamber is built with horizontal pavement by excavating the slope. A first portion of the chamber's perimeter—from the base to about the level of the admittance architrave—is built against the rock, while the remaining part rises from the contour of the slope. The top is covered with a mound of earth, the so-called tumulus (Fig. 5.24). The entrance to the chamber, called the *dromos*, is a long passage cut into the slope of the hill and tumulus and delimited by masonry walls of gradually raising height.

The doorway opening, called the *stomion*, is found at the intersection of the *dromos* with the chamber perimeter, and therefore interrupts the continuity of the masonry wall built around the chamber.

The stomion is flanked by piers supporting the stone architrave. The masonry blocks forming the masonry structure of the round wall are placed in successively smaller concentric rings to project progressively inwards and upwards to close the interior space from above (Fig. 5.25a). The intrados of the dome covering the chamber is ogival in profile (Fig. 5.25b). The lower part of the dome, built into the rock and in close contact with it, is inserted into the burial pit and becomes increasingly thick up to the height of the architrave over the doorway (Fig. 5.24). A filling of stones and compacted clay is frequently found in the joints on the external surface of the masonry to embed it within the surrounding rock.

The Treasury of Atreus provides a splendid illustration of the architectural arrangement: the dome is 14.50 m in diameter at its base. As can be seen in Fig. 5.26, which shows its last course, trapezoid shaped stone blocks are arranged in horizontal course. The dome structure is moreover solidified by smaller stones and

Fig. 5.24 Longitudinal section of a Mycenaean tholos

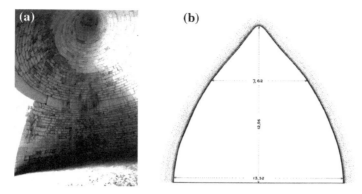

Fig. 5.25 a Tomb of Clytemnestra. The masonry; **b** Internal contour (from Mylonas 1966)

Fig. 5.26 Treasury of Atreus. Last stone course of the dome (from Blouet 1833)

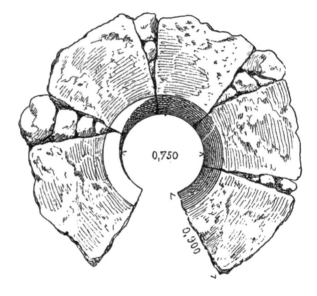

clay forcibly inserted between the blocks. Such details reveal the great care taken by the ancient builders to impart radial compression strength to the horizontal rings.

The masonry vault arrangement, with progressively inward projecting horizontal courses, is usually defined a *false dome*. By this definition, each one of the meridian slices into which the dome is ideally subdivided is statically independent from the others: only vertical actions are transmitted across them. Such behavior would be the same as in false arches, a prime example of which is the Lion Gate at Mycenae, shown in Fig. 5.3. However, thanks to the shrewd construction devices adopted by the ancient Mycenaean builders, tholoi exhibit the efficient static behavior of a true dome, as will be demonstrated in the next sections.

5.3.2 Statics of the Mycenaean Tholos

Let us consider, for instance, the tholos of the Treasury of Atreus, also known as the Tomb of Agamemnon. Recent research (Como 2006, 2007) has shown that the behavior of this typical tholos structure cannot be realistically described as that of a false dome (Cavanagh and Laxton 1981; Benvenuto and Corradi 1990). The following is based on the results of this research.

Figure 5.27 shows a meridian section of the tholos. Let us first consider the part of the slice emerging from the rock. The slice is formed by the superposition of concentric angular portions of courses progressively projecting inward. Each of the blocks making up the slice is loaded by its own weight and the weight of the overlying portion of the tumulus. According to the false dome model, each slice should be able to sustain itself, as it is statically independent of the others.

With reference to figure let us first consider the part of the slice emerging from the rock. The slice is formed by the superposition of concentric angular portions of courses progressively projecting inward. Each of the blocks making up the slice is loaded by its own weight and the weight of the overlying portion of the tumulus.

According to the false dome model, each slice should be able to sustain itself, as it is statically independent of the others. Let us look at a portion of the slice made up of all the rings overlying a generic horizontal plane a–a (Fig. 5.28). This portion can rotate around the toe O, representing the edge of the underlying course.

Each portion of the slice is subjected to both overturning and stabilizing couples, respectively represented by the moments of all the weights acting on the right and left of vertical line b–b. The portion of the slice will naturally be in equilibrium as long as the corresponding stabilizing moment is larger than the overturning moment. It has been shown that, due to the geometry of the rising rings, the slice is not at all self-sustaining (Como 2006, 2007).

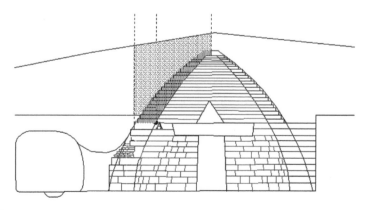

Fig. 5.27 A meridian slice loaded by the weights of the blocks and overlying tumulus

5.3 Mycenaean Tholos

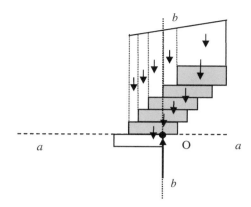

Fig. 5.28 Part of the slice made up of all the rings overlying horizontal plane a–a

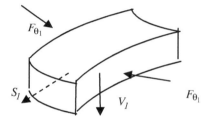

Fig. 5.29 Compressive forces $F_{\theta 1}$ acting on the *top* ring to produce force S_1

The slice is instead in equilibrium due to the horizontal resultant of the forces transmitted to the slice by the compressed rings: these compressive forces produce horizontal forces that stabilize it.

Figure 5.29 shows the top voussoir of the slice: the compressive forces $F_{\theta 1}$ acting on the sides of the voussoir produce the horizontal stabilizing force S_1. Not only vertical forces act on the slice, there are also the horizontal stabilizing forces transmitted by the compressed rings. The careful arrangement of the horizontal courses, with the insertion between the joints of small stones and clay acting as wedges locking the stone rings (Fig. 5.26), enable them to sustain radial compressions.

Near the base of the dome the situation is reversed. Whereas the slices tend to shrink and the upper rings are consequently compressed, the lower rings tend to widen and push against the rock, so as to activate the reaction of the rock backing and thereby ensure membrane equilibrium. The lower rings are thus unloaded (Fig. 5.30). Such a result is all the more remarkable in light of the fact that these stone constructions were built more than 3000 years ago, that is, about 1400 years before the Rome Pantheon.

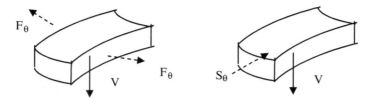

Fig. 5.30 Tensile forces in the lower rings, required to maintain the membrane stresses in the slice, are replaced by the reaction of the rock backing the chamber wall

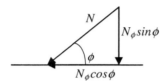

Fig. 5.31 Sliding force between the blocks

On the other hand, the horizontal arrangement of the courses represents the tholos' greatest weakness. The shear action of the horizontal component of forces N_ϕ can produce outward horizontal sliding of the courses, which is opposed by the friction between the blocks. If we denote ϕ as the angle between the meridian force N_φ and the horizontal, sliding does not occur if (Fig. 5.31)

$$N_\phi \cos \phi \leq f \cdot N_\phi \sin \phi \tag{5.70}$$

or

$$tg\, \phi \geq \frac{1}{f}. \tag{5.70'}$$

We can thus evaluate the values of the critical slope of the tholos contour as a function of the assigned values of the friction factor f. On average, we can assume $f = 1.125$ and obtain the $\phi \geq 41°\, 0.5$.

Figure 5.20b shows the upper portion of the inner, bottle-neck shaped contour of the Tomb of Clytemnestra: at the inflexion point, the slope ϕ is over 45°. This contour geometry is another ingenious device of the old Mycenaean builders. Humidity greatly reduces the friction between the courses and the tholos becomes more vulnerable. Only a radial arrangement of the blocks along a curved line following the meridian contour of the dome could avoid the risk of sliding. However, at the time such an arrangement was not yet technically feasible. Nonetheless, even today many *tholoi* are still in remarkably good static conditions.

5.4 Roman Concrete Vaults: Do They Push on Their Supports?

The ancient Romans built many different types of vaults, such as barrel, hemispherical, cross vaults, etc., both in Rome and throughout other regions of the Empire. Most of them were built of *opus caementicium* cast over wooden scaffolding or initial brick centering vaults (Adam 1984; Giuliani 1995). The use of this material conferred advantages in terms of both economy and speed of execution. On the other hand, it also suffered from the distinct disadvantage of producing considerably heavy vaults due to both the thickness of the masonry and the weight of the concrete, Their construction thus involved complex centering and the activation of large thrusts. Nowadays, even a naked-hands test of a piece of a Roman masonry immediately reveals the substantial effort required to separate the *caementa* from the mortar—so strong is the adherence between them. Such behavior is quite different from that of brick masonry, whose tensile strength is due solely to the relatively weak adhesion between brick and mortar, as discussed in Chap. 1. A natural question to pose is therefore: *do Roman vaults push on their supporting structures?*

In the presence of non-negligible tensile strength, a vault behaves as a solid, and the membrane solution can roughly represent the stress state in the vault under its own weight. In this case, hemispherical vaults do not push on their springing. However it is unlikely that the full internal cohesion of the material be reached by the time the centering was removed (Giovannoni 1925). Thus, the seeming solidity of vaulted concrete structures is often deceptive. The vault thrust will be activated upon stripping. Thrusts can also be activated by the cracking produced by an earthquake or differential foundation settlements. The slow penetration of water and humidity inside the cracked masonry mass gradually aggravates the situation. The presence of cracks on a vault's intrados is a certain sign of thrust activation. The presence of such cracking has been detected on the intrados of the ambulatory vaults of the Rome Colosseum and the dome of the Pantheon, the subject of the next section.

5.5 The Pantheon

5.5.1 Introductive Notes

The Pantheon, whose structure has, by good fortune, been preserved unaltered up to today, was built in a site sacred to the god Mars on the remains of two earlier temples. The first, built by Marcus Agrippa, was destroyed by a fire. The second, built by Domitian on the ruins of the earlier one, was shattered by lightning.

In the year 120 CE, Emperor Hadrian ordered another temple built on the same site—the structure we see today. By Hadrian's order the name of the first builder was inscribed on the bronze pediment to commemorate the first structure. Hadrian's rebuilding however modified the original orientation of the temple: its main axis is in fact along the east–west direction, towards the point where the planet Venus rises on April 1, a day sacred to Venus, the goddess whence *Gens Julia* (the house of Julius Caesar) descended (Lucchini 1996).

The building complex was originally composed of the forecourt, with its surrounding porticos, the *pronaos*, the *vestibule* and the *Rotunda*, the large domed circular chamber, all erected at the same time as a single unified whole or "*unicum*". The concept of the new temple was different from traditional ones because it presents an inner space accessible to worshippers. In earlier temples, people were instead forced to gather outside.

Stamps on the bricks have enable the temple to be dated precisely: it was built in 120–128 CE. Although not known with certainty, it was clearly a great artificer, perhaps Apollodorus of Damascus or the Emperor Hadrian himself, who designed and accomplished this extraordinary monument, considered one of Rome's most important buildings. The use of various materials and the adoption of differing methods of construction lend the building the character of a transition period work. Indeed, the temple's main component, the Rotunda with its cupola, is made of concrete, heralding future technical advances, while the trabeated pronaos, which nearly conceals the Rotunda completely, follows an architectural tradition that goes back to classical Greece (de Fine Licht 1968).

5.5.2 Structural Aspects of the Temple

The radius of the inner dome is precisely equal to its height, so it could hold a perfect hemisphere. The great dome, with an internal diameter of about 43.30 m and height of 21.65 m, rests on a masonry cylinder of the same height, so that the total interior height of the building is 43.30 m (Fig. 5.32).

These measurements have been a matter of some debate and will be discussed hereinafter. The massive circular wall, about 6.50 m thick, did not allow for windows to be opened, so a hole 9.0 m in diameter was left in the roof in the center of the dome, the *oculus*, which provides the only source of light.

The *pronaos*, larger than that of the Athens Parthenon, consists of an entablature and roof sustained by 16 monolithic corinthian columns, weighing about 50 t.

The columns' shafts are about 1.50 m in diameter and 12 m in height, equal to eight times its diameter, each with an *entasis*. The columns' total height, including the corinthian capital, reaches about 14 m. The corner columns, in conformity with the architectural principles laid down by Vitruvius, have shafts with slightly larger diameters than the others.

Of all the structural components of the temple, the Rotunda, is clearly the most prominent, important and extraordinary. The Rotunda is made up of the drum, in turn constituted by a cylindrical wall, and the dome. The drum wall is not a solid structure, but contains cavities and chambers and is open towards the interior with large niches and *exedrae*. Similar cavities are also presents at the base of the dome situated in correspondence to the exedrae and chambers of the drum.

5.5 The Pantheon

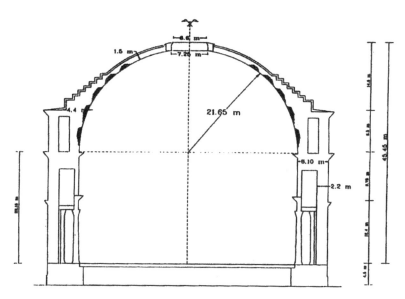

Fig. 5.32 Transverse section of the Rotunda (de Fine Licht 1968)

The interior of the dome presents the simple geometry of a hemisphere: we recognize the symbolic mark of Archimedes of a sphere inscribed in a cylinder.

The actual diameter of the drum, which is the same as the dome's at its springing, as before mentioned, has been a matter of some debate. The convexity of the floor, the internal niches, the columns ringing the inner edge of the wall, all combine to produce uncertainties in evaluating the true internal radius of the sphere. The measurements given in Fig. 5.27, from de Fine Licht, are not universally accepted. Another researcher, Wilson Jones (2000), identified the basic geometrical conception of the dome. The square inscribed at the base of the dome, along the axes of the internal columns, has sides whose lengths equal the width of the monument's front portico, measured between the axes of the outermost columns (Fig. 5.33). The diagonal of this square matches the diameter of the circle. The side length of the square measures 31.50 m and its diagonal $31.50 \times 1.4142 = 44.55$ m. The diameter of the sphere thus equals the length of this diagonal, that is, 44.55 m, i.e. 150 Roman feet, because $150 \times 0.297 = 0.44.55$. The length of 44.55 m may thus be considered an exact measure of the dome's diameter along the axes of the internal columns. The length of the internal diameter of about 43.30 m, given in Fig. 5.27 is thus only a bit shorter than the previous measure, taking into account the mean value of the internal columns diameter, about equal to 1.0 m.

The cylindrical drum continues for a distance of about 8.20 m above the dome's springer; the weight of this extra masonry serves to oppose the thrust of the dome. In this ingenious way, the builder was able to provide a counter-weight to strengthen the dome in its lowest portion.

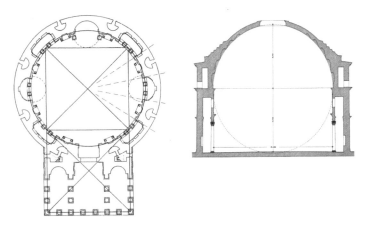

Fig. 5.33 Plan and section of the Pantheon with its geometrical modulus (Wilson Jones 2000)

The total height of the drum is 30.50 m. From the outside the dome seems to rise above the last cornice and looks like a shallow cupola with a saw-toothed profile created by seven steps rising from its base. The crown is topped by the oculus, the opening in the middle of the cupola, surrounded by a ring of bricks.

The dome's thickness falls from about 5.90 m at the base to nearly 1.50 m at the top, where the dome is topped with a compressed ring surrounding the oculus. The ring is about 1.60 m high and is made up of vertical tile-like bricks, called *bipedales* set radially in three circles one over the other. The extrados of the dome is covered by half-bricks, or *semilateres,* placed *in a herringbone* pattern and an upper impervious layer of *opus signinum* on which gilded bronze tiles were laid, though nowadays these have been substituted with a lead covering. A number of large relieving arches are incorporated into the wall masonry and can be seen in the brick façade as a frieze of alternating large and small arches (Fig. 5.34). The eight larger arches, arranged along the Rotunda's main and diagonal axes, lie above the vaults and the *exedrae* of the middle area. These arches are made up of two concentric courses of *bipedales* inserted about 2.0 m into the masonry to reinforce the concrete and direct stresses down towards the side piers. Studies carried out between 1829 and 1934 revealed that the dome did not present ribs, contradicting a famous etching by Piranesi. The dome intrados is divided into five orders of 28 hollow coffers set in the concrete. Apart from embellishing the vault, these also served to reduce its weight. The choice of 28 coffers was not by chance: Euclid considered it a *perfect* number, because it equals the sum of its divisors. It is moreover strictly linked to the lunar calendar. At the time, partitioning a circle into 28 equal parts was a complex geometry problem related to the problem of subdividing a circle into 7 parts, which had been studied by Archimedes (Martines 1989).

The wall of the Rotunda, sustaining the dome, stands on a massive foundation ring of concrete made of travertine fragments poured layer by layer into lime and *pozzolana* mortar, which over time has become hard as rock. The foundation ring,

5.5 The Pantheon

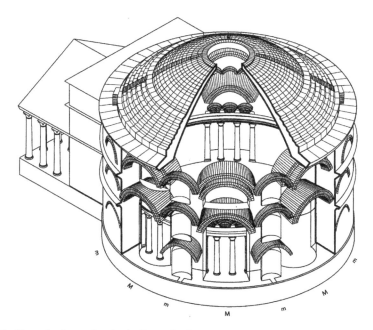

Fig. 5.34 The relieving arches in the Rotunda (Lancaster 2007)

similar to the Colosseum's, has a rectangular section 7.50 m wide and 4.50 m high and a cladding of *semilateres*, tightly bound to the concrete.

The cylindrical wall, from the foundation up to the first cornice is made of poured concrete and is about 12.50 m in height. Alternating layers of travertine and lumps of tuff were used as *caementa* in a lime and pozzolana mortar. This core is faced with about 60 cm thick bricks. The bricklaying shows fine horizontal joints about 1.5 cm wide and narrower vertical joints. Horizontal leveling courses of *bipedales* are distributed at intervals of about 1–1.5 m.

From the first cornice up to the dome springing, for a height of about 9.50 m, the masonry composition is similar to the lower part, except for the core, for which brick and tuff fragments were used as *caementa*. Further up, there is yet another distinct, 8 m-high zone that comprises the top part of the dome and is made of alternating layers of light tuff blocks (*tufo giallo*) and volcanic slag or scoria (Fig. 5.35). The arrangement and distribution of the cladding materials along the dome's height reflects a clear design principle: the heavier brick and travertine were used for the lower band of the drum, then the travertine was replaced by the lighter tuff. The same sort of materials arrangement is found within the dome. Bricks and then tuff are placed at the dome springer, then filling layers of decreasing weight follow in sequence: *capellaccio, tufo giallo*, pumice and volcanic slag (Lugli 1938).

During restorations performed in the years 1881–1882, significant cracks were discovered throughout the Rotunda's facade. Then in 1936, in-depth analyzes detailed the cracking patterns in the Pantheon's various types of masonry.

Fig. 5.35 The various materials of the dome masonry (Lancaster 2007)

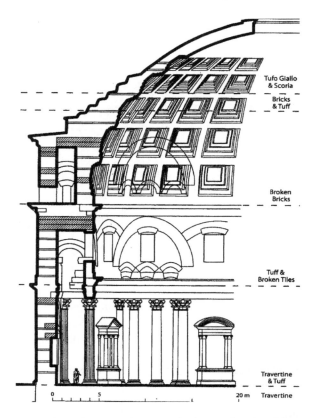

Over the centuries many attempts have been made to repair the cracks, the oldest of which probably appeared just after the temple's completion. In fact, a number of age-old foundation reinforcement operations have been clearly documented. Indeed, the inscription "*vetustate corruptum*" over the architrave of the *pronaos* testifies to one of these early reinforcement operations carried out in about 202. The

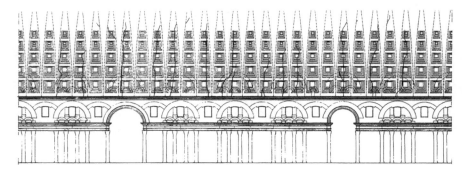

Fig. 5.36 Cracks detected by Terenzio on the dome intrados (de Fine Licht 1968)

5.5 The Pantheon

walls and the arches were also repaired using bricks stamped with dates corresponding to the rule of Septimius Severus (193–211). Figure 5.36 provides an illustration of the meridian cracks around the intrados of the entire dome as documented by Terenzio in 1934.

5.5.3 Thrust of the Dome

Let us now look at the thrust of the dome under the usual assumption of no-tensile strength concrete. The above mentioned cracking pattern attests to the activation of thrusts within the dome. We can also assume that slight widening of the drum at the dome springing has occurred, so as to activate the minimum thrust. Within this basic framework, we will follow the approach of Ferri and Pecci (1996–1997) to evaluate the dome thrust by taking a slice of 45° magnitude.

The slice is subdivided into 38 ashlars and the corresponding weights and centers evaluated, as indicated in the Table 5.1 and in the Fig. 5.37. The sum total of all weights is 2377.73 t, so the weight of the whole dome is about $W = 2377.73 \times 8 = 19.022$ t. The weight of the slice is concentrated particularly on the springing, thereby considerably reducing the thrust. The static approach is now applied to search for the minimum thrust. Figures 5.38 and 5.39 describe the pressure lines corresponding to various locations of the points of the curves at the

Table 5.1 .

Voussoirs	Weights (t)	Voussoirs	Weights (t)
1	13.09	20	39.50
2	11.15	21	41.80
3	12.60	22	47.70
4	14.66	23	74.20
5	16.83	24	82.10
6	19.17	25	91.40
7	21.65	26	116.10
8	21.05	27	116.10
9	18.12	28	116.20
10	28.94	29	105.70
11	27.39	30	103.20
12	23.18	31	127.87
13	28.10	32	127.87
14	35.00	33	80.80
15	27.00	34	80.80
16	30.80	35	143.58
17	26.00	36	143.58
18	34.00	37	143.58
19	45.40	38	143.58

220　　　　　　　　　　　　　　　　　　　　　　　　　　5　Masonry Vaults: Domes

Fig. 5.37 Weight distribution of various ashlars in the slice

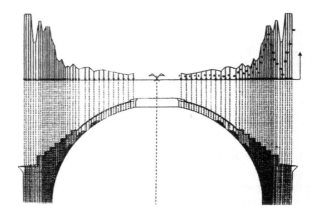

Fig. 5.38 Pressure curves C1, C2, C3

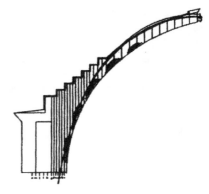

Fig. 5.39 Pressure curve D4, D5, D6

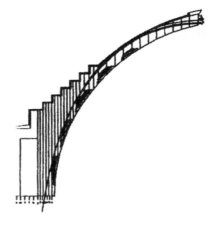

Fig. 5.40 Successive resultants of the forces acting on a slice of Rotunda

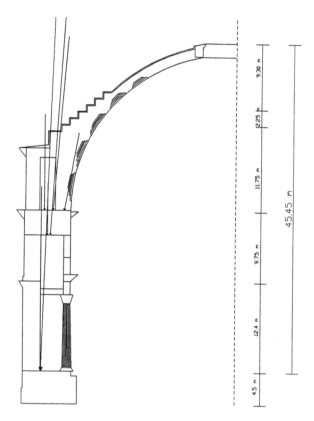

springing and key of the slice. Only below the springing do the pressure lines intersect the weight forces of the more external ashlars. In these sections the weight of the high cylindrical wall above the springing further deviates the pressure line. The minimum thrust, S_{min}, of all statically admissible ones, corresponds to the curve C_3 passing through the point located at the connection of the dome extrados with the upper central ring around the *oculus*. Curve C_3 is thus tangent to both the intrados, at the haunches, and the extrados, at the connection point of the dome to the upper ring. Such a pressure line allows a mechanism displacement to develop and produce a slight widening of the dome at the springing, due to the deformation of the drum. The estimated minimum thrust value of the splice considered, which is 1/8 of the entire dome, turns out to be 186.22 t. This value has also been checked via the kinematic approach, as it can also be determined as the maximum among all the kinematic values.

The drum's average radius, along half its thickness, is about $R_m = 22.27 + 3.25 = 25.52$ m and the thrust S_o per unit length of the average circumference at the dome springing equals $186.22 \times 8/(\pi \cdot 2 \cdot 25.52) = 9.29$ t/ml.

The thrust is about three times lower than the thrust of St. Peter's dome, which as will be shown later, is about to 30 t/ml. The ratio χ between the thrust and the weight of a single slice represents the *efficiency ratio* of a dome. In the case of the Pantheon we thus have

$$\chi = \frac{186.22}{2377.73} = 0.078$$

The value of this ratio, particularly low in comparison to other important domes, is indicative of the static efficiency of the Rotunda. This resulting value is a consequence of both the design geometry and the materials distribution of the dome. In this respect, the most significant of the geometrical features are the extension of the drum beyond the level of the dome springing, the gradual reduction in thickness of the vault towards the center and the widespread presence of coffers.

The extension of the drum beyond the level of the dome springing strongly deviates the thrust of the dome towards the vertical. The weight of the high drum reduces the eccentricity of the resultant of the forces acting at the base of the foundation structures (Fig. 5.40).

5.6 Brunelleschi's Dome in Florence

5.6.1 A Brief Account of the Cathedral's Construction

The dome of Santa Maria del Fiore in Florence is one of the greatest achievements of early Renaissance architecture. Construction of the cathedral was begun by Arnolfo di Cambio in 1302. Upon Arnolfo's death, the work stopped and recommenced only in 1331. Then in 1334 Giotto, the new master builder, was commissioned to build the bell tower.

Work on the church continued under various masters, including Andrea Pisano (1336–1349), Francesco Talenti, Giovanni di Lapo di Ghini and finally Neri di Fioravanti, whose design for the dome was chosen in a 1366 referendum. However, by the beginning of the 14th century, work on building the dome had not yet started and only on August 19th, 1418 was a special contest announced:

> Whosoever wishes to make some model or design for the vaulting of the main Dome of the Cathedral under construction by the Opera del Duomo—by armature, scaffolding, or any other means, or any lifting device pertaining to the construction and perfection of said cupola or vault—shall do so before the end of the month of September. If the model be used he shall be entitled to a payment of 200 gold Florins (modified from King 2000).

The contest specified that the dome, to be built on the already finished octagonal drum, had to have an external diameter of about 54 m and a height of more than 105 m from ground level. Two hundred florins was a good deal of money and so the competition attracted the attention of carpenters and masons from all across

Tuscany. In the final stages of the contest, the projects of both Ghiberti and Brunelleschi were chosen for consideration, though in the end, it was Brunelleschi's unprecedented plan, which called for building the dome without the use of centering, won. Construction of the dome was begun in 1420 under the direction of Brunelleschi himself and was concluded in 1434. The lantern was then added in 1461, after Brunelleschi's death in 1446.

5.6.2 The Supporting Pillars

The dome is sustained by four large pillars: two apsidal and two flanking the nave. Their heads reach the height of 28.00 m from floor level, assumed to be at 0.00 m. The two apsidal pillars, from the floor up to 20 m height, have hollow cross sections in the shape of an isosceles trapezium with side lengths of 10 and 5.0 m (Figs. 5.41, 5.42 and 5.43).

A schematic drawing of their cross section is shown in Fig. 5.44. The cross sections of the other two pillars flanking the nave are divided into two parts to create a wide passage between the aisles and transept. Beyond 20 m height, the sections of all four pillars become solid and are firmly connected to the drum. The masonry of the pillars is made up of thick stone facings tightly bound to an internal rubble and mortar core. Their foundations rest on solid gravel banks.

5.6.3 The Drum

The drum is octagonal in shape, resembling a crown. Its eight sides are supported by the four pillars with four pointed arches spanning them.

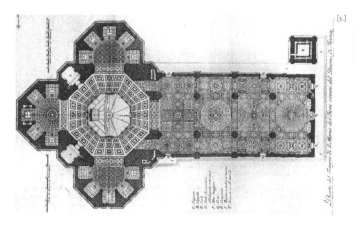

Fig. 5.41 Plan of the cathedral of Santa Maria del Fiore (Sgrilli 1733)

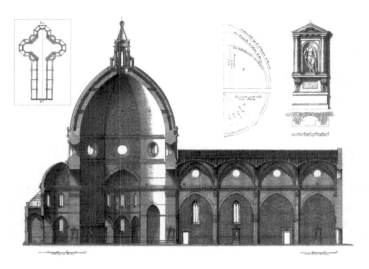

Fig. 5.42 Longitudinal section (Sgrilli 1733)

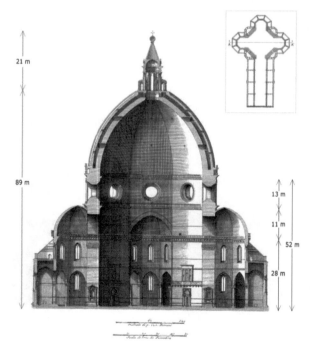

Fig. 5.43 Transverse section (Sgrilli 1733)

5.6 Brunelleschi's Dome in Florence

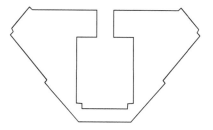

Fig. 5.44 A schematic drawing of the apsidal pier section

The drum is divided into an upper part, from 39 m height up to the springing of the dome at 52 m, and a lower mixtilinear portion where it joins the piers (Fig. 5.45). The following are the basic average measurements of the drum (Fig. 5.46):

(1) distance D between two opposite corners of the inner octagon, which is 77 *arms*, as indicated in a parchment by Gherardo di Riccardo di Prato (1 *arm* = 0.584 m, so $D = 77 \times 0.584 = 44.97$ m).
(2) drum thickness s_d = ZH along the corner direction, equal to 4.60 m.

From these figures, we obtain (Fig. 5.40):

Fig. 5.45 The drum

Fig. 5.46 Horizontal section of the drum

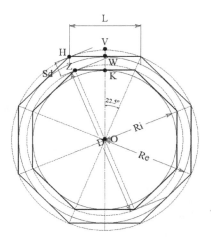

- R_i = OK = radius of the *internal circle*, tangent to the internal sides of the octagon = OZ · cos 22° 0.5 = D/2 · cos 22° 0.5 = 20.77 m
- R_e = OV = length of the radius of the *external circle*, in which the drum external boundary is inscribed = D/2 + s_d = 44.97/2 + 4.60 = 27.08 m
- L = 2 · HW = length of the *side of the external octagon* = 2 · OH · sin 22° 0.5 = 2 · OV · sin 22° 0.5 = 20.73 m
- L_{min} = internal length of the side of the octagon = 20.73 − 9.20 sin 22° 0.5 = 17.21 m
- L_{min} = average length of the side of the octagon = 0.5(20.73 + 17.21) = 18.97 m
- a_e = length of the *apothem of the external octagon* = OW = OH · cos 22° 0.5 = 25.02 m
- s' = thickness of *the crown* in which the octagonal drum is inscribed = KV = OV − OK = 27.08 − 20.77 = 6.31 m
- s'' = thickness of *the crown* inscribed *inside the drum* = OW − OZ = 25.02 − 44.97/2 = 2.53 m
- s'' = thickness of *the dome* at the midpoint of the drum side:
- KW = ZH · cos 22° 0.5 = 4.60 · cos 22° 0.5 = 4.25 m

The drum is lightened by eight *oculi* crossing the sides of the octagonal crown. The average diameter of the *oculus* measures about 5.80 m.

5.6.4 The Dome

Unlike most cupolas, which like the Pantheon are hemispherical in profile, the vault of Santa Maria del Fiore is a pointed dome (Fig. 5.47). The dome intrados, sectioned by vertical planes passing through opposite corners, is a circular arch. Each of these eight arches is called a "*pointed fifth*" and inscribes an angle of about 60°

Fig. 5.47 The dome of Santa Maria del Fiore (model at the "Opera del Duomo", Gizdulich (from Ippolito and Peroni 1997)

with the center of the circle. The radius of a *pointed fifth* is equal to 4/5 of the distance spanning the internal corners of the octagonal annulus, i.e. it equals $4/5 \cdot 44.97$ m = $4/5 \cdot 77$ *arms* (1 *arm* = 0.584 m) (Fig. 5.48). These eight arches, converging upwards in the center, form the groin ribs of the dome's intrados.

The dome is an octagonal cloister-vault made up of four interpenetrating barrel vaults. There are eight webs and their surface is produced by horizontal straight lines extending from the octagon sides. Each of the eight webs spans two adjacent *corners, or groin ribs*, the "*speroni d'angolo*" (Fig. 5.49). Each web is composed of two shells stiffened by two *median ribs*, called "*speroni mediani*", connected to the groin ribs by nine horizontal arches. This stiffening system connects the external and the internal shells tightly together, so that the composite vault behaves like a single solid dome.

As will be shown, the internal masonry pattern differs from the external shape of the dome and, surprisingly, presents the structure of a so-called dome of rotation. The dome can be divided into three sectors delimited by four galleries ("camminamenti" in Fig. 5.50), running horizontally into the masonry to provide access to interior walkways (Fig. 5.50).

Fig. 5.48 Geometry of the pointed fifth (From parchment of Giovanni di Gherardo da Prato 1426; da Ippolito Peroni 1997)

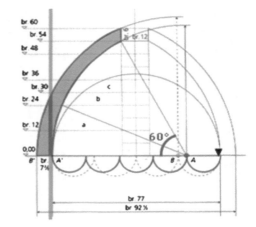

Fig. 5.49 A web with meridian groin ribs, two median ribs and nine horizontal arches connecting the median to the corner ribs

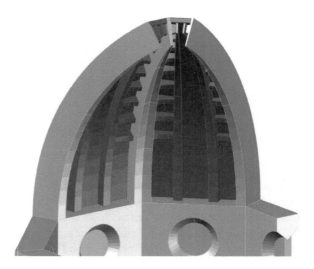

The first gallery is located at a height of 3.50 m above the level of the third ledge ("ballatoio" in Fig. 5.50), which, by convention, is assumed to be at height 0.00 m. The first, 3.50 m section rises from the springing just up to the height of the first gallery and is built of solid masonry. In this sector the bending of the dome is visible only on the intrados while from the exterior the extrados is seen to rise vertically. In this first sector the masonry, visible from the internal passages, is composed of blocks and mortar beds and has the same pattern as the underlying drum, with lateral joints oriented orthogonally to the octagon sides. The first so-called *catena di macigno,* or chain of stones) is located in this section.

Above the first sector, the dome divides into two shells connected by ribs. The masonry pattern in these sections reveals the ingenious strategies deviced by

5.6 Brunelleschi's Dome in Florence

Fig. 5.50 Transverse section of the dome with the two shells (Ippolito and Peroni 1997)

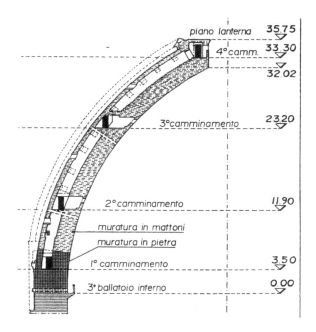

Fig. 5.51 The herringbone

Fig. 5.52 Herringbone pattern with converging vertical bricks shells (Ippolito and Peroni 1997)

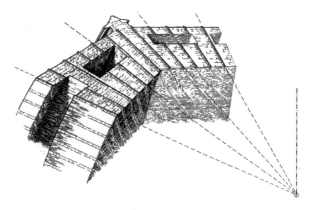

Brunelleschi: the *spinapesce* (*herringbone*) and the *corda blanda* (slack-line), the *connection horizontal arches*, as shown in Figs. 5.51, 5.52 and 5.56.

5.6.4.1 Constructional Devices of Brunelleschi

The Herringbone

With the curving of the dome, the flat bricks gradually begin to take on an inclined position in the meridian planes. Long bricks, placed vertically and converging in the center, wedge the increasingly inclined flat bricks in between, preventing them from sliding off while the mortar is still wet. As the masonry was raised, the vertical bricks had to be put in place one adjacent to the other. The diagonal and jagged lines of the vertical bricks form the so-called *herringbone pattern*.

The corda blanda

Brunelleschi's second device concerns the positioning of the bricks rows around the dome: they do not follow the rectilinear alignment of the octagon sides and so do not intersect at the groin ribs, but are instead positioned on conical beds along continuous curves with circular projections. These curves, called *slack lines*, are intersections of the dome with ideal cones with downward directed vertices. Each of these ideal cones shares a common vertical axis, which coincides with the dome's central vertical axis (Figs. 5.53 and 5.54).

A cone-shaped slack line is produced where each ideal cone intersects the octagonal shape of the rising vault. The presence of these slack lines was predicted by Mainstone (1977). Di Pasquale discovered these curved brick beds in 1977, through measurements made over external areas of the outer dome stripped of its tiles (Fig. 5.55). Through this strategy Brunelleschi avoided discontinuities in the brick rows at the corners It seems natural to assume that Brunelleschi was seeking to build a rotational dome, despite the octagonal external form.

5.6 Brunelleschi's Dome in Florence

Fig. 5.53 Intersections of the webs with the cones' surfaces

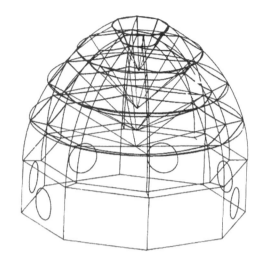

Fig. 5.54 Rows of bricks in the *slack-line* arrangement

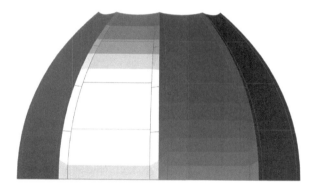

Fig. 5.55 Discovery of the conical brick beds beneath the horizontal rows of tiles (Di Pasquale 2002)

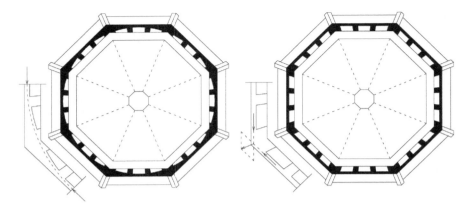

Fig. 5.56 The *left figure* shows the horizontal section of the outer shell together with the horizontal arches connecting the corner ribs with the adjacent median ribs. The *right figure* shows the same horizontal section but without the horizontal arches

The connection horizontal arches

The third constructional device conceived by Brunelleschi concerns the construction of the horizontal arches connecting each corner rib with the two adjacent median ribs, as shown in Fig. 5.56. This device was conceived directly by Brunelleschi in the year 1326 during the construction of the dome, with a change of the program of works (Mainstone 1977; Como 2012).

The construction of the dome went on with continuity circle by circle creating a kind of circular skeleton over which the external octagonal structure of the dome took shape. The purpose of Brunelleschi was to build the dome so that it contained within its thickness of its two shells continuous circular rings. This was possible without difficulty in the inner shell, whose thickness was included between 7 and 5 feet. The outer shell was, on the contrary, thinner: its thickness varied from 2 feet at its base narrowing to just 1 foot at the summit. It was not possible a circular dome within its thickness. Brunelleschi bypassed this difficulty. Connection arches were built on the inside of the dome's outer shell as shown in Fig. 5.56 the horizontal arches connecting the corner ribs with the two adjacents median ribs ensued the existence of continuous circular rings also in the outer shell. Nine horizontal arches were built at 8 foot interval. These arches served a vital function in building the cupola because, creating continuous masonry horizontal rings, prevented the inwards failure during the construction when the shell curving inwards, had passed a critical angle of overturning of webs, as it will shown in the next section.

5.6.4.2 Equilibrium of the Dome During Construction

The dome was built without centering, circular layer by circular layer, so its internal stress state changed continuously during construction.

Fig. 5.57 Equilibrium of a slice at critical height (Scarano 2007–2008)

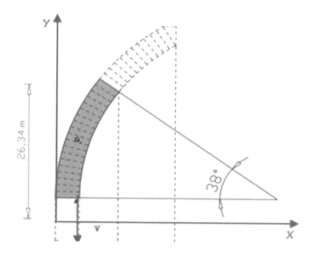

The segment of each web is self-supporting, as long as their height sufficiently small, when the resultant of the weights falls within the footprint of the base of the web. As construction proceeds, this resultant will become ever more eccentric, up to the point that it reaches the boundary of the base section, when the web would tend to tip over (Fig. 5.57). Upon further load increments, the rings, and particularly the top ring of the webs, will become compressed and will produce a balancing force, F, over each web or dome slice (Fig. 5.58). The overturning of the built webs tends to occur when the height of these webs reaches the height of about 26 m above the dome springing. At this height the rings will be thus fully engaged securing the equilibrium of the dome. The construction of the dome circle after circle, both in the inner and in the outer shell, positioning the brick rows along continuous curves on conical beds, allowed the engagement of these rings. This choice to build the cupola probably was suggested to Brunelleschi during a stay in Rome observing the Pantheon, with its open *oculus*.

Stability of the single outer shell during the construction has been studied by Mainstone (1977) and then by Como (2012). These studies showed the importance

Fig. 5.58 Equilibrium of the slice beyond the critical height

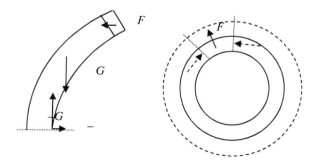

of the horizontal rings connecting the median to the corner ribs able to ensure the presence of a resistant ring also inside the thin outer shell.

5.6.5 Crack Patterns in the Dome

Documentation of the first cracks detected on the dome intrados date back to the year 1639, when meridian cracks on the intrados of the internal shell were first noticed. They crossed the *fresco* by Vasari and followed a jagged line along the herringbone brick pattern (Figs. 5.59, 5.60 and 5.61).

The largest cracks on the dome intrados were surveyed by Blasi and Ceccotti in 1984 (Fig. 5.62). Cracks have developed on the intrados along the meridians in the middle of the webs over the four cathedral pillars. No cracking has instead been detected along the groin ribs. The largest cracks, averaging 5–6 cm in width, run along the inner shell on the north- and south-eastern webs.

From the crown down to a co-latitude angle of about 25°, the dome webs over the pillars are intact, but are separated thereafter into slices by meridian cracks.

Long meridian cracks are also present in the drum, as detected since the end of the 17th century by G. Nelli, (see Fig. 13, p. 222, in the book edited by Giandebiaggi et al. 2009).

The dome thrusts out at its supports, which have given way slightly to produce a small increase in the span. The dome has thus undergone further deformation in order to adapt itself to the settling of its supports.

Hipped vaults, and more generally segmental vaults, under dead loads concentrate strong hoop tensile stresses at their angles, and cracks develop along the edges. Figure 7.574.4.65 in the next sections illustrates the typical cracking pattern of such vaults. No edge cracks are instead present on Brunelleschi's dome.

Fig. 5.59 Cracks running across the fresco of Vasari (Ottoni 2012)

Fig. 5.60 Old survey of a meridian fracture detected by *Giovanni Nelli, 13 giugno 1690* (Ottoni 2012)

Fig. 5.61 Draft of Vincenzo Viviani of two cracked webs (Ottoni 2012)

The crack pattern thus shows the presence of a stress state typical of a dome of rotation. It can thus be concluded that, in spite of its external geometry, this dome behaves like a dome of rotation by virtue of its smooth masonry structure. Similarly, some simple considerations can explain why cracks are not present on the webs sustained by the arches spanning the pillars.

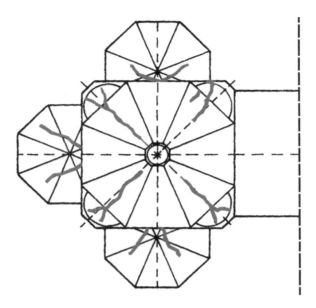

Fig. 5.62 The crack pattern according to Blasi and Ceccotti (1984)

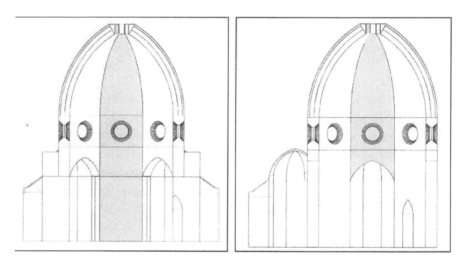

Fig. 5.63 Different behavior of webs sustained by pillars or by arches

Figure 5.63 shows the different positions of the webs: some are located above a pillar, others above the pointed arch. In the later case the arching effect, required to span the underlying opening, produces compression in the web, which opposes the tensile hoop stresses.

5.6.6 Thrust of the Dome

The occurrence of thrust is due to the presence of meridian cracks in the webs. The thrust (Fig. 5.64) distorts the initial membrane stress flow in the structure and produces eccentricities in the axial load on the piers.

At the same time, the meridian dome sections vary with the position of the vertical sectioning planes and the thrust, corresponding to the various slices, changes with the meridian section.

The thrust per unit length of the average circle at the dome springing will thus exhibit the undulating pattern sketched out in Fig. 5.65, where the middle sections of the webs correspond to the dotted lines. The two shells forming the dome are strictly connected by the ribs and horizontal arches, so the vault can be accurately described as behaving like a single solid structure despite its complexity.

A Valuation of the mean thrust in the dome

The small deformation of the drum produced by the thrusting action yields a small increasing in the dome span. The webs follow this deformation and, as described earlier, develop thrust that is the minimum of all statically admissible ones.

In this framework, we can evaluate the thrust by considering a dome slice *equal to 1/8 of the entire dome,* corresponding to a web supported directly by an underlying pier. The slice is sectioned into 31 voussoirs: the first corresponds to the voussoir at the crown.

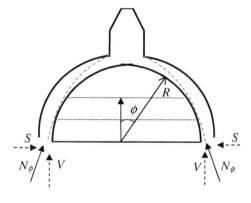

Fig. 5.64 Thrust due to the presence of meridian cracking

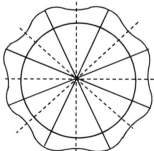

Fig. 5.65 Variation of the thrust along the sections: *dotted lines* indicate the web centres

Table 5.2 Dead loads

Voussoir	Vol. m^3	Weight (t)	X$_G$ (m)	Y$_G$ (m)
1	16.64	30.79	357.41	35.51
2	21.57	39.84	356.30	35.47
3	24.72	45.74	355..21	34.59
4	28.67	53.03	354.08	33.68
5	37.13	68.68	352.94	32.86
6	37.22	68.86	351.91	31.81
7	44.46	82.26	350.81	30.85
8	47.21	87.34	349.87	29.85
9	48.79	90.26	348.92	28.76
10	58.48	108.19	347.92	27.76
11	55.60	102.85	347.11	26.57
12	61.66	114.07	346.21	25.45
13	65.66	121.47	345.42	24.34
14	64.48	119.28	344.69	23.10
15	74.32	137.49	343.87	21.93
16	69.88	129.09	343.28	20.67
17	72.52	134.15	342.63	19.42
18	81.11	150.05	341.97	18.22
19	75.78	140.20	341.49	16.88
20	82.51	152.64	340.90	15.58
21	82.51	152.64	340.46	14.31
22	80.52	148.95	340.07	12.95
23	90.51	166.66	339.58	11.63
24	82.92	153.41	339.34	10.27
25	83.90	155.21	339.04	8.92
26	84.74	156.77	338.78	7.56
27	85.45	158.08	338.57	6.19
28	86.02	159.14	338.39	4.82
29	86.46	159.94	338.27	3.45
30	86.76	160.51	338.18	2.07
31	86.93	160.82	338.14	0.69

Table 5.2, lists the volume, weight, and position of the center of the various voussoirs. The average unit weight of the dome masonry has been assumed to be $\gamma = 1.85$ t/m^3. The distance y measures the height of the voussoir centre with the height 0.00 of the dome springing. The distance x gives the horizontal distance of the voussoir centre respect to a generic vertical axis. Also the share of the weight of the lanten, equal to 750/8 t, has been taken into account, as shown in Fig. 5.57. The construction of the pressure curve has been continued also inside the upper part of the drum as far as to the height of 39.0 m, taking into account of the corresponding weights. The thrust estimate has been performed by Scarano (2007–2008) (Fig. 5.66).

5.6 Brunelleschi's Dome in Florence

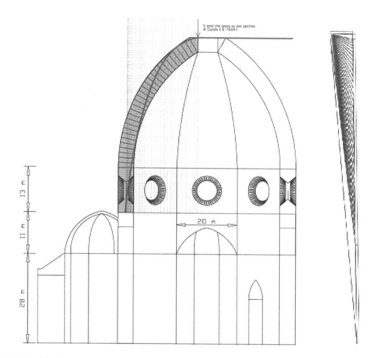

Fig. 5.66 Graphical valuation of the thrust of the dome slice (1/8 of the dome)

The pressure line corresponding to the minimum thrust state is plotted in Fig. 5.66. The curve passes through the extrados of the slice at the crown and is tangent to the intrados of the web near the back of the web. The pressure line intercepts the center of the internal side of the section of the pier head. The vertical load transmitted by the single slice to the pier head is about 3,800 t, including the share of the weight of the lantern, equal to 750/8. The total weight of the dome is $G_{cup} = 3,800 \times 8 = 30,400$ t, quite close to the value reported by others. Performing the necessary calculations gives us a horizontal thrust of 400 t transmitted by the slice (Fig. 5.57). Now, assuming a circumference at the dome springing equal to $\pi \cdot 0.5 \times (27.08 \times 2 + 20.77 \times 2) = 150.32$ m, averaging the lengths of the crown boundaries in which the drum is inscribed, the length of the base slice is $150.32/8 = 18.79$ m and the unit thrust per single meter of dome at its springing is $400/18.79 = 21.3$ t/ml.

5.6.7 Loads Transmitted to the Pillars

Vertical loads

- At the height of 52.00 m.

Fig. 5.67 A web overlying a pillar and two half-webs above the adjacent arches

The total vertical load V_{52} conveyed from the single dome slice is due to the weight of the web directly above the pillar and 1/8 the weight of the lantern. Thus we have $V_{52} = 3,800\ t$.
- At the height of 39.00 m.
- The weight of the corresponding drum upper band is added. At the same time the presence of the oculi within the drum deviates the force transmission and conveys the weight of the two halves of the adjacent webs to the pillar (Fig. 5.67).
- Evaluation of the weight G_{dup} of the upper band of the drum.
  The average lengths of the side of the octagonal crown and of the *oculus* respectively measure 19.0 m and 2.90 m. Thus,
- $G_{dup} = (19.0 \cdot 4.25 \cdot 13.0 - \pi \cdot 2.90^2 \cdot 4.25) \cdot 2.2 = 2062.4\,\text{t}$.
- Summing up the weight of the two halves of webs, equal to 3800 t, the overall weight conveyed at this height to the pillar is (Fig. 5.58):
- $V_{39} = 3,800 + 3800 + 2062.4 = 9662.4\,\text{t}$.

At the height of 28.00 m.
At this height, both the weight G_{dlow} of the portion of the lower band of the drum directly overlying the pillar and the vertical force transmitted by the adjacent pointed arches have to be added.
We now have $G_{dlow} = 19.0 \cdot 4.25 \cdot 11.0 \cdot 2.2 = 1954.15\,\text{t}$.

To evaluate the vertical load conveyed to the pillar by the two adjacent arches, we must first identify the load distribution acting on these arches. At this height these arches transmit to the pillar the shares of the weights of the drum upper band, together with their weights and that of the overlying masonry.

Fig. 5.68 Pointed arch and the above masonry subdivided into voussoirs

Figure 5.68 shows the pointed arch with the above masonry subdivided into voussoirs. Each half-arch is subdivided into 7 voussoirs and the overlying masonry into 6 voussoirs. Figure 5.68 also shows the three forces V_1, V_2 and V_3 representing the weights of the various masses of the parts of the upper band of the crown directly sustained by the arch. The unit weight of the masonry of the arch and drum has been assumed equal to 2.2 t/m^3. The thickness of the drum is 4.25 m.

We have, with some geometrical simplifying assumptions (Fig. 5.68):

$$V_1 = 13.0 \cdot 6.60 \cdot 4.25 \cdot 2.2 = 802.23.7 \, \text{t}$$
$$V_2 = 0.5 \cdot 2.90 \cdot 2.90 \cdot 4.25 \cdot 2.2 = 39.3 \, \text{t}$$
$$V_3 = 2.90 \cdot 2.60 \cdot 4.25 \cdot 2.2 = 70.5 \, \text{t}$$

where $\Sigma \, V_i = 912.04$ t.

Further, the distances of the forces V_i from the mid section of the pointed arch are

$$d_1 = 9.50 - 0.5 \cdot 6.60 = 6.20 \, \text{m}$$
$$d_2 = 2/3 \cdot 2.90 = 1.93 \, \text{m}$$
$$d_3 = 2.90/2 = 1.45 \, \text{m}$$

Table 5.3 reports the volumes, weights and center coordinates of the various voussoirs. The abscissas X give the distances of the centers from the arch key and the ordinates Y the vertical distances from a horizontal reference plane.

Table 5.3 Weights

Voussoir	Vol. m³	Weight (t)	X_G (m)	Y_G (m)
Upper masonry				
A	45.28	99.6	8.10	−268.16
B	35.81	78.8	6.65	−267.42
C	28.56	62.8	5.20	−266.86
D	22.94	50.5	4.75	−266.43
E	18.66	41.0	2.30	−266.09
F	15.51	34.1	0.80	−265.85
Half arch				
1	15.28	33.6	8.10	−277.57
2	17.46	38.4	7.15	−275.67
3	15.05	33.1	5.80	−274.37
4	13.62	30.0	5.15	−273.37
5	12.71	28.0	4.80	−272.59
6	12.11	26.6	2.60	−271.00
7	11.73	25.8	0.90	−271.54

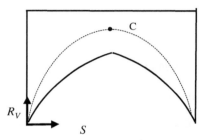

Fig. 5.69 The loads transmitted by the arch

Summing up all the vertical loads, we obtain $R_V = 1494.34$ t (Fig. 5.69).
Thus, the overall vertical load conveyed to the pillar at the height of 28.0 m is:

$$V_{28} = 9662.7 + 1954.15 + 1954.15 + 1494.34 \cdot 2 = 14052.24 \text{ t}$$

The weight of drum has usually been roughly estimated as equal to about 25,000 t. On the other hand, summing all the weights of the various parts of the drum, we get

$$G_{drum} = \left(G_{dup} + G_{dlow} + 2R_V\right) \cdot 4 = 4 \cdot (1954.15 + 2062.4 + 2 \cdot 1494.34)$$
$$= 28020.9 \text{ t}$$

Horizontal forces

The total thrust acting on the single pillar is obtained by summing (Fig. 5.70):

- the thrust S transmitted at the height of 39.00 m by the web directly above the pillar: $S = 400$ t.

5.6 Brunelleschi's Dome in Florence

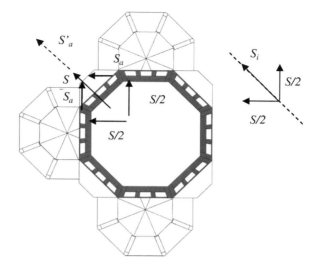

Fig. 5.70 Composition of thrusts acting on the single pillar

- the sum S_l of the thrusts acting at the height of 39.00 m on the two adjacent half-webs. The contribution S_l, is obtained as the vector sum of the corresponding radial half-thrusts, whence we obtain

$$S_l = S/\sqrt{2} = 283 \text{ t}$$

- the vector S'_a of the thrusts S_a transmitted at the height of 28.00 m to the pillar by the two adjacent arches. In order to obtain this last term, S'_a, we estimate the thrust S_a.

To evaluate the thrust of the pointed arch we assume that also the masonry above the arch also contributes to sustaining the loads. A pressure line thus develops inside the masonry, as sketched out in Fig. 5.69. The value of the thrust S_a can be thus obtained. We get

$$\begin{aligned} M_C &= -(99.6 \cdot 8.10 + 78.8 \cdot 6.65 + 62.8 \cdot 5.20 + 50.5 \cdot 4.75 + 41 \cdot 2.30 + 34.1 \cdot 0.80) \\ &\quad \times (33.6 \cdot 8.10 + 38.4 \cdot 7.15 + 33.1 \cdot 5.80 + 30.0 \cdot 5.15 + 28.0 \cdot 4.80 + 26.6 \cdot 2.60 \\ &\quad + 25.8 \cdot 0.90) - 802.0 \cdot 6.20 - 39.31 \cdot 1.93 - 70.5 \cdot 1.45 + 1494.34 \cdot 9.50 \\ &= S_a \cdot f \end{aligned}$$

Assuming $f = 10$ m, (Fig. 5.63) we obtain $S_a = 581.3$.

In conclusion at the height of 28.0 m the two adjacent pointed arches transmit to the pillar the thrust

$$S'_a = \sqrt{2} S_a = 822.04 \text{ t}.$$

5.6.8 Stresses at the Base of the Pillar

At the pillar head (h = 28.0).

$$N_{28} = 14052.24\,t$$

The load N_{28} is directed along the vertical axis passing through the center of the trapezium section of a single side of the octagonal crown. This section will sustain the load N_{28}, irrespective of the dimensions of the head pier section. The area of this section equals about 17.25 · 4.25 = 79.3 m² and the compression stress is 14,052.2/79.3 = 17.7 kg/cm².

At the pillar base

An axial load $N = 14{,}052.2$ t is transmitted to the pier head by the drum.

Pillar weight

The pillar cross section is hollow in the center up to the height of 20.0 m, while it is solid from the height of 20.00 m up to 28.00 m. The masonry structure of the pillar is made up of about 2.0 m thick stone facings, tightly bound to a rubble and mortar core.

Calculating the pillar weight from height 0.00 m up to 20.00 m:

Area of the pier section: $A = 226.5$ m²
Weight of this length of pier: 226.5 × 20.00 × 2.2 = 9966 t

From the height of 20.00 m up to 28.00 m, the pillar cross section A' is solid. So we have:

Area of the pier cross section: A' = 226.5 m² + 90 = 316 m²
Weight of the length of pier: 316 × 8.00 × 2.2 = 5562 t
Total pillar weight: $G_{pil\ l} = 9966\,t + 5562\,t = 15{,}528\,t$

The area of the base section with the distances of the center from its external sides (Fig. 5.71) are

Area: $A = 2{,}264.890$ cm²; $Y_{gs} = 609.5$ cm; $Y_{gi} = 957.3$ cm

Furthermore, the moment of inertia of the section with respect to the central axis x, orthogonal to its axis of symmetry y, is:

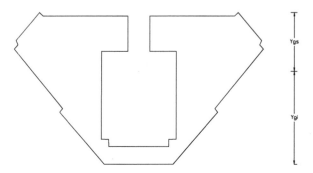

Fig. 5.71 Position of the pillar section centre

5.6 Brunelleschi's Dome in Florence

$$J_x = 3.98 \times 10^{11} \text{ cm}^4.$$

The axial load at the pillar base is obtained by summing the various vertical forces transmitted to the pillar:

$$N_{00} = 14{,}052.2 + 15{,}528 = 29{,}580.2 \text{ t}$$

We now evaluate the eccentricities of the various vertical loads conveyed to the pillar with respect to the center of the base section of the pillar.

The force of 14,052.2 t is conveyed across the side of the octagonal drum section at a distance of ½ 4.25 m = 2.125 m from the external boundary of the crown.

The corresponding eccentricity e' thus is = 6.09 − 2.125 = 3.96 m. The force of 15,528 goes across the center of the section. Thus the following equation gives the eccentricity of the resultant force N_{00}:

14,052.2 · 3.96 = 29,580.24 · e. Hence the load N_{00} = 14,052.2 t is acting on the inside at a distance $e^* = 1.88$ m from the center of the pillar base section.

The radial thrust $S + S_l = 683$ t transmitted by the dome webs to the pillar acts at a height of 39 m from the base, while the thrust $S'_a = 822.04$ t transmitted by the adjacent pointed arches acts at a height of 28 m. The bending moment at the pillar base is thus:

$M_{oo} = 683 \times 39 + 822.04 \times 28 = 49{,}654.12$ tm and is directed outside the pier. The composition of the bending moment M_{oo} with the eccentric load N_{00} yields a resultant eccentric load N_{00} with a total eccentricity $e^* = 1.88 − 49{,}654/29{,}580 = 1.88 − 1.67 = 0.20$ m. The position of the core section points can be on the other hand obtained as

$$d_{\sup} = \frac{W_i}{A} = \frac{J_x}{A \cdot Y_{gi}} = 183.56 \text{ cm}$$

$$d_{\inf} = \frac{W_s}{A} = \frac{J_x}{A \cdot Y_{gs}} = 288 \text{ cm} > e^* = 20 \text{ cm}$$

The section is wholly compressed. The maximum and minimum compression stresses are:

$$\sigma_{\text{int}} = \frac{N}{A} + \frac{M}{J_x} Y_{gi} = \frac{29{,}580000}{2{,}264890} + \frac{29{,}580000 \cdot 20}{3.98 \cdot 10^{11}} 957.3 = 13.06 + 1.42 = 14.48 \text{ kg/cm}^2$$

$$\sigma_{\text{int}} = \frac{N}{A} + \frac{M}{J_x} \sigma_{est} = \frac{N}{A} - \frac{M}{J_x} Y_{gs} = \frac{29{,}580000}{2{,}264890} + \frac{29{,}580000 \cdot 20}{3.98 \cdot 10^{11}} 609.5 = 12.15 \text{ kg/cm}^2$$

The estimated maximum compression stresses are only a small fraction of the high compression strength of the pillar masonry, which is constructed of regular marble blocks and good mortar. Over the centuries the dome has maintained a stable configuration and is an extraordinary example of the soundness and beauty of Renaissance architecture.

5.7 St. Peter's Basilica Dome by Michelangelo: The Static Restoration by Poleni and Vanvitelli

5.7.1 Dome Geometry

The story of the dome engineered by Michelangelo for St. Peter's Basilica in Rome is well-known to all, though the restoration works carried out in the early 18th are perhaps less so (Mainstone 1999; Benvenuto 1990; Di Stefano 1980).

Although the dome's construction was planned by Michelangelo, it was not to be completed until 1592, well after his death in 1564, under the supervision of Della Porta. Construction of the drum and sustaining pillars started much earlier and went on for many years. The four large pillars sustaining the drum and connected to each other by large arches at their heads were in fact built between 1506 and 1512 under the direction of Bramante.

The cross section of the four large piers can be inscribed into a square of about 19 m on a side. The span between them measures 22.60 m. The extrados at the crown of the four arches connecting the pillars reaches a height of about 50 m from their base.

The two eastern piers rest on solid marl and clay formations, but the other two overlie the remains of earlier Roman constructions. This nonuniformity in the foundations could be responsible for the differential settling that already began during their construction.

Some cracking in the arches were detected between 1514 and 1534, and called for some restoration and refurbishment operations before construction could be begun on the drum. Michelangelo himself ordered the pillars strengthened by eliminating the large niches and interior spiral staircase planned for in Bramante's original design. Unfortunately, Michelangelo never lived to see construction of the dome begin.

The large dome's structure is similar to Brunelleschi's in Florence. It is in fact made up of two interconnected shells, stiffened by 16 ribs. Figure 5.72 shows the dome in section as sketched by Vanvitelli and reported by Poleni (1748).

The main measurements defining the geometry of the dome and supporting drum have been obtained directly from Vanvitelli's drawings (Figs. 5.72, 5.73 and 5.74). Other measurements, made by Fontana (1694) and Beltrami (1929), differ somewhat from the reported values, though such differences have a negligible effect on the static analysis (Galli 1994–1995).

5.7 St. Peter's Basilica Dome by Michelangelo …

Fig. 5.72 Dome longitudinal section (Vanvitelli, da Di Stefano 1980)

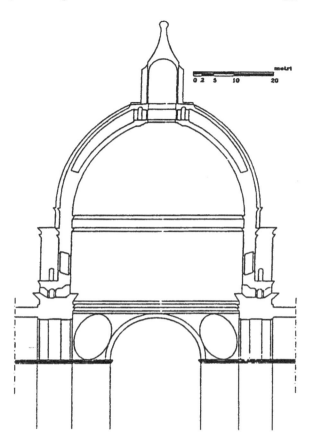

The thickness of the internal and external shells are respectively 2.00 and 1.00 m, while the total thickness dome varies from 3.00 m at the springing to about 5.00 m at the crown. The overall arrangement of the entire dome is that of an ogival spherical vault. The internal diameter of the dome at its base measures 42.70 m. The drum is composed of a 3.00 m thick cylindrical wall with an internal radius of 21.35 m. It is stiffened by 16 radial buttresses that arise from the drum for a length of about 4.50 m, are 3.00 m thick and reach a height of 14.50 m.

The elevation of the dome is reported relative to the height of the access gallery, F, that is, to the height of the floor of the passageway through the buttresses shown in Figs. 5.73 and 5.75.

The drum and buttresses are equal in height: 14.50 m. The *attico,* a circular vertical wall which extends the drum, rises from the top buttress up to the springing of the dome for a total height of 3.50 m. The drum and the attic, both of which are 3.0 m thick, thus reach a combined height of 18 m. Consequently, the dome

248   5 Masonry Vaults: Domes

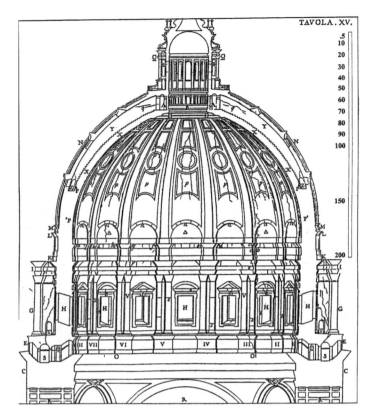

Fig. 5.73 Cracking detected by L. Vanvitelli (from Poleni 1748)

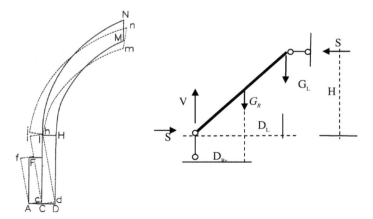

Fig. 5.74 The Three Mathematicians' model and the corresponding failure mechanism

5.7 St. Peter's Basilica Dome by Michelangelo ...

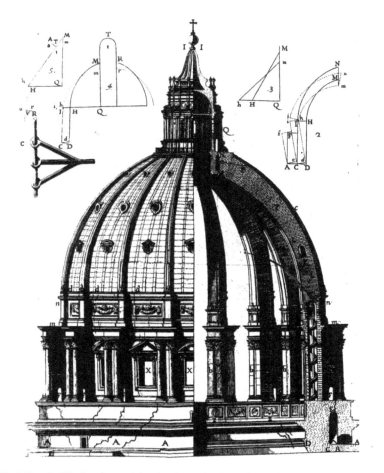

Fig. 5.75 A Vanvitelli's drawing and the simple mechanism of the "Three Mathematicians" (1742)

springing is found at a height of 18.00 m from the level of the access gallery (*F*, in Fig. 5.70). To sum up, some of the structures' fundamental measurements are:

- Distance *B* between the vertical edge of the lantern and the inner edge of the drum: $B = 21.35 - 3.5 = 17.85$ m;
- Internal diameter of the lantern: $D_L = 7.0$ m;
- Distance between the base and the extrados of the dome at the crown: $H = 28.20$ m;
- Ratio $H/B = 28.20/17.85 = 1.58$;
- Dome thickness at crown: 5.0 m;
- Distance between the base of the dome and the center of the dome section at the insertion with the lantern: $H' = 28.20 - 2.50 = 25.70$ m;
- Internal diameter of the drum and attic: 42.70 m;
- Thickness of the drum and attic: 3.0 m;

- Internal diameter of the drum/attic: 42.70 m;
- Height of the drum from the access gallery floor, F: 14.50 m.
- Distance between the interior edge of the drum and the inner edge of the dome section at the attachment of the lantern: 17.85 m.

In fact, calculating we get, $D_E = 42.70/2 - 3.50 = 21.75 - 3.50 = 17.85$ m.

It is interesting to point out that the so-called Rules of Fontana (1694), followed in late 17th century Roman construction, would have called for the drum to be thicker than the actual 3.00 m. However, the rules refer to unbuttressed domes. The masonry used for the dome is made up of bricks, travertine blocks and mortar beds, and it was laid with the support of wood scaffolds and centering. The two shells were built up between the ribs. Two iron ties encircling the dome were placed by Della Porta (Di Stefano 1980).

5.7.2 Damage to the Dome: Early Studies Assessing Its Safety

Many years after the dome's completion cracks began to develop and grow gradually over time. The first signs of damage were detected as far back as 1631 and more and more cracks appeared over the following years. In the mid-18th century, about 150 years after its completion, the dome exhibited widespread, serious damage and word spread throughout the scientific community.

Various descriptions and experts judgments were forthcoming, amongst which the dire account of Saverio Brunetti (Book II of his *Memorie, Stato dei difetti...*, Poleni 1748):

> ... the entire wall of the drum and the attic, together with the columns and buttresses, have rotated outwards, dilating the dome and lowering the lantern, ...

This description corresponds to the cracking pattern detected by L. Vanvitelli between 1742 and 1743 in an exquisite set of drawings published by Poleni in his "*Stato dei difetti.*" (e.g., Plate XV, shown in Fig. 5.73). In this figure long meridian cracks are clearly visible running along the dome intrados. They arise from the drum nearly up to the height of the ring connecting the crown to the lantern.

The sixteen buttresses were hard-pressed to contrast the thrusting action of the attic and drum: the strain is evidenced by large, diffuse sloping cracks across them. At the time, sheets of lead covered the exterior of the outer shell and the cracks could therefore visible on the dome's extrados.

The cracking pattern is similar to that of the failure mechanism shown Fig. 5.12, produced by excess weight at the center and low resistance of the dome at the springing. On the other hand, circular cracks accompanying the meridian fractures are not clearly visible in Vanvitelli's drawings.

A membrane stress state, with hoop tensile stresses acting along the lower rings, occurred first in the original undamaged dome. But the friction strength between the bricks rings, compressed along the meridians, slowly faded, probably because of humidity penetrating into the masonry mass.

The behavior of the dome gradually shifted from that of a rigid shell, stiffened by hoop stresses, towards that of a pushing dome, partitioned by long meridian cracks. It was in this latter state that Vanvitelli found the dome, a century and an half after its construction. Christianity's most revered place of worship seemed in serious trouble indeed. Alarm grew in Europe and in 1742 Pope Benedict XIV appointed a committee of scientists, known as *"The Three Mathematicians"*, composed of T. Le Seur, F. Jacquier, and R.G. Boscovich, to report on the condition of the dome. These scientists were well-known in the scientific community because they had previously published a commentary on *Principia*, Newton's revolutionary work which had appeared in a limited number of copies in 1697 and were published again in 1713.

The Three Mathematicians' initial assessment, published as the *"Parere"* (i.e., opinion) was that the dome was seriously damaged and would require extensive reinforcement operations.

A later report by the same authors, the so-called "Reflexions", confirmed their initial estimation. However, other scholars who collaborated in the analysis dissented from their opinion. To settle this dispute, Benedict XIV decided to seek the advice of a brilliant Italian scholar, Giovanni Poleni. Poleni, who was born in Venice in 1683, began to study the new disciplines of Cartesian mathematics and experimental Physics at an early age. A member of the Royal Society since 1710, Poleni attained the chair in Experimental Philosophy at the University of Padua, where he also directed a new materials testing laboratory. In 1743, at Pope Benedict's behest, Poleni analyzed the dome and came to the conclusion that its state was not nearly as dire as the Three Mathematicians had made out.

In 1743 Poleni prepared a first manuscript to explain his convictions regarding the origins of the damage and made a number of suggestions for improving the dome's safety. In this 1748 manuscript, published as the *"Memoirs"*, he also reported on the results of a static analysis of the dome performed by himself in his laboratory in Padua. This analysis was conducted in the wake of some recent results on the statics of masonry arches obtained by R. Hooke at Cambridge, recalled in Sect. 3.2. Poleni presented his proposal for restoration.

The *Memoirs* were received favorably by the pope, who then entrusted Poleni with carrying out the dome restoration in collaboration with L. Vanvitelli, the architect of the *"Opera di San Pietro"*.

5.7.3 The Three Mathematicians' and Poleni's Differing Opinions

According to historical accounts, the two discordant opinions regarding the dome's state and safety were heatedly debated (Mainstone 2003; Benvenuto 1990; Como 1997, 2008). The Three Mathematicians, backed by many other scholars, believed that the dome's failure was imminent and its restoration, involving profound architectonic changes to the entire monument, was required with the utmost urgency. Poleni instead sustained that the dome's state of safety was much less threatening. Moreover, Poleni was convinced that the so-called defects of the great dome could be repaired without any modifications to its architecture.

The Three Mathematicians' assessment that the dome was in danger of failure was based on their interpretation of the cracking pattern, which they, using a simple mechanical model, viewed as the start of a collapse mechanism. This model, drawn from a plate of their *Parere,* is sketched out in Fig. 5.74 and considers the combination of the dome with the attic and drum, together with the adjacent buttress. They reduced the complex dome—attic/drum—buttress system to the simple mechanism illustrated in Fig. 5.74.

The system was modeled as an inclined beam, HT, whose top T was free to move along the vertical and whose base H could move along the horizontal, as sketched to the right in Figs. 5.68 and 5.69.

The horizontal segment AD of the left-hand scheme in Fig. 5.68 represents the drum base and adjacent buttress, and the segment AF the external edge of the vertical buttress. The buttress and the drum/attic were bound together only very weakly, so the Three Mathematicians reasonably considered the buttress to have been detached from the drum wall. Such a mechanism describes the deformation of the damaged dome, with the drum and the attic rotating externally and the dome slices counter-rotated inward, with lowering of the lantern and dilatation of the dome. According to this mechanism, the dome slice HMNI rotates inward around the hinge H and produces counter-rotation of the drum/attic/buttress around A and C.

By applying kinematics to this scheme, the Three Mathematicians also evaluated the thrust of the dome. The principle of virtual work, first formulated in 1717 by Giovanni Bernoulli in a letter addressed to Varignon, was published in 1725. The Three Mathematicians were thus certainly aware of the principle and probably referred to it in evaluating the pushing work and the resisting forces along the assumed failure mechanism.

The restoration operations proposed by the Three Mathematicians were quite extensive: in addition to encircling the dome with new iron ties, they also wanted to thicken the buttresses and place new heavy statues on the top of them.

Poleni, on the contrary, did not accept the conclusions of the Three Mathematicians: he saw no correlation between the cracking of the dome and that of the attic and drum. He instead attributed the damage solely to defects in construction and the use of poor masonry. Even the finding of a broken old iron ring in the masonry was not ascribed to the dome's cracking, but instead to temperature changes or the effects of earthquakes.

Poleni's firm conviction stemmed from the results of a static analysis that he himself developed and performed. This analysis, though incomplete, proved to him that the dome was still safe, despite its defects. Poleni's analytical procedure was inspired by Hooke's theorem of the inverted chain examined in Chap. 2.

Accordingly, Poleni divided the dome into fifty slices, each subdivided into thirty-two "wedges", whose position and weight he then evaluated.

He then constructed a precise scale model of a dome slice in his Padua laboratory and considered two thin chains: one of equal small-sized rings (an ideal catenary), and another composed of thirty-two small lead balls, whose weights modeled the weights of the thirty-two wedges constituting a single dome slice, including the top wedge's share of the weight of the lantern atop the dome. The length of the chains was fixed so that their end sections could pass through the centers of the sections at the springing and the crown of the slice. The first step in the procedure was to determinate the equilibrium curve of the chains. Then by inverting these curves and marking them on the slice model, Poleni was able to verify their positions with respect to the curve of the wedges' centers.

The position of the curve of the homogeneous catenary differed considerably from the curve of the wedge centers: it effectively exited the slice section, though it could be made to fit within the masonry through some slight adjustments. The chain of lead balls, on the other hand, was contained within the slice and maintained a position quite close to the curve of the centers. Poleni wrote:

> Wedges as far as N tend to push themselves out and the upper ones are somewhat low,… and the form of the dome is not free from imperfections… these are of minor importance because our entire catenary is situated at the solid inside of the vault.…

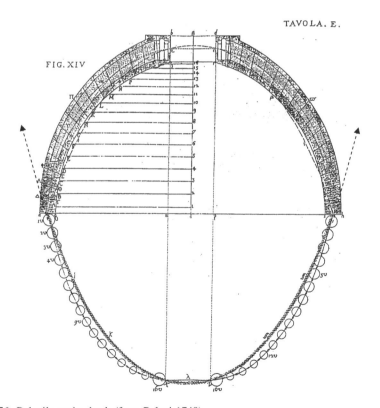

Fig. 5.76 Poleni's static check (from Poleni 1748)

Poleni thus concluded that the equilibrium within the slice was certainly admissible, because compression stresses alone should be able to sustain the loads: *"in short, the shape of the great vault was not bad at all"*.

Figure 5.76 shows the by now famous Poleni model with a sketch of his determination of the funicular curves with the vectors R tangent to the funicular at the springing. The horizontal component of R gives the pull of the chain and, consequently, according to the inverted chain model, the thrust in the dome. It is clear that Poleni's attention was focused exclusively on the equilibrium of the dome. In this way, Poleni was able to verify that the geometry of the meridian curve of the dome was actually admissible. On the other hand, he also discovered that, although the pressure line passed through the centers of the corresponding slice sections at the crown and springing, it drifted away from its axis. Thus, the shape of the meridian section of the dome was not the optimal one for the actual loads distribution: the dome would probably have been more stable had it had a more ogival shape. Notwithstanding, Poleni believed that the dome of St. Peter's was not at risk of failure and that the model proposed by the Three Mathematicians had to be flawed. The damage to the dome was due solely to defects in construction.

Poleni recommended two types of modifications, both of which had also been advised by the Three Mathematicians. He deemed both the insertion of four iron rings around the dome and the repairs to the cracks appropriate, but considered it unnecessary to thicken the buttresses, place statues at their top, or fill in the spiral staircase inside the pillars.

5.7.4 Poleni and Vanvitelli's Restoration Works

The dome was repaired and reinforced according to Poleni's recommendations and under the technical supervision of L. Vanvitelli.

The cracks in the dome were patched through the procedure know as *"scuci e cuci"* (literally, "unstitch and stitch"), which is still commonly used today. However, the most important measure was encircling the dome: in 1743 and 1744 five iron hoops were placed under tension at different heights around the dome and drum (rings A, B, C, D, E in Fig. 5.77).

Each hoop was put into position and then tensed by means of the ingenious connection equipment shown in Fig. 5.77. Once the various hoop segments were in place, they were tensed by beating in the wedges (a) shown in Fig. 5.72. In this way Poleni and Vanvitelli hooped the dome and partially closed the meridian cracks. A sixth hoop (Z) was placed in 1748 after Vanvitelli found that one of the two iron ties (hoop "u" in Fig. 5.78) originally emplaced during construction had broken. A number of other minor measures were taken after these operations and the dome has remained in good static condition since.

5.7 St. Peter's Basilica Dome by Michelangelo … 255

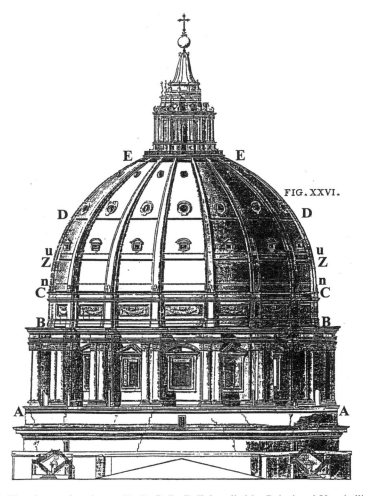

Fig. 5.77 The six new iron hoops (*A*, *B*, *C*, *D*, *E*, *Z*) installed by Poleni and Vanvitelli and the original ones (*u*, *n*) (Poleni 1748)

Fig. 5.78 The hoop connecting and tensing device (Poleni 1748)

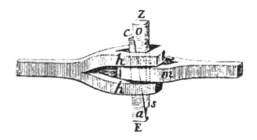

5.7.5 *Further Considerations on the Heated Debate*

The heated debate among scholars of the time and their contrasting opinions deserve further consideration. The next sections will therefore delve into the different studies performed by the Three Mathematicians and Poleni. Surprisingly, it will be shown that, in contrast to conventional opinion, the Three Mathematicians were actually less wrong than Poleni believed.

5.7.5.1 Remarks on the Poleni Opinion

As intimated earlier, Poleni's static analysis was in fact incomplete. He failed to take into account the presence of the attic and the drum, stiffened by buttresses. However, extending Poleni's approach to the complex system composed of the dome/attic/drum/buttresses would have required a knowledge of mechanics that had not been attained at the time. Indeed, Hooke's theorem of the inverted chain was insufficient to solve the problem: it would be necessary to resort to the general theorem of Limit Analysis of masonry structures, which were not formulated until 200 years later.

Essentially, Poleni omitted to control statics of the underlying supporting structures of the dome, as the attic and the drum and the cside counterforts.

The more accurate analysis of the problem, which will be covered in the following, will bring us to the conclusion that the evaluation of the Poleni thrust evaluation was correct but that the true state of the dome was not far from that estimated by the Three Mathematicians. Fortunately, the six iron hoops placed by Poleni proved fundamental to saving the dome.

5.7.5.2 Remarks on the Three Mathematicians' Model

As discussed in the foregoing, the Three Mathematicians formulated the simplified model illustrated in Figs. 5.74 and 5.79, whence they estimated the thrust in the dome.

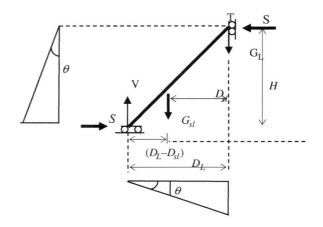

Fig. 5.79 The Three Mathematicians' evaluation of the thrust

From equilibrium of the rigid beam around the toe H, we have in fact the equation

$$SH = G_{sl}(D_L - D_{sl}) + G_L D_L \qquad (5.71)$$

whence the thrust S can be obtained immediately. Clearly, the model seems too simple: the entire slice of dome is assumed to rotate as a single rigid beam, without any hinges, when the buttress and drum/attic slice move outward. Surely, other mechanisms, with internal hinges, might be at work and need to be taken into account.

The Three Mathematicians evaluated the dome thrust assuming only one particular mechanism. Poleni obtained another estimate of the thrust. But what is the actual thrust of the dome? Only by knowing this thrust can we estimate the true safety conditions of the dome and verify whose opinion—Poleni's or the Three Mathematicians'—was really correct.

According to the cracking pattern detected by Vanvitelli, a spreading of the drum had occurred. The cracked dome, deforming through a mechanism, adjusted itself to the slightly increased diameter of its springing. Under such conditions the actual thrust of the dome is the minimum of all the statically admissible ones. The thrust obtained by the Three Mathematicians, which we will recalculate in the following, turns out to be lower than the actual thrust in the dome.

5.7.6 *Minimum Thrust in the Dome*

It is convenient to follow Poleni's partitioning and corresponding weight evaluations of the thirty-two wedges of the slice. In doing so, we shall refer to Fig. 5.74 from Poleni's *Memoirs (1748)*, where the various wedges are denoted as A, B, C, etc. The corresponding wedge weights are first evaluated using the same unit of weight as Poleni, the pound, *lb*, which corresponds to about 0.372 kg. The slice to be considered is 1/50 of the entire dome.

The weight of the lantern slice G_{slL}.

According to Poleni, weight G_{slL}, corresponding to 1/50 of the round angle, is: 81.6×10^3 lb, corresponding to 30.35 t. The weight G_L of the entire lantern thus equals $50 \times G_{spL} = 50 \times 81.6 \times 10^3$ lb $= 4{,}080 \times 10^3$ lb $= 1{,}517.74$ t.

The weight of the dome slice, G_{sl}:

The weight G_{sl} includes a share of the weight of the rib and the corresponding band of the outer and inner shells. According to Fig. 5.80, we have the following series of weights for the various wedges.

wedge A 89×10^3 lb
wedge B 88×10^3 lb
wedge C 87×10^3 lb
wedge D 85×10^3 lb
wedge E 82×10^3 lb

wedge F   79 × 10³ lb
wedge G   75 × 10³ lb
wedge H   71 × 10³ lb
wedge K   66 × 10³ lb
wedge L   60 × 10³ lb
wedge M   54 × 10³ lb
wedge N   48 × 10³ lb
wedge P   41 × 10³ lb
wedge Q   34 × 10³ lb
wedge R   27 × 10³ lb
wedge S   18 × 10³ lb

The horizontal axis passing through V indicates the head of the drum/attic, i.e. the level of the dome base. The alignment T, passing through point T in Fig. 5.80, is the vertical line passing through the center of the section joining the dome with the lantern, along which the load G_{spL} is conveyed.

The following distances are considered:

- distance D_L between the vertical line passing through the drum's internal edge (V in Fig. 5.75) and alignment T;
- distance D between the slice's center and alignment T;
- distance D' between the slice's center, including its share of the lantern weight, and alignment T;
- distance D_{GsIT} between the drum's internal edge and the slice's center, through which the weight G_{slTot} passes;

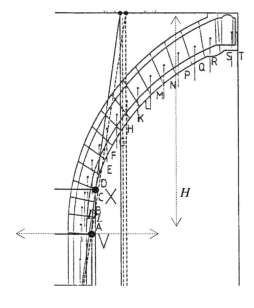

Fig. 5.80 The wedge weights and center positions (Memorie, Poleni 1748)

5.7 St. Peter's Basilica Dome by Michelangelo ...

- height H, with respect to the dome's base, of the extrados of the section joining the dome with the lantern.

According to the measures previously reported, it follows that $H = 28.20$ m. The same value is obtained by measuring this distance in Fig. 5.74, which is 5.94 cm, and calculating $5.94 \times 4.754 = 28.20$ m. Likewise, in Fig. 5.74, distance $D_L = 17.85$ m measures 3.76 cm, whence using $s = 1$ cm $= 4.75$ m, we have: $3.76 \times 4.75 = 17.85$ m.

Position of the slice's centre

Moments of the wedges' weights around point T in Fig. 5.75 (placed at the intrados of the dome-to-lantern connection section):

wedge A	89×10^3 lb $\times 4.15 = 369.35 \times 10^3$ lb $\times$ s
wedge B	88×10^3 lb $\times 4.14 = 364.32 \times 10^3$ lb $\times$ s
wedge C	87×10^3 lb $\times 4.05 = 374.10 \times 10^3$ lb $\times$ s
wedge D	85×10^3 lb $\times 3.95 = 335.75 \times 10^3$ lb $\times$ s
wedge E	82×10^3 lb $\times 3.85 = 315.70 \times 10^3$ lb $\times$ s
wedge F	79×10^3 lb $\times 3.65 = 288.35 \times 10^3$ lb $\times$ s
wedge G	75×10^3 lb $\times 3.45 = 262.50 \times 10^3$ lb $\times$ s
wedge H	71×10^3 lb $\times 3.20 = 227.20 \times 10^3$ lb $\times$ s
wedge K	66×10^3 lb $\times 2.95 = 194.70 \times 10^3$ lb $\times$ s
wedge L	60×10^3 lb $\times 2.65 = 159.00 \times 10^3$ lb $\times$ s
wedge M	54×10^3 lb $\times 2.30 = 124.20 \times 10^3$ lb $\times$ s
wedge N	48×10^3 lb $\times 1.95 = 93.60 \times 10^3$ lb $\times$ s
wedge P	41×10^3 lb $\times 1.60 = 65.60 \times 10^3$ lb $\times$ s
wedge Q	34×10^3 lb $\times 1.15 = 39.10 \times 10^3$ lb $\times$ s
wedge R	27×10^3 lb $\times 0.75 = 20.25 \times 10^3$ lb $\times$ s
wedge S	18×10^3 lb $\times 0.15 = 2.70 \times 10^3$ lb $\times$ s
Lantern slice	81.6×10^3 lb $\times 0.0 = 0.00$

The total moment is thus

$$M = 3210.92 \times 10^3 \text{ lb} \times 4.75 = 5673.69 \, \text{t} \times \text{m}.$$

Resultant of the wedges' weights + the lantern slice's weight:

$$G_{slTot} = 1004 \times 10^3 \text{ lb} + 81.6 \times 10^3 \text{ lb} = 1085.6 \times 10^3 \text{ lb} = 403.84 \, \text{t}$$

Distance D of the slice center from alignment T:

$$D = 5673.69/373.4 = 15.19 \, \text{m}$$

Distance D' of the slice center, including its share of the lantern weight, from alignment T:

$$D' = 5673.69/403.84 = 14.05\,\text{m}.$$

Distance D_{GslT} of the slice center, including its share of the lantern weight, from the internal edge of the drum:

$$D_{GslpT} = 17.85 - 14.05 = 3.80\,\text{m},\ H = 28.20\,\text{m}$$

Summing up the geometry of weights
According to the foregoing calculations, the total weight G_{slTot} of the slice of the dome, including its share of the lantern, evaluated up to the height of the extrados of the drum/attic, equals 403.84 t and acts at a distance, $D_{GslT} = 3.80$ m from the drum's inner edge.

Figure 5.81a sketches the outline of a generic mechanism v of a dome slice whose base undergoes a slight broadening. The point O indicates the position of the internal hinge. When point O falls on the intrados of the dome springing section, this mechanism corresponds to that envisioned by the "Three Mathematicians".

Figure 5.81b shows all the quantities involved. The mechanism is represented by the outwards horizontal settling $\Delta(v)$, along which the unknown thrust $S_{cin}(v)$ does work depending on the chosen mechanism v. The kinematic thrust corresponding to mechanism v is thus given by

$$S_{kin}(v) = \frac{\langle g, v \rangle}{\Delta(v)}. \tag{5.72}$$

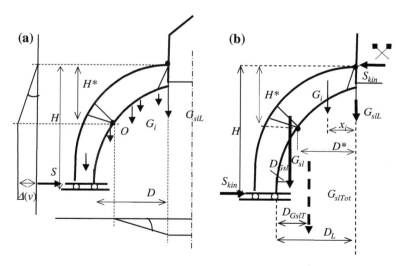

Fig. 5.81 The minimum thrust evaluated via the kinematic procedure

5.7 St. Peter's Basilica Dome by Michelangelo … 261

where g is the load distribution, i.e. the sequence of forces G_i representing the weights of Poleni's wedges (Fig. 5.74).

The minimum thrust of the dome will now be evaluated by applying the kinematic procedure set forth in Chap. 2. The unknown thrust, that is, the minimum of all statically admissible thrusts, can be obtained as the maximum of all the kinematically admissible ones

$$S_{Min} = Max\ S_{kin}(v) = Max \frac{\langle g, v \rangle}{\Delta(v)} \tag{5.73}$$

The work done by loads g is given by

$$\langle g, v \rangle = \theta \sum_i G_i(D^* - x_i), \tag{5.74}$$

where D^* indicates the horizontal distance of the hinge O from alignment T, and x_i the distance of the single force G_i from the same alignment, as in Fig. 5.75b. Finally, we have

$$\Delta(v) = \theta H^*. \tag{5.75}$$

where H^* is the vertical distance between the internal hinge O and the extrados of the section where the dome joins the lantern. The vertical displacements v_i of the application points of loads G_i, representing the weights of the various wedges, are then given by

$$v_i = (D^* - x_i)\theta \tag{5.76}$$

Thus, we have

$$S_{kin}(v) = \frac{\sum_i G_i(D^* - x_i)}{H^*}. \tag{5.72'}$$

The maximum kinematic thrust $S_{kin}(v)$ is found by varying the position of the internal hinge O along the intrados curve of the slice. In the following, five different mechanisms, numbered (0) to (4), are considered and the corresponding kinematic thrust $S_{kin}(v)$ evaluated for each. The highest value among the $S_{kin}(v)$ turns out to be for mechanism (4).

Mechanism(0)—The mechanism of the Three Mathematicians
Position of the internal hinge O: point V in Fig. 5.74, on the intrados of the dome springing, at the height of the top of the attic/drum.

The work of the loads includes the weights G_i running from wedge A to S, in addition to the corresponding share of the lantern weight.

Horizontal distance D^* of the hinge O from alignment T: $D^* = D_L = 17.85$ m

Vertical distance H^* of hinge O from the extrados of the section where the dome joins the lantern: $H^* = H = 28.20$ m.

Total work L_{tot} of loads G_i, including the work of the lantern slice, on the assumed mechanism: $L_{tot} = 1615.67$ t × m × θ

Work performed by the thrust: $S_{kin} \times H \times \theta = S_{kin}$ 28.20 θ t × m

From Eq. (5.52'): $S_{kin} \times 28.20 \theta = 1615.67$ t × m × θ

Kinematic thrust: $S_{kin} = 57.29$ t

In this case, the entire slice moves as a single rigid body, as assumed by the Three Mathematicians. Thus, $D^* = D_L$ and we get

$$\langle g, v \rangle = \theta \sum_i G_i(D_L - x_i) + G_L D_L \theta = \theta[G_{sl}(D_L - D_{sl}) + G_L D_L],$$

where, according to the definition of the centrer

$$D_L - D_{sl} = \frac{\sum_i G_i(D_L - x_i)}{G_{sl}}.$$

Consequently, Eq. (5.52') becomes

$$\{S_{kin}H - [G_{sl}(D_L - D_{sl}) + G_L D_L]\}\theta = 0.$$

and we return to (5.51), whence the thrust value $S = 57.29$ t determined by the Three Mathematicians.

Mechanism (1)

Position of the internal hinge O: point X in Fig. 5.74. According to this mechanism, the weight of wedges A, B and C do no work at all.

Horizontal distance D^* of the hinge (1) from alignment T: $D^* = 17.56$ m.

Vertical distance H^* of the hinge (1) from the extrados of the attachment section of the dome to the lantern: $H^* = 22.325$ m.

Total work of loads G_i, including the work of the lantern slice, along the assumed mechanism: $L_{tot} = 1579.556 \times \theta$ tm.

Work performed by the thrust: $S_{kin} \times H^* \times \theta = S_{kin} \times 22.325$ m × θ and $S_{kin} = 70.75$ t.

Mechanism (2)

Position of the internal hinge O: between wedges H and K in Fig. 5.74. According to this mechanism the weights of wedges from H to A do no work at all.

Distance D^*: D: $H^* = 13.3$ m.

$S_{kin} = 78.51$ t

Mechanism (3)

Position of the internal hinge O: between wedges E and F in Fig. 5.74. According to this mechanism the weights of wedges E to A do no work.

Distance D^*: $D^* = 16.62$ m.

Distance H^*: $H^* = 18.52$ m.

$S_{kin} \times 18,52\theta = 1444.78\theta$
$S_{kin} = 78 \, \text{t}$

Mechanism (4)
Position of internal hinge O: between wedges D and E in Fig. 5.75.
According to this mechanism the weights of wedges D to A do no work.
Distance D^* d: $D^* = 17.81$ m.
Distance H^*: $H^* = 20.42$ m.
Total work of the weights of the wedge and the lantern slice:
$L_{tot} = 1715.40 \times \theta$
Work performed by the thrust S_{kin}: $S_{kin} \times H^* \times \theta = S_{kin} \times 20.42 \times \theta$
Equation (5.52′): S_{kin} 20.425 $\times \theta = 1715.40 \, \theta$ and $S_{kin} = 84$ t
Other mechanisms, with internal hinges positioned elsewhere on the slice intrados, do not furnish larger thrust values than mechanism (4).

The *maximum* kinematic thrust value thus corresponds to mechanism (4) and the *minimum* thrust of all *statically admissible* states therefore corresponds to the kinematic thrust that results by positioning hinge O between wedges D and E, namely 225.77×10^3 lb = 84 t. This thrust is transmitted by a slice 1/50 of the dome's round angle. Given an average drum diameter of 45.70 m, and average circumference of 45.70 m $\times \pi = 143.57$ m, the length of the arch corresponding to the assumed slice is 143.57/50 = 2.87 m. The thrust of 83.98 t is thus transmitted along a length of 2.87 m, and the thrust per unit length of the drum equals 83.98 t/2.87 = 29.26 t/ml. The vertical load transmitted by the slice equals the total weight of the slice, or V = 108.6 $\times 10^3$ lb = 403.84 t. The position of this vertical force V at the base of the slice can be obtained by considering the equilibrium of the slice. Let x be the distance of force V from alignment T (Fig. 5.82). The condition of zero moment of all the forces around point T at the extrados of the section of the dome's connection with the lantern gives

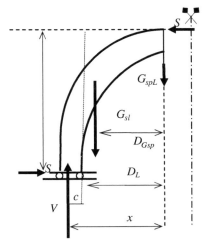

Fig. 5.82 Equilibrium of the slice

$$Vx - G_{sl}D_{Gsl} - SH = 0$$

or

$$403.84x - 373.4 \times 15.19 - 84 \times 28.20 = 0,$$

whence we obtain $x = 19.9$ m. Thus, the distance of force V from the inner edge of the drum is

$$c = 19.85 - 17.85 = 2.0 \text{ m},$$

and the thrust ratio, i.e., the ratio between thrust and vertical reaction, is

$$\chi = \frac{S}{V} = \frac{83.98}{403.84} = 0.207.$$

This value is much larger than the value $\chi = 0.106$ for the dome of Santa Maria del Fiore, let alone the even lower value, $\chi = 0.078$ calculated for the Pantheon dome.

Let us now compare the value of the minimum thrust with the value of the thrust obtained by Poleni using the pressure line determined via his ingenious approach of the inverted chain in Fig. 5.76. Figure 5.83 shows the vector, R, having the same direction as the tangent line to the inverted chain at the dome springer section. The vectors V and S_{Poleni} balance the force R.

The horizontal component of force R, i.e., Poleni's thrust, can be obtained directly from Fig. 5.70. The ratio S_{Poleni}/V is about $7.5/32 = 0.234$, and we get $S_{Poleni} = 0.234 \times 1004 \times 10^3$ lb $= 235 \times 10^3$ lb $= 87.4$ t, only about 1 % more than the 84 t minimum thrust evaluated in the foregoing. The value S_{Poleni} turns out to be slightly larger than the minimum thrust evaluated using modern methods. The thrust S_{Poleni} corresponds to an admissible pressure line wholly contained between the slice's extrados and intrados, but without touching them, i.e. without producing a mechanism, and, specifically, the mechanism (4).

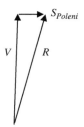

Fig. 5.83 The thrust corresponding to Poleni's inverted chain

5.7.7 Comparisons: St. Peter's Lower Static Efficiency

It is interesting to compare the weight and corresponding thrust of St. Peter's dome with others of about the same diameter, such as the domes of the Pantheon and Santa Maria del Fiore. What emerges from such comparison is the relatively low static efficiency of the dome of St. Peter's.

The total weight of the St. Peter dome equals $403.84 \times 50 = 20{,}192$ t and transmits a thrust of 29.3 t/ml. The Pantheon dome weights about 19,000 t and experiences a thrust of 10.30 t/ml, while the dome of Santa Maria del Fiore weights about 30,000 t and has a thrust of 21.3 t/ml.

5.7.8 Checking Safety of the Drum/Attic/Buttresses System

The attic is a round vertical wall overlying a series of large windows at the level of the buttresses. On its interior, the masonry bears a sequence of internal arches, just above the attic, that transfer the load over the underlying openings. As the drum is damaged by vertical cracks, as in Fig. 5.84, we assume that it works along vertical strips.

Let us now consider a strip of the drum 1/50 of the round angle together with its overlying attic, for a total height of 18.00 m. Figure 5.85 shows this drum strip together with the corresponding forces.

By way of definition, the quantities shown in the figure are:

S_{domT} thrust transmitted directly by the dome slice, acting at a height of about 18.00 m from the drum base;
V vertical load due to the weight of the dome slice alone;
C_3 weight of the slice of the attic/drum of thickness equal to the dome base;
c distance of the vertical force V from the drum's internal edge, as previously evaluated, $c = 2.0$ m;
h_1 height of the strip of the attic/drum, equal to 18.00 m.

Fig. 5.84 The drum/attic band

Fig. 5.85 Check of the vertical strip of the drum

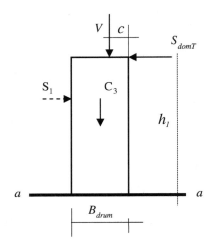

The dotted arrow in Fig. 5.79 represents the counter-thrust of the buttress, which was not considered in the foregoing analysis. Previous calculations gave us

$$S_{domT} = 84\,\text{t};\ V = 403.84\,\text{t}.$$

The total weight of the drum and the attic, according to Poleni, is 17.861 t, and the corresponding weight of the slice, i.e., the force C_3 in Fig. 5.86 is: $C_3 = 17.861 \cdot 1/50 = 357.2$ t.

Let us consider the base section of this slice drum/attic, which is 1/50 of the round angle: this base is found at the height of the floor of passageway F through

Fig. 5.86 Check of the drum strip including the buttress

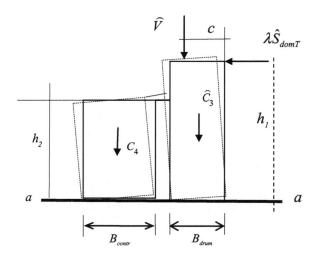

the buttress, as shown in Fig. 5.75. The total moment of all the *vertical forces* acting on the center of this base section, of thickness 3.00 m, is

$$M_{tot} = 84 \times 18.0 + 403.84 \times 0.55 = 1734.1 \text{ tm}.$$

It should be recalled that in the first analysis the possible counter-thrust of the buttress was not taken into account. The section, having a width of 2.87 m, equal to the arch length of the slice, bears the axial load

$$N = 403.84 + 357.2 = 761.04 \text{ t}$$

with an eccentricity

$$e = 1734.1/761 = 2.28 \text{ m} > B/2 = 1.50 \text{ m}.$$

The resultant of the axial force N and the bending moment M thus turns out to be outside the base section of the slice. The effects of the buttress counter-thrust must therefore be taken into account. Indeed it is the composite system drum/attic, coupled with the buttresses, of height h_2, that represents the structure resisting the dome thrust (Fig. 5.86). Failure of this system could occur via the mechanism foreseen by the Three Mathematicians, through equal rotations of the sliced drum/attic and buttresses, as in Fig. 5.79. Instead, we now consider thrust S_{domT} influenced by factor λ, as shown in Fig. 5.86.

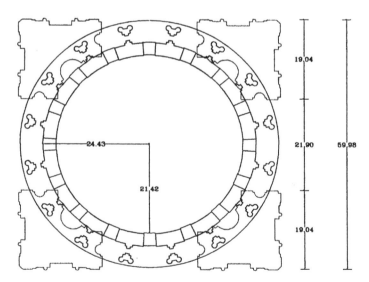

Fig. 5.87 Horizontal section of the drum

Given that the buttresses around the drum are sixteen in number, we must take 1/16 of the round angle as the width of the drum/attic slice associated to a single buttress (Fig. 5.87). The values of the corresponding weight and thrust considered above must now be multiplied by the factor 50/16 = 3.125. These values of the weight and thrust will be respectively indicated as $\hat{C}_3$ and $\hat{S}_{domT}$ to distinguish them from the other values corresponding to 1/50 of the round angle.

The weight of the buttress, indicated as C_4 in Fig. 5.80, is very difficult to evaluate because of its complex geometry. As a rough estimate, averaging its transverse section, we have $C_4 = 2.3 \times 14.50 \times 3.60 \times 5.60 = 672$ t, to account for the greater weight from the marble elements, the presence of large cornices, and so forth (Fig. 5.81). The thrust load multiplier λ is thus obtained by equating the resistant and pushing work done by the various forces along the mechanism, by which we have

$$\left[\hat{C}_3 \frac{B_{drum}}{2} + C_4 \frac{B_{contr}}{2} + \hat{V}(B_{drum} - c)\right]\theta = \lambda\theta\hat{S}_{domT}h_1,$$

whence we obtain

$$\lambda = \frac{\hat{C}_3 \frac{B_{drum}}{2} + C_4 \frac{B_{contr}}{2} + \hat{V}(B_{drum} - c)}{\hat{S}_{domT}h_1}$$

and

$$\lambda = \frac{3.125[357.2 \times 1.5 + 403.84 \times (3.0 - 2.0)] + 672 \times 2.8}{3.125 \times 84 \times 18.0} = 1.02.$$

Another simplified analysis, performed by Como (1999), yielded a safety factor λ of 1.06, only slightly higher than the just determined value of 1.02.

Note that we have assumed the drum to be cracked along its entire height. Such an assumption leads to a picture of the dome's static condition that is considerably more pessimistic than its actual condition. In any event, the cracking pattern detected by Vanvitelli included severe diagonal cracks in the buttresses (Fig. 5.73).

The resisting buttresses were seriously stressed by the dome and the equilibrium was extremely precarious. The particular geometry of the buttresses, with their external columns practically detached from the masonry wall, indicates that they were not originally designed as resistant elements to the dome's thrust.

Despite the imprecision of the rough estimate performed, it can nonetheless be concluded that the static conditions of the dome were actually quite critical, even close to failure.

5.7.9 Insertion of Iron Hoops

The most crucial reinforcing operation performed on the dome was undoubtedly encircling it with six iron hoops, as illustrated in Fig. 5.78, which shows their positions at:

- the buttress bases (A);
- the top of entablature of the main order of the drum (B);
- the dome springing (C);
- the dome's mid-height (D);
- the lantern springing (E);
- halfway between hoops C and D (Z).

Hoop Z was the sixth and last ring installed, when during the restoration in 1748 it was discovered that the old rectangular lower iron hoop (lines n–n and u–u in Fig. 5.71), emplaced during the dome's construction, had broken. During installation, the new hoops were heated and emplaced hot, so they could be tensed after cooling. The strength of the encircled dome has been greatly increased by the positioning of the hoops, which, when stretched during the displacement mechanism, perform plastic dissipation work.

Let R_c be the radius of the iron ring and w_r the radial displacement of the centers of the hoop sections consequent to the mechanism. The dilatation ε_ϕ occurring during hoop stretching, is thus

$$\varepsilon_\phi = w_r/R_c. \tag{5.77}$$

Let σ_o be the yield stress of the iron of the hoop, and A_c the area of the circle's cross section. The plastic dissipation D_{pl} is

$$D_{pl} = \sigma_o A_c R_c \int_0^{2\pi} \varepsilon_\phi d\phi = 2\pi \sigma_o A_c w_r \tag{5.78}$$

According to Poleni, the area of the rectangular cross section of the iron hoop (in the old unit of *oncia*) was "*once 3 per once 5*", which is about 51.9 cm². The failure strength σ_R of the iron of the hoop, according to Poleni's evaluations, is about 4,000 kg/cm² and the corresponding yield stress σ_o can be assumed equal to $\sigma_R/1.5$ and therefore 2,667 kg/cm². We can thus estimate the increase in dome strength gained by inserting a single hoop, for example, near the base of the dome.

For instance, by considering mechanism (4) considered above, we get, referring to Fig. 5.88, $w_c = H^* \theta$, where according to previous calculations, $H^* = 21.60$ m. The plastic work corresponding to the stretching of this iron hoop is thus

Fig. 5.88 The reinforcing effect of the new iron hoops on the strength of the dome

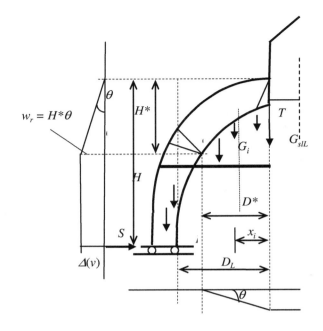

$$D_{pl} = 2 \times 2,667 \times \pi \times 2,160 \times \theta \, \text{kg} \times \text{cm} = 18,786 \times \theta \, \text{t} \times \text{m}.$$

The resisting work due to raising all the weights of the dome is $L_{tot} = 86,673 \times \theta$ t × m. The presence of the considered single hoop therefore increases the resistant work by about 20 % of the resistant work done by raising the weight of the entire dome during the mechanism development. The efficiency of the presence of the iron oopos depends on their position on the dome. In any case the installation of the six iron hoops has no doubt strongly increased the strength of the dome.

5.7.10 Conclusions

To sum up then, the attic/drum and buttresses were too weak to support the dome, given the nearly semicircular profile of its meridian section and the heavy lantern above. In particular, the rather frail geometry of the buttresses contributed little to its static stability, something probably overlooked by Michelangelo.

The Three Mathematicians realized the static inadequacy of the drum, albeit through too simplified a kinematic model. However, the repair and restoration work they proposed were too invasive and would have modified the monument's architecture. Poleni, on the other hand, never understood the actual static precariousness of the dome's state. However, although his static analysis was flawed, he, with the invaluable help of L. Vanvitelli, had six iron hoops installed to encircle the dome, which, in the end, were able to counterbalance the static deficiency of both the drum and buttresses. In brief, Poleni and Vanvitelli's restoration, carried out in full respect of the monument's architecture, turned out to be a great success indeed. To this day, not only is the wondrous Vatican dome still standing, but it has retained its original architecture.

References

Adam, J. P. (1984). *L'Arte di Costruire presso i Romani, materiali e tecniche, Biblioteca di archeologia* (Vol. 10). Milan: Longanesi & C.
Beltrami, L. (1929). *La Cupola Vaticana*. Vatican City: Tipografia Poliglotta Vaticana.
Benvenuto, E., & Corradi, M. (1990). *La statica delle false volte*, Ambrosi et al. (Eds.) *Architettura in pietra a secco*, 1° Semin. Intern., Fasano.
Blouet, G. A. (1833). Expedition scientifique de Morée. *Architecture et Sculture* (Vol. 2). Paris.
Bossut, C. (1778). Recherches sur l'equilibre des voutes, *Mémoires de l'Académie Royale des Sciences*, Paris.
Vanvitelli L., & Breve parere di, N. N. (1742). riportato in G. Poleni (2).
Cavanagh, W. G., & Laxton, R. (1981). *The structural mechanics of the Mycenaean Tholos Tomb*, BSA 76.
Como, M. (1997). *Un antico restauro statico della cupola di S. Pietro a Roma*. In C. Conforti (Ed.), *Lo specchio del cielo*. Milan: Electa.
Como M. T. (2006). Analysis of the statics of the Mycenaean tholoi. In M. Dunkeld et al. (Eds.), Proceedings of Second International Congress Construction History, Cambridge 2006.
Como M. T. (2007). *L'Architettura delle tholoi Micenee. Aspetti costruttivi e statici*. Naples: Università degli Studi Suor Orsola Benincasa.
Como, M. (2008). *Sulla storia del restauro statico della cupola di S. Pietro in Roma eseguito da Poleni e Vanvitelli*, Conv. Naz.le di Storia dell'Ingegneria, Cuzzolin Ed., Naples.
Como, M. T. (2012). Structural devices concerning the progressive outer shell construction in Brunelleschi's Dome. In: *CSH Conference*, Paris.
Di Pasquale, S. (2002). *Brunelleschi, La costruzione della cupola di Santa Maria del Fiore*. Venice: Marsilio.
Di Stefano, R. (1980). *La cupola di S. Pietro – Storia della Costruzione e dei Restauri*, Edizioni Scientifiche Italiane, Naples.
de Fine Licht, K. (1968). *The Rotunda in Rome, a study of the Hadrian's Pantheon*. Copenaghen: Jutland Arch. Soc., Publ. VIII.
Fontana, C. (1694). *Il tempio vaticano e sua origine*, Rome.
Galli, L. (1994–1995). *Sulla statica della cupola di S. Pietro ed interpretazione dei suoi antichi dissesti*, Supervisors: M. Como, Ugo Ianniruberto, a.y.
Giandebiaggi, P., Mambriani, C., & Ottoni, F. (Eds.). (2009). *Santa Maria del Quartiere in Parma*. Parma: University of Parma, Cariparma.
Giovannoni, G. (1925). *La Tecnica della Costruzione presso i Romani*, SEAI, Rome, ristampa 1994, Bardi, Rome.
Giuliani, F. C. (1995). *L'Edilizia nell'Antichità*. Rome: la Nuova Italia Scientifica.
Heyman, J. (1977). *Equilibrium of shell structures*. Oxford: Oxford University Press.
Heyman, J. (1995, 1997). *The stone skeleton*, Cambridge University Press.
Ippolito, L., & Peroni, C. (1997). *La cupola di Santa Maria del Fiore*. Florence: La Nuova Italia Scientifica.

Lancaster, L. C. (2007). *Concrete vaulted construction in imperial Rome*. Cambridge: Cambridge University Press.
Lucchini, F. (1996). *Pantheon, Monumenti dell'Architettura*. Rome: la Nuova Italia Scientifica.
Mainstone, R. (1977). Brunelleschi's Dome. *Architectural Review, 162*, 156–166.
Mainstone, R. (1999). *The Dome of St. Peter's, structural aspects of its design and construction and inquiries into its stability in structure in architecture: History, design and innovation*. Ashgate, Variorum Collected Studies series.
Mainstone, R. (2003). *Saving the dome of St. Peter's, Construction history* (Vol. 19).
Martines, G. (1989). *Argomenti di Geometria Antica. A proposito della cupola del Pantheon*. Quaderni dell'Istituto di storia dell'Architettura, 13.
Poleni, G. (1784). *Memorie istoriche della gran cupola del tempio vaticano e de' danni di essa, e de' ristoramenti loro, Padova 1748*. Rome: Ristampa della Biblioteca Facoltà di Arch. Univ. di Roma La Sapienza.
Scarano, M. (2007–2008). *Cupola di Santa Maria del Fiore, Costruzione e Statica*, Supervisors Como M.; Fabiani, F.M., a.y.
Sgrilli, B. S. (1733). *Descrizione e studi dell'insigne Fabbrica fi santa Maria dedl Fiore*. Firenze: Paperini.
Terenzio, A. (1934). *La Restauration du Pantheon de Rome*. In aa.vv. *La Conservation des monuments d'art et d'histoire*, Paris.
Wilson Jones, M. (2000). *Principles of Roman architecture*. New Haven and London: Yale University Press.

Chapter 6
Masonry Vaults: Barrel Vaults

Abstract This chapter deals with statics of masonry barrel vault, the simplest form of a curved roof covering spaces with a rectangular plan. They have an ancient origin but began to be extensively built in the Roman and Romanesque architecture. Barrel vaults are generally thin cylindrical solids of a given profile, supported at their boundaries. Unlike spherical shells, masonry cylindrical shells, with their zero Gaussian curvature, do not benefit from shape strength. Consequently, they are more prone to deformation and consequently to cracking. At cracking the vault reduces to a series of side-by-side arches. Lastly the cylindrical masonry barrel vault with semicircular profile and sustained by side walls is thoroughly studied under various loading conditions.

6.1 Introduction

The barrel vault, which belongs to the family of cylindrical shells, is the simplest form of a curved roof covering spaces with a rectangular plan. Many examples of barrel vaults are found in the world's architectural heritage, some, such as the renowned roof of the Sistine Chapel, are famous all over the world. Barrel vaults were used extensively in Roman architecture (Lancaster 2007).

Figure 6.1 shows the barrel vault of the Basilica of St. Sernin, one of the most beautiful examples of Romanesque architecture in France. In this case the vault is supported by arches erected transverse to the barrel. These arches can be built first and then used as the formwork for constructing the barrel, which can be completed bay-by-bay using a movable scaffold.

Barrel vaults are generally thin cylindrical solids of a given profile, supported at their boundaries, and thereby exhibit three-dimensional behavior. On the other hand, it seems evident that the static behavior of the barrel vault can be viewed as that of a series of side-by-side arches.

The capacity (or incapacity) of a vault's constituent materials to sustain tensile stresses is the decisive factor in its behavior. In the presence of tensile strength, the

Fig. 6.1 Barrel vault of the Basilica of St. Sernin in Toulouse (from wikipedia creative commons 2009)

stress state in a vault can be simply analyzed by means of shell membrane solutions, which can provide a good approximation of the stresses occurring in the vault before cracking. On the contrary, when cracking occurs, the behavior of the vault will actually correspond to that of a series of side-by-side arches, as will be shown in the following sections.

6.2 Membrane Stresses in Cylindrical Vaults

Unlike spherical shells, cylindrical shells, with their zero Gaussian curvature, do not benefit from shape strength. Consequently, they are more prone to deformation and, if made of no-tension material, cracking. Nonetheless, describing the uncracked state of the barrel vault, which is well-represented by membrane stresses, is useful not only for a better understanding of the transition from the uncracked to the cracked state, but also to validate the assumed simple model of side-by-side arches.

A cylinder is a geometric solid generated by moving a straight line along a curve while maintaining it parallel to its original direction. From this definition it follows that through any point on the cylinder a straight line may pass which lies entirely on its surface. Each of these lines is called a generatrix. For convenience, we shall assume that the generatrices are horizontal. All planes normal to the generatrices intersect the cylinder in identical curves which are called profiles.

Generatrices and profiles represent a natural net of coordinates lines. We can choose an arbitrary profile as the datum line and measure coordinate x along the generatrices from this line, positive in one direction and negative in the other.

A local reference system $Pxyz$ is defined at each point P on the surface with the x axis directed along the generatrices, that is, along the axis of the cylinder; the y axis along the tangent to the profile passing through P; and axis z having the direction of the outward normal to the surface.

If r is the profile's radius of curvature at P, the element of the arc ds can be expressed as $ds = r\, d\phi$, where ϕ, in the plane of the profile, is the angle between the normal to the profile at P and the vertical axis (Fig. 6.2).

Let us consider a shell whose middle surface is a cylinder. We cut from it an element bounded by two adjacent generatrices ϕ and $\phi + d\phi$ and by two adjacent profiles x and $x + dx$. (Fig. 6.3). The membrane forces acting on the four edges must all lie in tangential planes to the middle surface and may be resolved into normal and shear components as shown. The forces per unit length of section are N_x, N_ϕ, the normal forces, and $N_{x\phi} = N_{\phi x}$ the shearing forces. The load per unit area of the shell element is the weight g. The element of surface has area $dxds$ and the total load acting on the element is $gdxds$.

Figure 6.3 shows the components of the membrane forces acting on the element of the middle cylindrical surface of the vault. The equilibrium in the three directions x, y, z yields the equations (Flugge 1962; Belluzzi 1955; Heyman 1977):

$$\frac{\partial N_x}{\partial x} + \frac{\partial N_{\phi x}}{\partial s} = 0 \quad \frac{\partial N_\phi}{\partial s} + \frac{\partial N_{x\phi}}{\partial x} + g \sin \phi = 0 \quad N_\phi = -gr \cos \phi. \quad (6.1)$$

The force N_ϕ is independent of x. Moreover, the internal equilibrium directly determines force N_ϕ, which cannot therefore depend on the boundary conditions of the vault. Owing to the profile curvature, force N_ϕ balances the load component in

Fig. 6.2 Element ds of the profile

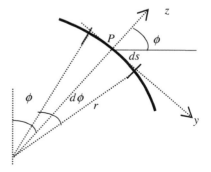

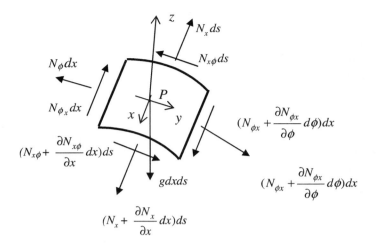

Fig. 6.3 Components of the membrane forces acting on the element of a cylindrical shell

the direction normal to the vault surface. When this normal becomes horizontal, $N_\phi = 0$. From Eq. (6.1), we have

$$N_{x\phi} = -Kx + A(\phi); \quad N_\phi = -gr\cos\phi; \quad N_x = \frac{x^2}{2}\frac{1}{r}\frac{\partial K}{\partial \phi} - x\frac{1}{r}\frac{\partial A(\phi)}{\partial \phi} + B(\phi). \tag{6.2}$$

with the position

$$K = (g\sin\phi + \frac{1}{r}\frac{\partial N_\phi}{\partial \phi}). \tag{6.3}$$

Force $N_{x\phi}$ varies linearly with distance x, while force N_x varies with the square of distance x. This follows from the fact that the membrane solution must also be able to represent the beam behavior along direction x. Since the total applied load is constant, the shear force and bending moment will respectively vary linearly or with the square of distance x.

Equations (6.2) and (6.3), which hold for the generic cylindrical shell, take on specific forms for the three profile cases considered. The profile equation can be expressed in terms of the change in its radius of curvature r with angle ϕ. Thus, we can write

$$r = r_o \cos^n \phi. \tag{6.4}$$

where r_o is the radius of curvature of the profile at the crown. When $n = 0$, Eq. (6.4) represents an arc of the circle; when $n = 1$, an arc of a cycloid; when $n = -3$ an arc of a parabola; and when $n = -2$, an arc of catenary.

6.2 Membrane Stresses in Cylindrical Vaults

Catenary profile
In this case the profile equation is

$$r = r_o \cos^{-2} \phi. \tag{6.5}$$

Thus, from the second equation in (6.2), we have

$$N_\phi = -g r_o \cos^{-1} \phi \tag{6.6}$$

and consequently

$$K = g \sin \phi - \frac{\cos^2 \phi}{r_o} \frac{\partial}{\partial \phi} (g r_o \cos^{-1} \phi) = g \sin \phi - g \sin \phi = 0. \tag{6.7}$$

The barrel vault is uniformly supported on its lateral walls, as shown in Fig. 6.4. The central section, defined by $x = 0$, belongs to a plane of symmetry and therefore there $N_{x\phi} = 0$. Consequently, the first of Eq. (6.2) yields $A(\phi) = 0$ and the shear force $N_{x\phi}$ in the vault cancels perfectly.

Transverse walls are present at the vault's end section, where, because these walls are unable to sustain loads directed orthogonally to their plane, the condition $N_x = 0$ holds. Consequently, from the third equation in (6.2), we get $B(\phi) = 0$ and once again in this case we have $N_x = 0$. The force N_ϕ, which is the only force different from zero, is given by (6.6).

The barrel vault transmits its weight uniformly to the side walls by means of the forces N_ϕ. The model of side-by-side arches fits the membrane behavior of the vault very well. Such result was in any event to be expected because the catenary is the funicular curve of the loads distribution acting along the profile.

Circular profile
In this case we have:

$$r = R \tag{6.8}$$

All three force components N_x, N_ϕ ed $N_{x\phi}$ are now acting in the vault. However, according to previous considerations, we still have $A(\phi) = B(\phi) = 0$ and

$$N_\phi = -gR \cos \phi. \tag{6.6'}$$

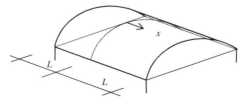

Fig. 6.4 Scheme of a barrel vault

Furthermore,

$$K = \left(g\sin\phi + \frac{1}{R}\frac{\partial N_\phi}{\partial\phi}\right) = 2\,g\sin\phi \quad \frac{\partial K}{\partial\phi} = 2g\cos\phi. \tag{6.9}$$

and the other two force components are

$$N_x = -(L^2 - x^2)\frac{g}{R}\cos\phi \quad N_{x\phi} = -2xg\sin\phi. \tag{6.10}$$

If the profile is a semicircle, at $\phi = \pi/2$ we have $N_\phi = 0$. The vault's weight cannot therefore be supported by the lateral walls. Only shear forces $N_{x\phi} = -2gx$ act on these walls and increase linearly along the generatrices up to the value $N_{x\phi} = -2gL$ at the transverse end walls (Fig. 6.5).

The forces N_x vary with the square of x along the generatrices and vanish at the end sections. The forces $N_{x\phi}$ transmit the vertical shear, represented by the weight, to the transverse end walls. The presence of these end walls ensures equilibrium of the vault.

The forces N_x and N_ϕ are always compressive at each point of the vault and vanish at the straight edges, in the neighborhood of the side walls. The straight edges of the vault are thus subjected to pure shear which produces tensile stresses along these edges in the diagonal direction (Fig. 6.6).

While reinforced concrete barrel vaults are fitted at their edges with suitable steel tie rods, masonry circular barrel vaults, lacking these, are prone to cracking along the longitudinal edges.

Fig. 6.5 Membrane stresses along the longitudinal and transverse edges

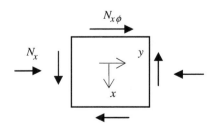

Fig. 6.6 Membrane stresses in the neighborhood of the end walls

6.3 Transition from the Uncracked to the Cracked State. The No-Tension Model of the Barrel Vault

Let us now refer to a common barrel vault with semicircular profile to analyze the transition from the uncracked to the cracked state.

We have seen that in the uncracked state the vault behaves like a longitudinal beam supported at its ends by transverse walls. Internal stresses are transmitted directly to masonry material along the straight edges and transverse walls. Cracks occur along the vault edges and the membrane stresses become inconsistent: both forces N_x and $N_{x\phi}$ must cancel out. Only forces N_ϕ do not, and the second equation in (6.1) cannot be satisfied by membrane forces. A bending moment takes place in the plane of the profile leading to a shear force T_ϕ. Owing to the moment M_ϕ, axial forces N_ϕ deviate from the midline of the vault section, the middle semicircle, and winding within the thickness of the vault, take on the profile of a catenary and load the side walls. This occurs at each transverse section of the vault and thus, upon cracking, the vault subdivides into a series of independent side-by-side arches. In order to avoid failure, the pressure line must be completely contained within the thickness of the vault. The thickness of a semicircular profile barrel vault must be greater than 1/10 its radius, at least, to be stable. The ensuing thrust must be sustained by the lateral walls (Fig. 6.7).

In the end, masonry barrel vaults behave as a series of side-by-side arches supported by lateral walls, which must be quite thick and cannot present wide openings. The static analysis of a masonry barrel vault is thus reduced to that of a masonry arch with equal profile and thickness.

Fig. 6.7 Stress transmission from the barrel vault to supporting walls

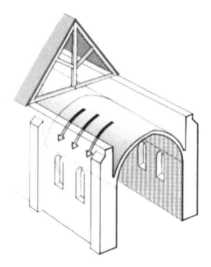

6.4 Systems Made up of Vaults and Walls

Composite structural systems can be found in various forms in architecture.

Figure 6.8 shows some common systems constituted by barrel vaults with side walls in various arrangements starting with the first, simplest scheme 1.1. In schemes 1.2 and 1.3 the side walls are lightened by means a series of small chapels covered by barrel vaults, an arrangement commonly found in churches. A series of wide piers runs along the nave flanked by the small vaults of the side chapels. These vaults also carry out the function of bearing the thrust of the main vault. Schemes 1.4 and 1.5 instead have two lateral aisles flanking the nave and the lateral vaults transmit the thrust of the central one. Case 1.6 illustrates the Gothic solution, in which the thrust is transferred by flying buttresses. The sequence of these various systems represents the evolution of church construction from the Romanesque to the Gothic style.

6.4.1 The Barrel Vault with Side Walls

6.4.1.1 Vertical Loads

Scheme 1.1 in Fig. 6.8 is the simplest system of a barrel vault with side walls. Such a system can easily fail if the walls are too slender, so throughout history numerous empirical rules have been formulated to establish safe dimensioning. At first, only vertical loads will be taken into account. Figure 6.9 defines the system's various geometric parameters, considering unit width in the direction orthogonal to the plane of the figure. The weights applied to the side walls are constant, while those

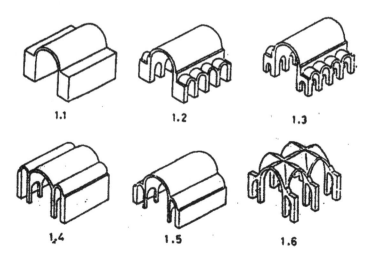

Fig. 6.8 Systems of barrel vaults and side walls (from Como, Lanni 1988)

6.4 Systems Made up of Vaults and Walls

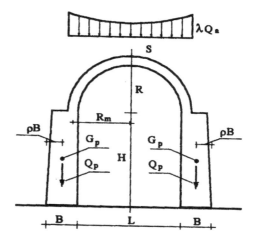

Fig. 6.9 Geometry of the barrel vault supported by side walls

on the vault increase proportionally according to the loading parameter λ. The failure mechanism may or may not be symmetrical. A symmetrical mechanism is illustrated in Fig. 6.10. Failure occurs when, despite the weight of the side walls, the axial load, because of the thrust of the vault, becomes so eccentric to reach the toe of the piers' base.

Precise analysis has shown that the failure mechanism is indeed symmetrical. The collapse load multiplier λ_{oV} of the vertical loads acting on the vault alone can be expressed as (Abruzzese, Como, Lanni 1995):

$$\lambda_{oV} = \frac{Q_p}{Q_a} \frac{\rho b}{k_V}. \qquad (6.11)$$

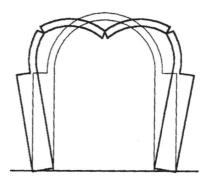

Fig. 6.10 Symmetrical failure of a composite vault/wall or arch/pier

where, with reference to Fig. 6.9, Q_p is the weight of a single wall, $\bar{Q}_a$ the weight of the whole vault, ρB the distance of the wall center G_p from the external toe, and k_V is a parameter representative of the geometry of the system:

$$k_V = 0.3 \frac{h+0.6}{s+0.4}\left[b+0.55-(h+s+1)\frac{b+0.2}{h+0.6}\right] - 0.2(b+0.02), \qquad (6.12)$$

where

$$h = \frac{H}{R} \qquad b = \frac{B}{R} \qquad s = \frac{S}{R}. \qquad (6.13)$$

The collapse load multiplier λ_{oV} of the vertical loads acting on the vault, expressed by (6.11), depends on the ratios between the various parameters representative of the geometry of the vaulted system but not on their absolute measures. This result is in agreement with the general statement proved at Sect. 2.7 about the theory of proportions of the past architecture

6.4.1.2 An Example

A round barrel vault with internal radius $R = 4.0$ m and thickness $S = 0.45$ m and unit maonry weight $\gamma = 1.6$ t/m^3 is supported by two side walls of width $B = 1.50$ m and height $H = 3.50$ m. Only the weight Q_a of the vault increases according to loading parameter λ. By applying (6.13) we evaluate the dimensionless geometric quantities s, b and h and get $s = 0.1125$; $b = 0.375$; $h = 0.875$. From (6.12), the quantity $k_V = 0.0507$.

Now, $\bar{Q}_a = 0.5 \cdot \pi(4.90^2 - 400^2) \cdot 1.6 = 9.557$ t, $Q_p = 1.5 \cdot 3.50 \cdot 1.6 \cdot 1.0 = 8.40$ t, $\rho = 0.50$, and we have $\lambda_{oV} = 3.25$. The vault/walls system will thus fail when the weight of the vault is increased more than three times of its initial value.

6.4.1.3 Horizontal Loads

Let us now consider the vault in the presence of horizontal forces, which can be used to represent the action of an earthquake. Figure 6.11 shows the geometry of the system under the action of both vertical and horizontal loads, these latter acting from left to right. The weights are maintained constant, while the horizontal forces increase according to parameter λ, for which it is assumed that a value of $\lambda = 1$ corresponds to horizontal actions equal to the vault weight.

Depending on the system's geometry and mass distribution, three different failure mechanisms can arise: local, semi-global or global.

A local mechanism develops wholly within the vault, the semi-global involves both the vault and right wall, while global failure affects the vault together with both

6.4 Systems Made up of Vaults and Walls

Fig. 6.11 A barrel vault supported by side walls under vertical and horizontal loads

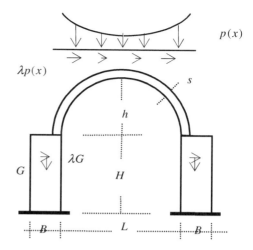

walls. This last type, which is quite uncommon and occurs in light vaults with slender walls, will not be considered in the following discussion, which will be limited to the first two mechanisms as they occur frequently in the common types of vault/walls systems. The semi-global mechanism occurs quite frequently because the horizontal forces act from left to right, so the vault thrust and the horizontal forces act concomitantly on the right side, while the opposite occurs on the other wall (i.e. the thrust opposes the horizontal actions).

Figures 6.12 and 6.13 respectively illustrate the local and semi-global mechanisms, showing the hinge positions (Como, Lanni 1988). Now let

$$u' \quad u'' \tag{6.14}$$

indicate two generic mechanisms, the first representing local deformation of the vault alone and the second the semi-global deformation involving both the vault and one wall.

Let also

$$M' \quad M'' \tag{6.15}$$

Fig. 6.12 Hinges in local failure of the vault alone

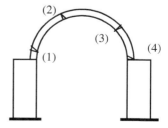

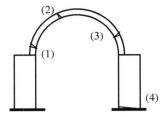

Fig. 6.13 Hinges in semiglobal failure of the vault Hinges of the semi-global failure

be the sets of all local and semi-global mechanisms. The failure horizontal load multiplier is determined by means of the kinematic theorem. We can consider

$$\lambda_c = \underset{M}{Min}\ \lambda'_c(u). \tag{6.16}$$

where $\lambda'_c(u)$ is the generic load multiplier corresponding to mechanism u, and M the set of all local and semi-global mechanisms

$$M = M' \cup M''. \tag{6.17}$$

The horizontal collapse multiplier is thus obtained as

$$\lambda_c = Min(\lambda'_o, \lambda''_o) \tag{6.18}$$

where λ'_o and λ''_o are the collapse multipliers restricted to the sets M' and M'' of the local and semi-global mechanisms.

6.4.1.4 Finding λ'_o in the Set of Local Mechanisms

The analysis follows the procedure set forth in Chap. 7 of the previous chapter—specifically, steps (a), (b), (c) and (d)—for a masonry arch subjected to horizontal forces (Fig. 6.14).

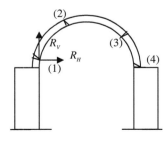

Fig. 6.14 Evaluating λ_c by imposing zero moment at hinges 2, 3 and 4

As for the arch, we recall that hinge (6.4) is located at the extrados of the right springer section and the distance between hinges (6.2) and (6.3) is approximately equal to $L/2$, where L is the internal span of the arch. These results allowing reducing the four unknowns to only two: x_1 and x_2, that define the positions of hinges (6.1) and (6.2) from the left springing (in this case of the vault). Function $\lambda_c(x_1, x_2)$ is defined by writing the equilibrium equations involving the unknowns λ_c, R_V, R_H. We will account for the fact that hinge (6.1) is very near the left springer. The multiplier λ'_o is obtained by numerically evaluating distances x'_1 and x'_2 when function $\lambda_c(x_1, x_2)$ reaches a minimum. Let us now examine the case of the semi-global failure mechanism.

6.4.1.5 Finding λ''_o in the Set of Semi-global Mechanisms

Once multiplier λ'_o has been obtained, we must verify whether this multiplier is also statically admissible with regard to the stresses inside the right wall. Let R'_V, R'_H. and M', respectively, be the three actions—vertical, horizontal and moment—that the vault, in the previously determined state of local failure, transmits to the right wall at the intrados of its springer section. We can now determine the value of multiplier λ_{op} of the horizontal forces that leads to failure of the wall according to the scheme in Fig. 6.15. The limit equilibrium of the wall yields the equation

$$-M' + R'_V B - R'_H H + \frac{GB}{2} - \lambda_{op} GH = 0. \qquad (6.19)$$

Thus if

$$\lambda_{op} \geq \lambda'_o. \qquad (6.20)$$

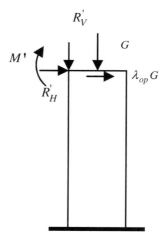

Fig. 6.15 The limit state for overturning of the right wall

the collapse of the system occurs through the development a local failure mechanism. On the other hand, if

$$\lambda_{op} < \lambda'_o. \qquad (6.21')$$

collapse involves both the vault and the right wall according to a semi-global mechanism u'', which is yet to be determined. Let us consider a generic semi-global mechanism, as sketched out in Fig. 6.16.

The evaluation of kinematic *multiplier* $\lambda_c(u'')$ is pursued by solving the equilibrium equations of zero moment at the mechanism hinges. Equation (7.56) in Chap. 7 regarding arches, must now be reformulated to yield the new equation for zero moment at the right toe of the right wall (Fig. 6.16), as follows:

$$-R_V(L+B) - R_H H + M_V(L) + R_p B - \lambda_c M_{oi}(L) - \lambda_c R_p H + G\frac{B}{2} - \lambda_c GH = 0, \qquad (6.22)$$

where R_p is the resultant of load p, and $\lambda_c R_p$ the resultant of load $\lambda_c p$.

Performing the numerical calculations reveals some regularities in the locations of hinges in the semi-global collapse mechanism. In fact, the same regularities found in the positions of hinges 2, 3 and 4 in local failure still hold. Consequently, in evaluating λ''_o, the only unknowns are the distances x_1 and x_2 of hinges 1 and 2 from the left vault springing (Fig. 6.16). The search for the collapse multiplier is thus reduced to the two-variable minimum problem

$$\lambda_o = \lambda''_o = \underset{M''}{\mathrm{Min}}\, \lambda_c(u'') = \mathrm{Min}\, \lambda_c(x_1, x_2) \qquad (6.23)$$

Now, if x''_1, x''_2 represent the values of variables x_1 and x_2, where the multiplier λ_c is at a minimum (6.23), the following simplifying condition holds

$$x''_1 > x'_1 \quad x''_2 > x'_2. \qquad (6.24)$$

Fig. 6.16 Evaluating λ_c by imposing zero moment at hinges 2, 3 and 4

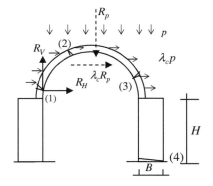

where x'_1 and x'_2 are the values of x_1 and x_2 corresponding to the local failure. Hinges 1, 2 and 3 move rightwards due to failure of the right wall. A simple expression for the failure multiplier of the horizontal force for round vaults supported by side walls has been obtained by Abruzzese, Como, Lanni (1995), as follows:

$$\lambda_{oH} = \left(\frac{\rho b}{\delta h} - \frac{Q_a k_1}{Q_p \delta h}\right)\left(1 + \frac{Q_a k_2}{Q_p \delta h}\right)^{-1}. \tag{6.25}$$

By using the same notation as above and denoting δh as the height of the center of the wall, we obtain

$$k_1 = \frac{b+0.2}{C-1.8}(0.61 - 0.47C) - 0.2(b+0.02);$$
$$k_2 = \frac{b+0.2}{C-1.8}[0.58 - (0.62+0.3s)C] + 0.2(h+0.3) \tag{6.25'}$$

$$C = \left[\frac{h+0.6}{b+0.2}(2+b) - h\right]\left(1 + s + \frac{h+0.6}{b+0.2}\right)^{-1}. \tag{6.25'}$$

Also in this case the strength of the system depends only on the ratios between the parameters defining the geometry of the system but not on their absolute values, confirming the result given at Sect. 2.7.

By way of example, let us consider the same vault/wall system described above. The dimensionless quantities s, b and h take the same values, namely, $s = 0.1125$; $b = 0.300$; $h = 0.875$; $\delta = 0.5$. According to (6.25') and (6.25"), parameters C, k_1 and k_2 are $C = 1.4548$; $k_1 = 0.0066$; $k_2 = 0.5238$. Thus,

$$\lambda_{oH} = \left(\frac{\rho b}{\delta h} - \frac{Q_a k_1}{Q_p \delta h}\right)\left(1 + \frac{Q_a k_2}{Q_p \delta h}\right)^{-1}$$
$$= \left(0.4286 - 2.3966\frac{0.0066}{0.4375}\right)\left(1 + 2.3966\frac{0.5238}{0.4375}\right)^{-1} = 0.1014.$$

The system thus fails when the horizontal forces reach a value equal to about the 10 % of its weight.

6.4.2 Tie Rod Reinforcing Vault/Walls Systems

Frequently vault/walls systems are reinforced with iron or steel tie rods. Figure 6.17 shows an example of a barrel vault reinforced by a tie rod with end plates. The increase in strength obtainable by fitting such a tie rod can be easily analyzed.

Fig. 6.17 The vault and side walls reinforced with tie rods

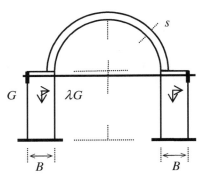

Let us first consider the presence of only vertical loads. The local mechanism, involving the presence of hinges in the vault alone, will not be considered. Referring to the mechanism shown Fig. 6.10, when both masonry walls outwards by angle θ, the resisting workdue only to the wall weight is $Q_p \rho B \theta$, where ρ indicates the eccentricity of the wall weight with respect to the external toe.

In the presence of the tie rod, it can be seen that the mechanism can take place only if the yield stress of the chain is attained. Thus, to the balance between the resisting and pushing work, we must add the work due to the plastic dissipation occurring in the tie rod. Referring now to half the system, we must add $T_o H \theta$, where T_o is the yield strength of the tie rod. We can consider a fictitious work of raising the wall weight that also includes the effects of the plastic dissipationl. We can write

$$L_{res,lat} = Q_p \rho B \theta + T_o H \theta = Q_p \rho B \left(1 + \frac{T_o H}{Q_p \rho B}\right) \theta = Q_p \rho * B. \quad (6.26)$$

and

$$\rho* = \rho \left(1 + \frac{T_o H}{Q_p B}\right). \quad (6.27)$$

The collapse multiplier can be thus be obtained by using (25), (6.25') and (6.25''), except that the fictitious eccentricity $\rho*B$ of the pier weight Q_p, where $\rho*$ is given by (6.27), is considered instead of ρB (Abruzzese, Como, Lanni 1985).

With reference to the vault/walls example considered above, let us consider the presence of a tie rod made of Fe 360 steel of circular section $\phi = 20$. The yield strength of the tie rod is thus $3.14 \times 2400 = 7.54$ t. From Eq. (6.27) we have $(T_o H)/Q_p B) = (7.54 \cdot 3.40)/(8.40 \cdot 1.50) = 2.035$, and the value of $\rho* = 1.5175$. From (6.27), by inserting the aforementioned tie rod and fixing it firmly to the side walls with suitable steel end plates, we get $\rho*/\rho = 3.03$. In conclusion, the limit vertical loads of the reinforced system increases by about 4.67 times, while for the unreinforced system the limit equilibrium is reached under loads any more than about 50 %. The results are similar in the presence of horizontal loads. For semiglobal

Fig. 6.18 The system vault/walls with reverse tie rod

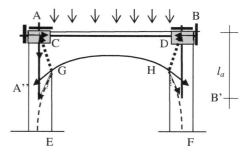

mechanisms, the limit horizontal strength can also be obtained by means of (6.25), except that coefficient ρ^*, given by (6.27), is considered in place of the actual ρ. Thus, we have

$$\lambda_{oH} = \left(\frac{\rho b}{\delta h}\frac{\rho^*}{\rho} - \frac{Q_a k_1}{Q_p \delta h}\right)\left(1 + \frac{Q_a k_2}{Q_p \delta h}\right)^{-1}$$
$$= \left(0.4286 \cdot 3.03 - 2.3966 \frac{0.0066}{0.4375}\right)\left(1 + 2.3966 \frac{0.5238}{0.4375}\right)^{-1} = 0.3262$$

In order to further counter the vault thrust, it is also possible to place a reverse tie rod at the vault extrados, though suitable supplementary reinforcements are naturally required. The idea is to transfer the thrust of the vault from its springing to the extrados. Figure 6.18 shows the solution to this problem: a vertical anchor rod ensures transmission of the vault thrust—a useful system for reinforcing vaults first suggested by Giuriani (Giuriani 1985).

References

Abruzzese, D., Como, M., Lanni, G. (1995) Some results on the strength evaluation of vaulted masonry structures. In: Brebbia, C.A., Leftheris, B. (eds.) Architectural Studies, materials and Analysis, STREMA 1995. Computational Mechanics Publ., Southampton, Boston.
Belluzzi, O. (1955). *Scienza delle Costruzioni* (Vol. III). Bologna: Zanichelli.
Como, M., & Lanni, G. (1988). *Sulla verifica alle azioni sismiche di complessi monumentali in muratura, Atti Convegno "Nel Cinquantenario della facoltà di Architettura di Naples, Franco Jossa e la sua opera"*. Facoltà di Architettura, Giannini editore, Naples: Istituto di costruzioni.
Flügge, W. (1962). *Stresses in shells*. Berlin: Springer Verlag.
Giuriani, E., Gubana, A. (1985) Extrados ties for structural restoration of vaults. In: Brebbia, C.A., Leftheris, B. (eds.) Architectural studies, materials and analysis, Computational Mechanics Publ., Southampton, Boston.
Heyman, J. (1977). *Equilibrium of shell structures*. Oxford: University Press.
Lancaster, L. C. (2007). *Concrete vaulted construction in imperial Rome*. Cambridge: Cambridge University Press.

Chapter 7
Masonry Vaults: Cross and Cloister Vaults

Abstract The aim of this section is the study of statics of cross and cloister vaults: they are related together in various respects, since from the point of view of their geometric generation. A brief introduction gives information about the historical development of these vaults: there are magnificent examples already in the Roman architecture. Membrane stresses in cross and cloister vaults are thoroughly studied. The existence of tensile stresses is a necessary condition for the membrane equilibrium and cracking is nearly inevitable. A study of sliced models of both the vaults is thus pursued. In the sliced model of the cross vault, firstly proposed by Heyman, webs transmit vertical and horizontal loads to diagonal ribs or to groins that, in turn, convey these loads to piers and flying buttresses. A dual model for the cloister vault, that cracks along the diagonals and transmits loads on the side walls, is firstly proposed. A detailed analysis of the cracks patterns is conducted for both the vaults at a their minimum thrust states. The chapter ends with the study of statics of some important examples of the vaults.

7.1 Geometric Generation of Cross and Cloister Vaults

Cross and cloister vaults are related in various respects, particularly from the points of view of their geometric generation.

Let us consider two equal, semicircular-profile barrel vaults on a square plan, and let us section each one with two diagonal vertical planes, as shown in Fig. 7.1. Such sectioning produces two different pairs of cylindrical webs. The pair of opposing webs A and the pair of opposing webs B have respectively the same profile and the same edges as the initial barrel vaults.

We can generate both the cross vault and the cloister vault by joining the four A webs or the four B webs, as shown in Figs. 7.2 and 7.3.

The cross vault is supported on its four corners, while the cloister vault rests on its four edge walls. The intersection of two equal, though not necessarily semi-

---

The original version of the chapter was revised: Partially processed figures 7.36 and 7.37 have been replaced. The erratum to the chapter is available at: 10.1007/978-3-319-24569-0_13

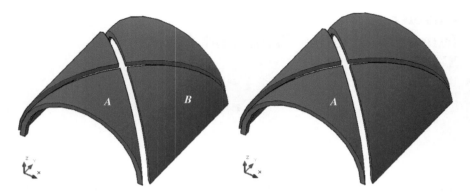

Fig. 7.1 Two semi-cylindrical barrel vaults with square plan and their diagonal sections

Fig. 7.2 Generation of a cross vault

Fig. 7.3 Generation of a cloister vault

circular, barrel vaults yields the simplest cross vault with a square in plan. More in particular, if the intersecting barrel vaults are semicircular in profile, the resulting cross vault will have rounded lateral arches and semi-elliptical diagonals.

The intersection of two barrel vaults having circular profiles, but different radii, yields a rectangular-plan cross vault. The geometry of the vault is thus modified: if the arch of the smaller span vault is round, the arch of the longer span vault must

7.1 Geometric Generation of Cross and Cloister Vaults 293

Fig. 7.4 Horizontal projection of intersection curves among the various webs composing the groin vault with rectangular plan

Fig. 7.5 Groin vault of the San Martino Carthusian monastery in Naples

necessarily be depressed. In the case of a rectangular plan, the intersection of the webs is a space curve.

The consequent cross or cloister vaults will present curvilinear groins in plan. The mismatch between the exact intersection curve and the approximate linear intersection is negligible, except when dealing with very elongated rectangular vaults (Fig. 7.4).

In this last case these space curved ribs can be made into an architectonic decoration. Cross vaults with rectangular plans and space curved ribs are uncommon. Figure 7.5 shows one unusual example of a cross vault with sinusoidal ribs in plan: the cross vault of the Carthusian monastery of San Martino in Naples.

7.2 Surface Areas and Weights of Webs and Lunes

Cylindrical webs with semicircular directrix

Figure 7.6, shows the cylindrical surface with semicircular directrix of diameter 2R with square plant generating both as the web as the lune according to the above discussion.

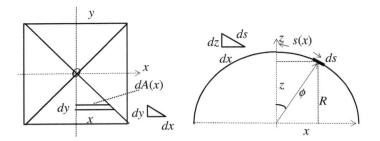

Fig. 7.6 Valuation of the current area $A(x)$

The radius R is the inner radius of the cylinder. The thickness of the web is denoted with t, a small quantity compared to R.

We consider the generic *web segment* defined by its distance x of its end section from the axis y passing through the vertex O. Making reference to the half of the web segment the surface element is

$$dA = s(x)dy = s(x)dx \qquad \text{(a)}$$

where, due to the assumed square plan, $dx = dy$. From Fig. 7.6 we have

$$ds \cos \phi = dx \quad x = R \sin \phi \qquad \text{(b)}$$

whence

$$\frac{ds}{dx} = \frac{1}{\cos \phi} = \frac{1}{\sqrt{1 - \sin^2 \phi}} = \frac{R}{\sqrt{R^2 - x^2}} \qquad \text{(c)}$$

e quindi il differenziale della lunghezza d'arco $s(x)$ vale

$$ds = \frac{R}{\sqrt{R^2 - x^2}} dx \qquad \text{(d)}$$

The length of the arc $s(x)$ is thus

$$s(x) = R \int_0^x \frac{d\xi}{\sqrt{R^2 - x^2}} = R \arcsin \frac{x}{R} \qquad \text{(e)}$$

7.2 Surface Areas and Weights of Webs and Lunes

We can valuate now the current area $A(x)$ of the surface of the half web as

$$A(x) = \int_0^x s(\xi)d\xi = R\int_0^x \arcsin\frac{\xi}{R}d\xi = R^2(\frac{x}{R}\arcsin\frac{x}{R} - 1 + \sqrt{1 - x^2/R^2}) \quad (f)$$

so that the current area of the whole web segment of width x is

$$U(x) = 2A(x) = 2R^2(\frac{x}{R}\arcsin\frac{x}{R} - 1 + \sqrt{1 - x^2/R^2}) \quad (g)$$

or, considering the dependance on the angle ϕ

$$U(\phi) = 2A(\phi) = 2R^2(\phi\sin\phi - 1 + \cos\phi) \quad (g')$$

For $x = R$ we obtain the surface area of the *complete web*. We have

$$U_c = U(R) = 2A(R) = R^2(\pi - 2) \quad (h)$$

Then, if the weight g for unit surface area of the web is $g = \gamma t$, where γ the unit weight of the masonry, the weight of the web segment of length x, is

$$G_w(x) = gU(x) = 2gA(x) = 2gR^2(\frac{x}{R}\arcsin\frac{x}{R} - 1 + \sqrt{1 - x^2/R^2}) \quad (i)$$

In particular, the weight of the *whole web* G_{web} is

$$G_W = G_w(x = R) = g(\pi - 2)R^2 \approx 1{,}142 gR^2 \quad (l)$$

Cylindrical lune with semicircular directrix

We consider the dual case of the cloister vault wit directrix of diameter $2R$, constant thickness $\underline{t}$ and square plant. With reference to Fig. 7.7, the directrix of radius R, is defined by the function

$$z(y) = \sqrt{R^2 - y^2} \quad 0 \le y \le R \quad (m)$$

and the arc element is given by

$$ds = \sqrt{dy^2 + dz^2} = R(R^2 - y^2)^{-1/2}dy \quad (n)$$

The length of the arc $s(y)$ of the directrix as from the crown can be immediately valuated as

$$s(y) = \int_0^y \frac{R}{\sqrt{R^2 - y^2}}dy = R\sin^{-1}\frac{y}{R} \quad (o)$$

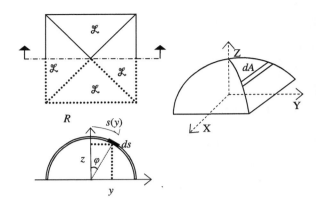

Fig. 7.7 Valuation of the surface area of the single lune

We can now obtain the surface area $F(y)$ of the segment of the lune. With reference to Fig. 7.9, the element of area is

$$dA = 2xds = 2yds \qquad (n')$$

and the surface area of F $F(y)$ of the segment of the lune is

$$F(y) = \int_0^y 2\xi \frac{R}{\sqrt{R^2 - \xi^2}} d\xi = 2R^2(1 - \sqrt{1 - y^2/R^2}) \qquad (p)$$

and the corresponding weight is

$$G_l(x) = 2gR^2(1 - \sqrt{1 - y^2/R^2}) = 2gR^2(1 - \cos\phi) \qquad (q)$$

In particular, for $y = R$ we have the surface area of the *complete lune*

$$F_L = F(y = R) = 2R^2 \qquad (r)$$

with the corresponding weight

$$G_L = 2gR^2 \qquad (s)$$

To check, summing up the weight of the two web and of two lunes, we have,

$$2G_W + 2G_L = 2g(\pi - 2)R^2 + 4gR^2 = 2g\pi R^2 \qquad (t)$$

the weight of the barrel vault from which we obtain the webs and the lunes, as shown in Fig. 7.1. Comparing (s) and (l) we get that the weight of the cloister vault is about 7/4 larger than the weight of the cross vault.

7.3 Historical Notes on Cross Vaults

The cross vault is highly efficient, both functionally and statically. In contrast to barrel vaults, wide windows can be opened in the side walls. For the sake of comparison, Fig. 7.8 shows the force transmission in a cross vault: it is clearly quite different from that of a barrel vault. Frequently, conspicuous diagonal ribs emerge from the intrados of the webs, while in other cases, internal ribs run along the groins within the masonry and are barely visible at the intrados as a crease along the diagonals. Some magnificent examples of cross vaults remain in Roman architecture: each bay of the cross vaults of the Diocletian Baths or the Basilica of Massentius in Rome span over 25.0 m.

Roman cross vaults, usually made of concrete, generally present rounded lateral arches and internal diagonal brickwork ribs. Centering was used to build both the lateral and diagonal arches, as well as to subsequently cast the concrete for construction of the webs. The concrete was laid in separate layers of *caementa* and mortar. The vaults were very heavy and transmitted large thrusts to their supporting structures. A thorough description of Roman vault construction techniques has been provided by Lancaster (2007).

Later Romanesque cross vaults were much smaller. They had pointed side arches and rounded diagonals. Like their earlier Roman counterparts, Romanesque cross vaults do not present outer ribs. The diagonal groins were built first, and only afterwards was the web added to the interior. Structurally, the Romanesque cross

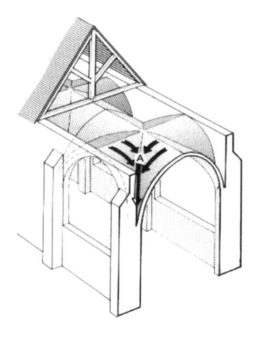

Fig. 7.8 Forces transmission in the groin vault

vault can be considered a single unitary system, with no differences between the webs and the ribs (Morabito 2004).

A significant change came during the Gothic period, whose builders diversified the ribs' functions from the webs', thus giving rise to the so-called *rib vaults*. The webs, which were much thinner and supported by ribs composed by ogival arches, emerged at the extrados of the vault. The vault's webs were fashioned to fit the boundary arches along both the groins and the four edges of the bay. Thirteenth century Gothic cross vaults, spanning as much as 10–15 m, are only about 20 cm thick, thereby producing limited amounts of thrust. Some Gothic cross vaults had extremely bold structural designs. For instance, the cross vaults of the Amiens Cathedral (Fig. 7.9) were built to a height of 42 m and those of *Saint-Pierre* in Beauvais to a height of 48 m. Figure 7.10a, b show sketches of quadripartite and sexpartite cross vaults.

In Gothic cross vaults the webs are sustained by diagonals ribs, which are in turn supported by corner pillars and a suitable system of buttresses. Thrusts are conveyed by the ribs to the flying buttresses, which, in turn, transmit them to the external buttresses. The curve of the pressure line is depressed in such vaults and a special filler is usually added in such a way as to reinforce the masonry near the springings.

One special type of cross vault is the so-called *star vault*, typical of Apulian architecture. These vaults are characterized by an incomplete intersection between the two orthogonal barrels generating the vault and present a central star-shaped band.

Fig. 7.9 Cross vaults of the cathedral of Amiens

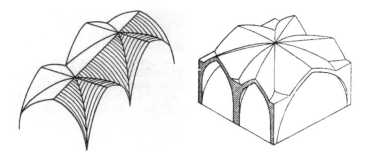

Fig. 7.10 a Quadripartite cross vault, **b** sexpartite cross vault (from Heyman 1997)

7.4 Statics of Cross Vaults

7.4.1 Initial Membrane Stresses

The study of the square in plan round cross vault of Fig. 7.11 provides sufficient information on the features of the membrane stress states occurring in cross vaults under their dead loads.

For sake of simplicity we will make reference to the cross vault with square plan having semicircular directrix of radius R. The vault is composed by four cylindrical webs intersecting each other along the diagonal groins. Each of the four webs is connected along the external side to a circular arch, called *formeret* and, along the diagonals, to the other webs. The webs are eventually sustained by diagonals ribs, which are in turn supported by corner pillars and a suitable system of buttresses (Fig. 7.12).

The local reference coordinates, defined at each point of the web, are the abscissa x, having the direction of the generatrices, and the straight line ϕ, tangent to the semicircular directrix. The axis x is the web axis. The edges of the single web are (Fig. 7.13):

- the semicircular boundary DBC, in contact with the *formeret* round arch
- the vertex O, the centre of the cross vault
- the two diagonal groins OC and OD, in contact with the two contiguous webs.

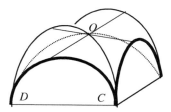

Fig. 7.11 Square in plan cross vault with semicircular directrix

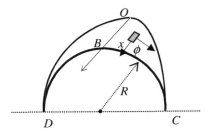

Fig. 7.12 A single web of the cross vault and the coordinates r and ϕ

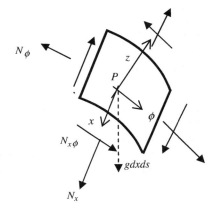

Fig. 7.13 Membrane forces on a cylindrical element inside the web

Now, consider this small element as contained within the web shown in Figs. 7.13 and 7.14.

In the above chapter we have seen that the internal equilibrium directly determines force N_ϕ, therefore independent on the boundary conditions of the vault. Force N_ϕ balances the load component in the direction normal to the vault surface and is

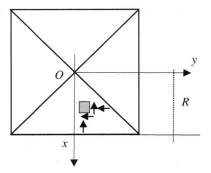

Fig. 7.14 Membrane stresses in the vault's web

7.4 Statics of Cross Vaults

$$N_\phi = -gR \cos \phi \tag{7.1}$$

The other forces N_x and $N_{x\phi}$ will obey the following equations that define the equilibrium of the small cylindrical element of vault along the direction x

$$\frac{\partial N_x}{\partial x} + \frac{\partial N_{\phi x}}{\partial s} = 0 \quad \frac{\partial N_\phi}{\partial s} + \frac{\partial N_{x\phi}}{\partial x} + g \sin \phi = 0. \tag{7.2}$$

Integration of Eq. (7.2), gives

$$N_{x\phi} = -Kx + A(\phi); \quad N_x = \frac{x^2}{2} \frac{1}{R} \frac{\partial K}{\partial \phi} - x \frac{1}{R} \frac{\partial A(\phi)}{\partial \phi} + B(\phi), \tag{7.3}$$

where

$$K = \left(g \sin \phi + \frac{1}{R} \frac{\partial N_\phi}{\partial \phi}\right). \tag{7.4}$$

With the reference axes coordinates Fig. 7.10, x is an axis of symmetry. Consequently, the shear forces $N_{x\phi}$ cancel at $x = 0$ and from the first of Eq. (7.3) above, we have

$$A(\phi) = 0, \tag{7.5}$$

Taking (7.1) into account together with (7.4) we get

$$K = 2g \sin \phi \tag{7.4'}$$

and the shear forces $N_{x\phi}$ become

$$N_{x\phi} = -2gx \sin \phi, \tag{7.6}$$

From (7.3) and (7.4'), the axial forces N_x along the generatrices will be

$$N_x = \frac{x^2}{R} g \cos \phi + B(\phi). \tag{7.7}$$

The side arches are unable to sustain the forces N_x orthogonal to their plane so that for $x = R$, $N_x = 0$ and

$$B(\phi) = -Rg \cos \phi \tag{7.8}$$

Whence we lastly obtain

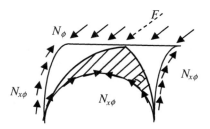

Fig. 7.15 Membrane forces acting in a groin vault. The lateral arches transmit the force distribution to the vault's side edges

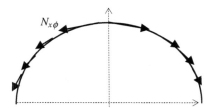

Fig. 7.16 Formeret arches hooping the sides of the vault

$$N_x = (x^2 - R^2)\frac{g}{R}\cos\phi. \tag{7.9}$$

The side arches cannot sustain forces orthogonal to their planes. Thus, taking Eqs. (7.1), (7.7) and (7.9) define the membrane forces acting within the webs of the considered cross vault (Fig. 7.11). The interplay among the actions transmitted by the webs to the diagonals ribs define the global equilibrium of the cross vault.

Figure 7.15 shows all the forces acting on a half groin vault. The compressive forces N_ϕ distributed along the web mid-sections equilibrate the horizontal components of the shearing forces $N_{x\phi}$. These shearing forces carry on a strong hooping action on the webs. The side arches, the *formeret*, behave as curved tie rods hooping the vault: they are subjected to tensile stresses (Fig. 7.16).

Thus, as shown for domes and barrel vaults, the existence of *tensile forces* is a necessary condition for the membrane state of stress in the vault. *Cracking* is thus inevitable and the masonry vault must reach a new internal equilibrium.

7.4.2 Transition from the Uncracked to the Cracked State

The *formeret* masonry arches cannot carry out the hooping action necessary for the membrane equilibrium. At cracking the shear forces $N_{x\phi}$ producing tension in these lateral arches will thus start to vanish in the neighborhood of the vault's edges and

7.4 Statics of Cross Vaults

then vanish along the whole web. The same will occur along the direction x, because, as forces $N_{x\phi}$ go to zero, from the first of Eq. (7.2) also the forces N_x will vanish. Thus the second equilibrium equation in (7.2) becomes

$$\frac{1}{R}\frac{\partial N_\phi}{\partial \phi} + g \sin \phi = 0, \qquad (7.10)$$

which cannot be satisfied with $N_\phi = -gR \cos\phi$. The membrane equilibrium along the direction y is thus *lost* and a variable bending moment M_ϕ must necessarily ensue, accompanied by a shear, T_ϕ, to replace the term $\partial N_{x\phi}/\partial \phi$ in Eq. (7.10).

With the vanishing of forces $N_{x\phi}$ and N_x, a new web equilibrium develops solely along the *direction* ϕ.

Moment M_ϕ causes the compressive force N_ϕ within the web thickness to be eccentric. Forces N_ϕ makes the web slices exert a thrusting action on the diagonal ribs. In turn, the diagonals activate a thrust S (Fig. 7.17) towards the corners: suitable structures such as flying buttresses are required to convey this thrust to the outer buttresses. Thrust S thus replaces the effects of the distribution of shear $N_{x\phi}$ along the edge arches in the membrane equilibrium. The webs split into a series of parallel arches supported by compressed diagonals.

The cross vault's transition from the uncracked to the cracked state is similar to the transition state occurring in masonry dome's. The resisting structure of the cracked cross vault is that described by the *Heyman model of the sliced vault* (1966, 1977), by which the webs are separated, or 'sliced', into a series of arches of variable span sustained by the diagonal ribs. The emergence of thrust causes a subsequent deformation of the vault's support structures, with a slight broadening of its span and probable increased cracking. The response of the sliced vault to this settlement brings about a state of *minimum* thrust. This model of the cross vault will be examined in the following. Particular attention will be devoted to determining the mechanisms able to describe the cracking patterns typically found on these vaults.

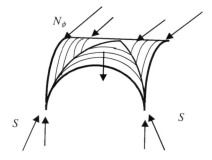

Fig. 7.17 Slicing of the webs and the source of thrust upon vault cracking

7.4.3 The Definitive Sliced Model

7.4.3.1 Force Transmission

As explained in the foregoing, the behavior of each web in a cross vault conforms closely to that of a barrel vault as described by the no-tension model, that is, like a series of independent parallel arches supported by longitudinal walls.

In the sliced webs of a cross vault the arch bands are sustained by the vault diagonals. Figure 7.18 shows the square plan ABCD of a simple cross vault. The sides AB, BC, CD and DA are the plane projections of the edge arches, called "formeret". They may be of various shapes, such as in rounded or pointed profiles.

Stress flow takes place in the direction ϕ along which the individual web slices transmit their loads in the vertical plane.

At the springings of each slice the actions transferred to the supporting diagonals interact with the actions transmitted by the slice of the adjacent web. These combine to produce a resultant force with vertical and horizontal components. The horizontal components produce a resultant force acting along the ribs in the diagonal plane, and each rib is thus loaded by both vertical and horizontal forces (Fig. 7.19).

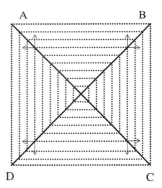

Fig. 7.18 Webs of a groin vault sliced into a series of arching bands of varying span supported by the diagonals

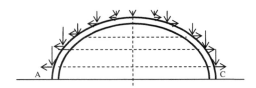

Fig. 7.19 Forces transmitted to the ribs by activated web arch bands

7.4 Statics of Cross Vaults

Whether there are actual physically distinct ribs or not, the actions of the cross vault will concentrate along the diagonals, which will in turn transmit the vertical and horizontal forces to their retaining structures. The ribs' resisting section, together with a portion of a web band, may be indeterminate, an issue of no particular import, given the low intensity of the stresses involved. The structures retaining the thrust accordingly undergo settling, which as a rule leads to slight widening of both the rib and web spans. A state of *minimum thrust* thus ensues in the vault.

7.4.3.2 Load Function

Let us firstly consider a square-plan cross vault with semicircular profile. In presence of a ribbed vault ribs have an external portion emerging from the vault intrados: conversely, ideal ribs can be considered present along the diagonals. Therefore, in any case, these ideal diagonal arch bands will be called ribs (Fig. 7.20).

Let L be the length of the side of the square plan of the vault and t the thickness of the webs, eventually including, in presence of a ribbed vault, also the ribs thickness. Let R_e and R_i indicate the radii of the external and internal round internal profile of the web; thus we have

$$R_e = \frac{L}{2} \quad R_i = \frac{L}{2} - t. \tag{7.11}$$

Figure 7.20 shows the plan of the cross vault.

The profile of the diagonal intrados of the vault is the ellipse of Fig. 7.21. Axes $O\xi\eta$ define the reference system of coordinates with the origin O placed at the central point of the alignment AB passing through the springings A and B.

Fig. 7.20 The plan of the square cross vault with its external profile

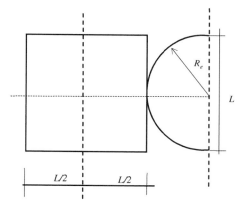

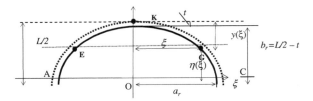

Fig. 7.21 The ellipse describing the intrados profile of the rib of the square in plan rounded cross vault

The equation of the ellipse

$$\frac{\xi^2}{a_r^2} + \frac{\eta^2}{b_r^2} = 1 \tag{7.12}$$

defines the contour of the diagonal intrados of the vault with respect to the reference axes $O\xi\eta$ of Fig. 7.21. The lengths of the ellipse's semi-axes, a_r and b_r, in Eq. (7.2) are

$$a_r = \frac{\sqrt{2}}{2}L - t \quad b_r = \frac{L}{2} - t \tag{7.13}$$

The weight of the web slices is defined by the function (Fig. 7.22)

$$g(x) = \gamma[(\sqrt{R_e^2 - x^2}) - R_i \sin(ar\cos\frac{x}{R_i})] \tag{7.14}$$

Fig. 7.22 The weight function $g(x)$ on the web

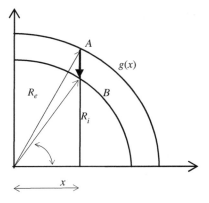

7.4.3.3 The Kinematic Approach

Square-cross vault undergoing horizontal settlement parallel to an edge
  Settlement mechanism in the ribs
  The settlement mechanism of the diagonal rib is sketched in Fig. 7.23.

We have to establish the positions of the hinges defining the settlement mechanism of the rib. The mechanism is symmetric and the extrados hinge is thus located at the key section K. Symmetrical hinges E and G cannot be located at the springing because statically incompatible: in this case, the pressure line, in fact, goes outside the arch. Hinges E and G are therefore placed *arbitrarily* near the haunches, between the springing A (or C) and the key section K of the arch.

By inspection of the plan of the vault sketched in Fig. 7.24, at first sight we could think that the diagonal rib, after the settlement, could take takes position A'KC'. In this way all the rib sections will undergo horizontal displacements *linearly increasing* with their distance from the centre K.

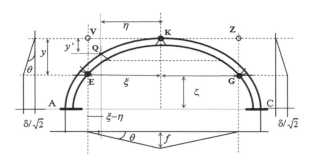

Fig. 7.23 Settlement mechanism of a diagonal rib

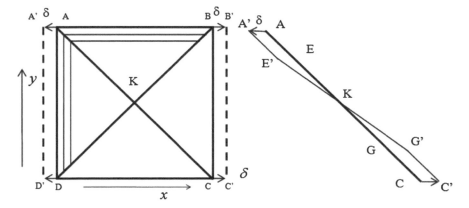

Fig. 7.24 Square-plan groin vault undergoing horizontal settlement parallel to its edge (direction *x*)

On the contrary, according to Fig. 7.23, that describes the settlement mechanism of the rib, all the sections of the arch, positioned between the internal hinge E as far as to the rib end, displace of a constant quantity. The rib will thus take in plan the broken configuration A'ÈKG'C' (Fig. 7.24) thereby the contact between the sections of the internal hinges E and G will occur only at the lower corner of the section.

Figure 7.24 shows the deformation of the diagonal rib. All the sections of the diagonal band included between the springings A (or B) and the hinges E (or G) will present constant horizontal displacement, equal to $\delta/\sqrt{2}$. Conversely the central rib portion EKG, of length 2ξ in plan, produces vertical and horizontal displacements linearly increasing respectively or with the distance from the hinge E or with the distance from the key section K.

The segment EK (or GK) of the rib rotates of the angle θ in the vertical plane around the centre V (or Z), so that the vertical displacement of the rib key K is

$$f = \xi\theta. \tag{7.15}$$

The end sides of the rib going from E to A, or from G to C, move only horizontally along the diagonal direction by a constant quantity given by

$$\frac{\delta}{\sqrt{2}} = \theta y. \tag{7.16}$$

Consequently, taking into account Eq. (7.12) and that

$$y(\xi) = \frac{L}{2} - \eta(\xi) \tag{7.17}$$

the horizontal settlement δ of the rib (or of diagonal arch band) along the diagonal direction, is given by

$$\frac{\delta(\xi)}{\sqrt{2}} = \theta\left(\frac{L}{2} - b_c\sqrt{1 - \frac{\xi^2}{a_c^2}}\right). \tag{7.16'}$$

Settlement mechanism in the webs.

A particular crack pattern results when the ribs have an external portion emerging from the vault intrados. The hinges located on the rib intrados, at sections E and G near the haunches (Fig. 7.7), form cracks that completely separate the webs. More in detail, the separation Δ that occurs at sections E and G of the rib extrados splits the overhanging webs. These cracks will thus traverse the entire thickness of the webs, as shown Fig. 7.25. Such cracks course along the vault parallel to the edges and are known as *Sabouret cracks*. They are generally wide enough that light can be seen filtering through them.

Figure 7.26, taken from Heyman's book "The Stone Skeleton", shows some typical cracking patterns of ribbed cross vaults. The drawing, from a study by

Fig. 7.25 Generation of Sabouret cracks

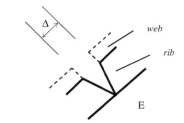

Fig. 7.26 Cracks on vault intrados (from Heyman 1997)

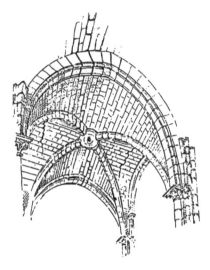

Abraham (1934), clearly shows the Sabouret cracks running parallel to the edges of the vault.

In any case, if emerging ribs are not present in the vault, alignments EF, FG, GH and HE will define line of cracks in the vault. These lines of cracks, produced by hinges E, F, G and H in the diagonal ribs, mark the subdivision of the webs into the two different regions respectively bearing the arch bands a and b defined above (Fig. 7.27).

Arches a are located in the outer zone of the webs, arches b in the inner. The mechanism of the arch bands a having the same direction of the settlement is defined by the position of internal hinge P on the arch intrados. The position of this hinge, in turn, depends on the ratio of thickness of the web to intrados radius, t_w/R_1 (Fabiani 2007; Como 2010).

Figure 7.28 illustrates the outcome of a particular position of P, that is, a continuation of the Sabouret fracture lines EH and EG. For the sake of simplicity, in the following analysis we will assume the same position of internal hinge P, in continuity with the lines EF and HG. The horizontal distance of hinge P from the centre M is thus $\xi/\sqrt{2}$.

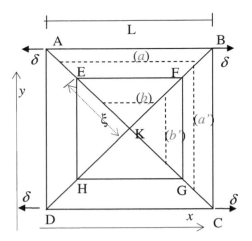

Fig. 7.27 The square cross vault undergoing horizontal settlement parallel to an edge

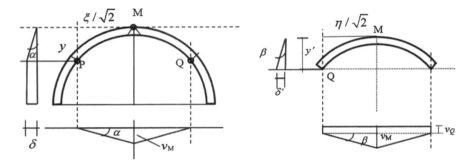

Fig. 7.28 Arch bands a (the outer band a is sketched in the figure) and the arch bands b

The vertical distance of P from the horizontal plane passing through the top K is the same as distance $y(\xi)$ of hinge E from the top hinge K in the diagonal arch band.

Arch bands a (Fig. 7.28)

The segment EK of the rib rotates of θ in the vertical plane around the absolute centre V (Fig. 7.23), and the horizontal displacement in the diagonal direction of the back end side of the rib is

$$\delta' = \frac{\delta}{\sqrt{2}} = y\theta \qquad (7.18)$$

7.4 Statics of Cross Vaults

The horizontal displacement at the springing of the band arch a, the same of the side settlement δ of the vault, as shown in Fig. 7.28, is

$$\delta = \alpha y \qquad (7.19)$$

Thus, equating the displacement δ given by (7.16) to the displacement (7.9) gives

$$\sqrt{2}y\theta = \alpha y \qquad (7.20)$$

and

$$\alpha = \sqrt{2}\theta \qquad (7.21)$$

The vertical displacement at the mid section of the rib (or the diagonal arch band) is

$$f = \xi\theta \qquad (7.22)$$

and the vertical displacements of all the points Q located at the extrados of the central portion of the rib (or the diagonal arch band) (Fig. 7.23) are

$$v_Q = (\xi - \eta)\theta \qquad (7.23)$$

where η indicates the horizontal distance of Q from the vertex K. Displacements (7.23) are the same of the displacements of the springs of the arch bands b. Particularly, from (7.23), when the point Q on the rib extrados is located along the vertical passing through the hinge E, we have $\eta = \xi$ and $v_Q(\eta = \xi) = 0$. The vertical displacement at the centre M of all the arch bands a, taking into account of (7.21) is

$$f_M = \alpha \frac{\xi}{\sqrt{2}} = \xi\theta \qquad (7.24)$$

and is the same of the rib mid section. The length of deformed side of the arch bands a have the same length $2\xi/\sqrt{2}$ and all present the following displacement function

$$v(x) = \theta\sqrt{2}(\frac{\xi}{\sqrt{2}} - x) \text{ per } 0 \leq x \leq \xi/\sqrt{2} \qquad (7.25)$$

Arch bands b (Fig. 7.28)

The component along the diagonal of the displacement δ_Q of springings of the arch bands b equals the horizontal displacement $y'\theta$ of the same point Q belonging to the rib extrados:

$$\frac{\delta_Q}{\sqrt{2}} = \frac{\beta y'}{\sqrt{2}} = y'\theta \qquad (7.26)$$

Hence

$$\beta = \sqrt{2}\theta \qquad (7.27)$$

The central displacement of any arch band b is obtained summing the vertical displacement v_Q of the springing with that due to the mechanism deformation of the band b. Thus

$$v_M = (\xi - \eta)\theta + \frac{\eta}{\sqrt{2}}\beta = (\xi - \eta)\theta + \frac{\eta}{\sqrt{2}}\sqrt{2}\theta = \xi\theta \qquad (7.28)$$

equals to the displacement of the mid section of the rib.

Arch bands a' (Fig. 7.27)

The arch bands a' having direction orthogonal to the settlement δ behave differently. Outer arch bands a' maintain their length and move rigidly, detaching each other. The inner arch bands b', *on the contrary,* move down because their bases, supported by the ribs or diagonal arch bands, move vertically with them. At the same time they detach each other. The arch band a', directed orthogonally to the settlement displacement δ, are supported by the sides of the rib and don't get down. No extension between springings of these arch bands occur that move rigidly.

Arch bands b' (Fig. 7.27)

The inner arch bands b', orthogonal to the direction of the settlement δ, are supported on the inner part of the rib and move vertically. No extension between the springings of these arch bands b' will occur. These bands translate only vertically due to the lowering $\delta_{Q'}$

$$\delta_{Q'} = (\xi - \eta)\theta \qquad (7.29)$$

At the centre $\delta_{Q'}(\eta = 0) = \xi\theta$ while along the alignment defined by $\eta = \xi$ and corresponding to the Sabouret crack we have $\delta_{Q'} = 0$.

Crack patterns

Crack pattern in the vaults reflects the geometry of the settlement mechanism examined.

Sabouret cracks define the frame separating the band arches a and b. Arch bands a parallel to the x axis, follow the settlement and deform with the same mechanism of ribs with the presence of two inner hinges located symmetrically at the intrados and a central hinge at the extrados. Arches a, indicated as a', orthogonal to the

7.4 Statics of Cross Vaults

settlement direction, dragged by rib deformation don't deform and detach each other moving outside. A consequent sequence of diffused thin cracks passing through the web, marks these detachments.

At the intrados of the vault a long transversal crack having direction orthogonal to the settlement together with the passing through Sabouret and detachment cracks will be visible (Fig. 7.29). This long central crack at the intrados, is the typical crack running at the intrados of the large cross vaults spanning on the nave of a cathedral. In this case the vault, rectangular in plan, settles in fact in the transversal direction. At the extrados the crack pattern will be similar except that there are no transversal cracks while will be visible diagonal cracks along all the vault and cracks along y in the outer zone of the vault.

Minimum Thrust Valuation

The evaluation of the minimum thrust will be performed by the kinematical approach. The minimum thrust μr_{min} of the vault can be thus obtained as the maximum of all the kinematical thrusts μr defined by application of virtual work equation:

$$4\mu r \cdot \delta(v) = \langle g, v \rangle \tag{7.30}$$

expressing the balance between the active work $\langle g, v \rangle$ of the dead loads g with the resisting work $4\mu r \cdot \delta(v)$ of the thrusts μr for any admissible displacement v settlement mechanism of the vault undergoing an horizontal settlement parallel to its edge (Fig. 7.30) (Como 2010).

Thus we get

$$\mu r_{min} = \max \mu r(v) = \frac{\langle g, v \rangle}{4\mu r \cdot \delta(v)} \quad v \in M \tag{7.31}$$

where M is the set of all the displacement mechanisms that admit settlement of the vault supports. The kinematical thrust μr is obtained by application of virtual work

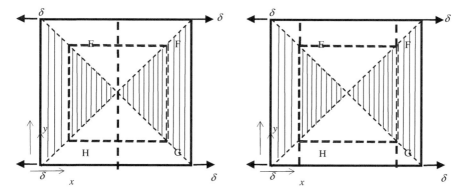

Fig. 7.29 Crack patterns at the intrados and at the extrados

Fig. 7.30 The thrust μr

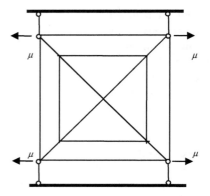

Eq. (7.30). The work of the weights acting on the webs can be obtained for the various considered arch bands.

The arch bands a, that are parallel to the settlement δ, behave equally and develop along a band of width $(L/2 - \xi/\sqrt{2})$ and all present the same following displacement function

$$v(x) = \theta\sqrt{2}(\frac{\xi}{\sqrt{2}} - x) \text{ per } 0 \le x \le \xi/\sqrt{2} \qquad (7.32)$$

The corresponding work done by the weights is:

$$L_g = 4\theta\left(L/2 - \xi/\sqrt{2}\right) \int_0^{\xi/\sqrt{2}} g(x)\sqrt{2}(\frac{\xi}{\sqrt{2}} - x)dx \qquad (7.33)$$

The arch bands b, parallel to the settlement δ, have length $2\eta/\sqrt{2}$, variable between 0 and $2\,\xi/\sqrt{2}$, and the displacement function

$$v(x) = \theta(\xi - \eta) + \theta\sqrt{2}(\frac{\eta}{\sqrt{2}} - x) \qquad (7.34)$$

The corresponding work of the weights is

$$L_g(\xi) = 4\theta \int_0^{\xi/\sqrt{2}} \left(\int_0^{\eta/\sqrt{2}} g(x,y)[(\xi - \eta) + \sqrt{2}(\frac{\eta}{\sqrt{2}} - x)]dx \right) d\eta \qquad (7.35)$$

7.4 Statics of Cross Vaults

There is no work of the weight for the arch bands a', *orthogonal* to the settlement. that don't get deformed. The arch bands b', having length $2\eta/\sqrt{2}$ and displacement function

$$v(x) = \theta(\xi - \eta) + \theta\sqrt{2}(\frac{\eta}{\sqrt{2}} - x) \tag{7.36}$$

The corresponding work is

$$L_g(\xi) = 4\theta \int_0^{\xi/\sqrt{2}} (\xi - \eta)(\int_0^{\eta/\sqrt{2}} g(x)dx)d\eta \tag{7.37}$$

Summing up all the various shares we have the total work of the weights g

$$L_{gTOT}(\xi) = 4\theta(L/2 - \xi/\sqrt{2}) \int_0^{\xi/\sqrt{2}} g(x)\sqrt{2}(\frac{\xi}{\sqrt{2}} - x)dx$$

$$+ 8\theta \int_0^{\xi/\sqrt{2}} (\xi - \eta)(\int_0^{\eta/\sqrt{2}} g(x)dx)d\eta + 4\theta \int_0^{\xi/\sqrt{2}} [\int_0^{\eta/\sqrt{2}} g(x)\sqrt{2}(\frac{\eta}{\sqrt{2}} - x)dx]d\eta \tag{7.38}$$

The displacement δ can be expressed in function of the rotation parameter θ according to (7.19), (7.20) and we can obtain the thrust function depending on the position ξ of the internal hinge of the rib

$$\mu r(\xi) = \frac{1}{\frac{L}{2} - b_c\sqrt{1 - \xi^2/a_c^2}} \left[(L/2 - \xi/\sqrt{2}) \int_0^{\xi/\sqrt{2}} g(x)\sqrt{2}(\frac{\xi}{\sqrt{2}} - x)dx \right.$$

$$\left. + 2 \int_0^{\xi/\sqrt{2}} (\xi - \eta)(\int_0^{\eta/\sqrt{2}} g(x)dx)d\eta + \int_0^{\xi/\sqrt{2}} (\int_0^{\eta/\sqrt{2}} g(x)\sqrt{2}(\frac{\eta}{\sqrt{2}} - x)dx)d\eta \right] \tag{7.39}$$

The same approach can be developed in the more general case of cross vault with rectangular plan. The maximum of this function, by varying the distance ξ, gives the minimum thrust of the settled cross vault. Numerical investigations will give in the next sections the values of the minimum thrust for various geometries of the cross vaults.

Square cross vault undergoing diagonal settlement

The cross vault, even though infrequently, can also settle diagonally. The crack pattern at the intrados will present two orthogonal central cracks as well as the frame of the Sabouret cracks.

Settlement mechanism in the ribs

The ribs present the mechanism as described in the previous Fig. 7.23. The vertical displacement at the mid section is still given by

$$f = \xi\theta. \tag{7.40}$$

Also in this case the sides of the ribs going from E to A, or from G to C, move only horizontally along the diagonal direction by the constant quantity δ, that is given now given by

$$\delta = \theta y. \tag{7.41}$$

and consequently is

$$\delta(\xi) = \theta(\frac{L}{2} - b_c\sqrt{1 - \xi^2/a_c^2}). \tag{7.42}$$

Settlement mechanism in the web
Arch bands a

For the reasons explained in the foregoing, sliced webs can be represented by arches of two different kinds: arches *a* and arches *b* as shown in the previous Fig. 7.27 and also shown in Fig. 7.32.

Arches *a* are located in the outer zone of the webs. These arch bands follow the displacements of the ribs, their mechanism being defined by the position of internal hinge P on the arch intrados. The position of this hinge, in turn, depends on the ratio of thickness to intrados radius, t/r_1.

Figure 7.31 illustrates the outcome of a particular position of P, that is, a continuation of the Sabouret fracture lines EH and EG. For the sake of simplicity, in the

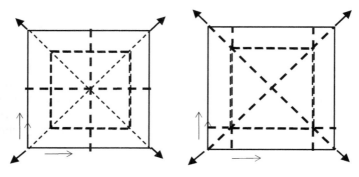

Fig. 7.31 Crack pattern at the intrados/extrados in presence of diagonal settlement

7.4 Statics of Cross Vaults

Fig. 7.32 Hinges alignments

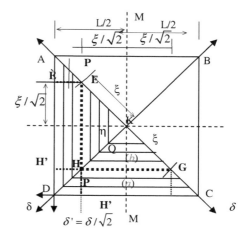

following analysis we will assume the same position of internal hinge P in continuity with the Sabouret lines EF and HG (Fig. 7.32).

Now, the horizontal distance of hinge P from the centre M is $\xi/\sqrt{2}$. The vertical distance of P from the horizontal plane passing through the top K is the same as distance y of hinge E from the top hinge K in the rib. The arch bands a, drawn along by the deformation of the rib, exhibit only horizontal displacements at their springing, all equal to (Fig. 7.33)

$$\delta' = \delta\sqrt{2}. \qquad (7.43)$$

Internal hinge P (or the opposite Q) is always located along the alignment HE in Fig. 7.17, and the horizontal displacement at the springing can be expressed as

$$\delta' = \alpha y. \qquad (7.44)$$

where α indicates the rotation of PM, and y the vertical distance (7.7) of P from M.

Fig. 7.33 The mechanism of the external arch band a

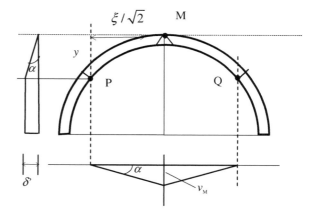

Whence

$$\alpha = \frac{\theta}{\sqrt{2}}. \tag{7.45}$$

Equation (7.35), albeit approximate due to the simplifying assumption regarding the position of internal hinge P, links the mechanism of arch band *a* with that of the ribs through a very simple relation and moreover provides a rough description of the web deformation. Let us now consider the arch band *a* with semicircular profile. The vertical displacement v_M of the central point M of this arch band is thus given by

$$v_M = \alpha \frac{\xi}{\sqrt{2}} = \frac{\theta}{2}\xi = \frac{f}{2} \tag{7.46}$$

and equals *half* the vertical displacement of the centre of the cross vault. The internal hinges P in all the arch bands *a* are found at the same position along alignment, EH, as far as the line of the Sabouret crack, HG, in Fig. 7.31.

At conclusion, the segments of the arch bands *a*, all having length $(L/2 - \xi/\sqrt{2})$ deform according the displacement section

$$v(x) = \frac{\theta}{\sqrt{2}}(\frac{\xi}{\sqrt{2}} - x) \text{ per } 0 \le x \le \xi/\sqrt{2} \tag{7.47}$$

The corresponding work done by the weights on the vertical displacement of all the arch bands *a* is:

$$L_g = 8\theta\left(L/2 - \xi/\sqrt{2}\right) \int_0^{\xi/\sqrt{2}} \frac{g(x)}{\sqrt{2}}(\frac{\xi}{\sqrt{2}} - x)dx \tag{7.48}$$

The arch bands *b* belong to the internal zone of the web: they are sustained by ribs at their springings Q. These springings Q undergo both vertical and horizontal displacements. Figure 7.34 shows point Q's location in the section where the arch band *b* joins the rib: the *diagonal* distance of K from Q is η.

Consequently, the distance of Q along the side direction is $\eta/\sqrt{2}$, as shown in Fig. 7.19. The vertical displacement v_Q of springing Q of arch band *b* is thus (Fig. 7.8)

$$v_Q = (\xi - \eta)\theta. \tag{7.49}$$

7.4 Statics of Cross Vaults

Fig. 7.34 The mechanism of the inner arch band b

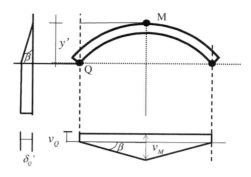

Displacement v_Q varies amongst arch bands b because η gradually decreases from Q to K. The outer horizontal displacement δ_Q of Q along the diagonal direction is

$$\delta_Q = y'\theta. \tag{7.50}$$

Thus, the horizontal displacement δ'_Q of the springing of arch band b along the direction of the groin vault edge is

$$\delta'_Q = \delta_Q/\sqrt{2}. \tag{7.51}$$

However, referring to the same Fig. 7.19, we also have

$$\delta'_Q = \beta y', \tag{7.51'}$$

where β is the rotation angle of the segment QM of the arch band. Consequently, the compatibility condition

$$\beta y' = \delta_Q/\sqrt{2} \tag{7.51''}$$

holds, and

$$\beta = \frac{\theta}{\sqrt{2}}. \tag{7.52}$$

Equation (7.52) links the deformation of arch bands b to the deformation of the ribs. The total vertical displacement of the centre M of arch band b, including the vertical settlement of its springings as well, is

$$v_M = (\xi - \eta)\theta + \frac{\theta}{\sqrt{2}}\frac{\eta}{\sqrt{2}} = \theta(\xi - \frac{\eta}{2}). \tag{7.53}$$

When point M approaches the groin vault centre, K, a progressive reduction in the span of arch bands b occurs and, at the limit, for $\eta \to 0$,

$$v_M \to \theta \xi = f. \tag{7.54}$$

Thus, for $\eta \to 0$, the vertical displacement of arch band b equals the vertical displacement of the rib key section. When, on the contrary, arch band b approaches the Sabouret line, HG, we have

$$\eta = \xi \tag{7.55}$$

and the central vertical displacement of arch band b becomes

$$v_{M,Sabouret} \to \theta \xi /2 = f/2, \tag{7.56}$$

which equals the vertical displacement of arch bands a. Thus concluding, the arch bands b of length $2\eta/\sqrt{2}$, varying between 0 and $2\,\xi/\sqrt{2}$, displace according the function

$$v(x) = \theta(\xi - \eta) + \theta\sqrt{2}(\frac{\eta}{\sqrt{2}} - x) \tag{7.57}$$

The work of the dead load $g(x)$ (or $g(y)$) is the same for all the arch bands b parallel to x or to y. Thus we have the corresponding work

$$L_g(\xi) = 8\theta \int_0^{\xi/\sqrt{2}} \left(\int_0^{\eta/\sqrt{2}} g(x)[(\xi - \eta) + \sqrt{2}(\frac{\eta}{\sqrt{2}} - x)]dx \right) d\eta \tag{7.58}$$

The total work of the loads g is:

$$\begin{aligned}L_g = 8\theta[&\left(L/2 - \xi/\sqrt{2}\right) \int_0^{\xi/\sqrt{2}} \frac{g(x)}{\sqrt{2}}(\frac{\xi}{\sqrt{2}} - x)dx \\ &+ \int_0^{\xi/\sqrt{2}} \left(\int_0^{\eta/\sqrt{2}} g(x)[(\xi - \eta) + \sqrt{2}(\frac{\eta}{\sqrt{2}} - x)]dx \right) d\eta]\end{aligned} \tag{7.59}$$

7.4 Statics of Cross Vaults

By using the virtual work Eq. (7.30), likewise to (7.39), we obtain the following equation giving the kinematical thrust

$$\mu r(\xi) = \frac{1}{\frac{L}{2} - b_c \sqrt{1 - \frac{\xi^2}{a_c^2}}} 2[\left(L/2 - \xi/\sqrt{2}\right) \int_0^{\xi/\sqrt{2}} \frac{g(x)}{\sqrt{2}} (\frac{\xi}{\sqrt{2}} - x) dx$$
$$+ \int_0^{\xi/\sqrt{2}} \left(\int_0^{\eta/\sqrt{2}} g(x)[(\xi - \eta) + \sqrt{2}(\frac{\eta}{\sqrt{2}} - x)] dx \right) d\eta] \quad (7.60)$$

The maximum of this function, by varying the distance ξ, gives the minimum thrust of the settled cross vault. Numerical investigations will give in the next sections the values of the minimum thrust for various geometries of the cross vaults.

7.4.3.4 The Static Approach

Evaluating the minimum thrust of the groin vault via the static approach involves the following steps:

- subdividing the webs into various orders of arch bands, splitting them into a series of voussoirs, and determining their weights and center positions;
- tracing the funicular of the loads acting on the voussoirs of the various arch bands at minimum thrust via trial and error. This evaluation must account that:

  for more depressed arches, the funicular curve passes through the top of the key section and on the intrados of the springings
  for more external arch bands, the funicular curve is tangent to the arch intrados towards the haunches, as shown in Fig. 7.35.

- calculating the vertical and horizontal actions transmitted by the arch bands to the ribs and, if these actions emerge from the web intrados, adding the rib weight, and lastly tracing the minimum thrust funicular curve.

The above procedure has been defined under the assumption of diagonal settlement of the vault.

The loads acting on the arch diagonal, consisting of vertical and horizontal forces, involve the presence of a *variable* thrust along the rib. The thrust at the key is lower than at the springings, though it is generally non-negligible.

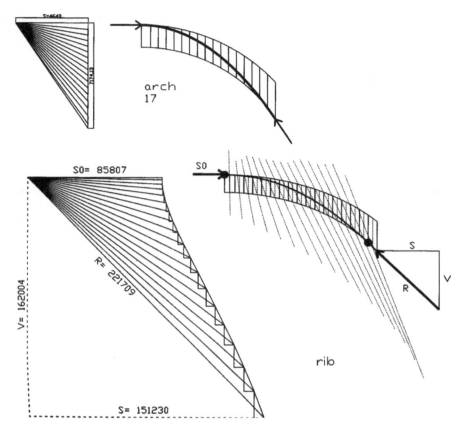

Fig. 7.35 Static evaluation of the minimum thrust for a rib (indicated as *costola*). The above funicular curve refers to the arch band designated as *arco 17*

7.4.3.5 Numerical Calculations

Lateral settlement assumption

The minimum thrusts of rounded cross vaults have been obtained by numerical investigations centred in the search of the minimum of the kinematical multipliers given by (7.29) and (7.50). The research has been extended also to the case of the rounded cross vaults with rectangular plan with the assumption of lateral settlements. In this case the side arches of shorter span are rounded and the other depressed.

Geometrical parameters are:

- the lengths a and b of the sides of the rectangular plan, with con $a > b$,
- the radii R_{1i} and R_{2i} of the intradoses, with $b = 2\,R_{1i}$
- the thickness t.

7.4 Statics of Cross Vaults

The outcomes of the numerical investigation have been the construction of two abacuses whose the first gives the ratio thrust/weight S/W as function of the ratio a/b and for different values of the thickness t. The ratio a/b varies between 1 and 2.5. All the curves of this abacus have been obtained by assuming the value of the radius of the rounded arch of minor span equal to $R_{1i} = 1.00$ m, i.e. with $b = 2 \times R_{1i} = 2.0$ m.

The single curve of the abacus has been traced with the assumed value of the thickness t. The ratio S/W takes a different value for a vault that has radius R^*_{1i} different from R_{1i} and then same thickness t. It is possible to show that *different vaults having the same t/R_{1i}, present the same ratio S/W*.

In order to prove this statement let us firstly a rounded masonry arch.

The weight W of the arch is proportional to the specific weight of the masonry, to the thickness t, to the radius al R_i of the intrados and to the width b. We can thus write

$$W = \chi \cdot \gamma \cdot b \cdot t \cdot r_i \tag{7.61}$$

where χ is a proportionality factor. The minimum thrust of a rounded arch can on the other hand given by (Como 2010)

$$S = \gamma \cdot b \cdot r_i^2 \bar{f}\left(\frac{t}{r_i}\right) \tag{7.62}$$

where the quantity $\bar{f}(t/R_i)$ is a suitable function of the ratio t/R_i. The ratio thrust/weight of the arch is thus given by

$$\frac{S}{W} = \frac{1}{\chi} \frac{r_i}{t} \gamma \cdot \bar{f}\left(\frac{t}{r_i}\right) \tag{7.63}$$

Thus, given two different arches respectively with radius and (r_i, t) and (r_i^*, t^*) such that

$$\frac{r_i}{t} = \frac{r_i^*}{t^*} \tag{7.64}$$

we have also

$$\frac{S}{W} = \frac{1}{\chi} \frac{r_i}{t} \gamma \cdot \bar{f}\left(\frac{t}{r_i}\right) = \frac{1}{\chi} \frac{r_i^*}{t^*} \gamma \cdot \bar{f}\left(\frac{t^*}{r_i^*}\right) = \frac{S^*}{W^*} \tag{7.65}$$

Consequently, two rounded arches having equal ratios t/R_i have also same ratios S/W. The same rule holds also for depressed arches, as can be easily proven, and for two cross vaults at minimum thrust state because in this condition the vaults slices into a sequence of arch bands. Conversely, if a given cross vault has radius R^*_{1i} and thickness t^*, we obtain from (7.27) the equivalent thickness

$$t = t^* R_{1i}/R_{1i}^* \tag{7.66}$$

of the rounded vault with the assumed radius $R_{1i} = 1.0$ m that has the same ratio S/W for any value of the ratio a/b. The various curves of the abacus of Fig. 7.36 that give the value S/W as function of the ratio a/b, correspond to different values of the equivalent thickness t given by (7.28).

In conclusion, given a rounded cross vault with rectangular plan $a^* \times b^*$ with the round side arch of minor span of length b^*, with $b^* = 2\, R_{1i}^*$ and with the ratio (a^*/b^*), by means (7.66) we can obtain the equivalent thickness t of the similar vault with $R_{1i} = 1.0$ m that has the same ratio S/W.

The same result holds in term of the ratio $V/(a*b*t)$ if V is the volume occupied by the vault so that if two cross vaults have equal ratios t/R_{1i} they have also equal ratios $V/(a*b*t)$. In brief

$$\frac{R_i}{t} = \frac{R_i^*}{t^*} \Rightarrow \frac{V}{abt} = \frac{V^*}{a^*b^*t^*}$$

The second abacus that gives the values of ratio $V/(a*b*t)$ as function of the ratio t/R_{1i} is given in Fig. 7.37.

The following example illustrates the proposed approach.

Evaluating the weight and the minimum thrust of the rounded cross vault, square nin plan, the following dimensions are

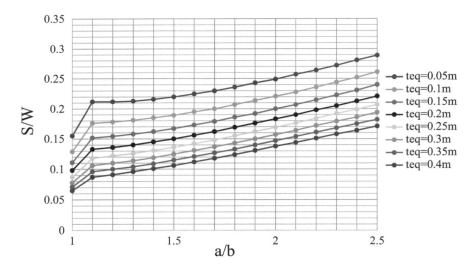

Fig. 7.36 Abacus giving the ratio S/W by varying t and ratio a/b for rectangular cross rounded vaults enduring parallel settlement

7.4 Statics of Cross Vaults

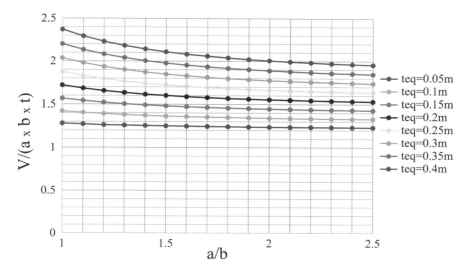

Fig. 7.37 The ratio V/(a*b*t) for different thicknesses t versus the ratio a/b for rectangular crossrounded vaults

$$a^* = 10.00 \, \text{m}; \quad b^* = 10.00 \, \text{m} \; ; \; t^* = 1.0 \, \text{m}$$

Thus Eq. (7.66) gives the equivalent thickness t of the vault having radius $R_{1i} = 1.0$ m that has the same ratio S/W of the assigned vault. Taking into account that $R^*_{1i} = 5.00$ m, we get

$$t = 1.00 \frac{1.00}{5.00} = 0.20 \, \text{m}$$

We determine the curve 10 of S/W the abacus of Fig. 7.20 corresponding to the equivalent thickness $t = 0.20$ m. From this curve, to the assigned value of $a^*/b^* = 1$ corresponds the value $S^*/W^* = 0.096$.

By using the abacus of Fig. 7.21 we evaluate the value of the ratio $V^*/(a^*b^*t) = 1.73$ corresponding to the curve of the equivalent thickness $t = 0.2$ and for the abscissa $a^*/b^* = 1$.

The volume (a^*b^*t) is given by $10.0 \times 10.00 \times 1.0 = 100$ mc so that $V^* = 1.73 \times 100 = 173$ mc. By assuming a specific weight of $\gamma = 1.7$ t/mc, the weight of the assigned cross vault is è $173 \times 1.7 = 294.17$ t so that the minimum thrust S of the vault is 0.096×294.17 t $= 28.24$ t $= 282.24$ kN.

By using directly the Eq. (7.39) we can obtain directly the minimum thrust by means the function of the kinematical thrust $\mu r(\xi)$, as function of the distance ξ of the internal hinge of the rib from the ket section, by using the program MAT.

We obtain the value $\mu r(\xi_{soluz}) = 269.39$ kN that we can check with the previous value of 282.23 kN evaluated by means the abacuses of figures Figs. 7.20 and 7.21; with a gap of the 5 %.

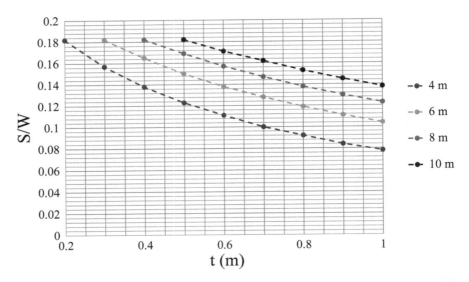

Fig. 7.38 Abacus giving the ratio *S/W* by varying *t* for the square cross rounded vaults enduring diagonal settlement

Diagonal settlement assumption

The case of the diagonal settlement has been considered in the numerical investigation only for the case of the round cross vault with square plan. The outcome of the analysis is the abacus of Fig. 7.38.

We compare the values of the minimum thrusts for the cross vault under side or diagonal settlement. Inspection of the above diagrams shows that the diagonal settlement yields a thrust about 30 % larger than the corresponding thrust of the vault suffering sideway settling. In brief, if S_L and S_D are respectively the thrust corresponding to the lateral and diagonal settlement, we can write

$$S_L = 0.7 S_D \qquad (7.67)$$

7.4.4 Are Ribs Necessary for Cross Vault Equilibrium?

Upon conclusion of the static analysis of groin vaults, it is interesting to take up an old, often hotly-debated question regarding the actual static function of ribs emerging from the intrados of cross vaults.

In this debate, scrupulously reconstructed by Di Pasquale (1996) some scholars, such as Viollet le Duc and Masson (1934), attributed a primary static function to the ribs, while others, such as Abraham (1934) and Sabouret, on the contrary, believed that the ribs served a purely ornamental function. In any event, in both the uncracked and cracked state, the forces in groin vaults concentrate along the diagonals. This stress concentration may be sustained by ribs emerging from the intrados of webs, as in Gothic rib vaults, or along the creases in the neighborhood of the groins within the thickness of the vault, as in the case of thick Roman vaults.

7.4.5 The Cross Vaults of the Diocletian Baths in Rome

7.4.5.1 Description of the Vaults

Originally, the Baths of Diocletian extended over an area of more than 135,000 m^2 with a rectangular plan of 376 m × 361 m (Fig. 7.39).

The main rooms—the *calidarium*, the *tepidarium* and the *frigidarium*—served the three main functions of the baths and were arranged along the building's minor axis. The dressing rooms, gymnasium, concert halls and libraries were located at the sides.

The vast rectangular hall, which today forms the transept of the church of *S. Mary of Angels*, was originally the center of the Baths, set along its minor axis between the tepidarium and frigidarium. The hall extends over a surface of 27 m by 90 m and is covered by three large groin vaults sustained by eight monolithic columns to reach a height of 28 m. Figure 7.40 shows an axonometric drawing of the building. Cross vaults were probably chosen in order to be able to arrange windows high up on the outer walls. An arching gridwork of brick ribs, as shown in Fig. 7.41, was erected before the concrete was cast. Different *caementa* were used in the construction of the various parts of the webs: tuff was placed near the springings, while lighter material, such as black pumice, were used near the crown. Various systems of buttresses, which anticipate the strategies of Gothic architecture and facilitate thrust transmission, are present in the Baths, as is clear from Fig. 7.41.

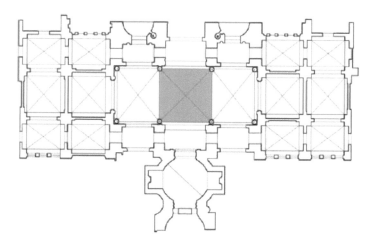

Fig. 7.39 Plant of the Baths of Diocletian

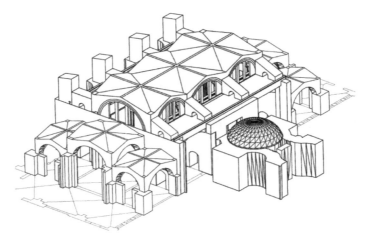

Fig. 7.40 Axonometric view of the covered *frigidarium* and *tepidarium*

Fig. 7.41 Internal ribs and concrete webs of the groin vault of the Baths of Diocletian (from Crema 1942)

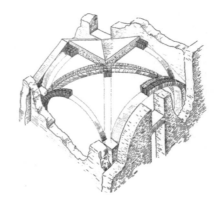

7.4.5.2 Thrust Evaluation via the Static Approach

Figure 7.42 shows the cross groin vault together with its dimensions. The thrusts and the vertical forces transmitted by the different web arch bands combine with the weights of the various rib voussoirs (Fig. 7.43).

A *diagonal settlement* mechanism has thus been assumed. According to the results shown in Fig. 7.44, obtained by Ciciotti (2006–2007), the minimum thrust of the rib equals 193 t. Figure 7.45 shows the pressure line in the rib: the hinge positions are identified by the tangent points of the funicular curve with the rib extrados and intrados shown in Fig. 7.45.

7.4 Statics of Cross Vaults

Fig. 7.42 The main groin vault with the diagonal rib

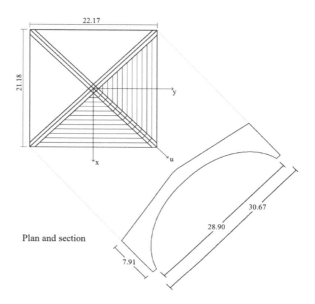

Plan and section

7.4.5.3 Thrust Evaluation via the Kinematic Approach

The kinematic approach has been applied to check the results obtained via the static theorem.

A first evaluation of the minimum thrust is performed by applying formula (7.60), assuming the mechanism defined by the hinge positions in Figs. 7.45 and 7.46. The corresponding value of the thrust is $\mu r = 1447.5$ tm/7.50 m $= 193$ t, equal to the value obtained via the static theorem (Fig. 7.44). A second value of the kinematic thrust has also been evaluated by assuming the mechanism defined by the hinge positions shown in Fig. 7.47.

The work of the weights acting on the rib and the work of thrust μr have been estimated for this latter mechanism: condition (7.30) gives $\mu r = 1035.98$ tm/ 6.04 m $= 171.5$ t, lower than the thrust corresponding to the mechanism assumed in Fig. 7.46. In this regard, it should be recalled that the minimum thrust corresponds to the maximum of all kinematic thrust multipliers.

The obtained values of the minimum thrust have been compared with the corresponding values given by the previous abacuses. The comparison can verify only the order of magnitude of the thrust because the geometry of the vaults of Diocletian Baths is non-exactly that of a rounded rectangular vault.

To use the abacuses of Figs. 7.36 and 7.37 we have to consider the occurrence of a lateral settlement. We will thus scale, according to (7.67), the valuated value of the thrust to obtain the thrust corresponding to the occurrence of the diagonal settling.

The vault can be assumed to be a rounded rectangular vault with a constant thickness $t^* =$ of 1.8 m and having an internal plan with the side lengths a^* and b^*. To obtain the lengths of the internal sides a^* and b^* with reference to Fig. 7.42:

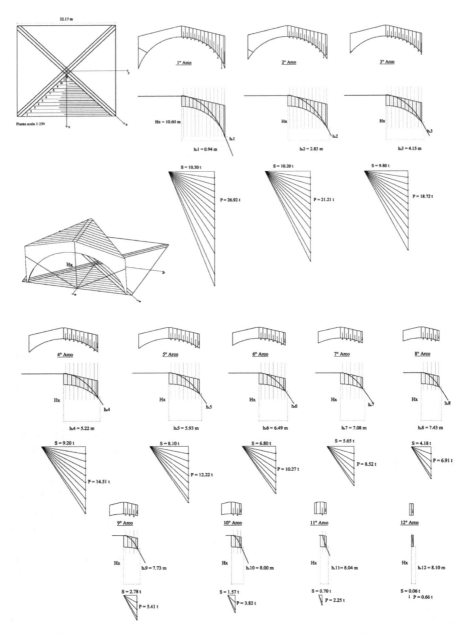

Fig. 7.43 Splitting of the web into 12 slices and evaluation of their minimum thrusts

7.4 Statics of Cross Vaults

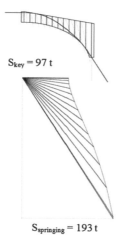

Fig. 7.44 Minimum thrust evaluation for a diagonal rib

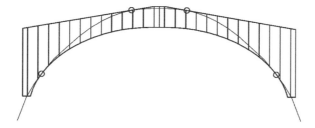

Fig. 7.45 Pressure line in a rib obtained via the static approach

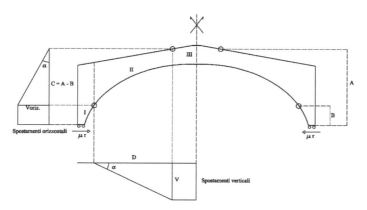

Fig. 7.46 Rib settlement displacements corresponding to the mechanism with hinge positions defined by the static approach in Fig. 7.45

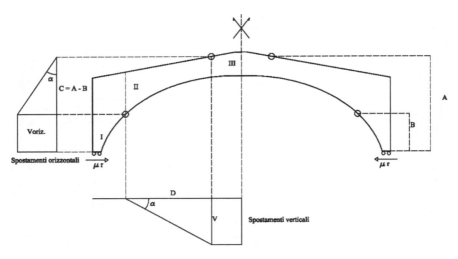

Fig. 7.47 Kinematic evaluation of the thrust assuming different hinge positions from Fig. 7.46

$a^* = 22.17 - (30.67 - 28.90) = 20.40$ m; $b^* = 21.18 - (30.67 - 28.90) = 19.41$ m and $a^*/b^* = 1.05$. Further $R_i = b^*/2 = 9.70$ m and application of (7.57) gives $t_{eq} = (1.5 \text{ m})(1.0 \text{ m})/9.70 = 0.185$ m.

From the above abacuses, the values are obtained as $S^*/W^* = 0.12$ and $V/a^*b^*t^* = 1.685$ corresponding to the ratio $a^*/b^* = 1.05$ and to the equivalent thickness $t_{eq} = 0.185$ m. Then, with $\gamma = 0.95$ t/m^3, these values are achieved:

- $V^* = 1.55 \cdot a^* \cdot b^* \cdot t^* = 1.685 \cdot 19.40 \cdot 20.40 \cdot 1.80 = 1201$ m^3 •

Weight = $W = V^* \cdot \gamma = 1201 \cdot 0.95$ ton = 1141 t; and Thrust $S = W^*0.12 = 0.12 \cdot 1141 = 137$ t.

The minimum thrust, obtained with the exact geometry of the vault and by application of the static theorem, evaluated in the direction of the side settlement is Eq. (7.67): Thrust$_{\text{static theorem}}$ 0.70 = 193 0.70 = 135.1 ton, near to the value of 137 ton obtained by using the previous abacuses. This result shows the utility of the proposed abacuses to calculate the value of the minimum thrust of rounded cross vaults.

7.5 Cloister Vaults

The cloister vault has been very commonly used to roof a variety of buildings. In fact, such vaults are closed, that is supported by four lateral walls topped by four cylindrical webs. They were built with the aid of wood centering by adding successive courses of masonry in directions parallel to the lateral walls.

7.5 Cloister Vaults

The earliest examples of cloister vaults date back to the first century BCE, some notable example being the roofs of the *Tabularium,* built in Rome in 78 BCE, Hadrian's Villa at Tivoli and those covering the octagonal rooms of the Baths of Antoninus (145–160 CE) in Carthage. The similarly octagonal room of *Domus Aurea* has a roof that begins as a cloister vault then gradually transforms into a hemispherical vault. Later examples are the roof of the Palatine Chapel in Aachen (13th–14th century CE).

With advances in construction techniques, cloister vaults even came to be used to roof the halls and rooms of aristocratic buildings. They are thoroughly analyzed in Palladio's *Four Books of Architecture* (1601) and in the treatise, *Universal Architecture,* by Scamozzi (1615). Castigliano's book for engineers (1879) is also a valuable source of information and practical guidelines for their construction.

7.5.1 Initial Membrane Stresses

The primary stresses occurring in a cloister vault prior to cracking is represented by the membrane solution (Fig. 7.48). As discussed above, determining the membrane stresses is useful in order to be able to formulate a model of the vault after cracking, i.e. the final resistant vault model. The stresses in the vault webs depend on their boundary conditions, which involve the interactions transmitted by the groins. Let us return to the equilibrium equations of cylindrical shells examined above and recalled here for convenience:

$$\frac{\partial N_x}{\partial x} + \frac{\partial N_{\phi x}}{\partial s} = 0 \quad \frac{\partial N_\phi}{\partial s} + \frac{\partial N_{x\phi}}{\partial x} + g \sin \phi = 0 \quad N_\phi = -gR \cos \phi. \quad (7.68)$$

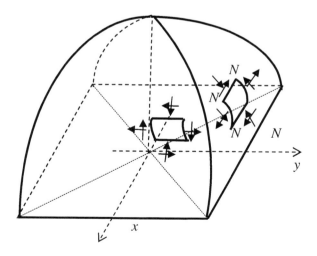

Fig. 7.48 Membrane stresses in webs of a cloister vault

By accounting for the expression for N_ϕ the second of these equations becomes

$$\frac{\partial N_{x\phi}}{\partial x} = -2g \sin \phi, \qquad (7.69)$$

which when integrated yields

$$N_{x\phi} = -2gx \sin \phi + C_1(\phi), \qquad (7.70)$$

and consequently

$$C_1(\phi) = 0. \qquad (7.71)$$

Shear forces $N_{x\phi}$ are thus

$$N_{x\phi} = -2gx \sin \phi. \qquad (7.72)$$

The shearing forces $N_{x\phi}$ are directed towards the centre of the lune. Figure 7.49 shows the distribution of forces $N_{x\phi}$ acting along a generatrix line.

Shear forces near the corners, where intersect two adjacent lunes, produce a diagonal resultant N_d that will be further determined (Fig. 7.50).

To complete the description of the membrane stresses it is necessary to evaluate the forces N_x that, as it will be shown, play a relevant role on the cracking of the

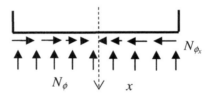

Fig. 7.49 The distribution of forces $N_{x\phi}$ acting along a generatrix line

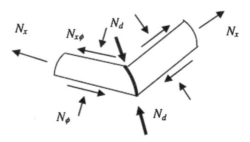

Fig. 7.50 Equilibrium at corners in the diagonal direction generates the diagonal force N_d

7.5 Cloister Vaults

vault. Let us therefore substitute the expression for $N_{x\phi}$ from the first equilibrium equation:

$$\frac{\partial N_x}{\partial x} = -\frac{1}{R}\frac{\partial}{\partial \phi}(-2gx \sin \phi) = \frac{1}{R}2gx \cos \phi, \qquad (7.73)$$

whence, by integration, we get

$$N_x = \frac{1}{R}gx^2 \cos \phi + C_2(\phi) \qquad (7.74)$$

Thus, in order to valuate forces N_x, we must know function $C_2(\phi)$. As we will show in the following, $C_2(\phi)$ is determined by determining the web interactions across the diagonals. It is thus necessary to valuate the force N_d. This force increases gradually moving down along the vault, as an increasing portion of the vault weight comes to bear. At the base the weight of the entire vault is sustained by diagonal forces N_d, because forces N_φ vanish at $\varphi = \pi/2$. To valuate force N_d, let us section the cloister vault with a horizontal plane π at a generic height Z (Fig. 7.51).

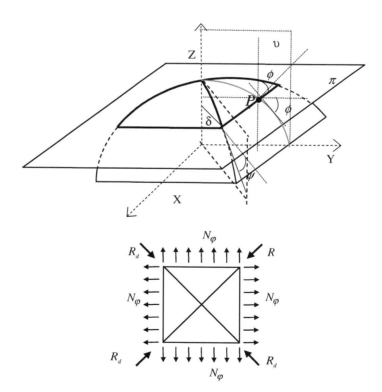

Fig. 7.51 Vertical equilibrium of the vault cap

Now consider two vertical planes, both passing through the vault center: one, which we denote as δ, passes through the diagonal of the vault, and the other, υ, is orthogonal to axis X in Fig. 7.51. The horizontal plane, π, defines the level ϕ at which forces N_d and N_ϕ are to be evaluated. Diagonal force N_d is contained within diagonal plane δ and makes an angle ψ with the horizontal (Figs. 7.51 and 7.52). The vertical plane, υ, orthogonal to axis X, sections the cylindrical surface along an arc of a circle whose tangent forms the angle ϕ with the horizontal at level ϕ .

The relation linking angles ψ and φ must be determined. To this end, we consider a unit segment 1 along the diagonal. This segment projects onto the vertical and horizontal directions to produce the segments a and b', thereby defining the triangle $1ab'$ in plane δ.

The projection of the triangle $1ab'$ onto the vertical plane υ is the triangle $1'ab$. Thus, we have

$$a = btg\phi = b'tg\psi. \tag{7.75}$$

Furthermore, the vault has a square plan and segments b and b' form an angle of $\pi/4$ on the horizontal plane, whence we get

$$b = b'/\sqrt{2}. \tag{7.76}$$

Substituting (7.76) into (7.75) gives

$$b'\frac{1}{\sqrt{2}}tg\phi = b'tg\psi, \tag{7.77}$$

whence

$$tg\psi = \frac{1}{\sqrt{2}}tg\phi. \tag{7.78}$$

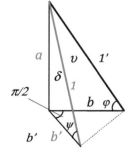

Fig. 7.52 Tangent to diagonal and circle arc in the vertical plane orthogonal to X

7.5 Cloister Vaults

The diagonal forces N_d can be determined accounting for the global equilibrium of the vault along the vertical direction. We have

$$4N_d \sin \psi + 4N_\phi 2x \sin \phi + G = 0, \tag{7.79}$$

Equation (g) at Sect. 7.2 gives the weight G_l of the single lune segment as

$$G_l(x) = 2gR^2(1 - \cos \phi)$$

Further, taking into account that $x = R \sin\phi$ from (7.79) we get

$$N_d = -\frac{2gR^2}{\sin \psi}[\cos \phi(1 + \sin^2 \phi) - 1]. \tag{7.80}$$

On the other hand,

$$\sin \psi = \frac{\sin \phi}{\sqrt{2 - \sin^2 \phi}} \tag{7.81}$$

and we finally obtain the expression for the compressive force along the vault diagonal

$$N_d = 2gR^2 \frac{\sqrt{2 - \sin^2 \phi}}{\sin \phi}[\cos \phi(1 + \sin^2 \phi) - 1], \tag{7.82}$$

whence it is moreover clear that

$$\lim_{\phi \to 0} N_d = 0 \qquad \lim_{\phi \to \pi/2} N_d = -2gR^2. \tag{7.83}$$

Figure 7.53 plots force N_d, as expressed by Eq. (7.82): tensions result at the top and compressions in the lower zone of the vault. At the base, as forces N_ϕ vanish, the entire weight of the vault is sustained by the diagonals.

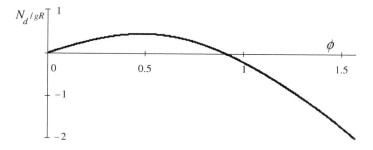

Fig. 7.53 Plot of N_d with varying φ

Now, in order to determine function $C_2(\varphi)$, and hence define forces N_x, it is convenient to analyze the equilibrium of a corner band of the vault with the two sides on the cylindrical surfaces belonging to two adjacent webs.

The width of the sides of this band is $Rd\phi$. The sides of this corner band straddle the corner and run from the middle of the webs as far as the corner. In this way, the contributions of forces $N_{x\varphi}$ acting on sides ds cancel due to symmetry. We now formulate the equilibrium condition of this corner band along the diagonal direction. The contributions of the various forces are considered separately in the following (Fig. 7.54).

- *Forces N_x*
  The two forces $N_x ds$, at $x = 0$, acting on the middle of the webs, yield the component acting internally along the diagonal and are consequently negative. By taking into account the expression for forces N_x evaluated at $x = 0$, we obtain the resultant diagonal force

$$-\sqrt{2}C_2(\phi)Rd\phi. \tag{7.84}$$

- *Forces N_φ*
  Here we evaluate the resultant of all forces N_φ acting along the edges of the corner band. They are constant along the edge.
  Firstly, along each side of the band we have both forces N_φ and the corresponding values due to the augmented coordinate φ. Accounting for the first side of the band, parallel to y, forces N_φ, acting along the side's two edges, yield a force acting along x on the horizontal plane

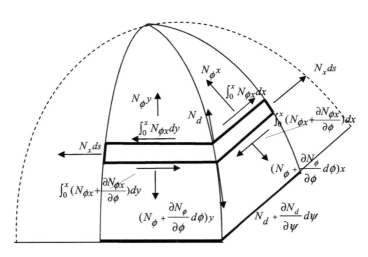

Fig. 7.54 Forces acting on the corner band

7.5 Cloister Vaults

$$-N_\phi y \cos\phi + [N_\phi y \cos\phi + \frac{\partial}{\partial\phi}(N_\phi y \cos\phi)d\phi] = \frac{\partial}{\partial\phi}(N_\phi y \cos\phi)d\phi. \quad (7.85)$$

Likewise for the other side of the band, parallel to x, we obtain the horizontal force acting along y

$$-N_\phi x \cos\phi + [N_\phi x \cos\phi + \frac{\partial}{\partial\phi}(N_\phi x \cos\phi)d\phi] = \frac{\partial}{\partial\phi}(N_\phi x \cos\phi)d\phi. \quad (7.86)$$

By summing the two forces vectorially, by accounting for $y = x$ we get the diagonal resultant acting outwardly

$$\sqrt{2}\frac{\partial}{\partial\phi}(N_\phi y \cos\phi)d\phi. \quad (7.87)$$

Thus, taking into account the expression for forces N_φ obtained above, and that $x = y = R\sin\phi$, we get the diagonal resultant

$$\sqrt{2}gR^2 \cos\phi(3\sin^2\phi - 1)d\phi. \quad (7.88)$$

- **Forces $N_{x\varphi}$**

  The resultant of all the forces produced by $N_{x\varphi}$ acting along the sides of the band parallel to y or x are evaluated as follows. Along each side of the band we have both forces $N_{x\varphi}$ and the values corresponding to the augmented coordinate φ. Concerning the first side of the band, the one parallel to x, the resultant of forces $N_{x\varphi}$, acting along the two edges of the band, yield a force along x

$$-\int_0^x N_{\phi x}dx + \int_0^x N_{\phi x}dx + \frac{\partial}{\partial\phi}(\int_0^x N_{\phi x}dx)d\phi = \frac{\partial}{\partial\phi}(\int_0^x N_{\phi x}dx)d\phi. \quad (7.89)$$

Likewise, considering the other side of the band, parallel to y, we obtain the horizontal force

$$\frac{\partial}{\partial\phi}(\int_0^y N_{\phi x}dy)d\phi. \quad (7.90)$$

Summing up these two forces vectorially, taking into account that $y = x$, we obtain a diagonal force acting outwardly

$$\sqrt{2}\frac{\partial}{\partial \phi}(\int_0^x N_{\phi x}dx)d\phi \qquad (7.91)$$

However, $N_{\phi x} = -2gx \sin \phi$, so

$$\sqrt{2}\frac{\partial}{\partial \phi}(\int_0^x N_{\phi x}dx)d\phi = \sqrt{2}\frac{\partial}{\partial \phi}[(-2g \sin \phi)\frac{x^2}{2}]d\phi$$

Whence, with $x = R \sin\phi$
we

$$\sqrt{2}\frac{\partial}{\partial \phi}\sqrt{2}\frac{\partial}{\partial \phi}(\int_0^x N_{\phi x}dx)d\phi = -\sqrt{2}gR^2\frac{\partial}{\partial \phi}(\sin^3 \phi)d\phi$$
$$= -3\sqrt{2}gR^2 \sin^2 \phi \cos \phi d\phi. \qquad (7.92)$$

- *Forces N_d*
  We now address the contribution of the diagonal forces N_d. The horizontal component of N_d is $N_d \cos \psi$. The rate of change along ψ of this force is thus given by

$$-N_d \cos \psi + N_d \cos \psi + \frac{d}{d\psi}(N_d \cos \psi)d\psi = \frac{d}{d\psi}(N_d \cos \psi)d\psi$$
$$= \frac{d}{d\phi}(N_d \cos \psi)d\phi. \qquad (7.93)$$

However,

$$\cos \psi = \frac{1}{\sqrt{1+tg^2\psi}} = \frac{1}{\sqrt{1+1/2tg^2\phi}} = \frac{\sqrt{2}\cos \phi}{\sqrt{1+\cos^2 \phi}} \qquad (7.94)$$

whence, taking into account the expression for N_d

$$\frac{d}{d\phi}(R_d \cos \psi)d\phi = 2gR^2 \frac{d}{d\phi}\{\frac{\sqrt{2-\sin^2 \phi}}{\sin \phi}[\cos \phi(1+\sin^2 \phi) - 1]$$
$$\frac{\sqrt{2}\cos \phi}{\sqrt{1+\cos^2 \phi}}\}d\phi. \qquad (7.95)$$

Summing up all the contributions of the forces considered, we arrive at

7.5 Cloister Vaults

$$-\sqrt{2}C_2(\phi)Rd\phi + \sqrt{2}gR^2\cos\phi(3\sin^2\phi - 1)d\phi - 3\sqrt{2}gR^2\sin^2\phi\cos\phi d\phi$$
$$+ 2gR^2\frac{d}{d\phi}\{\frac{\sqrt{2-\sin^2\phi}}{\sin\phi}[\cos\phi(1+\sin^2\phi) - 1]\frac{\sqrt{2}\cos\phi}{\sqrt{1+\cos^2\phi}}\}d\phi = 0.$$
(7.96)

The derivative of the function in parenthesis in (7.96) with respect to the variable φ has been evaluated by Ciampa (2007–2008) using the program *Mathcad*. This enables evaluating the constant

$$C_2(\phi) = gR\cos\phi[3\cos(2\phi) - 2] - N_d\frac{\sin\psi}{R\sin^2\phi}, \quad (7.97)$$

thereby yielding the same expression obtained by Tomasoni (2008) via a different approach. Equation (7.97) enables calculating forces N_x. Thus we have

$$N_x(\phi) = gR[\frac{x^2}{R^2} + 3\cos(2\phi) - 2]\cos\phi - N_d\frac{\sin\psi}{R\sin^2\phi}. \quad (7.98)$$

or, taking into account that $\sin\psi = \sin\phi/\sqrt{2-\sin^2\phi}$

$$N_x(\phi) = gR[\frac{x^2}{R^2} + 3\cos(2\phi) - 2]\cos\phi - \frac{N_d}{R}\frac{1}{\sin\phi\sqrt{2-\sin^2\phi}} \quad (7.98')$$

Function $C_2(\phi)$ represents the force N_x evaluated on the midline axis of the webs, that is at $x = 0$ (Fig. 7.55). At the base of the cloister vault we have tractions costantly

$$N_x(\phi = \pi/2) = 2gR^2 \quad (7.99)$$

Figure 7.55 shows the distribution of ther forces N_x at mide section of the lunes.

Fig. 7.55 Variation of forces N_x along the mid section of lunes

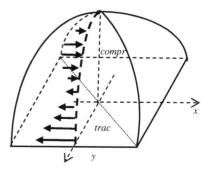

7.5.2 Cracking

The membrane description of the stresses in the vault indicates a strong hooping action produced by forces N_x. Figure 7.56 schematically illustrates the tensile forces acting on the vault corners. The corner tensions N_x are balanced by the diagonal compressions N_d. A strong pull action is produced at the vault corners and cracking of the masonry occurs first along the *diagonals* of the vault. Other cracks may then follow along lines parallel to the edges. Ultimately, diagonal cracks occur systematically along the groins of cloister vaults (Fig. 7.57).

Consequent to the diagonal cracking, forces N_x vanish, as do $N_{x\phi}$, by virtue of the first of the equilibrium equation (7.68). Without forces N_x and $N_{x\phi}$ the

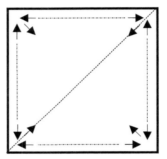

Fig. 7.56 Equilibrium at corners

Fig. 7.57 Typical cracking pattern at the corners of a cloister vault; Palazzo Martinengo delle Palle in Brescia (Tomasoni 2008)

compressive diagonal forces N_d also disappear at the corners, leaving only forces N_ϕ. These will necessarily incline towards the horizontal, withdraw from the vault's middle surface, and produce a shear component T_ϕ able to ensure equilibrium. Thus, the third equation in (7.68) no longer holds.

The web slices and resistant bands of the vault run along the coordinates lines ϕ as far as the perimeter walls. Forces N_ϕ transmit both the weight of the vault and the thrust onto these walls. The definitive resistant structure of the cloister vault is thus determined upon its cracking. The next section will take up the search for this new resistant system of the vault within the framework of the no-tension model of masonry material.

7.5.3 Definitive Resistant Model

Square plan vault

The definitive resistant model of cloister vaults can be immediately defined taking into account that the cylindrical bands of the webs at cracking transmit both weights and thrusts onto the perimeter walls. The cloister vault thus slices into two orders of arches supported by the edge walls.

For the sake of simplicity, we will first consider the case of a cloister vault with square plan and circular profile. We divide the vault webs into cylindrical slices of a specific width, as shown in Fig. 7.58.

A grid defines the curved axes of these two orders of slices and we can consider the loads represented by a series of vertical forces applied at the grid nodes. The intensity of each of these forces corresponds to the weight of the influence area of the node, represented by the gray square in Fig. 7.59. The loads W, corresponding to the various influence areas, have different intensities for the different areas of the vault surface involved. Each node is designated by two coordinates (i, j)

Fig. 7.58 Resistant system of the sliced cloister vault

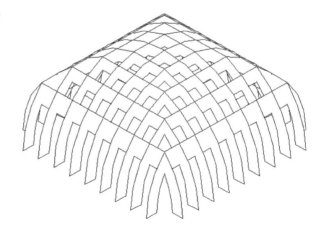

Fig. 7.59 Grid formed by the axes of the two orders of sliced arches constituting the vault and the corresponding loads

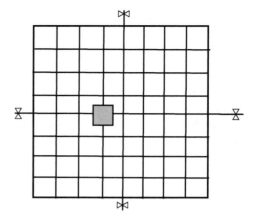

respectively representing the position at which the two parallel axes of the grid edges intersect at the node. The problem is to determine the distribution of load W_{ij} between the two arches i and j intersecting at node (i, j).

Two different arches cross at each node of the grid. In particularly, two circular arches pass through the vault center. These arches have the same axis and the same span. Load W applied at the center of the vault divides into equal parts.

Moving away from the center, different arches intersect at various nodes. For instance, one node may lie where a full arch intersects with another one with a flat central platband, as illustrated in Fig. 7.60. The two ideal arches are in equilibrium configuration and their pressure curves will always remain within the two arches. Specifically, the pressure line in the flat stretch of the arch will be represented by a very flat curve, hence the load acting on this stretch will be negligible.

All pairs of arches intersecting at each of the nodes belonging to the vault axes of symmetry, that is, the central and diagonal arches, are equal. The loads applied to these nodes are thus split equally into 50 % shares of the total. The location of these points on the axes of symmetry are indicated with small circles in Fig. 7.61. A circular arch segment will intersect with a flat arch segment in all the other nodes. At these nodes, the entire load Q_{ij} is transmitted only onto the curved stretch, or in other words, the load acting on any flat, or nearly flat segment, is zero (Fig. 7.62).

An example

The cylindrical cloister vault is question is subdivided into 9 × 9 arch bands, as shown in Fig. 7.63. The external and internal radii are $R_e = 7.50$ m; $R_i = 6.50$ m; the

Fig. 7.60 Two different types of arches in the sliced cloister vault

7.5 Cloister Vaults

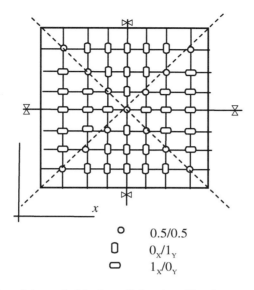

Fig. 7.61 Distribution of the vertical loads applied at the grid nodes

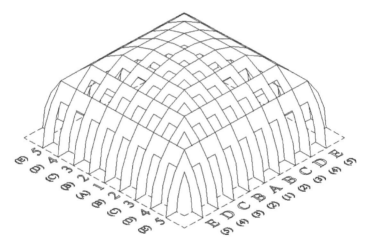

Fig. 7.62 Slicing the cloister vault into in 9 × 9 arch bands

thickness of the vault is therefore $s = 1.0$ m. The distance between adjacent arch bands is thus $i = 1.44$ m.

Figure 7.63 shows the various arch. The division of the loads is therefore immediate, as shown in the figure. Table 7.1 reports the distribution of the loads amongst the various arch bands. Each arch has been divided into six voussoirs whose weights are given in Table 7.2. The pressure lines within the various arch

Fig. 7.63 Pressure curves and minimum thrusts on the arch bands

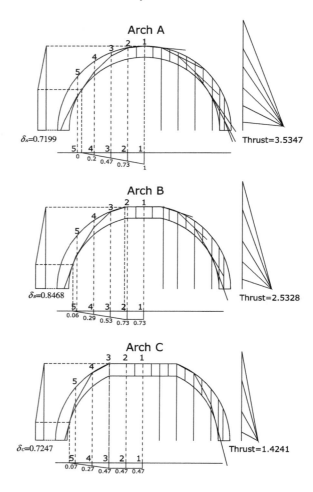

Table 7.1 Distribution of the vertical loads amongst the various arch bands

%	E	D	C	B	A
5	50	100	100	100	100
	50	0	0	0	0
4	0	50	100	100	100
	100	50	0	0	0
3	0	0	50	100	100
	100	100	50	0	0
2	0	0	0	50	100
	100	100	100	50	0
1	0	0	0	0	50
	100	100	100	100	50

7.5 Cloister Vaults 347

Table 7.2 Weights voussoirs

	Arch A	Arch B	Arch C
P_1	1.447	1.716	1.587
P_2	1.479	1.732	1.587
P_3	1.592	1.592	1.610
P_4	1.860	1.860	1.610
P_5	2.801	2.801	2.801
P_6	2.530	2.530	2.530

Table 7.3 Thrusts transmitted by the various arch bands

Arch	A	B	C	D	E
Thrust (t)	3.53	2.53	1.42	0.70	0.23

bands are shown in Fig. 7.62. It is a simple matter to determine the distribution of the thrusts along the perimeter walls (Table 7.3). The precision of the results can be improved by narrowing the grid in Fig. 7.64.

Rectangular-plan vault

The same principle of simplified load distribution also applies in the case of vaults with a rectangular plan, such as that illustrated in Fig. 7.64.

In brief, the principles of distribution of the loads amongst the arch bands are:

- no load can be sustained by flat arch segments;
- in nodes where curved and flat arch segments intersect, the entire load Q is sustained by the curved arch, even if this curved arch becomes flat immediately after the node;

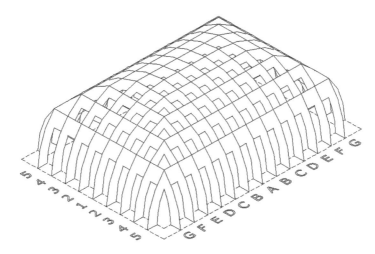

Fig. 7.64 Cloister vault with rectangular plan sliced into arch bands

Table 7.4 Load distribution amongst the various arch bands

%	G	F	E	D	C	B	A
5	50	100	100	100	100	100	100
	50	0	0	0	0	0	0
4	0	50	100	100	100	100	100
	100	50	0	0	0	0	0
3	0	0	50	100	100	100	100
	100	100	50	0	0	0	0
2	0	0	0	50	100	100	100
	100	100	100	50	0	0	0
1	0	0	0	0	50	100	100
	100	100	100	100	50	0	0

- in nodes where equal arches intersect, the load is allocated equally, i.e. 50 % each.

An example

The defining vault parameters are: $R_i = 4.5$ m; $s = 0.70$ m; distance between bands $i = 1.0$ m; $\gamma = 1$ t/mc.

The cylindrical cloister vault under consideration is subdivided into 9 × 9 arch bands, as shown in Fig. 7.62. Table 7.4 reports the distribution of the loads amongst the various arch bands. The upper number in each box indicates the percentage load assigned to the arch denoted by a letter, the lower number in the box is the load percentage assigned to the arch denoted by a number.

Once the loads have been allocated, the pressure lines of the various arches can be determined according to the minimum thrust assumption and the various thrusts evaluated.

Tables 7.5 and 7.6 report the thrusts transmitted to the perimeter walls by the various arch bands. The total thrust on the longer abutment walls, whose overall length is 13.00 m, equals 13.87 t, with an average value of 1.067 t/ml. The total thrust on the shorter abutment walls, 9.00 m in length, equals 6.62 t, with an

Table 7.5 Thrust on the long abutment wall

Arch	A	B	C	D	E	F	G
Thrust(t)	1.93	1.93	1.71	1.20	0.69	0.33	0.11

Table 7.6 Thrust on the short abutment wall

Arch	1	2	3	4	5
Thrust(t)	1.96	1.20	0.69	0.33	0.11

average value of 0.74 t/ml. The considerable difference between the central and lateral thrusts is due to the tendency of the vault to 'work' along the shorter span arches.

The calculations on the cloister vault presented here have been drawn from Ciampa (2007–2008).

References

Abraham, P. (1934). *Viollet le Duc et le rationalisme médieval*. Paris: Vincent Fréal.
Castigliano, C. A. (1879). *Théorie de l'équilibre des systèmes élastiques, et ses applications*. Turin: A.F. Negro.
Ciampa, D. (2007–2008). *Statica delle volte a padiglione in muratura*, supervisor. M. Como, a.y.
Ciciotti, E. (2006–2007). *Tecniche costruttive e Sistemi Statici della fabbrica delle Terme di Diocleziano trasformata nella basilica di S. Maria degli Angeli e dei Martiri*, Supervisors. M. Como e C. Conforti, a.y.
Como, M. (2010). On the collapse of the Beauvais Cathedral in 1284. Reasons and investigations. In Proceedings of the Second International Congress on Construction History, Cambridge, Vol. 1, edit. By the Coconstruction History Society.
Crema, L. (1942) La volta nell'architettura romana, L'Ingegnere, n. 9, Industrie Grafiche Italiane Stucchi, Milano.
Di Pasquale, S. (1996). *L'Arte del Costruire, tra conoscenza e scienza*. Venice: Ed. Marsilio.
Fabiani, F. M. (2007). *Analisi statica delle volte a crociera in muratura*. Tesi di dottorato: Dipartimento di Ingegneria Civile, Università di Roma Tor Vergata.
Heyman, J. (1966). The stone skeleton. *International Journal of Solids and Structures*, 2.
Heyman, J. (1995, 1997). *The stone skeleton*. Cambridge: Cambridge University Press.
Heyman, J. (1977). *Equilibrium of shell structures*. Oxford: Oxford University Press.
Lancaster, L. C. (2007). *Concrete vaulted construction in imperial Rome*. Cambridge: Cambridge University Press.
Morabito, G. (2004). *Caratteri e tecniche del costruire nell'Europa del Medioevo, in Europa, Civiltà del costruire: dodici lezioni di cultura tecnologica dell'Architettura*. Gangemi, Rome: a cura di G. Morabito.
Palladio, A. (1601). *I quattri Libri dell'Arcjitettura*. Venezia.
Scamozzi, V. (1615). *L'idea dell'Architettura Universale*. Venezia.
Tomasoni, E. (2008). *Le volte in muratura negli edifici storici*. Rome: Aracne.

Chapter 8
The Colosseum

Abstract This chapter concerns with the static behavior of the Colosseum, one of the most important monuments of the Roman architecture. After a description of the structure of the monument, the following topics are carried out:

(a) static analysis of the original configuration
(b) description of the damage heaped by the monument during its whole life and of the past restoration works
(c) analysis of the two conjectures concerning if the past earthquakes or the past dismantlement works were responsible for the damage.

The study is developed according to the above common static approach, based on the limit analysis and, in some cases, with the use of suitable non linear programs. The static analysis of the original configuration of the monument shows that the lack of radial constraints on the ring walls, partially weakened their static safety under vertical loads. This weakness was removed by the 19th century restoration works. The monument suffers little from the seismic action. The huge mass of the vast building, together with the presence of underlying soft soils, produces a strong mitigation of the intensity of the seismic waves that propagate from the bedrock up to the surface. At the same time the monument has a seismic strength far in excess of the possible maximum seismic forces that reached the monument throughout the whole of its history. The second conjecture, that of the dismantlement, is also taken into account in the chapter and it is shown that the demolition of at least two adjacent outside piers can produce collapse of an entire vertical strip of the outer wall.

8.1 The Original Colosseum Structure

The Colosseum, Rome's greatest amphitheater (Cozzo 1928), was commissioned by the Emperor Vespasian in 72 CE. It was built on the marshy site of an artificial lake on the grounds of Nero's palace, the *Domus Aurea*, in the valley between the Palatine, Esquiline and Caelian Hills.

The soil on which the Colosseum is built is made up mostly of stiff marine clays, gravels and tuffs, with a cover of soft alluvial sediments of a former tributary of the River Tiber (Sciotti 2004).

Prior to the construction of the amphitheater, a preliminary hydraulic assessment of the area was required to drain the waters that Nero had channeled into the valley to form the artificial lake. After the lake was drained, a huge concrete elliptical ring was erected as an ordinary masonry wall, constructed of *opus caementicium*—a mixture of rubble and pozzolanic mortar. Its transverse section was about 50 m wide and 13 m high (Fig. 8.1). Later, with subsequent fillings, this heavy ring became the foundation, on which the external structure of the Colosseum was built.

The monument has an elliptical plan, similar to the foundation ring, with external diameters of 188 and 155 m (Rea 2002) (Fig. 8.2). At its interior lies the arena, where the various events took place.

The plan of the arena is also elliptical in shape, its major and minor axes measuring 75 × 44 m, respectively. The *podium,* constructed of *opus latericium* and faced in marble, surrounded the arena.

The emperor's seat was set at the center of the podium, called the *suggestum,* while members of the senate and court occupied the rest of the podium. The arena had a wood pavement covered with yellow sand; the term *arena,* in fact, derives from the Latin word for the sand covering the grounds (La Regina 2001).

Beneath the arena, in the area known as the *hypogeum* (literally, the underground), were rooms that served various purposes related to the events. The rooms were divided by tuff and brick walls and were arranged in four symmetrical quadrants connected by two corridors at right angles to each other, running along the ellipse axes and two corridors parallel to the podium wall. Each quadrant had three straight corridors parallel to the main axis and a room whose plan formed a segment of the ellipse (Rea 2002; Luciani 1993) (Fig. 8.3). The large, 75 m-long and 4.30 m-wide central corridor was paved in *opus spicatum* with yellow and reddish brick. Not only did the corridor walls support the wood pavement of the arena, but they also served as the foundation structures for the hoists used during events.

Below the amphitheater's main gateways were four *cryptoportici,* built into the foundations to extend the area surrounding the building and connect the underground rooms with the area outside. The *cryptoporticus* below the southeastern entrance reached the *Ludus Magnus,* the large barracks housing the gladiators, while the cryptoporticus at the opposite side extended to the area of the Temple of Venus, where the backdrops were made. The external sides of the two *cryptoportici* contained the maneuvering rooms, whose walls were made mainly out of large undressed travertine blocks.

The main part of the Colosseum's external structure consists of the *cavea* (the seating area) and three ring walls with a total of 240 piers for each floor, connected at various levels by arches supporting the perimeter arcades. The piers were made of travertine voussoirs laid without the use of mortar. The façade is made up of three superimposed regular arcades of *fornices*; these were structural elements composed of a pier plus an arch with three column orders: Tuscan, Ionic and

8.1 The Original Colosseum Structure

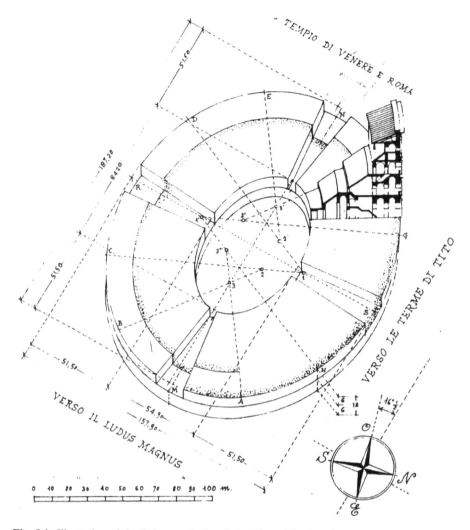

Fig. 8.1 Illustration of the Colosseum's foundation (from Moccheggiani Carpano 1977)

Corinthian (Rea 2002; Coarelli et al. 1999) (Fig. 8.4). The arches typically consist of an odd number of travertine voussoirs, so that each side of the arch would have been built up on its own and then closed with a key stone at the top (Conforto 1988, 1993).

The facade has an upper cornice, known as the *atticus*, capped by an entablature consisting of an architrave, frieze and cornice. Sturdy corbels in the frieze support the cornices. The Colosseum is 48.5 m high on the exterior, with four floors: the first three floors have round arches, while the attic bears forty rectangular windows.

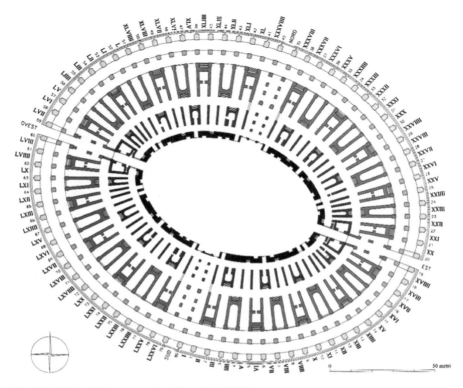

Fig. 8.2 Original Colosseum plan (from Rea 2002)

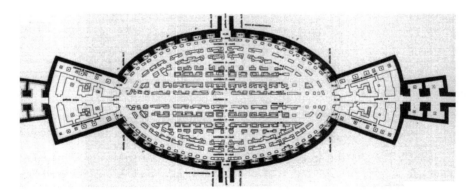

Fig. 8.3 The arena and its underground sections (from Luciani 1993)

The Colosseum was usually uncovered, though in the event of rain it could be enclosed by an enormous awning, called the *velarium*, which was maneuvered and fixed in place by two squads of sailors from the fleets of Ravenna and Cape

8.1 The Original Colosseum Structure

Fig. 8.4 Reconstruction of the original façade (from the Colosseum model by Lucangeli and Dalbono, La Regina 2001)

Misenum (La Regina 2001). To this end, the attic was fitted with a series of brackets, three for each span between the piers, for a total of two hundred and forty. Each bracket had a support to hold up an equal number of vertical struts that sustained the *velarium* above. The struts were passed through holes bored in the cornice and inserted into the bracket supports (Fig. 8.4).

The upper cornice was constructed with travertine blocks plastered on the inner side with a thick concrete cover (Fig. 8.5).

The *cavea* was formed by radial walls covered with concrete barrel vaults over which rows of stone seats were built. It was divided into five seating sections, each connected to one of five entrance ways corresponding to the external architectural orders. Stone was the main material employed to build the radial walls, which were made up of two outer layers of coursed ashlars, with an inner rubble and mortar fill. The height of the radial walls varied from a maximum of 23 m near the innermost perimeter ring wall, to a minimum of 14 m towards the internal edge of the cavea.

Originally, the Colosseum had a greater number and variety of (Lancaster 1998). Barrel vaulting was used for the perimeter arcade on both the first and second stories. This was possible thanks to the substantial heights involved and the presence of two more stories above, which applied high enough dead loads to counteract the outward thrusts.

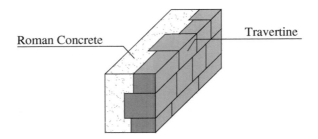

Fig. 8.5 The upper cornice built with travertine blocks and concrete cover

More stable cross, or groin, vaulting was used on the third story and the inner gallery of the second because of the lower vault heights involved. Judging from the traces that still remain on the inner part of the wall, the uppermost vault, which formed the attic floor, was unquestionably a groin vault. Stretches of the vault were interrupted by stairs providing access to the uppermost floor, the *maenianum summum ligneis*. All the circumferential arcades encircling the cavea between the ring walls (Fig. 8.6) are about 0.80 m thick at the keystone.

In constructing the vaults, Roman masons placed mortar and tuff rubble in horizontal layers on timber centering, beginning at the springers and gradually filling in the arch up to the keystone. There are some brick-faced concrete vaults which may have acted as provisional centering.

The construction of the foundation ring reveals that at the time concrete was being used in quantities that were unprecedented in the history of Roman architecture.

The design of the building must have been extremely precise, which in turn required exacting attention to detail during its construction (Conforto 1986; Rea 2002). To build the piers, the travertine blocks were carefully prepared and worked

Fig. 8.6 The first-story circumferential arcade (da Luciani 1993)

8.1 The Original Colosseum Structure

at the building site itself. To ensure a more efficient use of stone and save time, blocks of different sizes were used, making it necessary to lay down courses of different heights. The construction of the piers was carefully controlled and the blocks laid in an organized pattern, necessary to reach the pre-established springer heights of the circumferential arches (Como 2004). Bricks were also used extensively to erect the walls of the podium, the radial walls of the hypogeum, and the intradoses of the circumferential arcades, as well as to finish the lateral walls of the imperial gateway, where the intrados of the arch is made of *bipedale* bricks and decorated with stucco. Moreover, from the second order upwards, brick walls also connect the external travertine support piers, ensuring that the entire structure becomes gradually lighter towards the upper portions (Fig. 8.7).

The building site of the amphitheater required provisional timber scaffolding, which was carefully prepared during the planning stages: raising incredibly heavy travertine blocks several tens of meters off the ground was not an operation that could be left to chance. Roman carpenters built normal scaffolding on the overlying

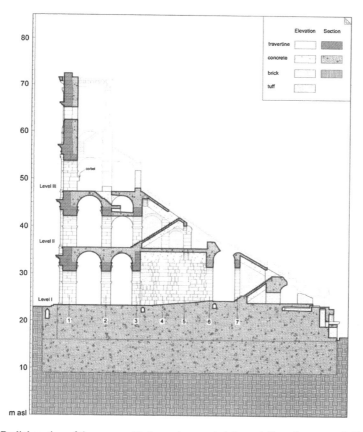

Fig. 8.7 Radial section of the cavea with the various materials used (from Lancaster, in Rea 2002)

levels, where the roof vaults of the second order and the *cavea* had not yet been built, and overhanging scaffolds where the *cavea* interrupted the walkway at ground level. This procedure was clearly unnecessary for the external facade, where the scaffolding could be added freely. The passageway of the third level, still bears a series of brackets protruding about 30 cm from the inner façade of the piers: these were used to support the overhanging scaffolding. When the building was finished, the brackets were left, as was normal practice in many Roman buildings.

8.2 Static Analysis of the Colosseum's Original Configuration

8.2.1 Pier Stresses

The first three elliptical rows of travertine piers, the passageway vaults and the attic constitute the most structurally problematic parts of the Colosseum. It has been discovered that the circumferential arcades were constructed at several different sites, in separate parts which were joined only subsequently and the centering dismantled at different times. This suggests that these vaults behave like a series of radial arches placed on the walls, joined together to form a single unit (Fig. 8.8).

Along the arcades intrados, where the heavy plaster is lacking, circumferential cracks are visible around the keystone area (Fig. 8.9). This proves that the vaults work radially and that they are in a state of minimum thrust. Indeed, the walls, and in particular the outer ring wall, which receives the thrusts of the external arcade, undergo small rotations and lateral deformations.

The values of these minimum thrusts can be evaluated by means of the funicular polygon of the forces for each perimeter arcade. The line of pressure is positioned

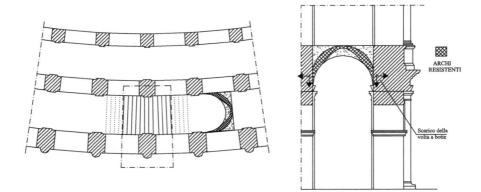

Fig. 8.8 The resistant system of the radial arch vaults

Fig. 8.9 Cracks visible at the key of the circumferential vaults

so that the vaults are in a condition of minimum thrust. As described above, the vaults are not all the same type throughout the building.

By way of example, in the following we calculate the thrusts in the first-order barrel vaults in corridor I (Fig. 8.10), whose characteristics are as follows:

Span (L) = 5.15 m
Thickness at the keystone (s) = 0.8 m
Length (H) = 6.5 m

The system of forces is obtained by dividing the element into blocks, to each of whose center of gravity is applied the value of the weight via a downwards vertical force. The funicular polygon is constructed by imposing the conditions of symmetry and minimum thrust. At the keystone, the pressure line is considered to be 15 cm from the vault extrados in order to take the floor thickness into account.

The piers of the outer wall are therefore exposed not only to the vertical and horizontal radial actions of the circumferential arcade, but to the imperfectly aligned thrusts of the circumferential arches, as well. The specific weights of the various materials used to evaluate the weights can be summarized as follows:

travertine (piers and attic): $\gamma = 2400$ kg/m^3
tuff (vaults): $\gamma = 1800$ kg/m^3

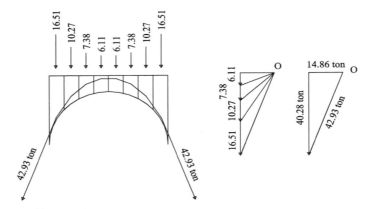

Fig. 8.10 Funicular polygon and radial pressure line for the circumferential arcades

concrete conglomerate (vaults): $\gamma = 1800$ kg/m^3
bricks (vaults and radial walls): $\gamma = 1800$ kg/m^3
wood (roofing): $\gamma = 600$ kg/m^3

The following calculations have been carried out previously (Lauri 1998–1999):

- the loads acting on the facade
- the thrusts in the plan
- the actions exerted by the circumferential arcades.

The weight of the outer wall is calculated by considering the weight of the attic, piers and parts of the circumferential arches. The calculations refer to the piers near the main axes, where the thrusting forces are greatest. The radius of curvature of the outer wall in this area is $R_{min} = 69$ m.

The opposing thrusts of two adjacent circumferential arches do not completely cancel one another out and thereby cause a *radial* action S_r which still tends to bend the external wall outwards. If the thrust of the circumferential arch is indicated by S_c and the in-plan curvature by ρ ($\rho = 1/R$), then (Fig. 8.11),

$$S_r = S_c \cdot \rho \cdot i, \tag{8.1}$$

with i the distance between the piers. The equivalent load, radially distributed on the external wall, is given by the equation:

$$q_r = S_c \cdot \rho. \tag{8.2}$$

In its original configuration the Colosseum had no tie-beams or radial chains: the thrusts S_r or, equivalently, the radial loads q_r, were exerted on the piers, in particular, those of the outer wall.

The ring walls can be represented by a series of single cells made up of a pier, together with the adjacent circumferential arcades, the architraves and the arches

8.2 Static Analysis of the Colosseum's Original Configuration

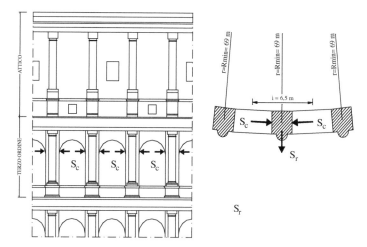

Fig. 8.11 Radial thrusts S_r

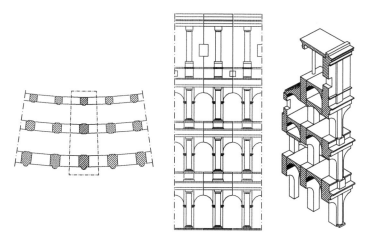

Fig. 8.12 Structural cell

below, all taken along the piers' half-spans (Fig. 8.12). The following calculations refer to this structural cell (Fig. 8.12).

Our attention shall be focused on the pier of the single cell, in particular on the pier of the outer wall (Fig. 8.13). Here the distributed loads (A_1) represent the vertical and radial loads transmitted by the arcades onto the architrave; thrust (B_1), acting within the circumferential arch, is due to the vertical load produced by the vault; thrust (C_1) is caused by the weight of the architrave and the arch itself; (D_1) is the radial component of thrust (B_1); (E_1) is the radial component of thrust (C_1); (F_1) is the axial load transmitted through the pier from the higher orders.

Fig. 8.13 The forces acting on a pier of the single cell

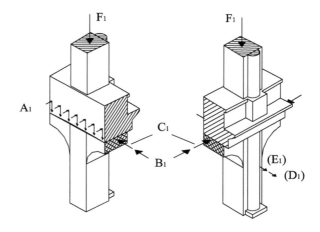

Table 8.1 Thrusting actions produced by the circumferential arches

Radial thrust	External wall	Middle wall	Internal wall
S_{r1}	2.84 t	1.41 t	1.00 t
S_{r2}	2.60 t	1.27 t	2.57 t
S_{r3}	6.70 t	1.68 t	–

The thrusting actions on the pier, directed radially outwards, and produced by the circumferential arches alone, are given in Table 8.1.

The greatest thrust corresponds to the third order of the external wall. This is due to the enormous weight of the attic, which produces a considerable thrust on the underlying arches and therefore a large component S_r. The additional radial forces exerted by the circumferential arcades have been calculated in a similar manner.

Although the intensity of these additional forces is moderate compared to those due to the circumferential arches, they have been added in order to evaluate the resulting actions on the piers. Figure 8.14 shows the set of all vertical forces acting on a pier of the external wall. It can be seen that at various stories the presence of offsets on the external side of the wall produce moments contrasting the action of the radial thrusts.

Table 8.2 summarizes the values of the axial load, the bending moment and shear acting at the base of the pier, where the thrusting forces are largest.

The internal pier undergoes very low stresses, because it is subjected to a centered axial load. The base section of the intermediate wall is also completely reactive. The external pier base section, on the contrary, is only partially reactive. To properly evaluate the stresses in the external pier, we should account for the fact that the pier is composed of large blocks set in a rather haphazard pattern. Nevertheless, in a first analysis which adopted a continuous no-tension model for the pier, it was found that only a portion of the geometrical section at the base of the half column-pier is compressed and that the maximum compression is about

8.2 Static Analysis of the Colosseum's Original Configuration

Fig. 8.14 Actions on the outer pier

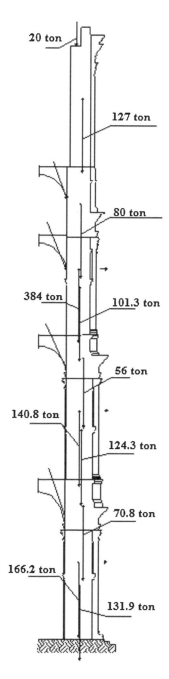

Table 8.2 Actions at the outer pier base

	External pier	Intermed. pier	Internal pier
N (t)	1543.5	952	488
M (t × m)	1350.7	47.6	–
T (t)	73.1	1.7	–

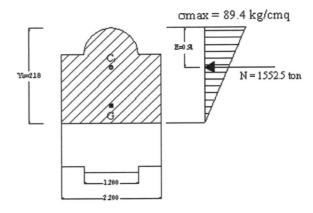

Fig. 8.15 Stresses at the base section of the external wall

96 kg/cm^2 (Fig. 8.15). This value is however considerably below the crushing strength of travertine, on the order of at least 300 kg/cm^2. It should be noted that this regards only the piers along the main axis.

In any event, the results reveal a certain weakness in the original resistant structure of the Colosseum. By contrast, the Roman amphitheater of Nimes, France, built in the 1st century CE, is smaller than the Colosseum, but its arches were constructed with transverse marble tie rods to oppose the outward bulging of the outer wall, as shown in Fig. 8.16.

Fig. 8.16 The transverse marble tie rods of the amphitheater of Nimes

8.3 Limit Analysis

8.3.1 Preliminary Remarks

The following section will examine the behavior of the Colosseum's structures from the standpoint of limit analysis, adopting the no-tension assumption according to the Heyman model. In this framework, the possible mechanisms by which the Colosseum's structures may be deformed should be kinematically compatible and should therefore not allow any sliding or material interpenetration.

The finite crushing strength of travertine has instead been taken into account only at the toe of the outer piers. The analysis firstly addresses the possibility of local failure, characterized by some piers being pushed outwards (Fig. 8.17). It is a simple matter to show that any local mechanism which allows for rotation of only some of the piers is incompatible (Fig. 8.18). A kinematically compatible mechanism (similar to the failure mechanism in domes with excess weight in the center) instead involves outwards rotation of *all* eighty piers along the three ring walls.

The circumferential arches of the wall crack. They are thereby split into blocks in the horizontal plane and follow the mechanism, continuing to transmit the thrust

Fig. 8.17 An incompatible mechanism

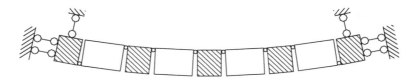

Fig. 8.18 Static scheme of a local mechanism involving 5 piers and 4 circumferential arches

due to the vertical loads (Fig. 8.19). The required condition (2.91), that the internal stresses offer no resistance to activation of the mechanism, is thus satisfied.

Hinges will develop in the arcades to allow it to follow the outward rotation of the external wall. With the circumferential vault straight in plan, a plane mechanism results, as shown in Fig. 8.20. For in-plan curvature of the vault, instead, hinges A, B, C are not cylindrical, but will follow a broken line and be accompanied by radial cracks (Fig. 8.21). Two possible failure mechanisms can ensue:

(a) involving rotation of the external wall alone (Fig. 8.20a);
(b) involving rotation of both the external and central walls as a whole (Fig. 8.20b).

Actually another, third mechanism, involving rotation of the internal wall as well is also possible, though it is however not considered herein. In fact, the resisting

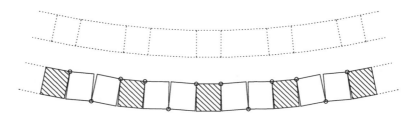

Fig. 8.19 Compatible mechanism involving the entire annular wall

Fig. 8.20 Radial displacements in the circumferential vault

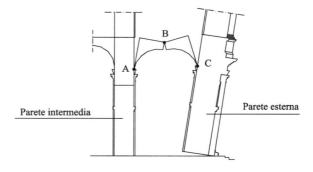

8.3 Limit Analysis

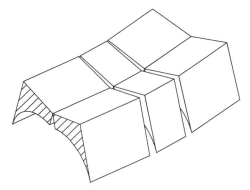

Fig. 8.21 Compatible mechanism of the circumferential vault

work done during development of this mechanism includes the extra resisting work from the lifting of the internal wall, which is not included in mechanism (*b*).

Thus, in the following the two possible mechanisms (*a*) and (*b*) will be analyzed. All calculations refer to the structural cell illustrated in Fig. 8.12.

Mechanism type (a)

This type of mechanism regards only the outer wall and the circumferential vault. The external wall rotates outwards, while the internal and central walls do not move.

The external circumferential vault between the central and outer walls cracks and three transverse hinges will thus form: two at the intrados of the abutments and one at the extrados near the crown. Figure 8.22 shows a possible mechanism that also involves crushing of the outer wall toe. The same figure illustrates the vertical and horizontal components of the displacement field. The mostly downward displacement of the circumferential arcades as the mechanism develops highlights the

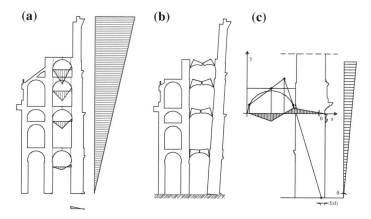

Fig. 8.22 a Vertical and horizontal displacements; **b** radial section of the mechanism; **c** centers of rotation and displacement field for the first-level inner vault

destabilizing effect of the loads acting on these vaults. All the displacements can be linked to parameter θ, that is, the rotation of the external wall. The relation between the displacements and rotation θ can be inferred from Fig. 8.20c, which shows the hinge positions, the vertical displacements of the vault at the first level, and the horizontal displacements of the external wall. In the figure, $f(d)$ is the distance of the hinge on the base of the external wall from its external edge; it has been assumed to be non-zero to account for masonry's finite compression strength.

Mechanism type (b)

All the circumferential vaults and both the external and the central walls are involved in this mechanism. The arrangement of the hinges in the inner vaults, which are located between the internal and middle walls, is similar to that in the external vault in mechanism type (a). Thus, three hinges will form on each vault: two at the intrados near the abutments and one at the extrados near the crown. With regard to the external vault, it is possible to distinguish two different hinge arrangements, which characterize distinct subtypes of this mechanism, denoted as (b_1) and (b_2), and are addressed separately in the following.

Mechanism type (b_1)

One first possible mechanism is defined by considering each external vault as behaving like a rigid strut. In this case, the two hinges in the vaults must be positioned so that the directions of the struts formed will intersect at the same point along the straight line passing through the hinges at the base of the walls (Fig. 8.23). Thus, only one hinge position in one of the external circumferential vaults (Fig. 8.24).

Mechanism type (b_2)

Another possible, alternative mechanism for the external vaults considers only a single vault behaving like a rigid strut: the other vaults will behave like those in the internal span. The problem is defining the level of the external span at which this strut vault is located. It can however be shown that the position of the strut vault that minimizes the collapse load is at the first level, as shown in Fig. 8.25.

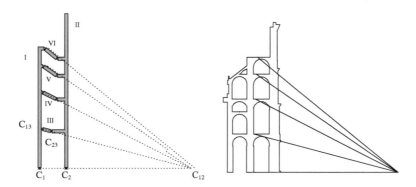

Fig. 8.23 Alignment of the centers of rotation to transform the structure into a mechanism can be chosen arbitrarily. The displacements corresponding to the mechanism deformation of the two walls and vaults are shown in Fig. 8.24

8.3 Limit Analysis

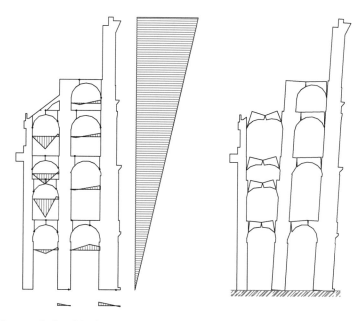

Fig. 8.24 **a** vertical and horizontal displacements; **b** radial section of the mechanism

Fig. 8.25 Mechanism type **b**: The position on the first level of the external vault that behaves as a simple strut

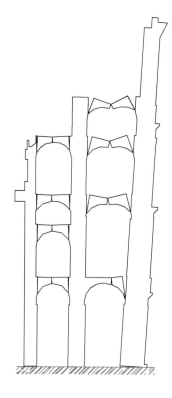

In this case, all displacements are expressed as functions of rotation θ of the central column.

8.3.2 The Collapse Load

The force system

The force system acting on the structure is made up of the following components:

- weight W of the walls;
- dead load p acting on the circumferential vaults;
- live load λq acting on the circumferential vaults;
- thrust S_c, which depends on the dead and live loads

Evaluation of the collapse load

By applying the kinematic theorem, the function λq is given by:

$$\lambda q = \frac{L_{dead}}{L_{live}}. \tag{8.3}$$

The different quantities of work are evaluated with reference to the structural cell shown in Fig. 8.12. Work L_{dead} is given by:

$$L_{dead} = L_p + L_W + L_{Sr(p)}, \tag{8.4}$$

where each term is defined as follows. The first term in (8.4) is given by the following expression:

$$L_p = \sum_{i=1}^{n} L_{p_i}, \tag{8.5}$$

where L_{pi} represents the work done by the dead load applied to each circumferential vault on the relative vertical displacements, which are illustrated in Figs. 8.15, 8.16, 8.17 and 8.18, respectively for mechanisms (*a*), (*b*$_1$) and (*b*$_2$). The sum is extended to n, which represents the number of vaults involved in the mechanism, hence n = 4 in the case of mechanism (*a*) and n = 8 for mechanisms (*b*).

The second term in (8.4) is:

$$L_W = \sum_{j=1}^{m} L_{W_j}, \tag{8.6}$$

8.3 Limit Analysis

with L_{Wj} the work done by the weight W_j of the central and external walls on the vertical displacements, which are illustrated in Figs. 8.15, 8.16, 8.17 and 8.18, respectively for mechanisms (*a*), (*b*$_1$), (*b*$_2$). The sum is extended to m, which represents the number of the columns in the mechanism, hence m = 1 in the case of mechanism (*a*), where only the external column rotates, and m = 2 for the mechanisms (*b*), where both the central and external columns rotate.

The last term in (8.4) is the work done by the radial components $S_r(p)$ of the thrusts $S_c(p)$ transmitted by the arches loaded by dead loads:

$$L_{Sr(p)} = \sum_{k=1}^{3} L_{Sr_k(p)}^c + \sum_{h=1}^{6} L_{Sr_h(p)}^e. \tag{8.7}$$

The first sum is related to the thrusts acting on the central vaults and is zero in the case of mechanism (*a*), while the second refers to the thrusts acting on the outer vaults.

Work L_{live} is given by:

$$L_{live} = L_q + L_{Sr(q)}. \tag{8.8}$$

The first term in (8.10) is given by the following expression:

$$L_q = \sum_{i=1}^{n} L_{q_i}, \tag{8.9}$$

where L_{q_i} represents the work done by the live load applied to each circumferential vault on the relative vertical displacements, which are illustrated in Figs. 8.15, 8.16, 8.17 and 8.18 respectively for the mechanism type (*a*), (*b*$_1$), (*b*$_2$).

The last term of (8.11) is the work done by the radial components $S_r(q)$ of thrusts $S_c(q)$ transmitted by the arches loaded by the live loads:

$$L_{Sr(q)} = \sum_{k=1}^{3} L_{Sr_k(q)}^c + \sum_{h=1}^{6} L_{Sr_h(q)}^e. \tag{8.10}$$

The first sum is related to the thrusts acting on the central vaults and is equal to zero in the case of mechanism (*a*), while the second sum refers to the thrusts acting on the external vaults.

The expression for λq clearly depends on the position of the hinges defining the mechanism. Thus, $\lambda q = \lambda q(d, d_1, x_i; i = 1...N)$, where x_i represents the abscissa of the i-hinge, and N is the number of hinges required to define the mechanism. The minimum of function $\lambda q(d, d_1, x_i; i = 1...N)$ yields the collapse load $\lambda_c q$ related to the set of the mechanisms analyzed, and is calculated under the constraint expressing the wall equilibrium in the vertical direction at a fixed value of masonry compression strength. With reference to the vertical loading conditions analyzed here, the resulting values of the collapse loads $\lambda_c q$ calculated via Eq. (8.3) (Coccia 2000–2001) for mechanisms (*a*) and (*b*) are:

mechanism (*a*): $\lambda_c q = 11.47$ kN/m^2;
mechanism (*b*$_1$): $\lambda_c q = 29.45$ kN/m^2;
mechanism (*b*$_2$): $\lambda_c q = 16.56$ kN/m^2.

These collapse load values seem rather low, a result that confirms the suspicion that the original external ring wall had a tendency to rotate outside because the vault thrusts tended to bend the wall outwards, despite the presence of the offsets on the external side of the ring wall.

The lack of radial constraints on the ring walls of the Colosseum in its original form compromised its static safety under live vertical loads. Other Roman amphitheaters were instead equipped with such constraints, for instance, the stone lintels on the Nimes amphitheater. The high upper cornice coursing around the Colosseum, whose tensile strength has been neglected in the analysis, may actually exert a strong binding action and increase the monument's strength. On the other hand, as for the ring vaults, different parts of the upper cornice were probably constructed separately at several different sites and joined together only later at different times. The presence in the cornice of radial sections weakened by joints cannot be neglected. The no-tension assumption, at first sight conservative, thus appears quite realistic and enables obtaining a safe lower bound of the monument's effective strength.

8.4 Damage and Subsequent Repairs

The damaged state in which the Colosseum had been reduced by the 18th century—nearly seventy centuries after its construction—was the result of alternating periods of use, disuse and abuse, during which many exceptional events, such as fires, earthquakes and stone pilfering, probably led to partial collapses and contributed to its steady decay. 'Work' was done on the Colosseum at various times, though not always with the aim of repairing it, so overall, a great deal of material has been definitively lost. Figure 8.26 shows the plan of the amphitheater first level in its current state. Comparing it to Fig. 8.1, it is immediately evident that the entire southern section of the monument—both the outer and middle walls with their piers—has been destroyed. The area just outside the arena is still well preserved around three quarters of its circumference, while 50 % of the outer wall is missing on the southern side and part of the interior wall has been rebuilt. Figure 8.27 shows a drawing by Gaspar Van Wittel, who portrayed the Colosseum as it appeared in the mid-18th century. It is unclear how the damage to the Colosseum progressed over time or how it reached the state shown in Fig. 8.27.

The most serious damage began when the circumferential arches were interrupted and the outer ring wall on the southern side began to break up. Whether this was the result of the earthquakes that struck Rome from the 4th to the 16th

8.4 Damage and Subsequent Repairs

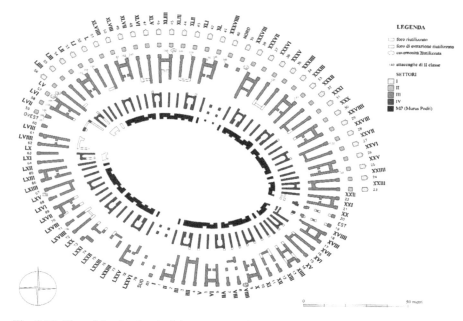

Fig. 8.26 Plan of the first level of the monument in its current state (from Rea 2002)

Fig. 8.27 Drawing by Gaspar Van Wittel showing the state of the Colosseum in the mid-18th century

centuries, as suggested by various authors (Croci 1993; Funiciello 1995; Cerone 2000), or whether it was caused by its being dismantled and stripped of its precious building materials (Conforto and Rea 1993) is not yet clear.

The historical accounts of the events that affected the monument and available documentation regarding the activities carried out within it has provided some relevant, though incomplete information (Conforto 1986; Croci 1990; Cerone 2000). In particular, from the 3rd century onwards, the historical sources indicate a series of events that caused damage to the building's structure, some of it quite extensive.

The earliest significant event to be documented is a serious fire that broke out in 217 CE. The first important restoration operations were carried out soon after—perhaps following the earthquakes in 217 and 233—under Emperor Alexander Severus. Serious damage to the monument were later caused by the 15th-century earthquakes, particularly in the years 443 and 508. An inscription engraved in a stone near the main entrance commemorates the restoration decreed by the prefect Decio Mario Venanzio Basilio to repair the damage caused by an *abominendi terrae motus*, i.e., the earthquake of 508. It was in this phase of the building's existence that, after a long period of inactivity, the last games were held in the Colosseum (519 and 523). It is thus likely that from this time on not all the 'work' done on the monument was aimed at restoring it, since its original function had ceased.

The progressive decline of Rome's population in the early 6th century reduced the number of seats needed in the amphitheater, and the first systematic work of dismantling began on the building's southern side. An epigraph documenting this activity is situated atop a pier on the second level between arches XIII and XV. One interpretation of this epigraph is that the senator *Gerontius*, by concession of the sovereign, was entitled to make use of that side of the monument as a stone quarry. Dated to the time of Theodoricus, the epigraph is situated in such a way that it could not have been visible from below unless the two perimeter porticoes had already been removed by then. It therefore seems reasonable to infer that the Colosseum was already being systematically dismantled and cannibalized for materials on its southeastern side (Rea 2002).

In the following centuries, historical sources and numerous archaeological finds point to a new phase in the building's use. On the one hand, it was in part utilized once again, though for activities quite different from beforehand. On the other, demolition continued and the monument was stripped of its precious building materials, travertine and marble in particular.

Written evidence from the late 10th century strongly suggests that several arcades and spaces under the steps were occupied by lime workers, who were attracted by the abundance of marble and travertine, which when burnt provide a rich source of lime. Thus, extensive fires were likely used to break up some piers and thus cause partial structural collapse.

A strong earthquake with epicenter in the Apennine mountains took place in 1349, as testified to by Petrarch: "*cecidit aedificiorum veterum neglecta civibus, stupenda peregrines moles*". According to several historical sources, arcades of the southern external ring collapsed. In the late Middle Ages, the monument was inhabited and during the 13th century, *Palazzo Frangipane*, amongst other buildings, was built in the southeastern sector, while a hospital was opened in 1381.

8.4 Damage and Subsequent Repairs

Until the end of the 17th century the building complex was used for a wide range of work activities, either in succession or at times even co-existing, such as a manure depot or saltpeter production for a nearby gunpowder factory. No work was done until Pope Pius VII (1800–1823), and later Leo XII (1823–1829), Gregory XVI (1831–1846) and Pius IX (1846–1878) presided over a lengthy process of restoration of the entire amphitheater.

The stability of the external wall was cause for concern, since only the 39 arches overlooking the Esquiline hill were still left standing and the balance of thrusts and counterthrusts in the circumferential direction had been interrupted. Stern, Palazzi and Camporesi constructed the southern buttress, which was completed in 1807 (Fig. 8.28), while Valadier added the buttress on the opposite side (Fig. 8.29). Despite the reinforcement work by Stern (1806–1820) and Valadier (1823–1826), the external wall continued to rotated outwards.

Therefore later, in 1850, Canina installed a series of three chains in correspondence to the thirteen central arcades (Fig. 8.30). In order to anchor the radial chains in the uppermost part of the external wall, 13 piers of the third level were reconstructed and 13 pairs of chains were installed on the upper floor of the second level. Once anchoring the wall structure had been completed, the chains were positioned in pairs on the same level as the vaults. The chains placed on the third-level floor were about 16 m in length and made of two pieces joined by hooks. These were linked together by passing horizontal iron bars through the chain links and then linking the bars themselves with a single piece of chain 9 m long linked by means of a vertical bar.

A similar system of iron supports were used to create anchorage on both the external facade and the intermediate pier using iron wedges and buffers. Canina's work of fitting the Colosseum with chains effectively completed the consolidation

Fig. 8.28 The Stern buttress

Fig. 8.29 The Valadier buttress

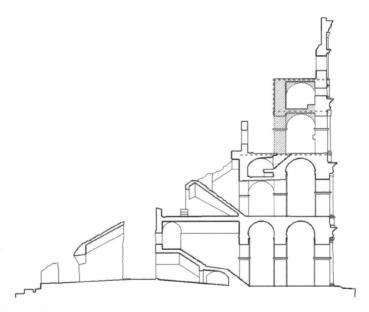

Fig. 8.30 The chains fitted by Canina at various levels together with new piers

measures. The entire central area of the northeast wall was stabilized for the first time and the efficacy of these measures can still be appreciated today, 150 years after they were carried out.

While the buttresses built by Stern and Valadier restore the balance of the thrusts of the circumferential arches, the chains installed restrain the external wall of the amphitheater in the radial direction. With these measures, the restored Colosseum, despite the damage it had suffered, was probably safer than it had ever been.

8.5 Possible Causes of the Damage

The causes of the damage suffered by the Colosseum over the centuries are still a matter for conjecture. As there in no evidence of differential subsidence of the foundations, the evidence points to either past earthquakes (Croci and Viscovitch 1993; Cerone 2000) or the systematic dismantling of the structure in the past (Conforto and Rea 1993; Como et al. 2006).

8.5.1 Seismic Excitability of the Monument: Effects of Soil-Structure Interactions

The seismicity of Rome is quite moderate. Nevertheless, during its over 2,500 years of history, the city has been struck by a sizeable number of earthquakes, many of which caused quite severe damage to its artistic patrimony. According to the Italian earthquake catalog, spanning more than two thousand years, the major effects are exclusively due to Apennine seismicity, which in some cases generated in Rome intensities of up to VII–VIII on the Mercalli intensity scale (MCS) (Guidoboni 1994). Local geological conditions, however, played a major role in transmitting the actions that caused the damage: buildings located over sedimentary fillings of the River Tiber are particularly subject to damage. As actually observed, during earthquakes, the surface ground motion in sediment-filled valleys may be significantly amplified and prolonged (Moczo et al. 1995).

The soils underneath the Colosseum, even if not completely known. are composed by soft alluvial deposits. Only on the North side there are stiffer Pleistocene soils.This non uniform soil condition could have further amplified the ground motion. (Moccheggiani Carpano 1977; Bozzano 1995; Jappelli et al. 2000; Funiciello et al. 2002; Sciotti 2004). It is thus reasonable to attribute to the foundation soil of the Colosseum the same maximum intensity level of the seismic motion recorded in Rome. The VII–VIII MCS degree corresponds to a range of horizontal seismic ground acceleration of 200–400 mm/s^2 (see, for instance, Como and Lanni 1981). A level of the maximum ground acceleration of the order of 400 mm/s^2 can be reasonably considered in examining the effects of earthquakes on the Colosseum's structure. One further important aspect must however be taken into account in examining the monument's seismic behavior: soil–structure interactions. The huge mass of the vast building, together with the presence of soft terrain,

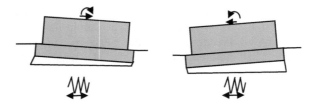

Fig. 8.31 L'interazione terreno struttura alla base del monumento

produces significant and complex interactions between the soil and the structure, as sketched in Fig. 8.31. This aspect of the problem has bee not yet considered in literature and deserves consideration.

A simple model of the structure, together with the assumed elastic connections to the underlying soil, is sketched out in Fig. 8.32. A rigid large block, representing the monument, of mass m and moment of inertia I_o about the horizontal transverse axis passing through its center of gravity G, can be displaced horizontally and rotate with respect to the rigid frame Oxy, considered to be integral with the soil. With respect to the soil the system has two degrees of freedom,

$$u(t), \quad \phi(t), \tag{8.11}$$

the horizontal displacement and the rocking rotation. The monument cannot be considered rigidly attached to the soil, but connected elastically by translational and rotational elastic springs of stiffness,

$$k_x, k_\phi. \tag{8.12}$$

One effect of the seismic action on the soil under the foundation—to which the block is connected elastically—is to impart the horizontal motion:

$$s(t). \tag{8.13}$$

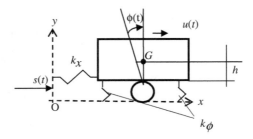

Fig. 8.32 Simple model of the interaction terrain-structure

8.5 Possible Causes of the Damage

Inertial forces arise and at each instant, t, the dynamic equilibrium of the block is governed by the following equations:

$$-m(\ddot{s} + \ddot{u} - h\ddot{\phi}) - k_x u = 0 \tag{8.14}$$

$$m(\ddot{s} + \ddot{u} - h\ddot{\phi})h - k_\phi \phi - I_o \ddot{\phi} = 0, \tag{8.15}$$

which can also be written in matrix form

$$M\ddot{\mathbf{D}} + K\mathbf{D} = \mathbf{M}\ddot{s}, \tag{8.16}$$

where

$$\mathbf{D} = \begin{bmatrix} u \\ \phi \end{bmatrix}; \ \mathbf{M} = \begin{bmatrix} m \\ -mh \end{bmatrix}; \ M = \begin{bmatrix} m & -mh \\ -mh & mh^2 + I_o \end{bmatrix}; \ K = \begin{bmatrix} k_x & 0 \\ 0 & k_\phi \end{bmatrix} \tag{8.17}$$

are respectively the displacement and mass vectors and the mass and the stiffness matrix of the system. The circular frequencies ω_I and ω_{II} of the free oscillations of the system are solutions to the algebraic equation

$$\omega^4 \frac{I_o m}{k_x k_\phi} - \omega^2 \frac{m}{k_x}\left(1 + \frac{k_x h^2}{k_\phi} + \frac{I_o k_x}{m k_\phi}\right) + 1 = 0. \tag{8.18}$$

The corresponding two eigenvectors are

$$\mathbf{Z}^{(I)} = \begin{bmatrix} 1 \\ \dfrac{m\omega_I^2 - k_x}{mh\omega_I^2} \end{bmatrix} \quad \mathbf{Z}^{(II)} = \begin{bmatrix} 1 \\ \dfrac{m\omega_{II}^2 - k_x}{mh\omega_{II}^2} \end{bmatrix}. \tag{8.19}$$

The first components of these vectors define translational displacements, while the second refer to rotations. Under the action of the seismic waves passing through the soil the overlying monument absorbs various amounts of energy depending on its oscillation periods. The values taken by the participation factors G_I and G_{II}, i.e., components of the inertia vector $\mathbf{G}$ of the dragging motion produced by the seismic action (Como and Lanni 1979), are

$$G_1 = \frac{1}{l_1}(Z_1^{(I)} M_1 + Z_2^{(I)} M_2) \quad G_2 = \frac{1}{l_2}(Z_1^{(II)} M_1 + Z_2^{(II)} M_2). \tag{8.20}$$

Factors G_I and G_{II} are the components along the vectors $\mathbf{Z}^{(I)}$ and $\mathbf{Z}^{(II)}$ of the inertia vector $\mathbf{G}$, representative of the dragging motion of the monument induced by seismic action $s(t)$, while $Z_1^{(I)}$, $Z_2^{(I)}$, $Z_1^{(II)}$, $Z_2^{(II)}$ are the translational and rotational components of the two eigenvectors $\mathbf{Z}^{(I)}$ and $\mathbf{Z}^{(II)}$. The constants M_I and M_{II} are the components of the mass vector $\mathbf{M}$, defined in (8.17), and l_1 and l_2, the generalized masses

$$l_i = \mathbf{Z}^{(i)T} M \mathbf{Z}^{(i)}. \qquad (8.21)$$

It is now necessary to calculate all the various geometric, inertial and geotechnical quantities involved, as follows.

Foundation weight. Area of the elliptical base annulus: 20,295.0 m^2; height of the foundation: 12.40 m; Foundation weight: W_f = 46,000.0 t; Distance between the center of gravity of the foundation annulus and the foundation plane: d_f = 6.20 m.

Weight of all the radial walls: W_s = 48,240.0 t; d_s = 20.60 m; Weight of the two podium walls: W_{mp1} = 11,374.0 t; W_{mp2} = 4,664 t; d_{mp1} = 15.60 m; d_{mp1} = 13.90 m; Weight of the staircase terraces: W_g = 63,848.0 t; d_g = 24.00 m.

Weight of the circumferential vaults and walls: Wp = 718,200.0 t; distance d_p = 25.00 m.

Total weight of the monument: W_T = 1,313,326.0 t.
Distance between the center of gravity and the foundation plane: d = 18.00 m.
Total mass of the monument: m = 133,876 t × s^2 × m^{-1}.
Moment of inertia I_o = 10,724,606t × m × s^2.
Elastic constraints between soil and foundation.

The soil is mostly composed of soft alluvial deposits: the corresponding value of the modulus of elasticity, E, ranges between = 1000 and 2000 t/m^2. Consequently, the shear modulus G ranges between 500 ^{2} and 1000 t/m^2.

The elliptical foundation annulus can be approximately represented by a circular annulus with external and internal radii equal to R_{me} = 171.5 m and R_{mi} = 59.5 m. The elastic sliding and rocking stiffnesses of a rigid annulus resting on the surface of an elastic half-space can be evaluated approximately by extending to this case the expressions obtained for a rigid circular footing (Hall 1967). We thus have

$$k_x = \frac{32(1-v)}{7-8v} = G(R_{me} - R_{mi}) \quad k_\phi = \frac{8}{3(1-v)} G(R_{me}^3 - R_{mi}^3) \qquad (8.22)$$

With the value of modulus G = 455 t/m^2, which results from assuming E = 1000 t/m^2 and v = 0.1, we get k_x = 236,709 t/m; k_ϕ = 6,516,349,570 tm. By doubling the value of G, we double the values of constants k_x and k_ϕ. and can now evaluate the periods of the two free oscillation modes of the monument. From Eq. (8.18), we obtain $\omega_I^2 = 1.747 r/s$; $\omega_{II}^2 = 614.686 r/s$, hence, the corresponding periods of free vibration are T_I = 4.75 s and T_{II} = 0.253 s. It should be noted that doubling the assumed values of the shear modulus, G, yields periods T_I = 3.35 s and T_{II} = 0.29 s. The oscillations modes, according to Eq. (8.19) and the first choice of G, are

$$\mathbf{Z}^{(I)} = \begin{bmatrix} 1 \\ -0,000659 \end{bmatrix} \quad \mathbf{Z}^{(II)} = \begin{bmatrix} 1 \\ 0,0554 \end{bmatrix} \qquad (8.23)$$

8.5 Possible Causes of the Damage

The first, very high-period mode is practically purely translational. The second mode, lower in period, is quasi-translational. The generalized mass factors, given by (8.11), are $l_I = 137{,}061 \text{t} \times \text{m} \times \text{s}^2$ and $l_{II} = 32.917 \text{t} \times \text{m} \times \text{s}^2$.

The seismic input is a dragging horizontal motion that nearly exclusively activates the first oscillation mode. The participation factors G_I and G_{II}, evaluated according to Eq. (8.20), are in fact: $G_I = 0.988$; $G_{II} = 0.012$. By using the common spectrum response equation defined by Italian seismic codes as

$$R(T) = 0.862/T^{2/3} \qquad (8.24)$$

we obtain R(4 s) = 0.34. Thus, while the maximum possible ground acceleration is of the order of 400 mm/s², the maximum acceleration that can effectively reach the monument is only about $0.4 \times 0.34 = 0.136$ m/s².

Because of the uncertainties in the geotechnical definition of the foundation soil, the period T_I of the horizontal oscillation mode has also been evaluated assuming larger values of the soil's elasticity modulus. By choosing the value E = 8000 t/m², eight times larger than that previously assumed, we get $T_I = 1.7$ s and the response factor becomes R(1.7) = 0.7, still below unity. As a consequence of the monument's large mass, even in the case of stiffer soil the spectral acceleration remains below the maximum possible ground acceleration and turns out to be $0.7 \times 0.4 = 0.28$ m/s².

Now, regarding the vertical seismic motion, the possible vertical ground acceleration due to an earthquake amounts to only small shares of g, thereby corresponding to vertical forces that would produce negligible effects.

The seismic shear waves that propagate from the bedrock up to the surface grow gradually as they traverse softer soils, but at the surface their action on the monument is mitigated due to the soil-structure interaction. The fundamental frequency of the monument is much lower than the predominant frequencies of the ground motion in consequence of the large mass of the monument and of the underlying layer of soft alluvial soil which interacts elastically with the structure.

The first dynamic mode of the monument thus involves only rigid translational oscillation. The higher modes, which produce deformation to the monument, are orthogonal to the first mode and consequently also to the ground motion. These higher modes do not participate in the motion, so that the high energy contained in the ground motion is only weakly transmitted to the monument. At the same time, the stiff foundation structure protects the monument from the distorting effects of differential subsidence, which certainly would occur in such soft soil.

8.5.2 Seismic Strength of the Monument

The seismic strength of the Colosseum's structure can be evaluated roughly via push-over analysis. The predominance of the translational mode suggests that the horizontal accelerations can be assumed constant across the various masses of the monument. A gradually increasing distribution of horizontal loads, linearly proportional to the masses, will thus represent the push–over loading condition quite closely.

Figure 8.33 shows the plan of the monument under the action of horizontal seismic forces. These forces act normally to the northeastern and southwestern façades. The southwestern half of the entire ring vault, which the horizontal seismic action pulls and forces to expand outward, detaches from other half, which, being an arched dam, is instead compressed by horizontal pushing loads. Only one half of the monument would thus fail under the action of an earthquake.

By considering the half of the whole ring vault pulled by the seismic loads, at first sight, we might come to the conclusion that collapse of this vault could effectively occur through the four-hinge mechanism illustrated in Fig. 8.34. However, deformation of the vault must be consistent with the assumptions of no sliding and no masonry interpenetration. When other parts of the building are considered, on the other hand, some difficulties arise. The ring vault is underlain by the piers and circumferential arches. As can be seen from Fig. 8.34, each pier must undergo movements in plan in the two orthogonal directions to rigidly follow the displacements of the circumferential vaults. Plane translation of the head of a pier corresponds to its being lifted.

Consequently, the rise experienced by the various piers would not always be the same, so the vaults must distort, with likely masonry interpenetration. Within the framework of the simple model of rigid in compression no-tension materials, compatible displacement mechanisms of the ring walls could thus not exist and the collapse load cannot be obtained by means of simple limit analysis applied to masonry structures.

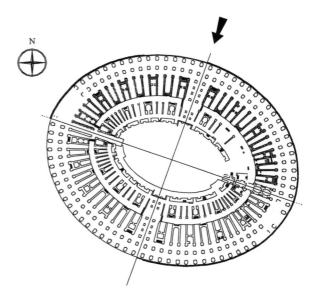

Fig. 8.33 Plan of the Colosseum and direction of seismic actions

8.5 Possible Causes of the Damage

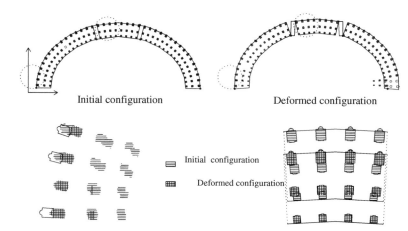

Fig. 8.34 Kinematic in-plan deformation of half the circumferential wall and deformation of half a circumferential vault and different piers' displacements

Fig. 8.35 Assumed uniaxial σ–ε diagram

Evaluation of the horizontal seismic strength of the monument has required the use of special nonlinear programs, in particular, DIANA (1990)—a finite-element structural analysis program that enables solving a wide range of nonlinear problems, and which is particularly useful in the presence of materials exhibiting different behavior under tension and compression. Relevant calculations have recently been made by Marasca (2004–2005). As for as the input data, we assume a uniaxial σ-ε diagram, with vanishing tensile strength for both the piers built with travertine voussoirs and the concrete of the vaults and upper cornices. The crushing strength of both travertine and concrete has been assumed to be 2,500 N/cm^2. In brief, with reference to Fig. 8.35, the assumed material parameters are:

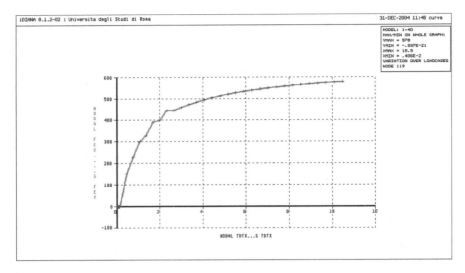

Fig. 8.36 Force–displacement diagram

$f_t = 0.3$ N/cm^2; $G_t = 0.00045$ Ncm/cm^2; $f_c = 2{,}500$ N/cm^2; $G_c = 400$ Ncm/cm^2; $E = 2 \times 10^6$ N/cm^2; $v = 0.2$, where f_t and G_t are respectively the tensile strength and the fracture energy. Likewise, f_c and G_c are the corresponding values for the compression.

While the vertical dead load is kept constant, the horizontal push-over loads are gradually increased until failure is reached. At each loading step the corresponding horizontal displacement at the top of the monument has been evaluated; the resulting force–displacement diagram is plotted in Fig. 8.36, which enables estimating the collapse load, as it occurs at the horizontal tangent point. The loading parameter λ—the ratio between the horizontal force and the weight—reaches the value $\lambda_o = 0.122$. Figure 8.37 shows the broadening of the wall at collapse.

According to the foregoing analysis, the Colosseum behaves as if it were isolated from the ground motion; the ratio between the possible horizontal seismic thrust and the weight is equal to about 0.028.

The very high horizontal strength of the monument is far in excess of the possible maximum seismic forces that could have reached the monument throughout the whole of its history.

The situation would be substantially different if the monument were already damaged, which may explain the partial failures that occurred during past earthquakes.

8.5 Possible Causes of the Damage

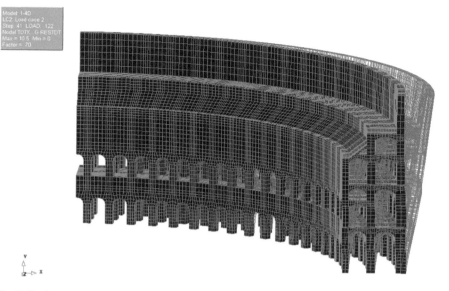

Fig. 8.37 Broadening of the outer wall at collapse

8.5.3 The Dismantling Hypothesis

Demolishing large parts of the Colosseum was certainly not an easy operation. It was probably carried out by provoking intentional failures, such as by burning the interiors of some of the travertine piers. The documented activities inside the monument of lime workers, who burnt marble and travertine to obtain lime, points to the use of fire to produce breakage of some piers and consequent partial failures of the monument. The demolition of one or two piers has been simulated via the program DIANA. The elements' weights are gradually applied to the structure deprived of the (demolished) piers on the first level.

According to the simulation, the destruction of one single pier does not lead to collapse—the monument would still remain standing after its demolition. Figure 8.38 shows the stress field in the monument after the razing of a single pier.

The demolition of two piers, on the contrary, would bring about collapse of the entire section of the outer ring wall overlying the two piers involved. Figure 8.39 shows the failure configuration of the monument after two lower-level piers of the outer ring wall have been demolished. Figure 8.40 instead shows the load–top vertical displacement diagram obtained by step-by-step calculations as the monument's initially very low weight is gradually increased.

Dismantling the monument by razing individual piers would therefore have required the simultaneous demolition of at least two piers. This could have been accomplished by boring holes in some nearby piers, inserting colophony in the

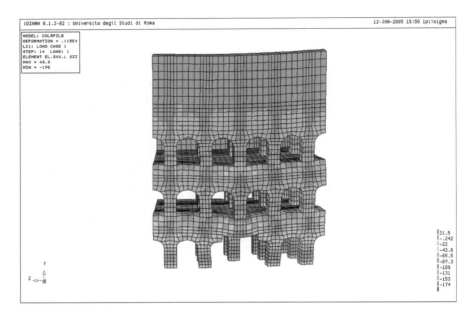

Fig. 8.38 Equilibrium configuration of the Colosseum structure after demolition of only one pier of the outer wall

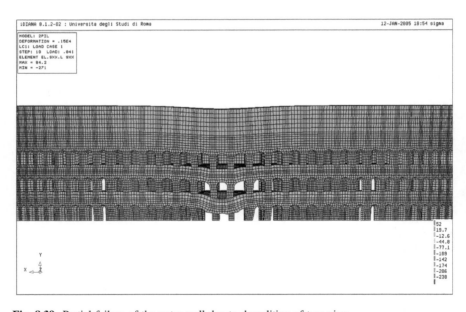

Fig. 8.39 Partial failure of the outer wall due to demolition of two piers

8.5 Possible Causes of the Damage

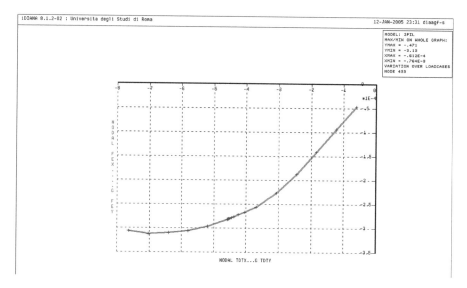

Fig. 8.40 Load–top vertical displacement diagram with gradually increasing monument weight

holes and igniting it. Historical documents indicate that dismantling operations were performed on the Colosseum as early as 16th century (Rea 2002).

8.5.4 Conclusions

Because of its enormous mass, the thick and stiff foundation structure and the soft underlying soil, the Colosseum is not very susceptible to seismic actions. Consequently, the damage that the monument suffered over nearly 17th centuries of its existence cannot be attributed solely, or even mostly, to earthquakes. Instead much of the severe damage wreaked on the Colosseum has come from the intentional demolition of many of its structures in order to reutilize its precious constituent materials. Systematic razing of the first-level piers of the external wall has been documented to as early as the 6th century. As soon as the continuity of the ring wall was broken, the external wall bulged outwards and, the consequent lack of connections between the radial and circumferential walls immediately led to an intrinsic weakness in the entire structure. Once cut off from the inner radial wall, the external circumferential wall became extremely vulnerable, and in such a seriously compromised state an earthquake would have produced devastating effects.

References

Bozzano, F. (1995). *Il* sottosuolo dell'area dell'Anfiteatro Flavio in Roma, *Geol. Appl. e Idrog.*, 30.
Cerone, M., et al. (2000). Analisi e documentazione dei dissesti strutturali ed individuazione delle situazioni a rischio, *Quaderni del Colosseo*, 1, Sovrintendenza Archeologica di Roma.
Coarelli, F., Gregori, G., Lombardi, L., Orlandi, S., Rea, R., & Vismara, C. (1999). *Il Colosseo* Electa, Milan.
Coccia, S. (2000–2001). *Analisi dei dissesti delle strutture del Colosseo*, supervisor M. Como, a.y. University of Rome Tor Vergata, Faculty of Engineering
Coccia, S., & Ianniruberto, U. (2004). Analisi della resistenza della parete dell'attico del Colosseo nella sua configurazione originaria in presenza di forze orizzontali, Atti *Conv. Naz.le ANIDIS*, Genova.
Coccia, S., Como, M., & Ianniruberto, U. (2000). Analisi Limite delle strutture del Colosseo, *Quaderni del Colosseo*, 1, Sovrintendenza Archeologica di Roma.
Coccia, S., Como, M., Ianniruberto, U., & Conforti, M. L. (2005). On the reasons of the Colosseum structural damage, *Intern. Seminar Theory and Practice of construction: knowledge, means, models*, Ravenna, 27–29.
Como, M. T. (2004). Sull'apparecchiatura dei pilastri del Colosseo, Atti del Convegno *Rileggere l'Antico*, Electa, Rome. (in press)
Como, M., & Lanni, G. (1979). *Elementi di Costruzioni Antisismiche*, (Ed.). Cremonese, Rome.
Como, M., et al. (2000). *Analisi Limite delle strutture del Colosseo*. Sovrintendenza Archeologica di Roma: Quaderni del Colosseo.
Como, M., Ianniruberto, U., Imbimbo, M., & Lauri, F. (2001). Limit analysis of the external wall of Colosseum: *Proceedings International Millennium Congress, More than Two Thousand Years in the History of Architecture*, vol. 1: session 1 and 2, UNESCO–ICOMOS, in partnership with the Bethlehem 2000 Project Authority.
Como, M., Coccia, S, Conforto, M. L., & Ianniruberto U. (2006). Historical Static Analysis of the Colosseum, *The Second International Congress on Construction History*, CSH Publ., Cambridge, Cambridge.
Conforto, M. L. (1986). L'Anfiteatro Flavio: costruzione ricostruzione e restauri, *Metamorfosi*, 3.
Conforto, M. L. (1988). Originalità del modello architettonico, in *Anfiteatro Flavio*, Rome.
Conforto, M. L., & Rea, R. (1993). *Colosseo: alcune considerazioni tecniche*. Manutenzione e Recupero nella Città Storica, Rome: Atti del Convegno ARCO.
Cozzo, G. (1928a). *Il Colosseo: L'Anfiteatro Flavio nella tecnica edilizia e nella storia*, Rome.
Cozzo, G. (1928b). *Ingegneria Romana*, Rome.
Cozzo, G. (1971). *In Colosseo. L'Anfiteatro Flavio nella tecnica, nella storia delle strutture, nel concetto esecutivo dei lavori*, Rome.
Croci, G. (1990). *Studi e ricerche sul Colosseo*, Sovrintendenza Archeologica di Rome.
Croci, G., & Viskovic A. (1993). Causes of the failures of Colosseum over the centuries and evaluation of the safety levels, *Proceedings Of the IASS–MSU International Symposium*, Istanbul, Turkey.
Diana User's Manual (1990). *Nonlinear Analysis*. Delft Netherlands: Frits C. de Witte and Wijtze Pieter Kikstra.
Docci, M. (2000). Ricerche sulla forma del Colosseo, *Quaderni del Colosseo*, 1.
Fanelli, L. (2000–2001). *Analisi numerica delle strutture del Colosseo*, supervisor M. Como, a. y. University of Rome Tor Vergata, Faculty of Engineering
Funiciello, R. et al. (1995). Seismic Damage and Geological Heterogeneity in Rome's Colosseum Area: are they related?, *Annali di Geofisica*, 38, 5–6.
Funiciello, R., et al. (2002). *La geologia della valle dell'Anfiteatro in Rota Colisei*. Rome: Electa.
Guidoboni, E. (1994). *Catalog of ancient earthquakes in the Mediterranean Area up to the 10th century*—I.N.G., Rome.

References

Hall, J. R. (1967). *Coupled rocking and sliding Oscillations of rigid circular footings*, Proceedings of the International Symposium on Wave propagation and dynamic properties of earth materials, Univ. of New Mexico Press, Albuquerque.

Jappelli, R., et al. (2000). *Restauro del Colosseo, Quaderni del Colosseo, 1*. Sovrintendenza Archeologica di Roma: Area Geotecnica.

Lancaster, L. C. (1998). Reconstruction the restorations of the Colosseum after the fire of 217, *Journal of Roman Archaeology*, vol. 11.

Lauri, F. (1998–1999). *Analisi Limite delle strutture del Colosseo*, supervisor M. Como, a. y. University of Rome Tor Vergata, Faculty of Engineering

Leli, G. (2001–2002). *Analisi Statica delle strutture perimetrali del Colosseo*, supervisors M. Como, U. Ianniruberto, F. Fabiani, a. y. University of Rome Tor Vergata, Faculty of Engineering

La Regina, A. (2001). *Sangue e Arena*. Milan: Electa.

Luciani, R. (1993). *Il Colosseo*, ed. Fenice, Rome.

Marasca, M. (2004–2005). *Problemi statici nelle strutture del Colosseo*, supervisor M. Como, a. y. University of Rome Tor Vergata, Faculty of Engineering

Mercuri, D. (2001–2002). *Analisi numerica delle strutture ad arco e dei pilastri del Colosseo*, supervisor U. Ianniruberto, a. y. University of Rome Tor Vergata, Faculty of Engineering

Moccheggiani Carpano, C. (1977). Nuovi dati sulle fondazioni dell'Anfiteatro Flavio, *Antiqua*, 7.

Moczo, P., Rovelli, A., Labàk, P., & Malagnini, L. (1995). Seismic response of the geologic structure underlying the roman Colosseum and a 2–D resonance of a sediment valley, *Annali di Geofisica*, 38, 5–6.

Molin, D., & Guidoboni, E. (1989). *Effetto fonti, effetto monumenti a Roma: I terremoti dall'antichità a oggi*. In: *Mediterranea. Storia–Archeologia–Sismologia*, edited by E. Guidoboni, SGA, Bologna, Italy.

Rea, R. (1996). *Anfiteatro Flavio*, Roma, (Ed.). Poligrafico dello Stato.

Rea, R., et al. (2002a). *Rota Colisei*. Rome: Electa.

Rea, R., Beste, H. J., & Lancaster, L. C. (2002). Il cantiere del Colosseo, *Boll. Istituto Arch. Germanico, Sez. Romana*, Verlag P.Von Zabern, Mainz Am Rheim, vol. 109.

Sciotti, M. (2004). Ricostruzione schematica dei terreni di fondazione del Colosseo, da *il Colosseo* Electa, Rome.

Chapter 9
Masonry Stairways

Abstract Statics of cantilevered masonry stairs, the so-called "*scale alla romana*", is the subject of this chapter. The flights of these stairs are cantilevered from a wall and connected by small vaults constituting the landings. The flight is composed of a long masonry vault having a depressed transverse sectional profile. For this type of structure the existence of an admissible equilibrium may appear paradoxical. A new resistant model of these stairs is proposed in the no tension context. The model is validated by numerical investigations and comparisons with tests.

9.1 Geometrical Features of Masonry Stairs: Cantilevered Stairs

There are many different types of masonry stairways. One very common type in Italy is the so-called "*scale alla romana*" (Roman stairs), whose flights are cantilevered from walls and connected by small vaults constituting the landings. In the following sections such an arrangement will be referred to as "*cantilevered stairways*."

Figure 9.1 shows the plan of this type of stairs. Usually there are three straight flights winding around an open well, four intermediate landings and a long floor-level landing, or stairhead, that provides access to the different quarters on each story. Figure 9.2 shows a view of the corner where the flight meets the landing composed of a quarter cloister vault.

The structure of the flights is composed of a long masonry vault having a depressed transverse sectional profile. They are built with bricks or stones laid in different arrangements (Fig. 9.3).

A thick layer of rubble and mortar is cast over the vaults, and the steps built on top of these bases (Figs. 9.3 and 9.4).

Fig. 9.1 Cantilevered stairway with open well and landings made up of quarter cloister vaults

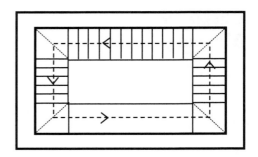

Fig. 9.2 Stair corner at the flight-to-landing intersection

Fig. 9.3 Typical section of a cantilevered flight

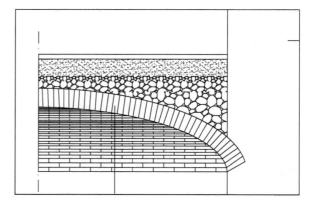

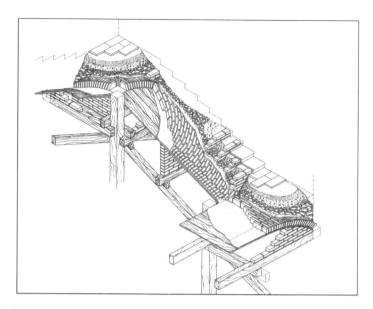

Fig. 9.4 Centering and scaffolding used to build cantilevered stairs (Giovanetti 1997)

9.2 Brick Layout

Figure 9.5 shows the typical brick pattern of a cantilevered stair with three flights and a long main landing. The stair angles around in the clockwise direction. Blocks —either stones or bricks—are laid along the axes of three-dimensional curves. The geometrical layout reveals the aim of the builder to keep the brick courses tight on the centering during construction.

Construction of a flight progresses bottom-up. The first course blocks are positioned parallel to the borders of the lower landing. Subsequent courses are then laid along arches in a gradual curve towards the well-hole, i.e. towards the flight's inner sides. As the construction progresses, these arches are made longer and longer. The positioning of the blocks is maintained up to reaching the stairwell edge.

The lowest, skewed band of a flight is built first. Then the central band follows, built with bricks laid following the curve of the last course of the lowest band. The addition of the upper band concludes construction of the flight.

Fig. 9.5 Brick layout of a typical cantilevered stairway (Formenti 1893)

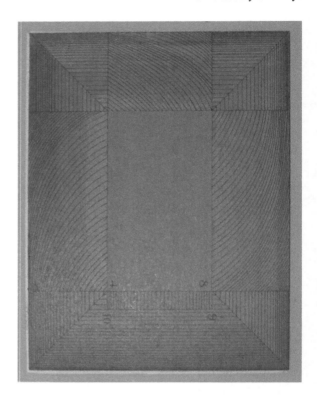

9.3 Other Types of Stairs

In some cases, the length of the stairhead does not allow for an open well. In such cases piers are constructed at the corners of the stairwell (Fig. 9.6). In the case of stairways with a central spine wall, the flights are composed of long barrel vaults (Fig. 9.7).

Fig. 9.6 Stairway with well on piers

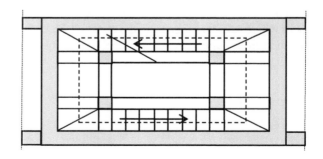

Fig. 9.7 Stair with spine wall

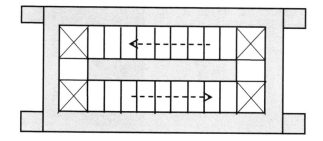

9.4 Paradoxical Static Behavior of Cantilevered Masonry Stairs

To anyone used to working with the framework of reinforced concrete structures, cantilevered masonry stairways must give the impression of extreme static instability. In such a perspective, the existence of an admissible equilibrium under loads may appear paradoxical. A widespread opinion holds that these structures are unsafe.

There are undoubtedly difficulties in formulating a consistent static model for these stairs taking into account the substantial incapacity of masonry to sustain tensile stresses. Their behavior has in fact long been the subject of in-depth study. Figure 9.8 shows a photograph of a gypsum model of a cantilevered stairway under loading tests (Lenza 1983). By increasing the loads, the first cracks occur transversely, at the middle of the free edge. Actually, it was expected that cracks would appear at the extrados, parallel to the axis of the flight, near the wall. No such cracks were however detected and the outcome is still a matter of debate (Baratta 2007).

9.5 Numerical Investigations on the Statics of a Single Cantilevered Masonry Flight

Various working hypotheses can be formulated. In place of the cantilever model, in which loads transmission occurs through bending and shear in the transverse direction, we could instead imagine longitudinal loads transmission, able to mobilize longitudinal resistant arches. However, as can be easily appreciated, even if such systems could actually be achieved, they would transmit very high thrusts, generally incompatible with ordinary staircase geometries.

The weight of the flight must be sustained by the wall in which it is embedded. The contribution made by the intermediate landings, made up of thin cloister vaults, is in fact negligible with respect to that offered by the wall. So we can assume that the weight of the flight is conveyed wholly to the wall. By this simplifying assumption, the long vault is also subjected to torsional actions and the problem

Fig. 9.8 Cracking pattern detected in the middle section of a gypsum model of a flight of stairs, by Lenza (1983)

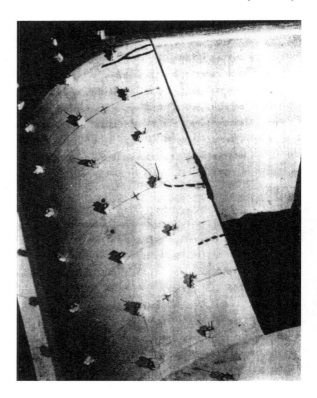

thus becomes understanding how transmission of such torsional loads can occur in the framework of no-tension behavior.

In this regard, one numerical investigation has been conducted (Soccolini 2008–2009) using the nonlinear program ATENA (Cervenka 2002), which can take into account both the presence of very weak tensile strength as well as the occurrence of cracks. The numerical analyses were conducted first considering a long horizontal vault cantilevered from a side wall. The vault section was a quarter circle with end constraints unable to sustain vertical loads. Figure 9.9 shows a section of the vault inserted into the wall.

Fig. 9.9 Transverse vault section assumed in the first numerical study

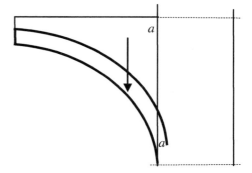

Fig. 9.10 Cracks at the extrados

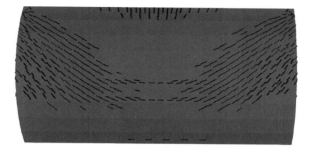

The vault is 12 cm thick, with internal radius $R = 1$ m. The length of the horizontal vault is 3 m. The loads per unit vault length, constant for the entire section, are:

masonry weight $g_p = 350$ kg/m,
rubble weight $g_r = 860$ kg/m,
live load: $q = 400$ kg/m.

The $\sigma - \varepsilon$ equation is linear, with a tensile strength of 1 kg/cm^2 and a compression strength of 200 kg/cm^2.

The loads are applied gradually through 31 successive steps. Figures 9.10 and 9.11 show the resulting cracking patterns at the extrados and intrados of the vault, respectively. The cracks are very thin and for the most part indicate the direction of the compressions. Figures 9.12 and 9.13 show the principal directions of stress respectively on the extrados and intrados.

The resulting transverse cracks, vertically cutting the external longitudinal edge of the vault (Figs. 9.11 and 9.12), match the cracks detected by Lenza (1983) in the previously mentioned tests on a gypsum model of a flight of stairs (Fig. 9.8) The same analysis was then conducted on an inclined flight. The trials regarded a vertical stairway height of 1.70 m and a linear flight length of 30 m. Figure 9.14 shows the resulting cracking pattern on the extrados. Comparing this figure with the

Fig. 9.11 Cracks at the intrados

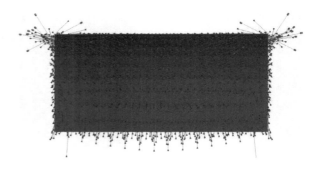

Fig. 9.12 Principal stress directions on the extrados

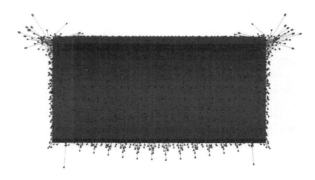

Fig. 9.13 Principal stress directions on the intrados

analogous Fig. 9.11 for the horizontal vault shows that the crack patterns are substantially similar, even if the influence of the inclination is noticeable. The results for the intrados cracking are the same.

Figures 9.13, 9.14 and 9.15 clearly reveal the occurrence of a longitudinal arching effect, with springings at the connection of the external vault edge with the

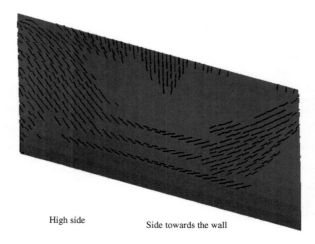

Fig. 9.14 The inclined flight. Cracks at the extrados

Fig. 9.15 generation of the torsional load on the transversal vault sections

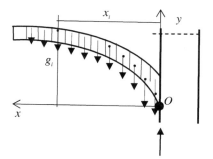

landings. Thus, the loads are *conveyed transversely* to the wall along vertical half-arches compressed by the pushing action exerted by longitudinal arches.

The cracking pattern confirms the occurrence of such longitudinal and horizontal arching. The resulting cracking patterns are similar to those occurring on the intrados near the key section of a masonry arch under vertical loads. Comparing the cracking pattern on the intrados and the extrados in Figs. 9.11 and 9.12 shows that the longitudinal flat arching arises mainly towards the vault extrados.

9.6 Resistant Model of the Horizontal Flight of Stairs with Side Landings

For the sake of simplicity, let us first examine the case of a vault with an horizontal axis. Any section of the the long vault is subjected to a *distributed torsional load*

$$m_T = \sum_i g_i x_i \tag{9.1}$$

as shown in Fig. 9.15.

No mechanism can be mobilized by the vault, rigidly fixed in the side wall and at the end landings. Thus the vault will certainly activate a resistant compression system inside. According to the results of the previous numerical investigations, we assume that the vault mobilizes a resistant system composed of (Fig. 9.16):

(a) a series of *longitudinal flat arches,* contained within the vault along horizontal planes, which transmit their thrusts to the intersection of the flight's free edge with the landings;
(b) a series of *transverse vertical half-arches* conveying the vertical loads to the longitudinal side wall.

The key point is the arising of compressions within these transverse half-arches, which enables them to transmit the vertical loads to the side wall. This compression

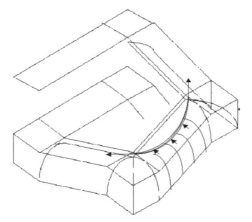

Fig. 9.16 Generation of the resistant system of a flight of stairs

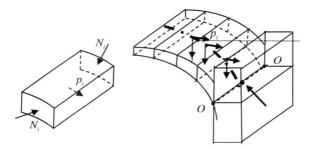

Fig. 9.17 Part of a flight subdivided into voussoirs, with the acting horizontal forces transmitted by the longitudinal arches, and the corresponding pressure line

is produced by the horizontal loads p_i conveyed by the longitudinal flat arches according to the scheme shown in Figs. 9.16 and 9.17.

Each of these longitudinal arches, denoted with A_i, is contained within an horizontal plane π_i crossing the profile of the flight section. We assume, for the sake of simplicity, that each of these arches is parabolic in profile, so that the horizontal load p_i is *constant* along the flight. In such a conception, the resistant system is able to sustain the torsion due to the misalignment between the resultant load g and the vertical wall reaction, which is assumed to pass through the point O, the toe of the flight section (Fig. 9.18).

Figure 9.19 shows the plan of the transversal vault with the assumed parabolic horizontal arches that convey the load p_i to the transversal arches.

The load p_i can be obtained by trial procedures so that the funicular curve of the loads p_i and of the weights g_i is all contained within the arch and passing through O.

9.6 Resistant Model of the Horizontal Flight of Stairs with Side Landings

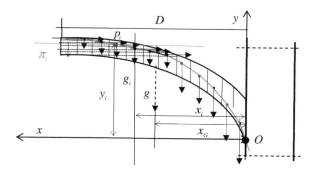

Fig. 9.18 The torsional equilibrium of the transversal vault sections

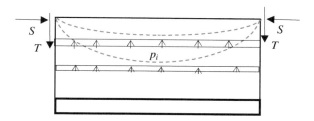

Fig. 9.19 Horizontal flat arches, interaction load p_i, total thrust and shear S, T

The stresses are only compressions, i.e. statically admissible. Acccording to Fig. 9.18, we have

$$\sum_i g_i x_i = \sum_i p_i y_i \qquad (9.2)$$

The resultant g of the loads g_i

$$g = \sum_i g_i \qquad (9.3)$$

has distance x_G from the toe O

$$x_G = \frac{\sum g_i x_i}{\sum g_i} \qquad (9.4)$$

The system, constituted by a single cantilevered flight with the two side landings fixed at rigid boundary walls, cannot become deformed by mechanisms. Masonry interpenetration tries to occur as soon as the loads are applied, so that only compression stresses can develop.

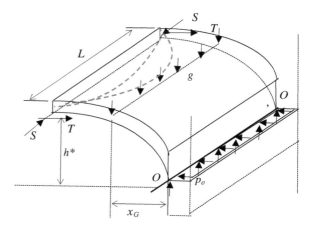

Fig. 9.20 The vault in its condition of global equilibrium

Following up now on the previous considerations we can affirm that for any distribution of weights g_i it will be possible to specify a distribution of constant horizontal forces

$$p_i \tag{9.5}$$

which, together with the weights g_i, give rise to a transverse pressure line wholly contained within the section of the flight and passing through the toe O.

Summing up all the loads p_i acting on the voussoirs constituting the vault, we obtain the horizontal load p_o representing the pushing load per unit length conveyed by the vault to the side wall (Fig. 9.20):

$$p_o = \sum_i p_i. \tag{9.6}$$

Uniform load p_o distributed along the length L of the vault is equilibrated by the two total shears T, the total transverse components of the action transmitted by the vault to the side landings (Fig. 9.19)

$$p_o L = 2T. \tag{9.7}$$

The two resultant shears T oppose the pushing load p_o exerted by the longitudinal wall. We can thereby obtain the arm h^* of the resisting global torque gLd (Fig. 9.19). For the global torsional equilibrium we thus obtain

$$gLx_G - 2Th^* = 0 \tag{9.8}$$

and

9.6 Resistant Model of the Horizontal Flight of Stairs with Side Landings

$$h^* = \frac{gLx_G}{2T}. \tag{9.9}$$

For each uniform load p_i there is one corresponding longitudinal arch of the series. The thrust S_i of these longitudinal arches A_i applied at the intersection of the landing with the free side of the flight is

$$S_i = \frac{p_i L^2}{8 f_i}, \tag{9.10}$$

where

$$f_i = D - x_i \tag{9.11}$$

is the corresponding sag of the arch A_i. The overall thrust is

$$S = \sum S_i = \frac{L^2}{8} \sum \frac{p_i}{D - x_i}. \tag{9.12}$$

The resulting internal stresses is both admissible and in equilibrium with the loads.

9.7 Determination of the Horizontal Forces P_i

The point of departure is to subdivide the section profile into a given number of voussoirs and then evaluate the corresponding weights g_i. The outer voussoir will be excluded because flat horizontal arches cannot form in the more external band.

It will be useful to trace the horizontal line a–a from the external corner of the intrados of the flight section: this line borders the lower plane of the part of the section where the lowermost horizontal arch forms. Horizontal segments are then traced above this line to indicate other planes where horizontal arches form.

A small settlement will occur due to a small widening of the wall cage under the action of the thrust of the flights. The research of the loads p_i will be thus conducted minimizing the thrust carried by transversal arches to the side wall. Through trial and error, the horizontal forces p_i can be found minimizing the resultant of the loads p_i. The loads p_i will be located mainly in the upper part of the profile, as sketched out in Fig. 9.21.

A pressure line can be traced to remain as high as possible within the section and pass through point O. Forces p_i in the lower part of the section can generally be neglected and an iterative procedure can be applied.

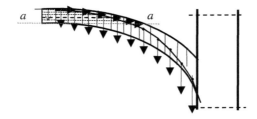

Fig. 9.21 Construction of the pressure curve within a transverse half-arch

9.8 An Inclined Flight of Stairs

Let us now consider the case of a flight of stairs whose axis is inclined by an angle ϕ with respect to the horizontal (Fig. 9.22). The section of the flight is the same as in the previous case of a horizontal axis, as are the reference axes in the section.

In the figure the upper broken line represents the external edge of the flight while the dotted line the internal edge, in contact with the wall. The overall torque is now lower and is expressed by

$$M_T = g\cos\phi \cdot L \cdot x_G \tag{9.13}$$

Shears T, misaligned with the pushing load p_o, balance the torque (9.13) according to the relation

$$2Th^* \cos\phi = g\cos\phi \cdot L \cdot x_G. \tag{9.14}$$

If the torque decreases, so does the arm of the resistant torque and

$$2Th^* = gLx_G. \tag{9.15}$$

Fig. 9.22 Torsional load on an inclined flight

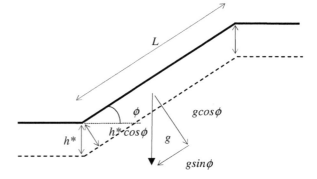

Both the pushing loads and the thrust transmitted by the series of horizontal arches to the side landings will have the same expressions (9.9) and (9.12) as for horizontal flights.

9.9 Cantilevered Stairways as a System of Flights and Landings

By using the model of the single flight with the two landings it is now possible to formulate the resistant model for the entire stairway. The play between the longitudinal flat arches and the transverse half-arches along two adjacent flights is shown in Figs. 9.23 and 9.24. It should first be noted that the vertical components of the thrusts of a pair of adjacent flights mutually cancel.

Figure 9.24 shows the connection between two adjacent flights and the inner landing. Dotted lines indicate the edges of the flights and the landing with the walls. In figure β is the angle formed by the inclined flight with respect to the horizontal, and subscripts u and l indicate quantities of the upper and lower flights, respectively. The actions conveyed to the horizontal edge of the landing, obtained by summing the thrust and the shear on the horizontal plane, are

$$(S_l \cos \beta - T_u) \quad (S_u \cos \beta - T_l) \tag{9.16}$$

When the flights are equal, we have

$$S_u = S_l = S, \quad T_u = T_l = T. \tag{9.17}$$

In this case, the forces acting on each of the two edges of the landing are (Fig. 9.25)

$$(S \cos \beta - T). \tag{9.18}$$

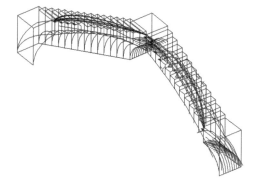

Fig. 9.23 Interplay between longitudinal and transverse half-arches of adjacent flights

Fig. 9.24 Canceling of the vertical components of the actions conveyed by two adjacent flights (*s*) and (*i*). Indicated thrusts and shears are transmitted by landings

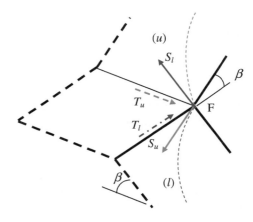

Fig. 9.25 Interactions between flights and actions on the external cage walls

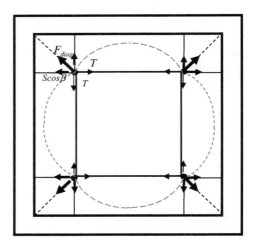

Figure 9.25 shows the plan of a staircase made up of four flights and four landings. For the sake of simplicity, the long stairhead has been omitted. The same figure shows the shear T and thrusts S acting at the *foci* of the stairs.

By considering the complex of the four flights, it can be seen that at each corner the sum of the shears T and of the thrusts S yields a resultant acting along the horizontal diagonal of the landing constituted by one quarter of a thin cloister vault. The overall compression force acting on the vertex of the diagonal is thus

$$F_{diag} = \sqrt{2}(S\cos\beta - T), \tag{9.19}$$

which, in turn, is conveyed to the corner of the staircase walls. This thrust is relatively moderate due to the contrasting effects of the thrust component, $S\cos\beta$, and the shear, T. Figure 9.26 shows an example evaluation of the actions on the

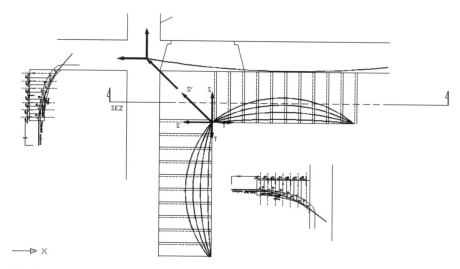

Fig. 9.26 Evaluation of thrust at staircase corners (Cartoni 2008–2009)

cage walls by the flights of a typical staircase. We understand the importance of a solid stair cage, with walls of suitable thickness.

References

Baratta, A. (2007). *Sulla statica delle scale in muratura alla romana, Ingegneri Napoli, 6*. Napoli: Legoprint Campania.
Breymann, G. A. (1926). *Trattato generale di Costruzioni Civili, 1*. Milan: Vallardi.
Cartoni, D. (2008–2009). *Sulla statica delle scale alla Romana*, M. Como. (University of Rome Tor Vergata, Rome, Italy).
Cervenka, V., & Cervenka, J. (2002). *Atena program documentation, User's manual for Atena 2D,* Prague.
Di Luggo, A. (2008). La scala nell'edilizia residenziale napoletana in *Applicazioni di geometria descrittiva e rilievo nell'Architettura "e-learning"*, Faculty of Architecture, University of Naples Federico II, Naples.
Franco, G. (1997). La scala tra valore simbolico e dimensione tecnica, *Costruire in laterizio*, n. 57.
Formenti, C. (1893). *La pratica del fabbricare*, atlante vol. 2, (ed.) Hoepli, Milan. D.M. 24.01.1986.
Galiani, V. (Ed.). (2001). *Dizionario degli elementi costruttivi*. Torino: UTET.
Giovanetti, F. (Ed.). (1997). *Manuale del Recupero del Comune di Roma*, DEI, Rome. D.M. 24.01.1986.
Heyman, J. (1995). *The stone skeleton, Structural Engineering of Masonry Architecture*. Cambridge: Cambridge University Press.
Lenza, P. (1983). Modelli di comportamento e direttrici di restauro delle scale in muratura realizzater con voltine a sbalzo, *Quaderni di Teoria e Tecnica delle Strutture*, Univ. di Napoili, Istit. Di Tecnica delle Coistruzioni, n. 531. D.M. 24.01.1986.
Soccolini, N. (2008–2009). *Ricerca di modelli resistenti delle rampe delle scale in muratura*, supervisor, M. Como. (University of Rome Tor Vergata, Rome, Italy).

Chapter 10
Piers, Walls, Buttresses and Towers

Abstract This chapter is addressed to the structural analysis under vertical loads of walls, piers and towers. For them, the non-linear interaction between the destabilizing effects of the axial loads and the masonry no-tension response can be very strong. Instability analysis of the masonry pier under an eccentric axial load is firstly studied in the wake of a relevant study of Yokel. The strong sensitivity of the pier strength to the eccentricity of the load is pointed out and comparisons are made with the case of reinforced concrete columns. Static analysis of building masonry walls is then examined. For them the presence of offsets of the wall thickness at the various stories plays a relevant role. Instability of towers whose behavior can be strongly influenced by foundation deformability, is analyzed at the end of the section. Special attention has been given to the stability analysis of the Pisa Tower, which recently underwent an outstanding restoration work.

10.1 Introduction

The topic of this chapter is the study of the statics of piers, walls and towers under vertical dead loads. Owing to their geometry, the behavior of such structures under vertical loads presents specific aspects whose analysis requires assumptions and approaches different from those considered so far. In fact, elastic masonry deformation, which is generally disregarded in arches and vaults because it yields negligible effects on their statics, in masonry piers and walls instead has important consequences on behavior.

The main aspect of the problem is the nonlinear interactions occurring between any changes in geometry and the no-tension response of the masonry: such interactions lead to high susceptibility of piers and walls to axial load eccentricities—far greater than that of reinforced concrete piers or steel columns to similar actions.

Furthermore, walls under vertical loads exhibit complex behavior, which is moreover strongly dependent on their state of conservation. The connections between walls and floors may in fact degrade over time, in which case, slow lateral deformation of the walls can significantly increase the axial load eccentricities and, consequently, their destabilizing effects. The cracking patterns in walls can furnish useful information for deciding on the most suitable repair and reinforcement operations to adopt. Similarly, nonlinear stability analysis is also required to study the equilibrium of a tower. Such analyzes, aimed at evaluating any tilting, can in many cases be performed taking into account the deformability of the foundation alone. The basic problem of evaluating the strength of eccentrically loaded piers and walls will be analyzed first in the next sections. Subsequently, many particular aspects of the statics of masonry piers, walls and towers will be examined and exemplified through some important case studies.

10.2 Piers

10.2.1 Strength of Masonry Piers Under Eccentric Axial Loads: Mechanical Aspects of the Problem

Eccentrically loaded masonry piers behave very differently from reinforced concrete columns. A regular series of cracks occurs in reinforced concrete piers, and the concrete between adjacent cracks bears the tensile stresses: this effect, usually called *tension stiffening*, attenuates the nonlinear response of the column.

In masonry piers, on the contrary, cracking spreads diffusely throughout wide areas of the structure, and the nonlinear effects are much more severe: they strongly reduce the pier's strength. Even small eccentricities of the axial loads can produce serious reductions in strength.

In the stability analysis of masonry piers it is usually assumed that the elastic strains vary linearly with distance from the neutral axis across the sections of the piers, if eccentrically loaded. The eccentricity e of the axial load P measures the distance of the point load, i.e. point C of application of load P, from the center of the section. The distance of point load C from the compressed edge of the section is indicated by u. For the sake of simplicity, we shall refer to rectangular pier cross sections. The section is wholly compressed (Fig. 10.1) only when the point load is included within the core of the section, i.e. when the eccentricity e of P falls within the interval

$$-t/6 \leq e \leq t/6. \tag{10.1}$$

This is the case of *low eccentricities*. In Fig. 10.1 the point load C is located at the edge of the core, i.e. $e = t/6$. In this case the neutral axis skirts the lower edge of the section: the stresses exhibit the triangular distribution shown in the figure. For the rectangular section we know that the core width equals $t/3$. Thus, with t the height of the section, the above-defined distance u is equal to $t/3$. Elastic flexural deformations must be taken into account in the analysis of an eccentrically loaded pier. Such deformations increase the axial load eccentricity and narrow the resistant areas of the pier sections (Fig. 10.2).

10.2 Piers

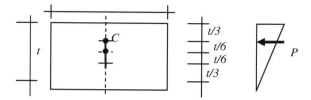

Fig. 10.1 Rectangular section with the positions of its core edge. The stress distribution corresponds to an eccentricity equal to half the core width

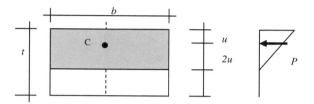

Fig. 10.2 The narrowed resistant section under a strongly eccentric axial load

With gradually increasing axial load intensity, this effect becomes more and more relevant and the pier may collapse. Figure 10.3 shows the effect of the flexural deformation and the crack distribution in a masonry pier of height h, with transverse rectangular sections of dimensions $b \times t$, eccentrically loaded at its end sections.

The hatched area in Fig. 10.3 indicates the resisting region of the pier and highlights the effect of the elastic bending deformations. Note that the extension of the cracked region would result to be greatly reduced if these bending deformations were neglectedSuch an evaluation of the strength of piers taking into account its flexural deformations was first performed by Yokel (1971).

This analysis refers to the pier's *compressed edge*. It is in fact simpler to refer to this edge, rather than to the pier central axis, which continuously changes position inside the resistant sections during loading.

The end sections of the pier are assumed to be hinged. The constraining effects of floors present at the head and base of piers or building walls, as well as the presence of flying buttresses and buttresses in cathedrals piers, justify this assumption.

10.2.2 Differential Equation of the Inflexion of an Eccentrically Loaded Cracked Pier

In this analysis the reference parameter is the distance u between the compressed edge and the axis of load P: owing to the lateral inflexion, this distance varies along

Fig. 10.3 Dramatic narrowing effect of the pier resistant zone due to flexural deformations

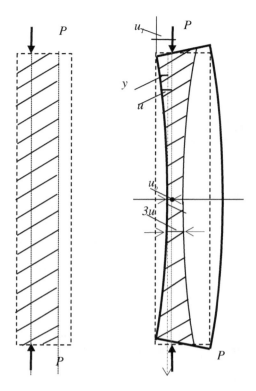

the height of the pier. The axial load P is applied at its end sections with eccentricity e falling within the interval $t/2 > e \geq t/6$, where t is the section width.

In Fig. 10.3 the distance of the axial load P from the generic edge of the pier section is indicated by u, while u_1 and u_o are the same distances but to the end and the mid-section of the pier, respectively. We thus have $u_1 = t/2 - e$. This *strong eccentricity* condition loading corresponds to positions of the point load of P at the pier end sections included between the edges of the section and its core. The axial load narrows *all* the resistant sections of the pier. (The case of *low eccentricities*, i.e. corresponding to eccentricities e included within the section core, with $t/6 \geq e \geq 0$, will be considered later.) Returning to the strong eccentricities, we must now evaluate the changes in the resistant sections' areas along the height of the pier due to its inflexion.

The straight line defining the direction of load P, represented by a dotted line in Fig. 10.3, passes at distance u from the compressed edge at any section of the pier. As shown in the figure, distance u gradually decreases from the end to the mid-section: the maximum distance u_1 is reached at the end sections of the pier, the minimum u_0 at the mid-section. The stresses have a *triangular distribution* at each transverse section of the pier: compression vanishes at the crack tip. On the opposite side, the compression is

10.2 Piers

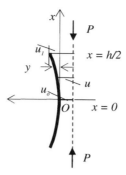

Fig. 10.4 Inflexion curve of the compressed edge with reference axes xy

$$\sigma_o(u) = \frac{2P}{3bu}, \qquad (10.2)$$

where $\sigma_0(u)$ is the *local* maximum stress in the section whose compressed edge is at distance u from the P axis.

The maximum compression stress of all the local maximum stresses $\sigma_0(u)$, indicated by σ_{max}, occurs at the pier mid-section, at distance $h/2$ from the end sections and equals:

$$\sigma_{max} = \frac{2P}{3bu_0}. \qquad (10.3)$$

Figure 10.4, shows the deformation of the pier's compressed edge: the origin of the reference axes is at the mid-section. Axis x is parallel to the direction of P and tangential to the deformed compressed edge at the origin O. Now let y be the distance between the line of the pier's compressed edges from axis x. Thus, at each section of the pier, we have

$$y = u - u_0 \qquad (10.4)$$

and, for $x = h/2$, $y = u_1 - u_o$. Figure 10.5 shows a small element of the pier between the external edge and the boundary of the cracked zone. The distance between the unloaded and the compressed edges of the element equals $3u$. The length of the element, from the unloaded side, is dl, whereas the length of the compressed edge is

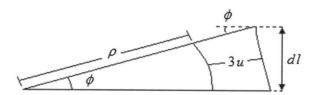

Fig. 10.5 Curvature of the compressed edge

$dl - \varepsilon dl$, with ε the strain in the external compressed edge of the element. A relative rotation ϕ occurs between the side sections of this element, as shown in Fig. 10.5.

This rotation ϕ can be obtained as

$$\phi = \frac{\varepsilon \cdot dl}{3u}. \tag{10.5}$$

The strain ε in the external compressed edge, on the other hand, can be evaluated according to (10.2), as

$$\varepsilon = \frac{\sigma_o}{E} = \frac{2P}{3bu}\frac{1}{E}, \tag{10.6}$$

where E is the elastic modulus of the masonry. Thus, taking (10.6) into account, the relative rotation ϕ becomes

$$\varphi = \frac{\varepsilon \cdot dl}{3u} = \frac{2P}{9Eb}\frac{dl}{u^2}. \tag{10.5'}$$

The length of the compressed side of the element, equal to $dl(1-\varepsilon)$, can be expressed in terms of the radius of curvature ρ of the curved compressed edge as

$$\rho\varphi = dl \cdot (1-\varepsilon) \tag{10.7}$$

and

$$\frac{1}{\rho} = \frac{\phi}{dl \cdot (1-\varepsilon)} = \frac{2P}{9Eb} \cdot \frac{1}{u^2 \cdot (1-\varepsilon)}. \tag{10.8}$$

The curvature of the function

$$y = y(x) = u(x) - u_o \tag{10.9}$$

is, as a rule, given by

$$\frac{1}{\rho} = \frac{\phi}{dl \cdot (1-\varepsilon)} = \frac{2P}{9Eb} \cdot \frac{1}{u^2 \cdot (1-\varepsilon)}, \tag{10.10}$$

whence

$$\frac{d^2y}{dx^2} = \frac{1}{\rho} \cdot [1 + (\frac{dy}{dx})^2]^{3/2} = \frac{2P}{9Eb} \cdot \frac{[1 + (\frac{dy}{dx})^2]^{3/2}}{u^2 \cdot (1-\varepsilon)}. \tag{10.11}$$

Strains ε are small quantities ($\varepsilon < 0.005$) and dy/dx is thus negligible with respect to unity; so all in all we can assume

10.2 Piers

$$\frac{[1+(\frac{dy}{dx})^2]^{3/2}}{1-\varepsilon} \approx 1. \tag{10.12}$$

The second derivative d^2y/dx^2, which represents the curvature of the compressed pier edge, is

$$-\frac{d^2y}{dx^2} = \frac{2P}{9Eb} \cdot \frac{1}{u^2}. \tag{10.13}$$

The factor $2P/(9Eb)$ is constant along x, so we can write

$$k_1 = \frac{2P}{9Eb}. \tag{10.14}$$

On the other hand, according to (10.9), we have

$$u = u_0 + y \tag{10.9'}$$

and the differential equation for the flexure of the eccentrically loaded cracked pier becomes

$$\frac{d^2y}{dx^2} = \frac{k_1}{(u_o + y)^2} \tag{10.15}$$

The following boundary conditions are associated to Eq. (10.15):

- at $x = h/2$, i.e. at the pier head, $y = u_1 - u_o$ (10.16)
- at $x = 0$, i.e. at the mid-section, $y = 0$. (10.16')

Integration of (10.15), satisfying the above boundary conditions (see Appendix), yields

$$P = \frac{9Ebu_1^3}{h^2} \cdot \alpha \cdot [\sqrt{1-\alpha} + \alpha \ln(\sqrt{\frac{1-\alpha}{\alpha}} + \sqrt{\frac{1}{\alpha}})]^2, \tag{10.17}$$

where

$$\alpha = u_0/u_1. \tag{10.18}$$

Note that the eccentricity e is given by

$$e = \frac{t}{2} - u_1. \tag{10.19}$$

Equation (10.17) expresses the relation between the flexure factor α and the eccentric axial load P.

10.2.3 Collapse Load

Let us now define a *reference critical load*

$$P_{eq} = \frac{\pi^2 E I_e}{h^2}, \qquad (10.20)$$

which is evaluated by means of the moment of inertia I_e of the resistant section at the pier head; this section has height $3u_1$. The moment of inertia I_e is thus given by

$$I_e = \frac{27 b u_1^3}{12}. \qquad (10.21)$$

In particular, when the point load at the head section is located at the edge of the core section, i.e. with $3u_1 = t$, the entire section is resistant and load P_{eq} matches the Euler load P_E. In fact, we obtain

$$P_{eq}(3u_1 = t) = \left(\frac{\pi^2 E}{h^2} 27 \frac{b u_1^3}{12}\right)_{u_1 = t/3} = \frac{\pi^2 E I}{h^2} = P_E, \qquad (10.22)$$

where

$$I = \frac{b t^3}{12}. \qquad (10.23)$$

Taking (10.20) and (10.21) into account, condition (10.17) becomes:

$$\frac{P}{P_{eq}} = \frac{4}{\pi^2} \cdot \alpha \left[\sqrt{1-\alpha} + \alpha \ln\left(\sqrt{\frac{1-\alpha}{\alpha}} + \sqrt{\frac{1}{\alpha}}\right)\right]^2. \qquad (10.24)$$

Equation (10.24) holds for $t/2 > e \geq t/6$, where e is the load eccentricity at the end pier sections. Condition (10.24), first determined by Yokel, expresses the dependence of the applied eccentric axial load P on the pier inflexion at its mid-section. The derivative of the function P/P_{eq} with respect to variable a is sketched out in Fig. 10.6. This derivative vanishes for

$$\alpha = \alpha = \bar{\alpha} = 0.6116. \qquad (10.25)$$

Function P/P_{eq} attains a maximum for $\alpha = \bar{\alpha} = 0.6116$. The flexure parameter a thus equals unity when the pier is not inflexed, i.e. for $P = 0$. With increasing P, the

10.2 Piers

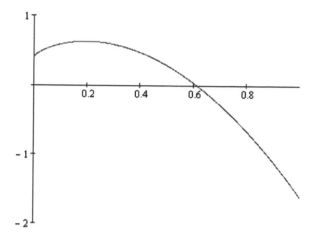

Fig. 10.6 The derivative of the function P/P_{eq} with respect to the variable a

pier starts to bend and parameter u_o decreases, as does factor α. Figure 10.7 gives a dimensionless representation of function (10.24) by assuming eccentricity $e = t/6$.

The point with coordinates ($\alpha = u_o/u_1 = 1$, $P/P_{eq} = 0$) represents the rectilinear configuration of the pier when the axial load is zero. On the contrary, the point with coordinates ($\alpha = u_o/u_1 = 0$, $P/P_{eq} = 0$) corresponds to a state of cracking so widespread as to produce an internal hinge in the pier and, consequently, vanishing of the axial load intensity strength, as shown in Fig. 10.8. In this case, equilibrium in the pier under zero axial load can be maintained only with $u_o/u_1 = 0$.

Using (10.19) we can trace the entire pier equilibrium path with gradually increasing P and the given eccentricity defined by parameter u_1. The equilibrium path runs from right to left along the diagram in Fig. 10.7. Initially, starting at $P = 0$, load P increases, while bending factor $\alpha = u_o/u_1$ decreases. The first branch of the equilibrium states describes this behavior for α varying from $\alpha = 1$ to $\alpha = 0.6115$, just where P attains its maximum. All the points belonging to this branch represent

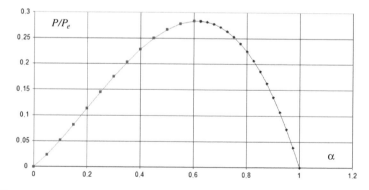

Fig. 10.7 Equilibrium states of the masonry pier loaded with eccentricity: $e = t/6$

Fig. 10.8 The pier at the state $P = 0$, $u_o = 0$

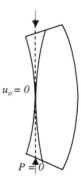

stable equilibrium states. At the point defined by $\alpha = 0.6115$, the axial load equals $P/P_{eq} = 0.285$, which is the maximum load that the pier can sustain: any further increment of P will now lead to collapse of the pier by *loss of equilibrium*.

In conclusion, collapse of the pier loaded axially with high eccentricity e, i.e. with $t/2 > e \geq t/6$, will come about under the axial load

$$P_{cr} = 0.285 \frac{I_e}{I} P_E, \qquad (10.26)$$

where, according to (10.21), the eccentricity e is present in the expression for I_e, the moment of inertia of the resistant section at the pier head. Alternatively, highlighting the dependence of the critical load on the eccentricity e at the end sections of the pier, from (10.26) we have

$$\frac{P_{cr}}{P_E} = 0.285 \left(\frac{3}{2} - \frac{3e}{t}\right)^3 \qquad t/2 > e \geq t/6. \qquad (10.26')$$

In particular, from (10.26'), when the eccentricity e equals $t/6$, i.e. when the load is applied at the core edge of the end sections, we obtain $P_{cr}/P_E = 0.285$. For larger eccentricities the reduction in the critical load with respect to the Euler load is greater. For $e = t/2$ the critical load vanishes altogether.

The values of P/P_{eq} corresponding to the descending branch of the curve in Fig. 10.7, i.e. to values of α within the interval $0.625 > \alpha > 0$, have limited physical significance: they correspond to the pier inflexion that can be maintained with a load below the critical one. They are all unstable equilibrium states.

10.2.4 Pier Strength with Variable Load Eccentricity

The foregoing results have been generalized by Frisch–Fay (1975) and De Falco and Lucchesi (2000, 2003) to consider the entire variability range of eccentricity. Figure 10.9 shows the collapse load of the pier versus the ratio e_L/d, in

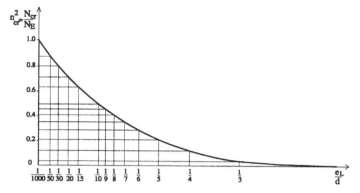

Fig. 10.9 Dimensionless strength P_{cr}/P_E of a masonry pier loaded eccentrically by varying the eccentricity factor e_L/d (De Falco and Lucchesi 2003)

dimensionless form, where e_L is the eccentricity and $d = t$ the section height, considering both weak and strong eccentricities. When the eccentricity vanishes, that is, when the ratio e_L/d is nearly 1/1000, the collapse load matches the Euler load P_E given by (10.22). As the eccentricity increases, the pier compression strength decreases: for $e_L/d = 1/10$ the pier strength is about 70 % of the Euler load, while the pier strength equals the above-cited strength of 0.285 P_E when e_L/d reaches the value 1/6.

10.2.5 Influence of the Pier Weight

The pier weight is represented by a series of uniformly distributed loads, w, equivalent, overall, to the total weight W, assumed to be non-negligible with respect to P. The combined action of load P, applied at the pier head, and distributed loads w is a significant and likely loading condition, frequent, for instance in the high piers of a cathedral.

Figure 10.10 shows a pier of height $2L$, hinged at its end sections, loaded by the force P at its head and by the distributed weight w. This case is also equivalent to that of a cantilever pier of height L under equal loads. Figure 10.11 shows the results obtained by La Mendola and Papia (1993).

The ordinates represent the dimensionless values of the maximum head load that the pier can sustain with the assumed ratio W/P and eccentricity ratio e/H, where H indicates the pier section height. Inspection of the diagrams in Fig. 10.11 reveals that the addition of weight W, *if relevant with respect to P*, in the presence of *large eccentricities*, increases the load P that the pier can sustain. In this case, we say that the weight W has a stabilizing effect.

Fig. 10.10 Pier loaded both by an eccentric force at its head and its own weight distributed along its length

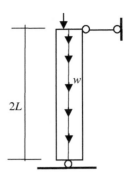

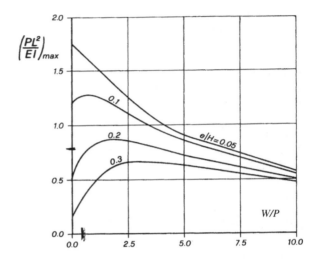

Fig. 10.11 Stabilizing or destabilizing effects of weight W on the magnitude of the head load P that a pier can sustain (da La Mendola and Papia 1993)

To the contrary, in the presence of *small eccentricities*, the effect of the weight has, as a rule, a destabilizing effect. Other load combinations have been considered, in particular, the case of a concurrent shear force together with an axial load applied at the pier head (Como and Ianniruberto 1995).

10.2.6 The Use of Nonlinear Programs in Stability Analysis of Piers

The use of nonlinear programs, such as ATENA (Cervenka 2002) or DIANA (Frits and Wijtze 1990), able to account for both material and geometrical nonlinearities, can be very useful for analyzing the static behavior of masonry piers with more complex geometry and load distributions. Both these programs assume low tensile strength, as in the $\sigma - \varepsilon$ diagram shown in Fig. 10.12.

10.2 Piers

Fig. 10.12 Tensile $\sigma - \varepsilon$ diagram assumed in the nonlinear programs

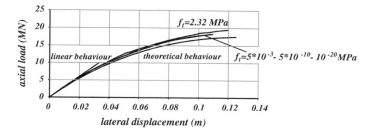

Fig. 10.13 Axial load-lateral displacement diagrams for eccentricity $e = L/6 = 16.7$ cm

The first example presented in the following can highlight the ability of these programs to accurately describe the behavior of masonry piers under eccentric loads. The example considers a pier of constant section loaded at its head by an eccentric load. The numerical results can thus be compared with those from Eq. (10.24). The pier has a square section with side length $L = 1$ m and height $H = 10$ m. The stress-strain diagram is of the type shown in Fig. 10.13, and discussed in Chap. 1. The program considers non-zero tensile strength that can be suitably reduced in the calculations.

The assumed elastic constants for the masonry are: elastic modulus, 3.032 E + 04 MPa; Poisson coefficient, $\nu = 0$; tensile strength, 5 E − 03 MPa; and compression strength, 2.5 E + 05 MPa. The example addresses two different eccentricities: $e_1 = L/6 = 16.7$ cm, and $e_3 = 30$ cm. The value of the head axial load, assumed initially $P = 3 \times 10^{-1}$ MN, increases gradually. Collapse comes about when

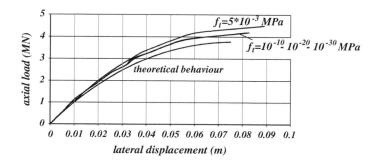

Fig. 10.14 Axial load-lateral displacement diagrams for $e = 30$ cm

the tangent becomes horizontal on the axial load—lateral displacement diagram, as shown in Figs. 10.13 and 10.14. The corresponding collapse loads turn out to be only slightly higher than the collapse loads resulting from application of Eq. (10.24).

10.2.7 Influence of Mortar Creep on the Behavior of an Eccentrically Loaded Pier

10.2.7.1 Simplified Model of a Visco-Elastic No-Tension Pier

The creep of mortars, examined in Chap. 1, can significantly, albeit slowly, increase the destabilizing effects of axial load on masonry piers or walls. Thorough study of the problem can be performed by modifying the previously examined Yokel formulation in order to account for the creep deformation of mortar, examined in Sect. 1.13.2.1. Such an approach is however very complex and only some simplified solutions to the problem can be obtained in practice. One simplified creep model of an eccentrically loaded, elastic no-tension pier considers the presence of a single central viscous, no-tension voussoir, as shown in Fig. 10.15. A load P is applied with eccentricity e with respect to the center G of the section at height h from its base of width L (Fig. 10.16). The pier rotates by a small angle θ under the action of the eccentric axial load P. Owing to pier rotation θ, the eccentricity of P becomes $(e + h\theta)$. The distance u of the axis of P from the external edge of the base section is thus

$$u = (L/2 - e - h\theta). \tag{10.27}$$

We can assume that *only* the central voussoir exhibits elastic no-tension behavior. Thus, the equilibrium equation of the inclined pier, in the case of small eccentricity, i.e. with $(e + h\theta) < L/6$, is

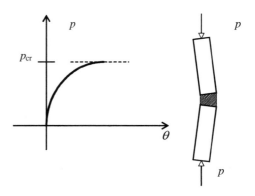

Fig. 10.15 Simplified creep model of an elastic no-tension pier

10.2 Piers

Fig. 10.16 Definition of the central viscous no-tension voussoir

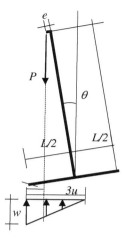

$$p = \frac{9}{2}(\gamma - \theta)^2 \theta, \qquad (10.28)$$

where

$$p = \frac{P}{Eh^2}, \quad \gamma = \frac{L}{2h} - \frac{e}{h}. \qquad (10.29)$$

The critical state is thus reached when

$$\frac{dp}{d\theta} = -\frac{9}{2}2(\gamma - \theta)\theta + \frac{9}{2}(\gamma - \theta)^2 = 0, \qquad (10.30)$$

i.e. when the critical rotation

$$\theta_{cr} = \frac{\gamma}{3} \qquad (10.31)$$

is attained. The dimensionless elastic critical load is therefore

$$p_{cr} = \frac{2}{3}\gamma^3. \qquad (10.32)$$

Creep deformation of the central voussoir gradually increases the rotation of the pier. By using the memory function considered in Sect. 1.13.2.1, in the framework of the viscosity model given by Krall (1947) in place of (10.30), we obtain the following equation governing the evolution of the pier equilibrium state over time:

$$3u(t)\theta(t) = \lambda\varepsilon_{tot}(t) = \lambda[\frac{2P}{3bEu(t)} + \alpha\beta\int_{t_1}^{t} e^{-\beta(\tau-t_o)}\frac{2P}{3bEu(\tau)}d\tau]. \qquad (10.33)$$

Substituting the expression for distance, $u(t)$, into Eq. (10.33) gives the relation between the rotation $\theta(t)$ and the dimensionless eccentric load p

$$\begin{aligned}\xi = &1 - \frac{9}{2}\frac{1}{\alpha p}[\gamma^2(\theta - \theta_o) - 3/2\gamma(\theta^2 - \theta_o^2) + 2/3(\theta^3 - \theta_o^3)] \\ &- \frac{1}{\alpha}\ln\frac{(\gamma - \theta)}{(\gamma - \theta_o)},\end{aligned} \qquad (10.34)$$

where the variable ξ is

$$\xi = e^{-\beta t}, \qquad (10.35)$$

and where for the sake of simplicity we have assumed $t_o = t_1 = 0$. In Eq. (10.34), θ_o and $\theta(t)$ respectively indicate the initial and generic rotations, i.e., the rotations occurring at the initial time $t = 0$ and at time t. The asymptotic pier rotation, for $t \to \infty$, i.e. when $\xi \to 0$, is indicated by θ_∞, hence, from Eq. (10.34), with $\xi = 0$ we obtain

$$\begin{aligned}0 = &1 - \frac{1}{\alpha}\frac{9}{2p}[\gamma^2(\theta_\infty - \theta_o) - 3/2\gamma(\theta_\infty^2 - \theta_o^2) + 2/3(\theta_\infty^3 - \theta_o^3)] \\ &- \frac{1}{\alpha}\ln\frac{(\gamma - \theta_\infty)}{(\gamma - \theta_o)}.\end{aligned} \qquad (10.34')$$

Thus, Eq. (10.34') furnishes the asymptotic rotation θ_∞ produced by load P.

10.2.7.2 Critical State: Comparison with the Solution Obtained via the Delayed Modulus Approach

The critical state of the pier is reached when

$$\frac{d\xi}{d\theta} = 0. \qquad (10.36)$$

Condition (10.36) defines the presence of *simultaneous* equilibrium states at the same time t. From (10.36) and by using (10.34), we obtain

$$-2\theta_{cr}^3 + 5\gamma\theta_{cr}^2 - 4\gamma^2\theta_{cr} + (\gamma^3 - \frac{2}{9}p_{cr}) = 0. \qquad (10.37)$$

The critical state is attained at $t \to \infty$ if both the rotation $\theta = \theta_{cr\,\infty}$ and load $p_{cr\,\infty}$ satisfy the equation

10.2 Piers

$$-2\theta_{cr,\infty}^3 + 5\gamma\theta_{cr,\infty}^2 - 4\gamma^2\theta_{cr,\infty} + (\gamma^3 - \frac{2}{9}p_{cr,\infty}) = 0 \qquad (10.37')$$

as well as Eq. (10.34'), i.e.,

$$0 = 1 - \frac{1}{\alpha}\frac{9}{2p_{cr,\infty}}[\gamma^2(\theta_{cr,\infty} - \theta_o) - 3/2\gamma(\theta_{cr,\infty}^2 - \theta_o^2) + 2/3(\theta_{cr,\infty}^3 - \theta_o^3)]$$
$$- \frac{1}{\alpha}\ln\frac{(\gamma - \theta_{cr,\infty})}{(\gamma - \theta_0)}. \qquad (10.34')$$

The quantities $\theta_{cr,\infty}$ and $p_{cr,\infty}$ in Eq. (10.37') are unknowns, while the unknowns in Eq. (10.34") are $\theta_{cr,\infty}$, θ_o and $p_{cr,\infty}$. The last equation required for the solution is obtained from Eq. (10.30), which gives the rotation θ_o at initial time $t = 0$ under load $p_{cr,\infty}$, which yields

$$\frac{2}{9}p_{cr,\infty} = (\gamma - \theta_o)^2\theta_o. \qquad (10.30')$$

Substitution of (10.30') into (10.37') and (10.34") yields

$$x(1 - 3/2x + 2/3x^2) - x_o(1 - 3/2x_o + 2/3x_o^2)$$
$$= x_o(1 - x_o)^2[\alpha + \ln\frac{(1 - x_0)}{(1 - x)}] \qquad (10.38)$$

$$(1 - x_o)^2 x_o = -2x^3 + 5x^2 - 4x + 1, \qquad (10.39)$$

where

$$x_o = \theta_o/\gamma, \qquad x = \theta_{cr,\infty}/\gamma. \qquad (10.40)$$

Solution of Eqs. (10.38) and (10.39) for assigned values of the creep factor α furnishes the values $\theta_{cr,\infty}$ and θ_o. The asymptotic critical load $p_{cr,\infty}$ is thus obtained by substituting the expression for θ_o into (10.30').

Table 10.1 reports the values of x and x_o obtained by solution of Eqs. (10.38) and (10.39) for the assumed values of creep factor α. Table 10.2 shows the values of the asymptotic rotation $\theta_{cr,\infty}$ and the initial rotation θ_o, together with the dimensionless asymptotic critical load and $p_{cr,\infty}$ associated with the above solutions for x and x_o.

Table 10.1 Solutions of Eqs. (10.38) and (10.39) according to the assumed values of creep factor α

$\alpha = 0$	$x_o = 1/3$	$x = 1/3$
$\alpha = 1$	$x_o = 0.102$	$x = 0.390$
$\alpha = 2$	$x_o = 0.067$	$x = 0.415$
$\alpha = 3$	$x_o = 0.050$	$x = 0.430$
$\alpha = 4$	$x_o = 0.040$	$x = 0.441$

Table 10.2 Asymptotic critical loads according to the assumed values of creep factor α

α = 1	$\theta_{cr,\infty} = 0.390\gamma$	$\theta_o = 0.102\gamma$	$p_{cr,\infty} = 0.370\gamma^3$	$p_{cr,\infty,del} = 0.335\gamma^3$
α = 2	$\theta_{cr,\infty} = 0.415\gamma$	$\theta_o = 0.067\gamma$	$p_{cr,\infty} = 0.262\gamma^3$	$p_{cr,\infty,del} = 0.222\gamma^3$
α = 3	$\theta_{cr,\infty} = 0.430\gamma$	$\theta_o = 0.050\gamma$	$p_{cr,\infty} = 0.203\gamma^3$	$p_{cr,\infty,del} = 0.167\gamma^3$
α = 4	$\theta_{cr,\infty} = 0.441\gamma$	$\theta_o = 0.040\gamma$	$p_{cr,\infty} = 0.166\gamma^3$	$p_{cr,\infty,del} = 0.133\gamma^3$

The asymptotic critical loads and $p_{cr,\infty,del}$ have also been obtained by direct substitution of the delayed elastic modulus

$$E_\infty = \frac{E}{1+\alpha} \tag{10.41}$$

into the elastic no-tension solution (10.30).

The critical loads have thus been obtained as

$$p_{cr} = \frac{1}{1+\alpha}\frac{2}{3}\gamma^3. \tag{10.42}$$

These values are also reported in the last column of Table 6.2. The elastic modulus (10.41) defines the ratio

$$E_\infty = \frac{\sigma}{\varepsilon_{tot,\infty}} \tag{10.43}$$

between the acting stress, which is constant over time, and the asymptotic total strain $\varepsilon_{tot,\infty} = \varepsilon_{el} + \varepsilon_{visc\,\infty}$, which is the sum of the elastic and asymptotic viscous strain. The delayed critical asymptotic loads $p_{cr,\infty\,del}$ are *approximate* solutions to the problem of the critical load evaluation of the no-tension creep model of a pier. Although these approximate values $p_{cr,\infty\,del}$ are consistently lower than the corresponding exact values $p_{cr,\infty}$, they also approximate them quite closely, as is evident in the last two columns of Table 10.2. This outcome highlights that, despite the complexity of the problem, the simplified critical loads obtained using the delayed elastic modulus can represent the actual critical loads with sufficient approximation. The collapse of the Beauvais cathedral in 1294 will be taken up in the Chap. 8 as an example application of the delayed modulus approach to creep buckling.

10.3 Building Walls

10.3.1 Introductory Remarks

Figure 10.17 shows the plan of a masonry building with different arrays of longitudinal and transverse walls. Figure 10.18 shows a section of another common historic building, from foundation to roof. The walls present offsets along the

10.3 Building Walls

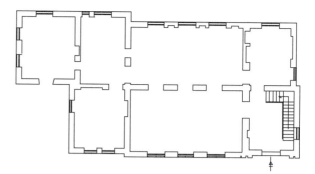

Fig. 10.17 Plan of an historic masonry building

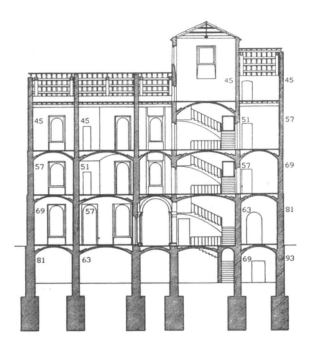

Fig. 10.18 Section of an historic masonry building (Giuffrè 1990)

vertical due to the varying wall thickness along the height—an arrangement which is justified by the considerably greater axial loads on the floor levels (Fig. 10.19). While the offsets on the internal walls are symmetrical, those on the external walls were generally made only on the inner side, in order to give buildings smooth vertical facades. Offsets were moreover frequently used as supports for floors.

In modern masonry buildings, ring beams running atop the walls at each floor efficiently oppose the transverse flexure of the facade walls and represent stiff transverse constraints on the walls. Historic and older buildings, to the contrary, lack such ring beams.

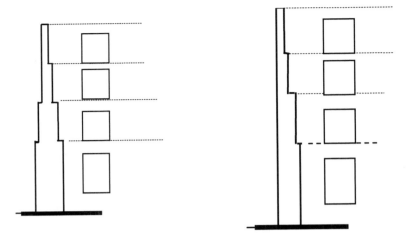

Fig. 10.19 Sections of inner (*left*) and outer (*right*) walls with symmetrical and asymmetrical offsets due to varying thickness along the height

Only a weak connection is offered at the floor levels by the friction between the floor's steel or wooden beams in contact with the walls at their supports. The connections between the different walls may thus be quite precarious. Particularly at corners, the connections between the stones or bricks may be degraded and even lacking, despite interpenetration between them at the wall intersections. The walls of old buildings are often visibly damaged, and studying the cracking patterns can furnish useful information about the causes of such damage.

10.3.2 Crack Patterns in Buildings Under Vertical Loads

Vertical loads clearly represent the most significant and long-lasting actions on historic buildings: the weight acts constantly and is responsible for most of the damage occurring in such buildings. Due to their varying thickness along the height, the external walls are subjected to eccentric axial forces and present a latent tendency to bulge outwards (Fig. 10.20). In historic buildings this tendency is opposed only by the connections between the different constituent walls, which may be more or less efficient. Moreover, these connections may be weakened by the presence of openings near the wall intersections or by cracking. Sometimes, chains were used to firmly connect the different arrays of walls. Figure 10.21a shows a typical cracking pattern in a transverse wall, near the joint to the facade, caused by rotation around the toe of the facade wall.

Cracks develop on the band of masonry overlying an opening and course vertically upward, increasing in width as they spread. Such cracks are due to the horizontal tensile stresses occurring in this band due to the outwards rotation of the

10.3 Building Walls

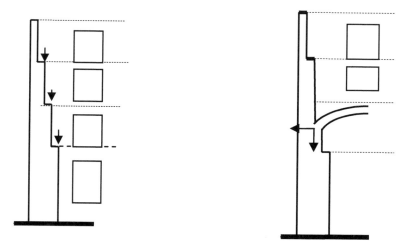

Fig. 10.20 Facade walls. Opposing and concurring actions of floor loads, wall weights and vault thrusts

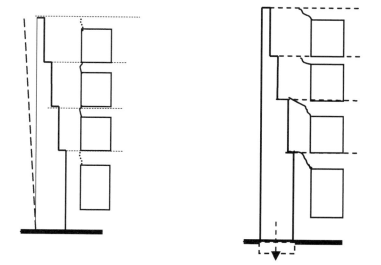

Fig. 10.21 Detachment cracks due to rotation (*left*) or subsidence (*right*) of a facade wall

wall. In some cases horizontal cracks may also appear in the floors near their connections to the façade wall. Figure 10.22 shows the cracking pattern in the transverse wall from subsidence of the facade. Typically, such cracks are inclined by about 45° from the vertical and are caused by subsidence of the façade wall, which produces shear stresses on the masonry band overlying the opening in the transverse wall (Fig. 10.22).

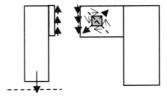

Fig. 10.22 Wall cracks above an opening due to subsidence of the facade wall

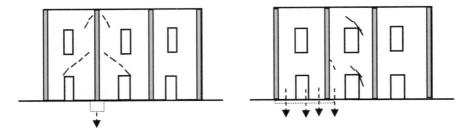

Fig. 10.23 Facade wall cracks due to subsidence of transverse walls

Figure 10.23 show analogous situations in the façades of buildings whose interior transverse walls wall have subsided. When the connection is weakened or lost, the external walls can undergo outward flexions and strong out-of-plumb rotations. The main aim of any restoration work is to re-establish firm connections between the disjointed walls.

10.3.3 Stresses Due to Vertical Loads

10.3.3.1 Weak Diffusion of Point Loads in Walls

Only two bricks of the lowermost course are engaged—numbers 8 and 9, as illustrated in Fig. 10.24.

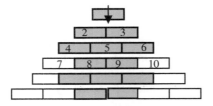

Fig. 10.24 Transmission of vertical load P across various courses of a wall

10.3 Building Walls

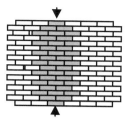

Fig. 10.25 Weak dispersion of a point load

Any lower course would again exhibit only two blocks engaged, and so on. A necessary preliminary operation for checking the safety of a building is to perform an evaluation of the compression stresses acting on the walls The bases of bricks 2 and 3 (in the second course from the top) are only partially compressed, but they, in turn, engage the underlying three bricks very differently. The two outer blocks, 4 and 6, are in fact subjected to strong eccentric compression.

The weak load diffusion described in Chap. 2 within the no-tension framework is thus also confirmed by regarding the wall as composed of bricks and weak mortar beds (Fig. 10.25). Consequently, evaluating the stresses due to vertical loading calls for working in terms of *vertical* bands.

10.3.3.2 Static Schemes for Vertical Wall Bands

Evaluating the stress along vertical wall bands of masonry buildings is a complex problem with various levels of uncertainties particularly dependent on the state of the connections between the walls making up the overall building structure.

In old buildings, these connections may be very weak and the walls, particularly of façades, behave much like vertical cantilever beams. In such cases, façade walls can frequently wind up out of plumb. Modern masonry buildings instead apply advanced systems for firmly connecting walls together. At each floor level reinforced concrete ring beams encircle the walls and are connected by slabs to the floor structure. This interconnected system of walls and slab floors produces a stiff three-dimensional cell structure. In some cases, historic buildings, if suitably stiffened, may also present such firmly connected structures. In the event that an efficient connection system has been fitted to a building, the static behavior of the vertical wall bands can generally be represented by the scheme of a continuous vertical beam with horizontal constraints at the floor level, as illustrated in Fig. 10.26b.

The floor offsets, produced by the varying wall thickness along the height, activates the floor couples due to the axial load misalignments, as shown in Fig. 10.27. Figure 10.26a shows a vertical beam, representing a wall band of the façade wall, connected at the floor levels to horizontal constraints, which act to oppose bending of the wall band. Due to the different heights of each story and the

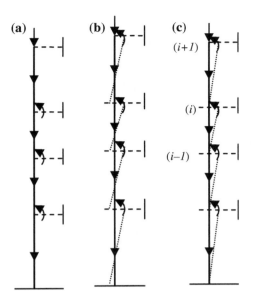

Fig. 10.26 Vertical wall bands with floor couples due to axis misalignments. Bending moment diagrams for actual (**a**) and building code schemes (**b, c**)

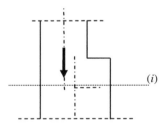

Fig. 10.27 Misalignments between wall axes along the height

different thicknesses of the walls between floors, the corresponding continuous beam will have varying spans and sections.

Figure 10.26b shows a possible diagram of the bending moment along the wall. Note the possible sign inversion in the bending moment diagram at the floor levels. Within this framework, it is useful to cite the simplified approach provided for by the Italian Building Code for evaluating stresses in vertical wall bands. This approach assumes that the wall is hinged at the wall base of each floor, as shown in Fig. 10.26c. Such an assumption can be justified by considering that, in modern masonry buildings, the presence of the ring beams at each floor interrupts the continuity of the wall, and thereby hinges the wall segments at their base on each floor.

The horizontal forces transmitted to the floors are shown in Fig. 10.28. At level i, the axial load N_i transmitted by the upper wall is *centered*.

10.3 Building Walls

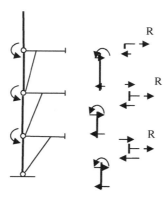

Fig. 10.28 Evaluation of horizontal floor constraints

Evaluation of the eccentricity at the head of each wall segment, that is, between levels i and $i-1$, will consider, together with the centered axial load N_i transmitted by the overlying wall, all other forces transmitted to level i by the beam floors in their actual positions, as shown in Fig. 10.29.

Thus, the eccentricity e of the axial load at the head of the wall segment between levels i and $i-1$ can be obtained as

$$e = \frac{N_i d_{i-1} + (\Sigma V_i) d_{vi}}{N_i + \Sigma V_i}. \qquad (10.44)$$

The presence of eccentricities in the axial load on the vertical wall bands may induce non-negligible destabilizing effects that must be adequately accounted for. A simplifying procedure for this purpose is presented in the next section.

10.3.3.3 Simplified Stress Analysis of Wall Bands to Account for the Destabilizing Effects of Axial Loads

The Norme Tecniche sulle Costruzioni (2005) furnishes a useful simplifying approach to account for the destabilizing effects of axial loads on walls and piers.

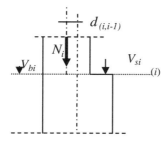

Fig. 10.29 Bending moment valuation at the head of the wall between floors $(i-1)$ and (i)

This procedure, which is a modified form of the common ω approach to buckling checks of steel columns, adequately accounts not only for the slenderness ratio λ of the wall, but also for the eccentricity e of the axial load. According to this approach, the mean compression stress in the wall section must be lower than an admissible reference stress $\bar{\sigma}_m$, that is,

$$\sigma = \frac{N}{\Phi A} \leq \bar{\sigma}_m, \qquad (10.45)$$

where N is the working axial load acting on the considered wall section, A the geometrical area of the wall section and Φ a suitable strength reduction coefficient that depends on the eccentricity ratio

$$m = 6e/t \qquad (10.46)$$

and the wall slenderness $\lambda = h_o/t$, where h_o is the inflexion length of the wall, which is in turn given by

$$h_o = \rho h, \qquad (10.47)$$

with h the distance between stories, and ρ the side constraint factor. As per Code provisions, this last is taken to be

$$\rho = 3/2 - h/a \quad 0.5 \geq h/a \leq 1.0; \quad \rho = 1/[1 + (h/a)^2] \quad 1 > h/a, \qquad (10.48)$$

where a indicates the distance between the constraining transverse walls.

Table 10.3 provides the eccentricity ratio factor Φ for various values of wall slenderness and load eccentricity. These values have been obtained through previous analyzes on the destabilizing effects of axial load eccentricities in masonry piers and walls. According to Italian Building Codes, the admissible mean reference stress $\bar{\sigma}_m$ can be obtained via the characteristic masonry compression strength f_k as

$$\bar{\sigma}_m = f_k/5. \qquad (10.49)$$

Table 10.3 Values of eccentricity ratio Φ

h_o/t	$m = 0$	$m = 0.5$	$m = 1$	$m = 1.5$	$m = 2$
0	1.00	0.74	0.59	0.44	0.33
5	0.97	0.71	0.55	0.39	0.27
10	0.86	0.61	0.45	0.27	0.15
15	0.69	0.48	0.32	0.17	–
20	0.53	0.36	0.23	–	–

They reduce the resistant area of the of the compressed wall section as a function of the wall slenderness and eccentricity factor m

Masonry compression strength, f_k, can also be evaluated by considering the strengths of the individual constituent stones or brick elements and mortar, as discussed in Chap. 1.

10.3.3.4 Composite Sections

Composite sections, in which a brick or stone facing covers an inner core of rubble and mortar, are very common in piers and masonry walls. Evaluation of the stresses in such composites must account for the different deformation capacities of the facing and core. Figure 10.30 shows a rectangular composite pier section, for which

$$E_p, \quad E_n \tag{10.50}$$

are the elasticity moduli of the different masonries in the pier facing and core, respectively. Let N be the axial load acting on the entire section, directed along the pier axis.

The mean compressive stresses, σ_p and σ_n, respectively acting on the facing and internal core are unknown. A first relation linking σ_p and σ_n is the following equilibrium equation

$$N = \sigma_p A_p + \sigma_n A_n, \tag{10.51}$$

where A_p and A_n are the areas of the facing and core, respectively. The second relation is the compatibility equation equating the facing and core strains:

$$\varepsilon_p = \varepsilon_n = \varepsilon \tag{10.52}$$

This condition depends on the bond between the bricks and the core, as well as on the connections potentially existing between the facing and the core. By substituting Eq. (10.52) into Eq. (10.53), and accounting for the elasticity relations

$$\sigma_p = E_p \varepsilon_p, \qquad \sigma_n = E_n \varepsilon_n, \tag{10.53}$$

we obtain the following expressions for the stresses in the facing and the core:

$$\sigma_p = \frac{N}{A_p + n_{n,p} A_n}, \qquad \sigma_n = n_{n,p} \frac{N}{A_p + n_{n,p} A_n}, \tag{10.54}$$

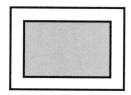

Fig. 10.30 Pier section composed of brick facing and inner rubble core

with

$$n_{n,p} = \frac{E_n}{E_p}. \tag{10.55}$$

Note that the destabilizing effects of axial load eccentricities, particularly when dealing with slender piers or walls, are adequately accounted for.

An example
Let us consider the composite section with the following characteristics:
$b = 1.50$ m; $d = 3.00$ m; $s = 0.30$ m;
$E_p = 5000$ MPa; $E_n = 1500$ MPa;
$N = 500$ t.
Thus, we have $n_{n,\,p} = 0.30$ and

$$\sigma_p = \frac{500}{2.34 + 0.30 \times 2.16} \, 16.73 \,\text{kg/cm}^2, \quad \sigma_n = 5.0 \,\text{kg/cm}^2$$

10.3.3.5 Mortar Creep Effects

Creep deformations of mortars influence the stress distribution in the facing and core of a composite pier. In evaluating this influence we can assume that creep effects will be greater in the core than in the masonry facing. Thus, considering both creep and shrinkage of the core, recalling the formulation described in Chap. 1, the compression stresses in the core and facing

$$\sigma_n(t), \quad \sigma_p(t) \tag{10.56}$$

are both functions of time t. The compatibility condition is now reformulated as

$$\varepsilon_{ep}(t) = \varepsilon_{en}(t) + \varepsilon_{vn}(t) + \varepsilon_{rs}(t), \tag{10.57}$$

where $\varepsilon_{en}(t)$, $\varepsilon_{vn}(t)$ e $\varepsilon_{sn}(t)$ are respectively the elastic, viscous and shrinkage strains of the core, and $\varepsilon_{ep}(t)$ the elastic strain of the facing. The elasticity conditions can now be expressed as

$$\sigma_p(t) = E_p \varepsilon_{ep}(t), \quad \sigma_n(t) = E_n \varepsilon_{en}(t). \tag{10.58}$$

Considering the creep deformation of the core, Eq. (10.57) becomes

$$\frac{\sigma_p(t)}{E_p} = \frac{\sigma_n(t)}{E_n} + \frac{\alpha \beta}{E_n} \int_{t_i}^{t} e^{-\beta(\tau-t_o)} \sigma_n(\tau) d\tau + \varepsilon_{sn}(t) \tag{10.59}$$

10.3 Building Walls

where, t is the current time and τ any given past time. The equilibrium equation at any time t yields

$$N = \sigma_p(t)A_p + \sigma_n(t)A_n. \tag{10.60}$$

Derivation of condition (10.59) with respect to time t gives

$$\frac{d\sigma_p}{dt} = \frac{E_p}{E_n}\frac{d\sigma_n}{dt} + \alpha\beta e^{-\beta(t-t_o)}\sigma_n(t) + E_p\frac{d\varepsilon_{sn}}{dt}, \tag{10.61}$$

where, recalling Eq. (10.4) in Sect. 1.11.4,

$$\frac{d\varepsilon_{rn}}{dt} = \varepsilon_R \beta e^{-\beta(t-t_o)}. \tag{10.62}$$

The derivative of condition (10.62) with respect to time t can also be written

$$\frac{d\sigma_p}{dt} = -\frac{A_n}{A_p}\frac{d\sigma_n}{dt} \tag{10.62'}$$

Thus, the following equation in the unknown $\sigma_n(t)$ is obtained simply by substituting (10.62') and (10.62) into (10.61):

$$\left(\frac{A_n}{A_p} + \frac{E_p}{E_n}\right)\frac{d\sigma_n}{dt} + \alpha\beta e^{-\beta(t-t_o)}\sigma_n(t) + E_p\varepsilon_R\beta e^{-\beta(t-t_o)} = 0. \tag{10.63}$$

With the position

$$\xi = e^{-\beta(t-t_o)} \tag{10.64}$$

and accounting for

$$\frac{d\xi}{dt} = -\beta\xi, \tag{10.64'}$$

Equation (10.65) becomes

$$-p\frac{d\sigma_n}{d\xi} + \alpha\sigma_n(\xi) + E_p\varepsilon_R = 0 \tag{10.64''}$$

where

$$p = \left(\frac{A_n}{A_p} + \frac{E_p}{E_n}\right) \tag{10.65}$$

The solution to (10.64″) is the function

$$\sigma_n(\xi) = A e^{\frac{a}{p}\xi} - E_p \frac{\varepsilon_R}{\alpha}. \tag{10.66}$$

At the initial time $t = t_i$, that is, at $\xi = \xi_i = e^{-\beta(t_i - t_o)}$, we have

$$\sigma_n(\xi_i) = A e^{\frac{a}{p}\xi_i} - E_p \frac{\varepsilon_R}{\alpha} = \sigma_{no}, \tag{10.67}$$

where σ_{no} is the purely elastic solution, given by (10.56). Thus, we obtain

$$\sigma_n(\xi) = \left(\sigma_{no} + \frac{\varepsilon_R}{\alpha}\right) e^{\frac{a}{p}(\xi - \xi_i)} - E_p \frac{\varepsilon_R}{\alpha}, \tag{10.68}$$

or more directly

$$\sigma_n(t) = \left(\sigma_{no} + E \frac{\varepsilon_R}{\alpha}\right) e^{\frac{a}{p}[e^{-\beta(t-t_o)} - e^{-\beta(t-t_i)}]} - E_p \frac{\varepsilon_R}{\alpha}. \tag{10.69}$$

Similarly, Eq. (10.60) gives the compression stress in the facing:

$$\sigma_p(t) = \frac{N}{A_p} - \sigma_n(t) \frac{A_n}{A_p}. \tag{10.70}$$

At the limit, for $t \to \infty$, the stresses in the core and facing reach their asymptotic values:

$$\sigma_{n\infty} = \left(\sigma_{no} + E_p \frac{\varepsilon_R}{\alpha}\right) e^{-\frac{a}{p} e^{-\beta(t_i - t_o)}} - E_p \frac{\varepsilon_R}{\alpha}, \quad \sigma_{p\infty} = \frac{N}{A_p} - \sigma_{n\infty} \frac{A_n}{A_p}. \tag{10.71}$$

If load N is applied to the pier at the same time t_i it takes the mortar to cure, we have $t_i = t_o$ and obtain

$$\sigma_{n\infty} = \left(\sigma_{no} + E_p \frac{\varepsilon_R}{\alpha}\right) e^{-\frac{a}{p}} - E_p \frac{\varepsilon_R}{\alpha}, \quad \sigma_{p\infty} = \frac{N}{A_p} - \sigma_{n\infty} \frac{A_n}{A_p}. \tag{10.72}$$

We can now re-evaluate the previous example considering the effects of creep and shrinkage of the core mortar. In this case, quantity p, defined by (10.18), is

$$p = \left(\frac{2.16}{2.34} + \frac{5}{1.5}\right) = 4.256.$$

By assuming $\alpha = 3$, we obtain $a/p = 0.705$. Considering now that load $N = 500$ t will act on the pier 1 year after the mortar has cured, we have $e^{-1} = 0.368$, and hence

$$e^{-\frac{z}{p}e^{-\beta(t_i-t_0)}} = e^{-0.705 \cdot e^{-1}} = e^{-0.705 \times 0.368} = e^{-0.259} = 0.771.$$

The asymptotic compression stress in the core is thus

$$\sigma_{n\infty} = (\sigma_{no} + E_p \frac{\varepsilon_R}{\alpha})e^{-\frac{z}{p}e^{-\beta(t_i-t_0)}} - E_p \frac{\varepsilon_R}{\alpha} = (5.02 + 50{,}000 \times \frac{0.3 \times 10^{-3}}{3})0.771$$
$$- 50{,}000 \times \frac{0.3 \times 10^{-3}}{3} = 7.725 - 5 = 2.725 \text{ kg/cm}^2.$$

The compression in the facing, on the contrary, is greater than the value resulting from ignoring creep effects:

$$\sigma_{p\infty} = \frac{500}{10 \times 2.34} - 2.725 \frac{2.16}{2.34} = 21.37 - 2.51 = 18.85 \text{ kg/cm}^2$$

This result, compared with the elastic solution, reveals the appreciable effect of mortar creep.

10.4 Buttresses

Figure 10.31 shows a masonry buttres under the vertical load Q and the thrust S to it conveyed by a vault or a flyer (Fig. 10.31).

The buttress has generally a stocky shape because with its weight has to balance the action of the thrust S. The weight G of the buttress is, as a rule, dominant with respect to the load V. In the next the limit thrust producing the overturning of the buttress will be obtained. The vertical load applied to the buttress will be the weight G and the vertical lod V: the thrust will be considered affected by the λ.

The activating of the overturning mechanism will be accompanied by the formation of a detachment fissure staring from the base of the buttress. Only a part of the weight of the buttress e mobilized in the opposition to the collapse.

Fig. 10.31 Scheme of the buttress under the loads conveyed by a vault

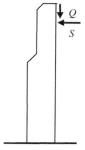

10.4.1 Geometry of the Detachment Inclined Crack (Ochsendorf)

Let us consider the loaded at its head by the thrust λS and the vertical load λQ. At the overturning state the pressure line, indicated as the curve a–a of Fig. 6.38, will pass through the left toe O of the buttress and at the section $m - n$ the curve a–a intercepts the core point P the section. Starting from the section $m - n$ and proceeding downward, the sections of the buttress will gradually crack and larger parts of the sections will be ineffective.

The curve b–b of Fig. 10.32 is the locus of points that trace the passage from the ineffective to the effective parts of the sections. At the failure, the part $rsnO$ that overturns, will detach from the part npO that remain on the ground. Only the weight of the part $rsnO$ of the buttress partecipates to define the detachment curve b–b. The element ABCD shown at the right of the figure, defines a small segment of thickness dy of the effective band of the buttress. Equilibrium of the element ABCD gives

$$W\frac{x}{3} + \frac{x}{2}dW - W\frac{x}{3} - d(W\frac{x}{3}) - \lambda H dy = 0 \qquad (10.73)$$

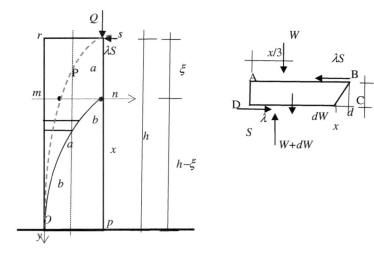

Fig. 10.32 Equilibrium in the cracked sections

10.4 Buttresses

from which we get

$$Wdx = \frac{x}{2}dW - 3\lambda Hdy \tag{10.73'}$$

But

$$dW = gxdy \tag{10.74}$$

and, substition of (10.75) into (10.73') gives

$$W = (g\frac{x^2}{2} - 3\lambda H)\frac{dy}{dx} \tag{10.75}$$

Derivating this last (10.75) with respect to the variable x gives

$$\frac{dW}{dx} = gx\frac{dy}{dx} + (g\frac{x^2}{2} - 3\lambda H)\frac{d^2y}{dx^2} \tag{10.75'}$$

On the other hand

$$\frac{dW}{dx} = gx\frac{dy}{dx} \tag{10.75'}$$

and the (10.171) becomes

$$(g\frac{x^2}{2} - 3\lambda H)\frac{d^2y}{dx^2} = 0 \tag{10.76}$$

Condition (10.76) ha sto be satisfied for any value of the x. Hence

$$\frac{d^2y}{dx^2} = 0 \tag{10.76'}$$

Consequently

$$\frac{dy}{dx} = \text{cost} \tag{10.76'}$$

The detachment fissure is thus rectlinear (Ochsendorf 2002).

10.4.2 Buttress Side Strength

Once obtained this result we can proceed to obtain the collapse multiplier λ_o of thew thrust S. At the failure the part OCBAO of the buttress, that overturns around the toe O, subdivides into two parts: the lower triangular zone OEC and the upper part

Fig. 10.33 Equilibrium in the cracked buttress at the limit overturning state

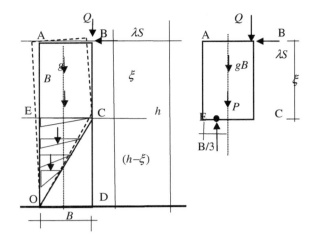

ABCE (Fig. 10.33). The line EC separates these two zones whose the upper has all sections effective and the lower, whose sections become gradually more and more ineffective. The section EC defines the passage from the cracked and the uncracked sections. The resultant of all the loads acting over the section EC will pass through the core point P of EC. Equilibrium around the point P of the upper part of the buttress gives on the other hand

$$\lambda S \xi - Q \frac{2B}{3} - g \frac{B^2}{6} \xi = 0 \qquad (10.77)$$

Solution of (10.77) gives the height ξ of the upper part of the buttress whose sections are all compressed. Ie With the position

$$G = gBh \qquad (10.78)$$

where

$$\beta = \frac{B}{h} \quad X = \frac{\xi}{h} \quad \Sigma = \frac{S}{G} \quad Z = \frac{Q}{G} \qquad (10.79)$$

we obtain

$$X = \frac{4Z\beta}{6\lambda\Sigma - \beta} \qquad (10.80)$$

The rotational equilibrium around the toe of the active part of all the buttress gives

10.4 Buttresses

$$\lambda Sh - QB - g\frac{B^2}{2}\xi - g\frac{B^2}{6}(h-\xi) = 0 \qquad (10.81)$$

or

$$\lambda Sh - \frac{B}{6}(6Q+G) - \xi G\frac{\beta}{3} = 0 \qquad (10.82)$$

The collapse multiplier is the obtained by substituting (10.80) into (10.82) and taking into account of (10.79). We get

$$\lambda S - \frac{\beta}{6}(6Q+G) - \frac{4Q\beta}{6\lambda S - G\beta}G\frac{\beta}{3} = 0 \qquad (10.83)$$

from which we obtain the algebraic equation of second degrre in λS

$$6(\lambda\Sigma)^2 - 2(\lambda\Sigma)\beta(1+3Z) + \frac{\beta^2}{6}(1-2Z) = 0 \qquad (10.84)$$

Discrminant of (10.84) is

$$\Delta = 4\beta^2 Z(9Z-2) \qquad (10.85)$$

and the solution $\lambda\Sigma$ of (10.84) takes the form

$$\lambda\Sigma/\beta = \frac{1}{6}[(1+3Z) + \sqrt{9Z^2 + 8Z}] \qquad (10.86)$$

Particular cases.
We consider first the case Q = 0 (Fig. 10.34). From (10.86) we have

Fig. 10.34 The case Q = 0

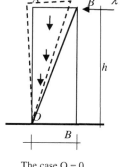

The case Q = 0

Fig. 10.35 The case G = 0

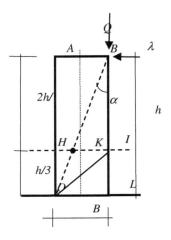

The case G = 0

$$\lambda S = \frac{\beta G}{6} \tag{10.87}$$

let us consider now the case $G = 0$ (Fig. 10.35). From (10.86), we get

$$\lambda S/\beta = \frac{1}{6}[(G+3Q) + \sqrt{9Q^2 + 8GQ}] \tag{10.88}$$

and, foir $G \to 0$

$$\lim_{G \to 0} \lambda S/\beta = \frac{1}{6}(3Q+3Q) = Q \tag{10.89}$$

Further, from (10.81)

$$\lim_{G \to 0} X = \lim_{G \to 0} \frac{4Q\beta}{6\lambda S - \beta G} = \frac{4Q\beta}{6Q\beta} = \frac{2}{3} \tag{10.90}$$

In this case the active part of the buttress has the contour ABIO of Fig. 6.41. The trianglular ineffective part is indicated as ILO. From Fig. 6.41 we have $tg\alpha = B/h$ and

$$HK = \frac{B}{h}HO = \frac{Bh}{h3} = \frac{B}{3} \tag{10.91}$$

The resultatnt of Q and λS passes throught the core point K, having distance $B/3$ from the edge AO. It can be useful to valuate the *reduced equivalent* width B' of the "*solid*" buttess, i.e. that cannot be cracked, having the same weight G and the same

10.4 Buttresses

Fig. 10.36 The equivalent solid buttress

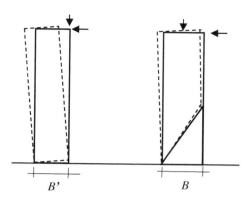

height h and that is under the same loads Q and λS, having the same overturning strength that of the actual buttress that, on the contrary, cracks (Fig. 10.36). Thus this *equivalent* width is given by the equation

$$\frac{\beta'}{\beta} = \frac{1 + 3Z + \sqrt{9Z^2 + 8Z}}{6(Z + 1/2)} \qquad (10.92)$$

where

$$\beta' = \frac{B'}{h} \qquad (10.93)$$

applications of (10.94) will be given.

10.5 Towers

10.5.1 Introductory Remarks

One particular class of masonry buildings, whose height predominates over their widths, are towers in all their forms, including bell-towers, minarets, and so forth. *Menhir*, sacred monuments built with gigantic stones, are the most ancient tower-like structures. Of the few livable *menhirs*, the tower in the Wall of Jericho is the most ancient, having been built about 9000 years ago; it is circular in section with a diameter of 8.0 m. Thousands of years later the Sumerians developed constructions called *ziqqurats*—terraced temples with square plans. Only a few traces of these remain today, mainly as the ruins of their foundations. In Italy the *Nuraghi*, built about 3000 years ago, are the most ancient surviving examples of tower constructions.

Towers have often dotted the landscape of many towns, marking their entrance gates or serving the function of watchtowers or lighthouses. They were also used as symbols of family power, as in San Gimignano, or as church bell towers and mosque minarets. Their height is clearly the characterizing feature of towers, the Lighthouse of Alexandria being one of the highest masonry towers ever built: it was reportedly 120–140 m in height. Built by Greeks in about 305 B.C.E., it collapsed in 1326, though nowadays some traces remain in the Castle of Qaitbay in Egypt.

10.5.2 Crack Patterns in Masonry Towers

Tower structures have to withstand winds and earthquakes. The action of the wind is probably the most critical because, due to their height, towers have high periods of oscillation and can thereby absorb seismic actions. In general, high compression stresses must be borne by tower masonry walls, particularly at the lower levels. Such stresses can cause the expulsion of stones and local failures in any irregular masonry. The presence of out-of-plumb walls makes matters worse, because it produces actions orthogonal to the wall plane.

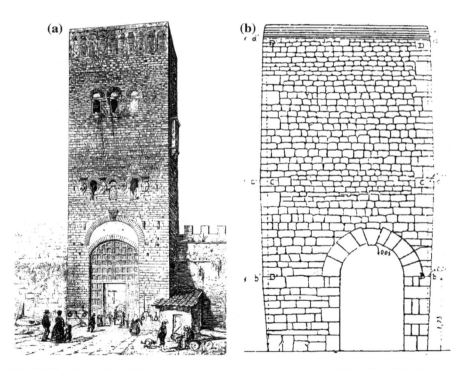

Fig. 10.37 **a** Typical cracking pattern in a masonry tower: the tower of San Niccolò in Florence (from Como 2000); **b** dislodged masonry

10.5 Towers

Fig. 10.38 Failure of the Bell Tower of St. Mark's in Venice in 1902, from a photomontage by L.H.N. Dufour (De Fez 1992)

The vertical walls of towers vary considerably in thickness: the tower shaft tends to open and the walls bulge outwards. Thus, vertical cracks frequently occur along the perimeter walls, particularly near window openings and in the upper part of the tower. Figure 10.37 shows a typical cracking pattern in an old, damaged tower, together with a sketch showing the dislodged masonry.

Figure 10.38 gives a photomontage of the failure of the Campanile di Venezia occurred in 1902.

Surveying cracking patterns can be a difficult task because of towers' heights. Nowadays, thanks to so-called *dynamical structural identification techniques*, tower frequencies and oscillation modes can be measured and different measurements compared over time, thereby enabling continuous monitoring of the evolution of any damage. One further hazard to towers is lightening.

Humidity can collect in cracks, which makes the structure a good conductor of electricity. Thus, if the tower is struck by lightening, the instantaneous increase in temperature of the humid air present in the cracks gives rise to an actual explosion within the masonry. This explains why old steeples and towers are frequently heavily damaged by lightening bolts. A clear lesson to be learned from this is that tower masonry must always be maintained in good condition.

In bell towers, the vibrations set up by the bells themselves may be a source of damage, hence checks of the vibrations induced by the motion of bells is frequently performed on such structures. Typical restoration works include masonry refurbishment and ringing with steel ties at various levels to restore the connections between the walls.

Serious static problems frequently depend on towers' foundations, which produce stress on narrower soil areas in comparison to ordinary buildings. Differential settling frequently occurs with consequent appreciable inclination and rotation of the tower. So-called *leaning stability* analysis is a problem particular to the statics of towers resting on deformable soils (Hambly 1985; Como 1993). The research on determining a suitable foundation model for towers is addressed in the following sections.

10.5.3 Plastic Model of the Tower Foundation

There are a number of different models for foundations that seek to describe their static response. These include both linear and nonlinear elastic approaches (see for instance, Hambly 1985; Napoli 1992), though plastic or visco-plastic models may be more appropriate. A tower involves the presence of high stresses in the foundation. Tilting of the tower, occurred during its construction, thereby confirming the strong stresses at their bases, moreover suggests that the underlying soil is at the plastic state.

Simple plastic and visco-plastic foundation models, following the basic approach of Meyerhof (1951), will thus be covered in detail in the next sections. Now let us consider a foundation plinth resting on the soil (Fig. 10.39) under the action of a centered axial load N and moment M, such that their resultant remains applied internally to the plinth. By gradually assigning increments to N and M, at a given point in the loading path, the plinth undergoes significant subsidence due to the plastic deformations occurred in the underlying soil.

By assuming different ratios between M and N (i.e., different eccentricities of N with respect to the center of the base section of the plinth), we can apply different loading paths and trace the locus Y of the points (M, N) in the plane M, N defining the attainment of the plastic state, also called the limit state. A typical interaction

Fig. 10.39 Yield locus of a foundation eccentrically loaded and supported by non-cohesive soil

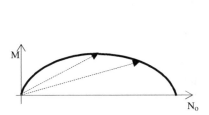

10.5 Towers

locus Y between the vertical load N and moment M, is drawn in Fig. 10.33, which outlines the various soil limit states.

For the sake of simplicity, let us assume that the foundation is rectangular with width a and transverse length b, and, according to Meyerhof (1951), let us also assume that *in the plastic state* a constant pressure p_o pushes on the compressed soil. In particular, if the foundation is centrally loaded, the limit load $N_o = p_o ab$ represents the ultimate centered vertical load, that is to say, the ultimate bearing capacity of the foundation under a centered vertical load. We can moreover assume (Fig. 10.34) that the supporting soil section of the plinth, eccentrically loaded, is subjected to a constant distribution of limit pressure p_o and engages only a band of limited width, equal to $(a/2 - x)$. Thus, the plinth equilibrium in the vertical direction yields

$$(\frac{a}{2} - x) = \frac{Na}{N_o}. \tag{10.73}$$

At the same time, rotational equilibrium yields

$$M = \frac{Na}{2}(1 - \frac{N}{N_o}). \tag{10.74}$$

Equation (10.74) describes the plastic state in terms of N and M and indicates the interaction locus in the plane M, N, as sketched out in Fig. 10.40. Generally, the interaction locus is represented by an equation of the type

$$f(M, N, N_o) = 0. \tag{10.75}$$

Thus, for a rectangular foundation, taking (10.74) into account, we have

$$f(M, N) = M - \frac{Na}{2}(1 - \frac{N}{N_o}) = 0. \tag{10.76}$$

The stress vector representing the loads acting on the foundation can be expressed by a two-component vector

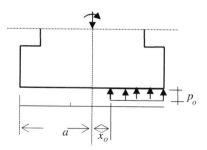

Fig. 10.40 Plastic model of an eccentrically loaded foundation

$$\sigma = \begin{bmatrix} M \\ N \end{bmatrix}, \quad (10.77)$$

disregarding the negligible shear. We can now evaluate the response of the foundation to an increase in stress. The corresponding plastic deformation increment is given by the two-component vector

$$d\varepsilon = \begin{bmatrix} d\theta \\ dv \end{bmatrix} \quad (10.78)$$

where $d\theta$ is the plastic rotation and dv the increase in the plastic settlement. When the loading path reaches a point P on locus Y, a plastic strain increment $d\varepsilon$ of the plinth occurs and has both components $d\theta$ and dv. These plastic strain increments develop both when the loading point P remains fixed on the locus as well as when P moves along it. To define the strain increment, we can thus move the loading point P along Y by applying a small increment $d\sigma$ tangent to Y. The plastic strain increment $d\varepsilon$ will occur without any work by $d\sigma$, in compliance with the basic principles of the Theory of Plasticity (Fig. 10.41).

If we move along the yield locus, the plastic deformation occurs without energy expense, hence

$$d\sigma \cdot d\varepsilon = 0. \quad (10.79)$$

Condition (10.79) indicates the orthogonality of $d\varepsilon$ to the boundary of locus Y (Fig. 10.35). The plastic strain increment $d\varepsilon$ is thus given by

$$d\varepsilon = \lambda \frac{\partial f}{\partial \sigma} \quad \lambda > 0 \quad \text{se } df > 0, \quad \lambda = 0 \quad \text{se} \quad df \leq 0, \quad (10.80)$$

where $f(\sigma)$ is given by Eq. (10.80) and df is its differential.

At the centered axial loading point ($N = N_o$, $M = 0$) the interaction locus presents a vertex, and the corresponding strain rate $d\varepsilon$ is a vector having any direction within the angle α and will include both plastic settlement and rotation, as shown in

Fig. 10.41 Normality rule for a plastic strain increment

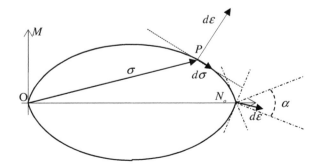

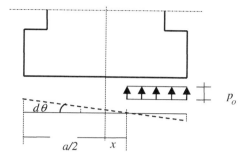

Fig. 10.42 Rotation of the foundation around the neutral axis

Fig. 10.49. This result explains the possible occurrence of a *sudden tilting* of the tower during its construction, when its weight, centrally applied on the foundation, reaches the limit value N_o. In particular, from (10.80), for a rectangular foundation we have

$$\frac{\partial f}{\partial M} = 1 \quad \frac{\partial f}{\partial N} = a\left(\frac{N}{N_o} - \frac{1}{2}\right) \tag{10.81}$$

and

$$d\theta = \lambda \quad dv = (-x)\lambda = -x\,d\theta. \tag{10.82}$$

According to (10.82), the plastic strain increment is thus produced by a rotation $d\theta$ of the foundation base section around the neutral axis corresponding to the current loading condition. This property holds for any type of foundation (Fig. 10.42).

10.5.3.1 Subsequent Yield Loci of Plastic Hardening Soils

Loose or weakly consolidated soils actually become stronger as loading progresses. The behavior of the foundation soil of tilting towers can often be explained by the presence of such soils. Soil strain hardening occurs as the soil deformations increase and a sequence of subsequent yield loci develops, as shown in Fig. 10.43.

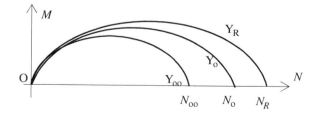

Fig. 10.43 Subsequent yield loci for a strain hardening foundation

The first plastic state is represented by the first locus Y_{oo}, while the final collapse state of the soil corresponds to the final locus Y_R. The plastic strain rate can thus be expressed via the following associated flow rule;

$$d\varepsilon = \frac{1}{\chi}\frac{\partial f}{\partial \sigma} df \quad \text{if } df > 0 \text{ and } d\varepsilon = 0 \quad \text{if } df \leq 0, \qquad (10.83)$$

where $\chi(\sigma)$ indicates the *strain hardening function* of the soil. This function will be determined for a tower foundation in the next section.

10.5.3.2 The Moment–Plastic Rotation Equation for the Tower Foundation

With reference to the case of a tower foundation, it should be noted that increasing the tower's tilt (Fig. 10.44) causes an increase in moment M, while the axial load N remains practically constant. A shear force acting on the foundation also occurs, though it is so small as to be negligible.

The strain hardening function $\chi(\sigma)$ can be considered to depend solely on the moment M via the function

$$\chi(M) \qquad (10.84)$$

This function represents softening behavior, that is, a gradual reduction in its tangent modulus as the rotation increases. It will also be able to describe *failure* of the foundation under the ultimate value of moment M_R, that is, the occurrence of

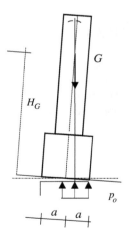

Fig. 10.44 A tilting tower

10.5 Towers

unbounded values of plastic rotations under the failure moment M_R. To construct this function, note that with N = cost, we have

$$df = dM, \tag{10.85}$$

and the corresponding rotation increment, with the assumption of a rectangular base foundation, is
In particular, for a rectangular plinth

$$d\theta = \frac{1}{\chi(M)} dM, \tag{10.86}$$

and the strain hardening function is given by

$$\chi(M) = \frac{dM}{d\theta}. \tag{10.87}$$

This strain hardening function, $\chi(M)$, defines the rotational tangent modulus and can be expressed as

$$\chi(M) = K_{\theta o} \frac{M_R - M}{M_R}, \tag{10.88}$$

where $K_{\theta o}$ is the initial tangent modulus, that is, the derivative $dM/d\theta$ at $M = 0$. With this position we thus get

$$d\theta = \gamma_{oR} \frac{dM}{M_R - M}, \tag{10.89}$$

where

$$\gamma_{oR} = \frac{M_R}{K_{\theta o}} \tag{10.90}$$

The quantity γ_{oR} can be defined as the *foundation deformability factor*. From (10.89) it can be seen that when the acting moment approaches the failure moment M_R, the rotation increment rises without limit. Integration of Eq. (10.89) gives (Como 1993)

$$M(\theta) = M_R(1 - e^{-\frac{\theta}{\gamma_{oR}}}). \tag{10.91}$$

Equation (10.91) describes the moment—rotation law of a rigid foundation resting on strain hardening soil. Upon unloading, permanent rotations occur, as shown in Fig. 10.45.

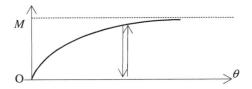

Fig. 10.45 Moment–rotation diagram

10.5.4 Stability of Leaning Towers

We assume that sudden, nonuniform soil settlement has taken place, with a consequent initial rotation θ_{oi} of the tower.

The tower will consequently undergo an additional instantaneous rotation θ due to the supervening eccentric position of weight G. The foundation base is thus loaded by the stress components (Fig. 10.46):

$$N = G\cos(\theta_{oi} + \theta), \quad M = GH_G \sin(\theta_{oi} + \theta), \quad T = G\sin(\theta_{oi} + \theta). \quad (10.92)$$

We also assume that in the past the tower has never undergone tilting rotations larger than the current one. During the loading history, the actual yield locus Y will thus never be contained within larger loci. The loading point $P(M, N)$, with components N and M given by (10.94), is thus located over the yield function Y (Fig. 10.47). Rotational equilibrium of the tower gives

Fig. 10.46 Equilibrium of a leaning tower after the initial rotation θ_{oi} of the foundation

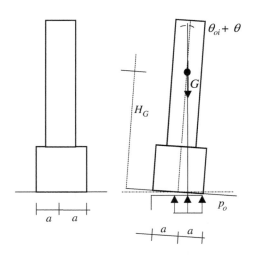

10.5 Towers

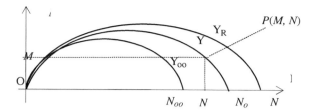

Fig. 10.47 Loading state of a tilted tower foundation

$$GH_G \sin(\theta_{oi} + \theta) - M_R(G)(1 - e^{-\frac{\theta}{\gamma_{oR}(G)}}) = 0. \quad (10.93)$$

Equation (10.93) equalizes the overturning moment given by the second of (10.92) to the resisting moment (10.91). Equation (10.93) can be symbolically expressed as

$$g[G(\theta), \theta] = 0 \quad (10.94)$$

and shows the dependence of weight G on the rotation θ.

By differentiating this condition, we get

$$\frac{\partial g}{\partial G}\frac{dG}{d\theta} + \frac{\partial g}{\partial \theta} = 0 \quad (10.95)$$

and then obtain

$$\frac{dG}{d\theta} = -\frac{\partial g}{\partial \theta}\bigg/\frac{\partial g}{\partial G}. \quad (10.96)$$

The denominator term in (10.98) does not vanish. The equilibrium of the tower becomes critical if

$$dG/d\theta = 0. \quad (10.97)$$

In the critical state, increments in the tower's rotation occur, in fact, without any further increments in the weight G. Specifically, Eq. (10.97) gives

$$\frac{dG}{d\theta} = \frac{\partial g}{\partial \theta} = GH_G \cos(\theta_{oi} + \theta) - K_{\theta o}e^{-\theta/\gamma_{oR}} = 0. \quad (10.98)$$

The rotational equilibrium is preserved in the critical state as well. Condition (10.98) will thus be associated with rotational equilibrium condition (10.93). From these equations, the two unknowns—the critical weight G^*_{cr} and the critical rotation θ^* of the tower—can be determined. Equations (10.93) and (10.98) yield the following condition

$$tg(\theta_{oi} + \theta^*) = \gamma_{oR}(e^{\theta^*/\gamma_{oR}} - 1), \qquad (10.99)$$

which, for small values of $(\theta_{oi} + \theta^*)$, gives

$$\theta^* \approx \sqrt{2\gamma_{oR}\theta_{oi}}. \qquad (10.100)$$

The critical weight of the tower $G^*_{cr}(\theta_{oi})$ can be obtained from Eqs. (10.100) and (10.98). We thus have

$$\frac{G^*_{cr} H_G}{K_{\theta o}} = \frac{e^{-\theta^*/\gamma_{oR}}}{\cos(\theta_{oi} + \theta^*)}. \qquad (10.101)$$

As θ_{oi} becomes smaller and smaller, θ^* also vanishes and the critical weight becomes

$$\lim_{\theta_{oi} \to 0} G^*_{cr} = G_{cro}, \qquad (10.102)$$

Where

$$G_{cro} = \frac{K_{\theta o}}{H_G}. \qquad (10.103)$$

The critical weight G_{cro} represents the critical weight of the *initially vertical* tower. Substitution of (10.103) into (10.101) gives the explicit expression for the critical weight of a leaning tower (Como 1993):

$$\frac{G^*_{cr}}{G_{cro}} = \frac{e^{-\theta^*/\gamma_{oR}}}{\cos(\theta_{oi} + \theta^*)} \qquad (10.104)$$

The critical tower weight has also been evaluated by Nova and Montrasio (1995).

Equation (10.104) shows that even a small initial rotation θ_{oi} of the tower can produce a large reduction in the critical weight G^*_{cr} with respect to the value G_{cro} of

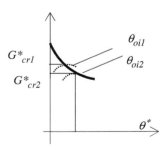

Fig. 10.48 Variation of the tower critical weight G^*_{cr} with initial rotation θ_{oi}

Fig. 10.49 Application of a counter weight on a leaning tower

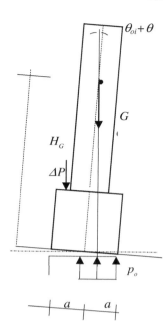

the critical weight of a vertical tower (Fig. 10.48). Such a critical condition is unstable.

We can define a factor, s, that expresses the safety of a tower in its rotated configuration. This safety factor is given by

$$s = \frac{G^*_{cr}}{G}, \qquad (10.105)$$

where G^*_{cr} is the critical weight of the tower that has undergone an initial subsidence rotation θ_{oi}. As long as coefficient s is not too small, the leaning tower will remain in its tilted position in a stable state, as has occurred, for instance, for the Garisenda tower in Bologna. To the contrary, if s is near unity, the tower equilibrium is quite uncertain. In such cases, small changes in the loads or in environmental conditions can lead to failure of the tower. Furthermore, creep deformations of the soil may lead to slowly increasing tilting of the tower.

Various systems can be used to improve the safety of towers. Some aim to reduce their inclination. The following sections will address these issues, with a particular focus on stability analysis of the renowned Leaning Tower of Pisa and the work carried out to stabilize it.

10.5.5 Counter Weights to Stabilize Leaning

Given the attempts made to stabilize a number of leaning towers in this fashion, it is interesting to evaluate the response of such structures to the application of counter weights. To this end, let us consider the scheme in Fig. 10.48, where an additional weight ΔP has been placed on the base of a leaning tower on the side opposite its inclination.

The plastic response of the foundation is different from the elastic case. According to the elastic foundation model, application of load ΔP on the side opposite the tilt would certainly reduce the tower's inclination. The stress acting on the foundation, before application of the additional load ΔP, is localized at point A of the interaction domain Y, corresponding to an assigned level of hardening N_o. The coordinates of this point, A, in the plane M, N are

$$N \approx G \quad M = GH_G \sin(\theta_{oi} + \theta) \tag{10.106}$$

According to the elastic model, the foundation response is represented by an increase in subsidence, together with a negative increment, $d\theta$, that is, counter rotation of the tower. According to the plastic model, instead, the foundation strain rate will, by the normality rule, be directed along the external normal at A to the interaction locus Y. Consequently, if $df > 0$, the rotation rate $d\theta$ will be positive and a further increase in the tower's inclination will occur (Fig. 10.50). The differential df of the yielding function f at A is, on the other hand, given by

$$df = \left(\frac{\partial f}{\partial M}\right)_A dM + \left(\frac{\partial f}{\partial N}\right)_A dN, \tag{10.107}$$

which, in the simple case of rectangular foundations, by accounting for (10.81), yields

$$df = -\Delta Ne + \Delta N\left(-\frac{1}{2} + \frac{N}{N_o}\right) = \left(-e - \frac{a}{2} + \frac{Na}{N_o}\right)\Delta N. \tag{10.108}$$

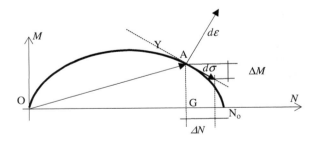

Fig. 10.50 Response $d\varepsilon$ of a plastic foundation to stress increment $d\sigma$

10.5 Towers

Let

$$e_L \tag{10.109}$$

be the *limit* eccentricity value, which corresponds to a zero increment in plastic rotation of the foundation, that is, when $df = 0$. Hence, we get

$$e_L = a\left(\frac{N}{N_o} - \frac{1}{2}\right). \tag{10.110}$$

Moreover, accounting for (10.110) and that

$$2H_G(\theta_{oi} + \theta) = x + \frac{a}{2}, \tag{10.111}$$

the limit eccentricity becomes

$$e_L = a\left[\frac{1}{2} - \frac{2H_G}{a}(\theta_{oi} + \theta)\right]. \tag{10.110'}$$

Consequently, we can write

$$df = (e_L - e)\Delta N, \tag{10.110''}$$

and we can thus conclude that only when

$$e > e_L \tag{10.112}$$

will $df < 0$ (Fig. 10.50). For instance, in the case of a rectangular foundation of width a, with $2H_g/a = 2.5$, $(\theta_{oi} + \theta) = 5°$, it can be seen that an eccentricity value $e = a/4$ is insufficient to ensure $e > e_L$.

10.5.6 Evolution of Tower Tiling

10.5.6.1 Soil Creep Effects on Leaning Tower Equilibrium

Leaning towers frequently attain a tilting configuration that remains unchanging over time. In such cases, the foregoing analysis can be deemed suitable for checking their stability. In other cases, however, such as for instance, the leaning Tower of Pisa, a structure may undergo slow but progressive increases in tilting. The equilibrium state of the tower evolves over time and can either stabilize asymptotically or deteriorate until it reaches failure. The evolution of a tower's rotation depends on the behavior of the foundation soil, whose response changes over time. There are various reasons for this behavior: periodic variations in the height of the water table linked to particular soil features, as creep of the solid particles of the soil itself, amongst others. Moreover, the interactions occurring between the time-dependent foundation soil response and the tower's tilting are extremely important. A small, uneven settling of the foundation causes an increase in rotation that, in turn, leads to greater axial load eccentricity on the foundation and consequently slow further tilting of the structure, and so on. A simple visco-plastic foundation model can describe this behavior. To this end, accurate geotechnical techniques are available to define the visco-plastic soil parameters involved in the various cases.

10.5.6.2 Visco-Plastic Model of Foundations

A visco-plastic model of the foundation can explain variations in a tower's inclination over time. A simple visco-plastic model considers the strain rate $\dot{\varepsilon}$ expressed as the sum of the plastic and viscous shares

$$\dot{\varepsilon} = \dot{\varepsilon}^p + \dot{\varepsilon}^v, \tag{10.113}$$

where the plastic strain rate $\dot{\varepsilon}^p$ is given by (10.80) and the viscous rate $\dot{\varepsilon}^v$, according to a common rheological equation, is given by

$$\sigma = K_v \dot{\varepsilon}^v \tag{10.114}$$

The stress vector, σ, of components M and N is given by (10.77), where K_v is the foundation viscous stiffness matrix:

$$K_v = \frac{1}{\alpha} e^{\beta(t-t_o)} \begin{bmatrix} k_{\theta\theta} & 0 \\ 0 & k_{vv} \end{bmatrix}. \tag{10.115}$$

In expression (10.115), t_o indicates the initial time, corresponding to completion of the tower, when it is assumed that the load began to act. The constant α is a factor expressing the intensity of the viscous deformation of the foundation, β a scale factor assumed equal to 1 century^{-1}, and $k_{\theta\theta}$, k_{vv} are positive quantities defining the viscous behavior of the foundation. The assumed viscous constitutive equation conforms to the formulation of creep deformation discussed in Chap. 1. Equation (10.114) can be written in the more explicit form:

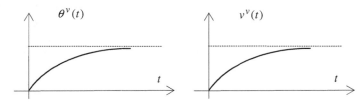

Fig. 10.51 Evolution laws of the subsidence and rotation of a purely viscous foundation under constant centered vertical load and constant couple

10.5 Towers

$$\begin{bmatrix} M(t) \\ N(t) \end{bmatrix} = \frac{1}{\alpha} e^{\beta(t-t_o)} \begin{bmatrix} k_{\theta\theta} & 0 \\ 0 & k_{vv} \end{bmatrix} \begin{bmatrix} \dot{\theta}_v \\ \dot{v}_v \end{bmatrix}. \quad (10.115')$$

Hence, we get

$$M(t) = \frac{1}{\alpha} e^{\beta(t-t_o)} k_{\theta\theta} \dot{\theta}_v, \qquad N(t) = \frac{1}{\alpha} e^{\beta(t-t_o)} k_{vv} \dot{v}_v \quad (10.116)$$

or

$$\dot{\theta}^v = \frac{\alpha}{k_{\theta\theta}} e^{-\beta(t-t_o)} M(t), \qquad \dot{v}^v = \frac{\alpha}{k_{vv}} e^{-\beta(t-t_o)} N(t). \quad (10.116')$$

For a better understanding of the mechanical significance of the various parameters involved according to the proposed visco-plastic model, as a first step let us analyze the slow subsidence of a foundation under *constant* loads, G or M, resting on a purely viscous soil, whose behavior is defined by Eq. (10.115'). Thus, from Eq. (10.115'), we get

$$\dot{\theta}^v = \frac{\alpha}{k_{\theta\theta}} e^{-\beta(t-t_o)} M, \qquad \dot{v}^v = \frac{\alpha}{k_{vv}} e^{-\beta(t-t_o)} N. \quad (10.116'')$$

Integration of (10.116") gives (Fig. 10.51)

$$\theta^v(t) = \frac{\alpha}{\beta k_{\theta\theta}} M[1 - e^{-\beta(t-t_o)}], \qquad v^v(t) = \frac{\alpha}{\beta k_{vv}} G[1 - e^{-\beta(t-t_o)}]. \quad (10.117)$$

The asymptotic values of the foundation rotation and settlement are thus

$$\theta^v_\infty = \frac{\alpha}{\beta k_{\theta\theta}} M, \qquad v^v_\infty = \frac{\alpha}{\beta k_{vv}} G. \quad (10.118)$$

The quantities

$$\frac{\beta k_{\theta\theta}}{\alpha} \qquad \frac{\beta k_{vv}}{\alpha} \quad (10.119)$$

represent the rotational and vertical asymptotic viscous stiffness of the foundation (Fig. 10.51).

10.5.6.3 Slow Tilting of Towers Resting on Visco-Plastic Foundation

Let us now consider a tower on visco-plastic soil, defined by the model discussed above. The tower undergoes an initial tilt θ_{oi} due to differential subsidence of the foundation at the onset of its construction. The initial equilibrium configuration of the tower is thus rotated: this initial inclination causes load G to become eccentric and slowly, over time, produces visco-plastic strains and further tilting. The slow displacement of the tower will be characterized by a predominating rotational component $\theta(t)$, which occurs under nearly constant axial load. Here, $\theta(t)$ now represents the *entire* rotation of the tower at time t, including the initial tilt as well as the visco-plastic share.

In light of previous results, we can disregard the contribution to the displacement of the small changes in the axial load during progressive tilting.

The rotation rate at time t is due to the strong interactions between the two plastic and viscous portions:

$$\dot{\theta}(t) = \dot{\theta}^p(t) + \dot{\theta}^v(t). \tag{10.120}$$

Creep rotation in fact produces rotation increments and, consequently, increases in the moment $M(t)$ acting on the foundation. This monotonically increasing moment, in turn, produces further increments in the plastic rotation and so on. The hardening of the soil, which slowly reduces the magnitude of the viscous rotation rates, conflicts with the increasing moment $M(t)$, whence new, additional rotation arises. Once this process has been initiated, either the tower's movement will slowly stabilize or it will progress fatally towards failure. In this context, from Eq. (10.120), and by accounting for (10.89) and the first of Eq. (10.117), we get

$$\dot{\theta} = \frac{M_R}{K_{\theta o}} \frac{\dot{M}}{M_R - M(t)} + \frac{M(t)}{k_{\theta\theta}^v} \alpha e^{-\beta(t-t_o)}. \tag{10.121}$$

The moment $M(t)$ acting on the foundation is given by

$$M(t) = G H_G \sin\theta(t). \tag{10.122}$$

Thus, taking into account that

$$\dot{M}(t) = G H_G \dot{\theta} \cos\theta(t), \tag{10.123}$$

from Eq. (10.121), we have

$$\dot{\theta} = \frac{M_R}{K_{\theta o}} \frac{\dot{\theta}\cos\theta(t)}{\mu_R - \sin\theta(t)} + \frac{\alpha}{k_{\theta\theta}^v} e^{-\beta(t-t_o)} G H_G \sin\theta(t), \tag{10.124}$$

where

$$\mu_R = \frac{M_R}{G H_G}. \tag{10.125}$$

10.5 Towers

With the position

$$\phi(t) = \mu_R - \sin\theta(t) \tag{10.126}$$

and taking into account that

$$\dot{\phi} = -\dot{\theta}\cos\theta(t), \tag{10.126'}$$

Equation (10.124) gives

$$-\frac{M_R}{K_{\theta o}}\frac{\dot{\phi}}{\phi} + \frac{\mu_R\alpha}{\chi^v_{\theta\theta}}e^{-\beta(t-t_o)} - \frac{\alpha\phi}{\chi^v_{\theta\theta}}e^{-\beta(t-t_o)} = -\frac{\dot{\phi}}{\cos\theta}, \tag{10.127}$$

The factor (10.128) where

$$\chi^v_{\theta\theta} = \frac{k^v_{\theta\theta}}{GH_G} \tag{10.128}$$

represents the dimensionless rotational viscous stiffness of the foundation. With the change in variable

$$\xi = e^{-\beta(t-t_o)}, \tag{10.129}$$

and accounting for

$$\phi = \phi[\xi(t)] \quad \dot{\phi} = -\phi'\beta\xi \quad ()' = \frac{d(\ )}{d\xi}, \tag{10.130}$$

we get

$$\frac{M_R}{K_{\theta o}}\beta\frac{\phi'}{\phi} + \frac{\mu_R\alpha}{\chi^v_{\theta\theta}} - \frac{\alpha}{\chi^v_{\theta\theta}}\phi = \beta\frac{\phi'}{\cos\theta}. \tag{10.131}$$

Thus, by taking position (10.126) into account, after some manipulations we obtain

$$\frac{\beta\chi^v_{\theta\theta}}{\alpha}\frac{\left(\frac{1}{\cos\theta} - \frac{\gamma_{oR}}{\mu_R - \sin\theta}\right)}{\sin\theta}(-\cos\theta d\theta) = d\xi \tag{10.132}$$

Separation of variables gives

$$\xi = -\frac{\beta\chi^v_{\theta\theta}}{\alpha}\int\frac{1 - \frac{\gamma_{oR}\cos\theta}{\mu_R - \sin\theta}}{\sin\theta}d\theta + k \tag{10.133}$$

and by integrating we get

$$\xi = 1 + \frac{\beta \chi_{\theta\theta}^{\nu}}{\alpha}[\ln(tg\frac{\theta}{2}) - \frac{\gamma_{oR}}{\mu_R}\ln(\frac{\sin\theta}{\mu_R - \sin\theta})] + k. \tag{10.134}$$

Let assume, for the sake of simplicity, that

$$t_o = 0 \tag{10.135}$$

Then $\xi(t_o = 0) = 1$, and the initial conditions are

$$\theta = \theta_i \text{ at } \xi = 1. \tag{10.136}$$

Thus, we obtain the explicit formulation of function $\xi = \xi(\theta)$:

$$\xi = 1 + \frac{\beta \chi_{\theta\theta}^{\nu}}{\alpha}[\ln(\frac{tg(\theta_i/2)}{tg(\theta/2)}) - \frac{\gamma_{oR}}{\mu_R}\ln(\frac{\mu_R - \sin\theta_i}{\mu_R - \sin\theta}\frac{\sin\theta}{\sin\theta_i})] \tag{10.137}$$

that describes the evolution of the leaning of the tower over time.

10.5.6.4 Critical State: Critical Time

Different, but *simultaneous,* equilibrium configurations occur at the critical state. Thus, in the critical state

$$\frac{dt}{d\theta} = 0, \tag{10.138}$$

or, taking (10.129) into account,

$$\frac{d\xi}{d\theta} = 0, \tag{10.138'}$$

However, from (10.138) and (10.137), the critical condition becomes

$$\frac{1}{\cos\theta} - \frac{\gamma_{oR}}{\mu_R - \sin\theta} = 0, \tag{10.139}$$

which gives the equation for the critical rotation

$$\gamma_{oR}\cos\theta_{cr} + \sin\theta_{cr} = \mu_R, \tag{10.140}$$

Fig. 10.52 The critical state

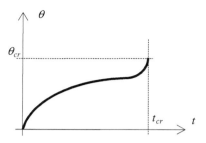

which, in turn, furnishes θ_{cr} with the assigned values of μ_R and γ_{oR}. Substituting this value of θ_{cr} into Eq. (10.137) yields the critical time. The condition,

$$\xi_{ct} \in [0, 1], \tag{10.141}$$

indicates whether or not the critical state can be actually be attained. Figure 10.52 shows the plot $\theta(t)$ of the tower rotation versus time. At the critical state, defined by the coordinates, critical time t_{crit}, and critical rotation θ_{crit}, the tangent to the curve $\theta(t)$, according to (10.138), becomes vertical. The impending critical state is signaled by the change in sign of the derivative $d\theta/dt$, from negative to positive values, implying *acceleration* of the motion.

We have applied the proposed visco-plastic model to analyzing the slow rotation of a leaning tower by assuming the values of M_R and of $K_{\theta o}$ considered in the next section. Figure 10.53 shows the plot of the tower inclination versus time, expressed in centuries, for various values of the dimensionless foundation asymptotic viscosity stiffness. All possible behaviors of such a tower can be described by varying the different quantities involved. Small values of the dimensionless foundation viscous stiffness, which also signify high, heavy towers, led to leaning failure.

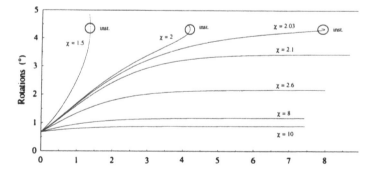

Fig. 10.53 Plot of the rotation versus time (in centuries) of a leaning tower with an initial tilt, resting on a visco-plastic foundation for different values of the dimensionless asymptotic viscous stiffness

10.5.7 The Leaning Tower of Pisa

The Tower of Pisa (Fig. 10.54) was designed as a cylindrical belfry that would stand about 56 m high. The width of its walls varies from 4.09 m at its base to 2.48 m at the top. The Tower is a hollow cylindrical shaft with eight stories, including the bell chamber. The external and internal diameters at the base are about 15.5 and 10.4 m, respectively. The bottom story is made up of 15 marble arcades, while each of the next six stories contains 30 arcades surrounding the inner cylinder of the tower. The top story is the bell chamber itself, with 16 arcades. The inner and outer surfaces are faced with marble and the annulus between these facings is filled with rubble and mortar within which extensive voids have been found.

Construction of the Tower, the Cathedral's bell tower, had a troubled history. Construction was begun in 1173, probably under the supervision of Bonanno Pisano. However, the Tower already began to sink after construction had progressed to the second floor in 1178. This was due to a mere 3 m-deep foundation, set in weak,

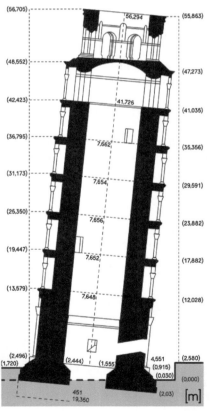

Fig. 10.54 The Leaning Tower of Pisa (From Wikipedia, Creative Commons 2009)

deformable subsoil. The work was suspended after this first foundation subsidence. During this period, the Republic of Pisa was almost continually engaged in battles with Genoa, Lucca and Florence, which allowed time for the underlying soil to settle.

About a century later, in 1272, construction was resumed under Giovanni di Simone and got as far as the as the sixth story. In an effort to compensate for the incline, the added floors were built out-plumb, by building one side higher than the other. However, construction was halted once again in 1284, when the Pisans were defeated by the Genoans in the battle of Meloria. The 7th floor with the upper bell chamber was finally completed in 1319 by Tommaso, son of Andrea Pisano.

The ground underlying the Tower consists of three distinct layers. The first layer, called horizon A, is about 10 m thick and consists primarily of soft estuarine deposits of sandy and clayey silts laid down under tidal conditions. The second layer, called horizon B, consists of soft, normally consolidated marine clay, known as *Pancone* clay, which extends to a depth of about 40 m.

This material is very sensitive and loses much of its strength if disturbed. The surface of the *Pancone* clay is dished beneath the Tower, revealing that the average subsidence is between 2.5 and 3.0 m. The third layer, called horizon B, is dense sand which extends to considerable depth.

The water table lies between 1 and 2 m below the ground surface in horizon A. The continuous long-term tilting of the Tower could be explained by continuous variation of the water-table level that produced, by racketing, incremental plastic deformation of the solid structure of the soil. The soil subsided considerably and the high deformability of the Tower foundation may, on the contrary, be mainly due to the high compressibility of the *Pancone* Clay (Burland 1998, 1999).

By 1992 the tower was leaning by an angle of about 5° towards the south. Precise measurements (begun in 1911) showed that during the 20th century the inclination of the Tower was increasing inexorably each year and the rate of tilting had doubled since the mid-1930s.

In 1990 the tilting rate was about 6 arc-seconds per year, equivalent to a horizontal movement of about 1.5 mm per year at the top.

The diagram in Fig. 10.55 shows the history of the tower's rotation according to Burland (1998).

The acceleration in the tilting from the year 1272–1360 has been attributed to soil consolidation during the first suspension of its construction. Another considerable increase in rotation occurred in 1838, when A. Della Gherardesca excavated a walkway around the foundations, the so-called "*catino*", to facilitate access to the tower. This work resulted in an inrush of water on the south side, since the excavation here reached below the water table, and eventually in a increase in the tower's inclination of more than half a degree.

Crucial operations to stabilize the tower through sub-excavations were performed in the years 2000–2001, as shown in Fig. 10.56 (Jamalkowski et al. 2003).

The sub-excavation technique involved gradual removal of small quantities of soil from the side opposite the incline. This technique, engineered by Terracina (1962), had first been applied successfully to stabilize the cathedral of Mexico City (Tamez et al. 1997). A large number of corkscrew drills were inserted at a shallow

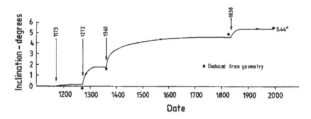

Fig. 10.55 Tower rotation in time (from Burland 1998)

Fig. 10.56 Inclined drill for soil extraction (Burland 1998)

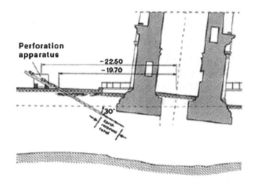

angle into the earth beneath the tower to remove soil from beneath the raised side of the tower. The progressive rotation was arrested, and the tower was straightened by 45 cm to about four meters off-center—returning it to its position in 1838.

Experimental research carried out at the Imperial College of London (Edmunds 1993) has shown that the sub-excavation technique is strongly influenced by the extension and localization of the region of sub-excavated soil. It is interesting to point out that this study envisaged the existence of a critical depth for the region sub-excavated on the side opposite the inclination. The incline of the tower would actually worsen if the soil extraction were to continue beyond this depth, though the reasons for this seem to be not yet clear (Como et al. 2001).

Some technical data on the Tower of Pisa
Weight (Lancellotta 1993; Desideri et al. 1997): G = 14450 t.

Tilt of the tower prior to the stabilization operations, $\theta = 5° 40'$. Height of the tower center of mass with reference to the foundation: $H_G = 22,515$ m.

Initial tower inclination upon its completion: $\theta_{0i} = 0° 40' = 0° 0.667 = 0.0116$ rad. This value has been determined by evaluating the axis corrections attempted during the last stage of its construction. The tower, in fact, has a slightly curved 'banana' shape due to the addition of the last stories out-plumb, but at an angle to the lower stories, in an attempt to correct the tilting caused by settling of the foundation soil.

The ultimate resistant moment of the foundation and the initial rotational stiffness, according to Lancellotta (1993): $M_R = 60,000$ tm; $K_{\theta 0} = 550,000$ tm/rad.

10.5 Towers

Critical weight evaluation according to (10.104).

We shall now apply the foregoing formulations of the foundation resistant moment and the initial rotational stiffness, as drawn from Lancellotta (1993), to evaluate the critical weight of the tower.

Using the given values of $K_{\theta 0}$ and M_R, we calculate the factor γ_{oR}, which from (10.90), gives us: $\gamma_{oR} = 0.109$. The critical weight of the tower, corresponding to the its initial vertical position, from (10.105), is: $G_{cr.o} = 24{,}428$ t.

We shall consider an initial inclination due to subsidence (according to Lancellotta) of $\theta_{oi} = 0° 40' = 0.0116$ rad, and from Eq. (10.100), the subsequent inclination due to foundation deformation is about $\theta^* = 0.050$ rad $= 2° 882'$.

Thus, from Eq. (10.104), the critical weight of the tower turns out to be $G^*_{cr} = 15472$ t. The corresponding critical rotation of the tower would be $\theta_{oi} + \theta^* = 0° 40' + 2° 882' = 3° 55'$, less than the 5.5° measured before the recent stabilization work.

Let us now assume the following values for the ultimate moment and the initial foundation stiffness: $M_R = 90{,}000$ tm; $K_{\theta 0} = 500{,}000$ tm/rad, together with a somewhat larger value of the initial inclination: $\theta_{oi} = 0° 50' = 0° 0.83' = 0.015$ rad.

From (10.90), we get $\gamma_{oR} = 0.180$ and from (10.103), $G_{cr.o} = 22{,}207$ t. By applying Eq. (10.100), the tilting of the tower after the initial settlement turns out to be $\theta^* = 0.073$ rad $= 4° 21'$. The corresponding critical weight is lower, that is, $G^*_{cr} = 14{,}863$ t, only slightly lager than the tower's actual weight. The total inclination of the tower becomes $4° 21' + 0° 83' = 5° 04'$, not too different from the rotation detected before its stabilization. Finally, the *safety factor*, evaluated in terms of the ratio between the critical and the actual weight of the tower, is very low, only about $14863/14450 = 1.03$. Further studies could take into account the creep deformations of the soil. These evaluations, albeit approximate, reveal the precariousness of the tower's state back in 1990 and moreover highlight the relevance of the stabilization works carried out.

10.5.8 Stability of Other Leaning Towers

The previous visco-plastic approach has been applied by. Lancellotta (1993) in the study of the asymptotic stability of the Ghirlandina Tower of the Dome of Modena (Fig. 10.57). Actually the tower is leaned in the direction South—West of $1° 14' 16''$. The studies have shown that the Tower is stable

10.5.9 Cracking of Leaning Towers. Heyman Collapse Analysis

In presence of sufficient tilting, an inclined fissure (Fig. 6.65) starting from the base can occur in a leaning wall and its stability can be compromised. In fact, part of the

Fig. 10.57 The Ghirlandina Tower of the Dome of Modena

wall remains attached to the base and its weight cannot any more explicate the stabilizing action on the shaft of the wall. The same occurs in a leaning tower.

The first problem is to define the profile of the fracture; then it will determined the maximum inclination of the tower under which the equilibrium becomes unstable.

Proceeding downward from the top, the weight of the upper part of the shaft becomes more and more eccentric. so that, at defined distance a from the top, the eccentricity reaches the end E of the section core. This section, distant a from the tower head, defines the starting point of the detachment fissure running as far as the base of the wall (Fig. 10.58).

With the assumption of rectangual cross sections, the scheme at the right of Fig. 10.59 shows the upper uncracked part of the shaft of the tower rotated of the angle ϕ. The end section of this part, far of a from its head, has the left hand corner K free of stresses. These are linearly varying along this section that is therefore compressed by a vertical force $W = \gamma Ba$, eccentric of $B/6$ respect to the centre of the

10.5 Towers

Fig. 10.58 Curved fissure detaching a masonry volume from the shaft of the tower

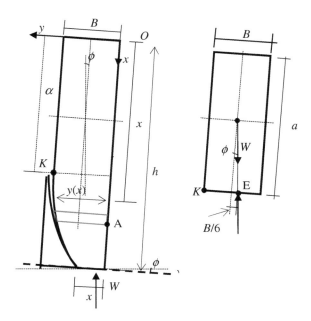

Fig. 10.59 Equilibrium of the element dy of the cracked segment of the tower shaft

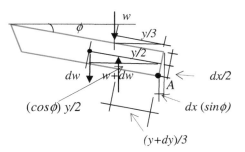

section. The height a of this upper part of the tower is connected to the width B and to the rotation ϕ by means the equation

$$\operatorname{tg}\phi = \frac{B}{6}\frac{2}{a} = \frac{B}{3a}, \tag{10.168}$$

Figure 10.59 shows an element of length dy of the cracked shaft of the tower.

Consider now a cracked section of the lower shaft of the tower and let y the width of its compressed part. Examine the downward evolution of the cracking starting from $x = a$.

A first *unknown* of the problem is thus the dependance of this *width y* of the compressed part of the section from its distance x from the head of the tower. A *second unknown* is then the dependance on the distance x of the weight w of the shaft of the tower that loads the considered section.

With the reference system of axes shown in Fig. 10.58, at the section passing through the point K of the cracked shaft we have $y = B$. The weight that loads this section is $w = \gamma B a$ and has an eccentrricity $B/6$, assuming constant the thickness of the shaft in the direction orthogonal to the plan of the figure.

In the cracked segment the width x of the section is connected to the rate of change of the weight w according to the equation

$$y = \frac{dw}{dx}, \qquad (10.169)$$

With reference to Fig. 10.59, the rotational equilibrium around the vertex A of the segment of length dx of the cracked shaft gives

$$dw\frac{y}{2}\cos\phi + w\frac{y}{3}\cos\phi - wdx\sin\phi - w\frac{y}{3}\cos\phi - w\frac{dy}{3}\cos\phi - \frac{y}{3}dw\cos\phi = 0 \qquad (10.170)$$

from which, dividing for $dx\,\cos\phi$, we get

$$\frac{y}{6}\frac{dw}{dx} - w \cdot tg\phi = \frac{w}{3}\frac{dy}{dx} \qquad (10.171)$$

On the other hand, from (10.169), we have

$$\frac{dy}{dx} = \frac{dy}{dw}\frac{dw}{dx} = y\frac{dy}{dw} \qquad (10.172)$$

and the (10.71) gives

$$\frac{y^2}{6} - w \cdot tg\phi = \frac{w}{3}y\frac{dy}{dw} \qquad (10.173)$$

Let us consider now the non dimensional quantities

$$X = \frac{x}{a}, \quad Y = \frac{y}{B}, \quad W = \frac{w}{aB} \qquad (10.174)$$

so that the (10.173), taking also account of the (10.168), becomes

$$Y^2 - 2W = W \cdot 2Y\frac{dY}{dW} \qquad (10.175)$$

10.5 Towers

This last equation, by making the substitution

$$z = Y^2 \tag{10.176}$$

with

$$\frac{dz}{dW} = 2Y \frac{dY}{dW} \tag{10.176'}$$

becomes

$$z - 2W = W \cdot 2Y \frac{dY}{dW} = W \frac{dz}{dW} \tag{10.175'}$$

and may be simplified in

$$\frac{dz}{dW} = \frac{z}{W} - 2 \tag{10.177}$$

Equation (10.177) has solution

$$z = W(C - 2 \ln W) \tag{10.178}$$

or, taking into account of (10.176),

$$Y^2 = W(C - 2 \ln W) \tag{10.179}$$

where C is a constant of integration. From Fig. 6.65 it may be noted that the fissure starts at the point K of coordinates $x = B$, $y = a$. To these determinations of x and y correspond the values $X = 1$, $Y = 1$ and $W = 1$, because with $x = B$, $y = a w = Ba$. But, with $Y = 1$ and $W = 1$, (10.179) requires that $C = 1$ and Eq. (10.179) becomes

$$Y^2 = W(1 - 2 \ln W) \tag{10.179'}$$

Thus, from this last Eq. (10.179') and the (10.169), i.e. with $Y = dW/dX$, we have

$$\frac{dW}{dX} = \sqrt{W(1 - 2 \ln W)} \tag{10.180}$$

or

$$X = \int_{1}^{W} \frac{dW}{\sqrt{W(1 - 2 \ln W)}} + A \tag{10.181}$$

where A is the second constant of iuntegration. This last has to be equal to 1 because when $W = 1$, $X = 1$. Thus we get

$$X = 1 + \int_1^W \frac{dW}{\sqrt{W(1 - 2\ln W)}} \tag{10.181'}$$

Equation (10.181') can be integrated with the change of variable

$$4t^2 = 1 - 2\ln W \tag{10.182}$$

or with

$$W = e^{1/2} e^{-2t^2} \tag{10.182'}$$

In fact, with (10.182) and from (10.179'), we have

$$Y^2 = 4Wt^2 \tag{10.179'}$$

and

$$Y = 2t e^{1/4} e^{-t^2} \tag{10.179'}$$

At the limit, when the width of the compressed zone vanishes, i.e. when $Y \to 0$, from (10.179''') we have $t = 0$. On the contrary, when the considered section is at $X = 1$, i.e. the section where cracking starts, there is $W = 1$ and $Y = 1$. Consequently, from (10.179) we have $t = \frac{1}{2}$.

The cracked segment of the tower, with the change of variable (10.182) is thus defined for $0 \le t \le 1/2$. Thus, from (10.182) we can write equivalently

$$W = e^{1/2} e^{-2t^2}, \quad 0 \le t \le 1/2 \tag{10.183}$$

and we obtain a first result: the non dimensional weight of the whole cracked shaft, at the limit of equilibrium, when its base is reduced to zero, is

$$W(X = 0) = W(t = 0) = e^{1/2} \tag{10.183'}$$

From (10.183) we have also

$$dW = -4t e^{1/2 - 2t^2} dt \tag{10.184}$$

and taking into account of (10.182) and (10.182') Eq. (10.181') can be written as

$$X = 1 + 2 e^{1/4} \int_t^{1/2} e^{-t^2} dt \tag{10.180'}$$

Fig. 10.60 The curved fissure as farv as to the hinge at the toe of the tilted tower

or

$$X = 1 + \sqrt{\pi}e^{1/4}[erf(1/2) - erf(t)] \qquad (10.185)$$

where *erf(t)* is the *error fuction* (Murray R. Spiegel, Mathematical Handbook, Schaum's Outline Series, McGraw Hill Book Company, New York, 1968)

$$erf(t) = \frac{2}{\sqrt{\pi}} \int_0^t e^{-u^2} du \qquad (10.186)$$

Concluding, we have

$$W = e^{1/2 - 2t^2} \quad Y = 2te^{1/4}e^{-t^2}; \, X = 1 + \sqrt{\pi}e^{1/4}[erf(1/2) - erf(t)] \qquad (10.187)$$

Figure 10.60, taken from Heyman (2002), gives the shape of the curved cracking of the tilted tower.

The maximum non dimensional height of the tower, for a given α, can be obtained taking into account that in this case

$$Y = 0. \qquad (10.188)$$

From the third of (10.187), with $x = h$, i.e. with $X = H = h/a$

$$H = h/a = [Y]_{t=0} = 1 + \sqrt{\pi}e^{1/4}erf(1/2) \qquad (10.189)$$

But $erf(1/2) = 0.5205$ and

$$H = 1 + 1.1846 = 2.1846 \qquad (10.190)$$

The height of the tower is $h = Ha$. Thus with (10.168) riesce

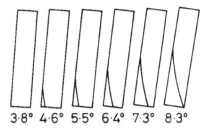

Fig. 10.61 Changes of the cracked zone at the base of the tower with increasing tilting, valuated at the ratio $h/B = 5$ (from Heyman 1992)

$$H/h = 1/a = B/aB = 3tg\phi/B \quad (10.191)$$

and the critical rotation of the tower at the limit of equilibrium is

$$(tg\phi)_{\lim} = \frac{0.7282}{(h/B)} \quad (10.192)$$

For instance, with a ratio $h/B = 5$, we have $(\phi)_{\lim} = 8.3°$.

The rotation of the tower is frequently valuated by means the *out-of-plumbness* of the top of the tower with respect to its base. The limit *out-of-plumbness* $F_{p\ Lim}$ is thus given by

$$F_{p\,Lim} = H \sin \alpha \approx H tg\alpha = 0.728B \quad (10.193)$$

that can be compared with respect to the critical *out-of-plumbness* corresponding to the case of the tower considered as a rigid uncracked block. Figure 10.61 gives for a tower with rectangular section and with a ratio $h/B = 5$, the increasing of the cracked zone for $h/B = 5$, as far as the overturning state.

The case of an hollow square section is considered in Fig. 10.62.

A tower with an hollow section is less dependant on the cracking than a tower with a solid section. The core of the hollow section is in fact much wider than the core of a rectangular section and are required eccentricities only a bit less than the double of B/6 to activate the cracking.

Fig. 10.62 The case of a tower with an hollow square section

10.5 Towers

The tubular section of the shaft of the Tower of Pisa can be considered alike to an uniform square thin walled section. Taking in account of the obtained results, we can say that the cracking unstability is not cause for concern and stability of the Tower depends only on the elasto-plastic—viscous response of its foundation.

Appendix

Yokel Integration of the Differential Equation for the Flexure of a Eccentrically Loaded Cracked Pier

By means of Eq. (10.15), we obtain

$$\frac{d}{dx}(\frac{dy}{dx})^2 = \frac{2dy}{dx} \cdot \frac{d^2y}{dx^2} = \frac{2k_1}{(u_0+y)^2} \cdot \frac{dy}{dx}$$

and

$$(\frac{dy}{dx})^2 = 2k_1 \int \frac{dy}{(u_0+y)^2} = -2k_1 \frac{1}{u_0+y} + c_1.$$

For $y \to 0$, we have $dy/dx \to 0$ and consequently $c_1 = 2k_1/u_0$,

and

$$(\frac{dy}{dx})^2 = 2k_1(\frac{1}{u_0} - \frac{1}{u_0+y}) = \frac{2k_1}{u_0} \cdot \frac{y}{u_0+y},$$

hence

$$\frac{dy}{dx} = \pm\sqrt{\frac{2k_1}{u_0}} \cdot (\frac{y}{u_0+y})^{1/2}. \qquad (10.194)$$

With the position

$$k_2 = \sqrt{\frac{2k_1}{u_0}} = \frac{2}{3}\sqrt{\frac{P}{Ebu_0}}, \qquad (10.195)$$

equation (10.194) becomes:

$$\frac{dy}{dx} = \pm k_2(\frac{y}{u_0+y})^{1/2}$$

or

$$dx = \pm \frac{1}{k_2} (\frac{u_0 + y}{y})^{1/2} \cdot dy.$$

Integration yields

$$x = \pm \frac{1}{k_2} \int (\frac{u_0 + y}{y})^{1/2} dy$$

and, consequently

$$\int (\frac{u_0 + y}{y})^{1/2} dy = \sqrt{y(u_0 + y)} + \frac{u_0}{2} \int \frac{dy}{\sqrt{y(u_0 + y)}}$$
$$= \sqrt{y(u_0 + y)} + u_0 \ln(\sqrt{y} + \sqrt{u_0 + y}),$$

hence

$$x = \pm \frac{1}{k_2} \left[\sqrt{y(u_0 + y)} + u_0 \ln(\sqrt{y} + \sqrt{u_0 + y}) \right] + c_2.$$

By applying boundary condition (10.15) (i.e., by accounting for the fact that at $x = 0$, $y = 0$), we get:

$$c_2 = -\frac{u_0}{k_2} \ln \sqrt{u_0}$$

and

$$x = \pm \frac{1}{k_2} [\sqrt{y(u_0 + y)} + u_0 \ln \frac{\sqrt{y} + \sqrt{u_0 + y}}{\sqrt{u_0}}] \quad (10.196)$$

Substituting (10.196) into (10.195) of k_2 furnishes:

$$x = \pm \frac{3}{2} \sqrt{\frac{Ebu_0}{P}} \cdot [\sqrt{y(u_0 + y)} + u_0 \ln \frac{\sqrt{y} + \sqrt{u_0 + y}}{\sqrt{u_0}}].$$

By considering that for $x = h/2$, $y = u_1 - u_0$, the solution is expressed in terms of u_1:

$$\frac{h}{2} = \pm \frac{3}{2} \sqrt{\frac{Ebu_0}{P}} \cdot [\sqrt{u_1(u_1 - u_0)} + u_0 \ln \sqrt{\frac{u_1 - u_0}{u_0}} + \sqrt{\frac{u_1}{u_0}}],$$

and hence

$$P = \frac{9Ebu_0}{h^2} \cdot [\sqrt{u_1(u_1 - u_0)} + u_0 \ln(\sqrt{\frac{u_1 - u_0}{u_0}} + \sqrt{\frac{u_1}{u_0}})]^2.$$

With the position

$$\alpha = u_0/u_1, \tag{10.197}$$

we finally arrive at

$$P = \frac{9Ebu_1^3}{h^2} \cdot \alpha \cdot [\sqrt{1 - \alpha} + \alpha \ln(\sqrt{\frac{1-\alpha}{\alpha}} + \sqrt{\frac{1}{\alpha}})]^2. \tag{10.198}$$

References

Burland, J. B. (1998). The enigma of the leaning of the tower of Pisa, *The sixth Spencer J. Buchanam Lecture,* Texas A&M University, December 9, 1998.
Burland, J. B., Jamiolkowski, M., & Viggiani, C. (1999). The Restoration of the Leaning Tower of Pisa: Geotechnical Aspects. *Workshop on the Restoration of the Leaning Tower of Pisa,* Pre-print Vol. 1, Pisa 1999.
Cervenka, V., & Atena, J. (2002). Program documentation, *User's manual for Atena 2D,* Prague 2002.
Como, M. (1993). Plastic and visco-plastic stability of leaning towers. In G. Ferrarese (Ed.), *Fisica matematica e Ingegneria: rapporti e compatibilità, Conv. Intern. in memoria di G. Krall.* Isola d'Elba, 10–14 Giugno 1993, Pitagora ed., Bologna.
Como, M., & Ianniruberto, U. (1995). Sulla resistenza laterale di pilastri caricati assialmente e costituiti da materiale elastico–non resistente a trazione, *XII Congresso Naz. AIMETA '95,* 3–6 Ottobre, Naples.
Como, M. T. (2000). *Il Restauro dei Monumenti a Torre in muratura, Restauro, 152–153.* Naples: Edizioni Scientifiche Italiane.
Como, M., Ianniruberto, U., & Imbimbo, M. (2001). A rigid plastic model of the under–excavation technique applied to stabilize leaning towers. In P. B. Lourenço, P. Roca (Eds.), *Constructions.* Guimarães.
De Falco, A., & Lucchesi, M. (2000). Stability of no-tension beam-columns with bounded compressive strength. *Proceedings of IASS–IACM 2000, Fourth International Colloquium on Computation of Shell and Spatial Structure,* June 4–7, Chania Crete, Greece.
De Falco, A., & Lucchesi, M. (2003). Explicit solutions for the stability of no-tension beam-columns. *International Journal of Structural Stability and Dynamics, 3*(2), 195–213.
De Fez, A. (1992). *Il consolidamento degli edifici.* Naples: Liguori.
Desideri, A., Russo, G., & Viggiani, C. (1997). La stabilità di torri su terreno deformabile. *Rivista Italiana di Geotecnica,* 1/97.
de Witte, F. C., Kikstra, W. P. (1990). *Diana user's manual, nonlinear analysis.* Delft, Netherlands.
Edmunds H. E. (1993). The use of under excavation as a means of stabilising the leaning Tower of Pisa: Scale and model tests, *MSc thesis, Department of Civil Engineering, Imperial College of Science, Technology and Medicine,* London.
Frisch-Fay, R. (1975). Stability of masonry piers. *International Journal of Solids and Structures, 11,* 2.

Giuffrè, A. (1990). *Letture sulla Meccanica delle Murature Storiche*. Rome: Facoltà di Architettura dell'università di Roma La Sapienza.

Hambly, E. C. (1985). Soil buckling and the leaning instability of tall structures. *The Structural Engineer, 63A*(3).

Jamiolkowski, M., Burland, J. B., & Viggiani, C. (2003). The statbilisation of the leaning Tower of Pisa, *Soil and Foundations, 43*(5).

Krall, G. (1947). Statica dei mezzi elastici cosiddetti viscosi e sue applicazioni. *Acc. Naz.le dei Lincei*, fasc. 3–4, Rome.

La Mendola, L., & Papia, M. (1993). Stability of masonry piers under their own weight and eccentric load. *Journal of Structural Engineering, 119*, 6.

Lancellotta, R. (1993). The stability of a rigid column with non–linear restraint. *Geotechnique, 2*.

Meyerhof, G. G. (1951). The ultimate bearing capacity of foundations. *Geotechnique, 2*(3).

Napoli, P. (1992). *Modellazione numerica della interazione struttura suolo*. Politecnico di Turin: Atti del Dpt. di Ingegn. Strutturale.

Norme Tecniche sulle Costruzioni, Ministero delle Infrastrutture e dei Trasporti, 2005.

Shrive, N. G., & England, G. L. (1981). Elastic, creep and shrinkage behavior of masonry. *International Journal of Masonry Construction, 1*(3).

Tamez, E., Ovando, E., & Santoyo, E. (1997). Under excavation of Mexico City's Metropolitan Cathedral and Sagrario Church. In *Proceedings of 14th International Conference on Soil Mechanics and Foundation Engineering*, Vol. 4.

Terracina, F. (1962). Foundations of the Tower of Pisa. *Geotechnique, 12*, 4.

Yokel, F. Y. (1971). Stability and load capacity of members with no tensile strength. In *Proceedings of A.S.C.E.*, 87, ST7.

Chapter 11
Gothic Cathedrals

Abstract Aim of this chapter is the study of statics of Gothic cathedrals, splendid achievements of engineering and architecture of the Middle Age. Some brief notes give information about times and places of their construction and an introductory analysis describes the structural elements constituting their stone skeleton. Two main static problems are then discussed: evaluation of the wind strength of the whole cathedral structure and analysis of the slender piers instability. The critical wind velocity for the cathedral of Amiens is determined by direct application of the Limit Analysis, via kinematical approach. The failure occurred at the cathedral of Beauvais in the past 1284 is studied in the second part of the chapter. The failure is generally attributed to the effects of foundations settlements. The chapter, conversely, inquires the possibility that the collapse could be due to piers instability, due to their exceptional height and slenderness. The question is examined in deep, analyzing, in particular, the effects of creep of the mortars on the piers strength. It is shown that the instability of the slender masonry piers, with their axial loadings eccentricities and the mortars creep effects, could be considered really responsible of the 1284 failure.

11.1 Introduction and Some Historical Notes

During the transition from the Romanesque to the Gothic period many radical changes came about in architectural style, especially in the construction of churches. Gothic cathedrals best highlight the structural originality of the architecture of the time. Large masonry masses were the hallmark of Romanesque constructions and only small windows opened in the perimeter walls of Romanesque churches. In contrast, the construction of a structural skeleton, unrelated to the masonry masses, was the first innovation. Large openings could thus be built into the walls and would eventually give rise to the stained glass and rose windows so typical of Gothic churches.

The main architectural elements of Gothic structures are pointed arches, ribbed cross vaults, flying buttresses and slender piers, some of which had already been used in Romanesque and, even earlier, in Roman architecture. Simultaneous application of all these elements to form a harmonious, unified whole was first achieved by French architects of *Ile de France,* who redefined the concept of cathedral. A prelude to Gothic architecture can be discerned in the mid-12th-century rebuilding of the apse of the Basilica of Saint-Denis, where French kings were once entombed, under the supervision of Pierre de Montereau and Suger: the broad windows under the arcades of the ogival vaults, supported by slender columns, herald the typical vast interior space of later Gothic cathedrals. Gothic architecture is typified by the cathedrals of Noyon (1160), Notre Dame in Paris (1163), Laon (1170), Saint Remy at Reims (1162 and 1181) and Soissons (1190). Somewhat later, construction of the cathedrals of Chartres (1195–1260) and Amiens (1220–1269), with ribs surrounding the piers starting at their bases and rising to join the vault ribs, gave rise to the Gothic *rayonnant,* which the pioneering naves of Saint Denis had originally introduced. Gradually, Gothic architecture spread throughout Europe and over subsequent centuries cathedrals were built in this magnificent style in nearly every major city.

The "golden age" of cathedral construction began with the inception of the choir of the Basilica of Saint-Denis, in about 1140, and continued up to 1284, the year the Beauvais Cathedral collapsed while still undergoing construction. With the Beauvais design, daring Gothic architecture had reached the extreme limit of static stability.

The principles of mechanics applied by cathedral builders were centered on the use of levers and the composition of forces. Mason lodges jealously treasured their knowledge of these principles, which were passed on during long years of training in workshops and at building sites.

The designers of Gothic cathedrals were at once architects and engineers. The thirty-three tables, in architect Villard de Honnecourt's notebook, drawn up in about the year 1235, document the techniques used in building sites and for the construction of Gothic cathedrals. Geometry was the sole basis of design: the construction codes, in fact, set out strict rules for proper geometrical proportions (Fig. 11.1).

11.2 Brief Notes on the Construction Techniques

The stones used by cathedrals builders originated in nearby, frequently marl, quarries. However, marl does not offer high compression strength, so the stones had to be cut precisely to present perfectly flat surfaces and thereby improve contact with the mortar beds. Lime mortars were used. A wall consisted of two, 20- to 30-cm thick outer layers or facings made of high-quality coursed ashlars, and a rubble and mortar inner filling. The two outer layers were connected by larger through-the-thickness stones in order to connect the two facings at regular intervals.

11.2 Brief Notes on the Construction Techniques

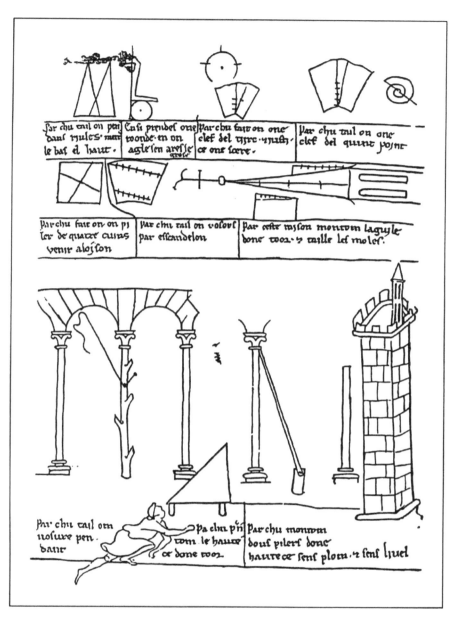

Fig. 11.1 A illustration from the notebook of Villard de Honnecourt (Heyman 1995)

As a rule, foundations were rather undersized: patrons did not enjoy seeing their money 'wasted' on underground structures that no one would ever see. Thus, nearly all the cathedrals built during the period suffered from differential subsidence, but

Fig. 11.2 Interior of the Cathedral Saint-Étienne in Bourges

were able to freely follow the settling and maintain an admissible equilibrium. Gothic cathedrals piers are very slender (Fig. 11.2), though they are well connected to flying buttresses and vaults and, as a rule, well braced. Buckling is the most common hazard. The aisles of Gothic cathedrals are covered by stone vaults, which are, in turn, covered by wooden trusses. The masonry vaults served to protect the interiors from fire. Figure 11.3 shows the progressive evolution of the transverse sections of three cathedrals built between the years 1190 and 1220: the cathedrals of Soissons, Chartres and Amiens.

The heights of the extrados of the nave vault increased gradually with improvements in constructional techniques—from 30 m of the Soissons cathedral and 34 m at Chartres, up to 42 m in the Cathedral of Amiens.

The challenge facing Gothic builders was to attain a nave vault height equal to the 48 m of the Beauvais Cathedral, the last cathedral to be built during France's golden age of Gothic architecture.

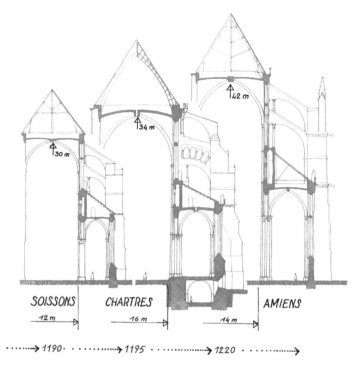

Fig. 11.3 Transverse sections of cathedrals built in successive times: Soissons, Chartres and Amiens

Pointed or ogival, arches were the most commonly used. Pointed arches mark the evolution of vaults spans. Ogival arches offer the advantage of reaching the same height despite the different spans (Fig. 11.4). A typical cathedral section is illustrated in Fig. 11.5, where the flying and the external buttress are indicated with their French terms.

The thrust of the nave vault of a major cathedral with lateral aisles is transmitted from above the aisle roofs by the flying buttresses to the external buttresses and thence to the ground. In such cases the flying buttresses work as simple props. The weight of the flying buttress yields a curved pressure line and the profile of the flying buttress must thus be that of a flat arch. With its ends fixed it is indeformable and the degree of the bearable thrust is limited only by the masonry crushing strength. Major cathedrals possess a double prop system and so only the lower flying buttresses convey the thrust of the vault. When the cathedral is exposed to the actions of the wind, the upper flyers, which are situated on the windward side, sustain the roof trusses and therefore work just as the lower flyers. The vertical buttresses at the outer end of the flyers were often capped with pinnacles that

Fig. 11.4 Ogival arches of the same height but different spans. Comparison with rounded arches

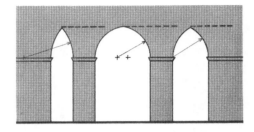

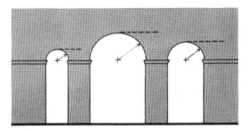

Fig. 11.5 The function of the flying buttress (arc boutant)

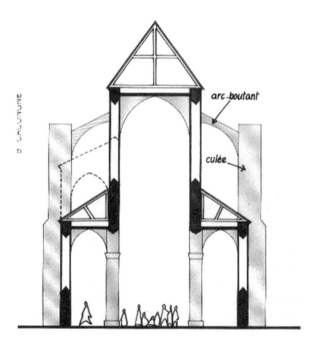

provide additional vertical loading to help resist the lateral thrust transmitted by the flyer. In the absence of wind, the thrust on the upper flying buttresses is near minimum. Figure 11.6 shows a detail of a flying buttress on the cathedral of Amiens.

11.3 Relevant Static Problems

Fig. 11.6 Flying buttresses of the Cathedral of Notre Dame at Amiens

11.3 Relevant Static Problems

Two main problems inherent to Statics of Gothic Cathedrals are talked in the chapter: the wind strength analysis of their transversal structure and stability analysis of the high piers flankin the nave.

The landscape of the plains of France is dotted with the distant profiles of cathedrals rising above the more common buildings of many cities and towns. It does not take much imagination to picture the force of the winds blowing through these plains to impinge on the cathedrals' façades and upper structures. Nor is it difficult to understand the enormity of the challenges facing architects of the time in building masonry structures able to oppose the actions of the fierce winds and prevent the buckling of such high piers. The study another time again will prove the high strength of the stone skeleton of the cathedrals and the skill of the past engineers and architects.

The study of the behavior of the slender Gothic piers is then developed carrying on an analysis of the old failure of the Beauvais Cathedral occurred in the far 1284.

11.4 Transverse Wind Strength

11.4.1 Wind Action on the Transverse Segment of the Cathedral: The Assumed Typology of the Cathedral of Notre Dame D0Amiens

The analysis considers a transverse segment of the cathedral having length equal to the longitudinal span of the piers flanking the nave. A wide cross vault, sustained by upper windowed walls, spans the central nave. These walls, in turn, are sustained by longitudinal arches spanning between piers bordering the nave.

Fig. 11.7 Plan of the Amiens Cathedral

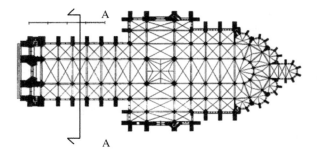

The study is conducted on a model of cathedral similar to the Cathedral of Amiens whose plan and transversal section is shown in Figs. 11.7 and 11.8.

The heads of the piers and the springings of the central vault are both connected to large external buttresses through two orders of flyings. The lateral aisle vaults are sustained by piers and by external side walls. Figure 11.9 shows the geometry of the considered sectional modulus with the corresponding measure lengths.

Cathedral structures are rigid in both the longitudinal the transverse directions. Wind action is essentially static and can be represented by a suitable distribution of horizontal forces. Figure 11.10 shows the typical distributions of wind velocity and pressure acting on the vertical projection of a transverse section of a cathedral. Wind pressure is proportional to the square of the velocity, thus explaining why,

Fig. 11.8 Front view of the Amiens Cathedral

11.4 Transverse Wind Strength

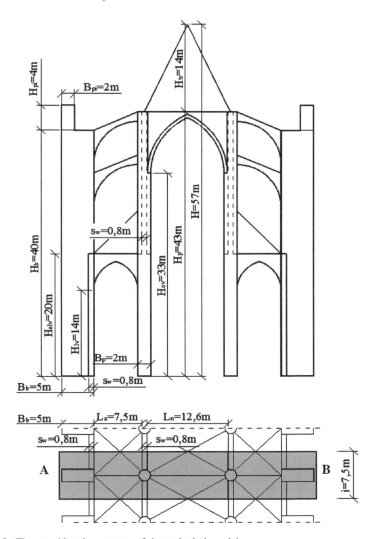

Fig. 11.9 The considered geometry of the cathedral model

while wind velocity increases less than linearly with height, pressure instead increases as a nearly linear function of height. Air velocity reduces on the windward face, speeds up on the lateral sides and generates eddies on the leeward side. Wind pressure is thus positive on the windward side, while it is negative on the leeward side, due to the suction produced by the airflow separation.

The reference velocity V_{ref} is generally measured at a height of 10 m from the ground, given that at ground level the velocity falls to near zero. A simple formula, drawn from Canadian building codes, expresses the variation in wind velocity $V(z)$ with height z, relative to the reference velocity V_{ref}:

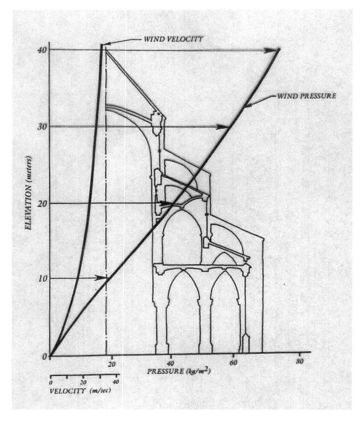

Fig. 11.10 Wind velocity and pressure distributions acting on the transverse section of a cathedral (from Wilipedia)

$$V_{wind}(z_m) = V_{ref}\left(\frac{z}{z_{ref}}\right)^\alpha, \tag{11.1}$$

where, in urban areas, $\alpha = 0.35$–0.40. For instance, a wind stream with a velocity of 50 km/h, or 13.9 m/s, at a height of 10 m from the ground, will at 40 m height, have a velocity $(4)^{0.35} = 1.62$ times larger than V_{ref}, that is, about 24.2 m/s.

According to the Italian code, the wind pressure on a unit surface orthogonal to the wind direction can be calculated as

$$p_v = q_b c_e c_p, \tag{11.2}$$

11.4 Transverse Wind Strength

where

$$q_b = 1/2 \rho V_b^2, \tag{11.3}$$

and:
- ρ standard air density taken to be 1.297 kg/m³;
- V_b wind velocity in m/s;
- c_e exposure factor, varying with height; this variation can be evaluated via Eq. (11.1),
- c_p aerodynamic factor, which can be assumed equal to 0.8 on the windward side and −0.4 on the leeward side.

By way of example, the wind pressure of an air stream with an average velocity of 50 km/h acting over an orthogonal surface on the windward side at a height of 10 m from the ground is

$$p_{v,10m} = \frac{0.5 \cdot 1.297}{9.81} \cdot \frac{s^2}{m} \frac{kg_f}{m^3} 13.9^2 \frac{m^2}{s^2} \cdot c_e \cdot 0.8 = 10.2 \frac{kg_f}{m^2}$$

On the leeward side the wind pressure is instead negative and equal to

$$p'_{v,40m} = -5.1 \frac{kg_f}{m^2}.$$

Wind pressure increases greatly with increasing wind velocity: for instance, a 100 km/h wind at 40 m from the ground produces a pressure on the windward side of

$$p_{v,40m} = \frac{0.5 \cdot 1.297}{9.81} \frac{s^2}{m} \frac{kg_f}{m^3} (27.77 \cdot 1.62)^2 \frac{m^2}{s^2} \cdot 0.8 = 107 \frac{kg_f}{m^2},$$

while on the leeward side it is

$$p'_{v,40m} = -53.5 \frac{kg_f}{m^2}.$$

Wind produces significant actions on cathedral structures, particularly on the piers and buttresses. However, cathedral geometries generally provide adequate strength even for exceptionally strong winds, as will be shown in the following.

Figure 11.11 shows the assumed distribution of the wind pressure acting on the transverse section of the cathedral: λ is the load multiplier. We have assumed the approximate wind pressure distribution on the cathedral walls, increasing linearly with the height. The collapse wind pressures $\lambda_{cr} p$ and $\lambda_{cr} p'$ will be obtained via determination of the failure multiplier λ_{cr} by application of Limit Analysis.

Fig. 11.11 Wind pressure distribution on the transverse section of a cathedral

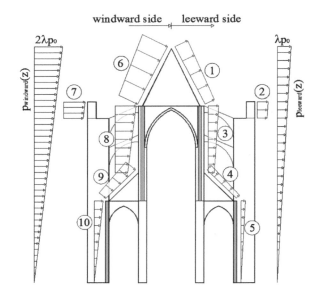

Four main groups of actions can be identified, as indicated in Fig. 11.11:

- *actions 1 and 6* are referred to the wind pressure acting on the roof truss, inclined of an angle of about 64.5° to the horizontal axis;
- *actions 2 and 7* are referred to the wind pressure acting on the pinnacles;
- *actions 3 and 8* are referred to the wind pressure acting on the upper internal walls;
- *actions 4 and 9* are referred to the wind pressure acting on the aisle roof, inclined of an angle of 45° to the horizontal axis;
- *actions 5 and 10* are referred to the wind pressure acting on the outward walls from the ground to the intrados of the aisle roof.

11.4.2 Dead Loads: Vertical Forces and Horizontal Vault Thrusts

11.4.2.1 The Assumed Masonry Unit Weights

The masonry pier has a regular brickwork with stone blocks bounded by horizontal mortar beds and vertical staggered joints. On the average the unit weight γ_p is assumed equal to 20 kN/m^3.

The walls and the outside buttresses are composed by two outer skins of good coursed ashlar and a solid rubble fill. On the average the assumed unit weight γ_b is 19 kN/m^3.

11.4 Transverse Wind Strength

The masonry of arches and vaults is a regular brickwork with stone elements. On the average the assumed unit weight γ_v is 21 kN/m³.

An average weight equal to 3 kN/m² on the inclined surface, including tiles, lead sheets, underlying wooden structures and trusses, has been assumed for the cumulative weight of the roof above the main vault. A reduced weight of 2 kN/m² has been assumed for the roof above the lateral aisles. An average weight of 1 kN/m² has been assumed for the two service wooden floors underlying the roofs.

11.4.2.2 The System of Vertical Dead Loads

The loading pattern relative to the dead loads is shown in Fig. 11.12. For sake of simplicity, the loads are shown only on the right half section.

The system of dead loads (Fig. 11.9) consists of the following actions:

- G_p, weight of the central pier and the upper wall aligned to it, lying in the width of the modulus;
- G_{pi}, weight of the pinnacle;
- G_b, weight of the lateral buttress;
- G_w, weight of the outward wall;
- G_{cr}, weight of half central truss;
- G_{lr}, weight of the lateral truss.
- G_{cv}, weight of half vault of the central nave;
- S_{cv}, resulting thrust of the vault of the central nave;
- G_{lv}, weight of half vault of the lateral aisle;
- S_{lv}, resulting thrust of the vault of the lateral aisle.

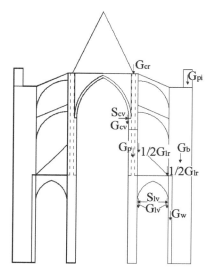

Fig. 11.12 Cross section of the Cathedral showing the considered dead loads

The internal pier is 43 m high and it is composed of two segments: the first segment rises from the ground to a height of 33 m up to the springer of the central vault. It has a circular cross section with diameter $B_p = 2$ m. The second segment rises from the height of 33 m and reaches the base of the roof trusses. Its length is equal to 10 m and its section has a smaller diameter $b_p = 1.5$ m.

The weight of the pier is:

$$G_{pier} = \left(\pi \frac{B_p^2}{4}\right) H_{cv} \gamma_p + \left(\pi \frac{b_p^2}{4}\right) (H_p - H_{cv}) \gamma_p = 2.427 \times 10^3 \text{ kN} \qquad (11.4)$$

where, as shown in Fig. 11.6:

- H_p is the total height of the pier, equal to 43 m;
- H_{cv} is the height of the lower segment of the central pier;

The weight of the wall aligned to the pier is given by:

$$G_{aw} = (H_p - H_{lv}) \cdot s_w \cdot (i - B_p) \cdot \psi_{gl} \cdot \gamma_b = 1.212 \times 10^3 \text{ kN} \qquad (11.5)$$

in which:

- H_{lv} is the height of the springing of the vaults of the lateral aisles, equal to 14 m;
- s_w is the thickness of the wall, assumed equal to 0.8 m;
- i is the longitudinal span between the piers, equal to 7.5 m;
- ψ_{gl} is a corrective factor which takes into account the percentage of windowed surface, taken equal to 0.5.

The overall weight of the pier is:

$$G_p = G_{pier} + G_{aw} = 3.639 \times 10^3 \text{ kN} \qquad (11.6)$$

The weight of the side buttress is:

$$G_b = s_b \cdot H_b \cdot B_b \cdot \gamma_b = 7.6 \times 10^3 \text{ kN} \qquad (11.7)$$

in which:

- s_b is the thickness of the buttress in the longitudinal direction, assumed equal to 2 m;
- B_b is the thickness of the buttress in the transversal direction, equal to 5 m;
- H_b is the height of the buttress, assumed equal to 40 m.

The weight of the pinnacle is:

$$G_{pi} = s_b \cdot H_{pin} \cdot B_{pin} \cdot \gamma_b = 320 \text{ kN}$$

where the height H_{pin} and the width B_{pin} of the pinnacle are equal to 4 and 2 m, respectively.

The weight of the outward wall is:

$$G_w = s_w \cdot H_{elv} \cdot (i - s_b) \cdot \psi_{gl} \cdot \gamma_b = 836 \, \text{kN} \qquad (11.8)$$

where:

- s_w is the thickness of the wall, assumed to be 0.8 m;
- $(i - s_b)$ is the length of the wall in the longitudinal direction, equal to 5.5 m;
- H_{elv} is the height of the wall from the ground to the top, taken to be 20 m;

Assuming that the trusses of the nave have an angle of about 64.5° to the horizontal axis, the half weight of the roof can be so evaluated:

$$G_{cr} = \left(\sqrt{\left(\frac{L_{tr}}{2}\right)^2 + H_{tr}^2} \right) i \cdot q_r + \frac{L_{tr}}{2} \cdot i \cdot q_{sr} \approx 400 \, \text{kN} \qquad (11.9)$$

where:

- q_r is an uniformly distributed load, perpendicular to the roof, having magnitude equal to 3 kN/m² (Heyman 1997);
- q_{sr} is the weight of the service wooden roof equal to 1 kN/m².
- L_{tr} is the distance between the center of gravity of the two piers of the nave, equal to $L_n + s_w$ = 13.4 m
- H_{tr} is the height of the trusses, equal to H − H_p = 14 m.

The trusses of the lateral aisles have the shape of an isosceles rectangular triangle, so that the dead load conveyed by the roofing is:

$$G_{lr} = \sqrt{2} L_a \cdot i \cdot q_{rl} + L_a \cdot i \cdot q_{sr} = 215.35 \, \text{kN} \qquad (11.10)$$

where L_a is the net width of the aisle, taken equal to 7.5 m and q_{rl} is an uniformly distributed load, perpendicular to the roof, having magnitude equal to 2 kN/m².

11.4.2.3 The Thrust of the Main Cross Vault

The thrust of the main vault plays a fundamental role on the wind strength of the cathedral because of its possible concurring pushing action with the wind. The vault is laterally closed by two side upper walls and by the overlying roof. The thrust of the vault is thus due only to the dead loads.

The thrust of the vault has been evaluated as the minimum among all the statically admissible ones. The structures supporting the thrust of the vault, i.e. the flying and the side buttresses, in fact settle lightly sideways in the plane of the

Fig. 11.13 The settled vault roofing the nave

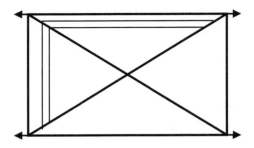

doubleau arches. The vault adapts itself to the settlement through a compatible three-dimensional mechanism and a state of minimum thrust takes place in it.

The behaviour of the vault is studied in line with the so called sliced model, firstly introduced by Heyman (1966): after cracking the webs are divided into a sequence of separate arches, each one characterized by a different span length, which convey the loads on the diagonal ribs (Fig. 11.13).

The evaluation of the minimum thrust of the main pointed cross vault has been performed according to the kinematic theorem of the Limit Analysis.

According to the kinematic approach the minimum thrust is *the maximum* of all the kinematic multipliers, $\lambda(\mathbf{v} \in \overline{M})$, varying $\mathbf{v}$ in the set of all settlement mechanisms $\overline{M}$, (Como 1998, 2013) where M is the set of the admissible displacement mechanisms of the vault related to the settlement along the direction perpendicular to the main nave (Coccia &.

In the considered case study the vault of the main nave has a rectangular plan 12.6 m long and 7.5 m wide. The thickness of the emerging ribs is equal to 0.5 m whereas the webs are 0.3 m thick. The obtained horizontal thrust F and the vertical action transmitted to the piers by each one of the two ribs converging to the point A or B are equal to 134 and 270 kN respectively.

The horizontal resulting thrust S_{cv} acting on the single pier can be obtained through a vector addition of the two components F: these last are inclined of an angle of about 31° with respect to the transverse axis, getting:

$$S_{cv} = 2 \cdot 134 \cdot \cos 31° \approx 230 \,\text{kN} \tag{11.11}$$

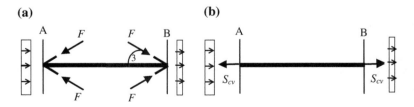

Fig. 11.14 The forces F transmitted by the diagonal ribs and the resulting horizontal thrust S

As shown in Fig. 11.14, the thrust of the vault and the wind action are concurring or contrasting respectively on the leeward and windward sides.

The overall vertical load conveyed to the pier by the two adjacent ribs is:

$$G_{cv} = 2 \cdot 270 \, \text{kN} = 540 \, \text{kN} \tag{11.12}$$

These results are in good agreement with the values of thrust and weight tabulated by Ungewitter (1901), for quadripartite vaults, varying the length and width of the vault, its rise from springing to crown, the thickness and materials of vault webs.

11.4.2.4 The Thrust of the Aisle Cross Vault

The smaller pointed cross vault spanning the aisle has a square plan with side 7.5 m. The thickness of the ribs is equal to 0.5 m whereas the webs are 0.3 m thick. The thrust is evaluated with the same kinematic approach and it is equal to 73.5 kN. The vertical action transmitted to the piers by each one of the ribs converging to the point A and B is 148 kN. Once again, two adjacent ribs convey their loads on the pier and on the lateral buttress. The horizontal resulting thrust on the single pier can be obtained through a vector addition of the two components, which present an angle of 45° to the transversal axis:

$$S_{vl} = 2 \cdot 73.5 \cdot \cos 45° \approx 104 \, \text{kN} \tag{11.13}$$

The overall vertical load conveyed to the pier and to the lateral buttress by the two neighbouring ribs is:

$$G_{vl} = 2 \cdot 148 = 296 \, \text{kN} \tag{11.14}$$

Similarly, to the previous case, assuming a rise of 5.3 m, the height to span ratio is equal to 0.7. The unit weight and thrust are estimated using the tabulated 2:3 ratio, so that $V_o = 13 \, \text{kN/m}^2$ and $H_o = 4 \div 4.3 \, \text{kN/m}^2$.

According to this approach, the weight G_{vl} and the thrust S_{vl} are respectively:

$$V = \frac{1}{2} L \cdot B \cdot V_o \cdot \frac{\gamma_{real}}{\gamma_{rubble}} = \frac{1}{2} 7.5 \cdot 7.5 \cdot 13 \cdot \frac{21}{24} \approx 320 \, \text{kN} \tag{11.15}$$

$$H = \frac{1}{2} L \cdot B \cdot H_o \cdot \frac{\gamma_{real}}{\gamma_{rubble}} = \frac{1}{2} 7.5 \cdot 7.5 \cdot (4 \div 4.3) \cdot \frac{21}{24} \approx 99 \div 106 \, \text{kN} \tag{11.16}$$

11.4.2.5 The Thrust of the Flying Buttresses

The cathedral has a double prop system. The lower flying buttresses constrain the springings of the vault of the main nave and transmit its thrust to the outer buttresses. The upper flying buttresses sustain the high trusses struck by wind and those situated on the leeward side work just as the lower flyers conveying the wind actions from roof trusses to the side buttress (Fig. 11.12). In this force transmission the weight of the flying buttress is neglected because small respect to other loads.

11.4.3 Valuation of the Lateral Wind Strength of the Cathedral: The Assumed Mechanisms

The following analysis is performed according to the kinematic approach and it is referred to a plane problem, considering a transverse segment of the Cathedral, characterized by a width equal to the longitudinal distance i between the piers of the central nave (Fig. 11.9).

Various failure mechanisms are possible. Specifically, three different groups of failure mechanisms have been taken into account. The first group considers the failure of the single roof truss, the failure of the windward pinnacle or the shear failure of the top wedge of the side buttresses.

All the other failure mechanisms affect different structural components but always engage at least one of the two side buttresses.

The second group thus considers the failures involving the leeward side of the Cathedral, pressed by the concurrent action of the thrust of the main vault and the wind pressure.

The third group considers the global failure of all the structural components of the cathedral, belonging both to the windward and the leeward sides. In this last case, the thrust of the vault isn't worth of interest because during the mechanism the central nave undergoes only a rigid translation, without any change of its span length.

A load multiplier λ affects the conventional unit wind pressure p_o acting at the top of the Cathedral. The minimum value λ_o among all the multipliers λ, corresponding to the various mechanisms, represents the wind strength of the Cathedral.

The lateral buttress is the main resistant element of the cathedral. It is composed by a series of stones placed roughly along horizontal courses. In the overturning failure, a lower region of the buttress fractures and becomes ineffective, thereby reducing the side strength of the element. According to the middle third rule, Ochsendorf (2002) proved that the fracture line is straight for buttresses characterized by a rectangular cross section and subjected to their weight and an inclined force at their top section, as shown in Fig. 11.15. This fracturing effect cannot be neglected when the weight of the buttress is dominant with respect to the other involved vertical loads. On the contrary, this fracturing effect has not been taken into account for the piers and the walls because their dead weight is non-dominant

11.4 Transverse Wind Strength

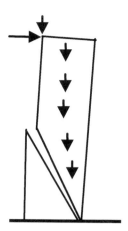

Fig. 11.15 The lateral buttress with its typical straight fracture line

with respect to all the other acting vertical loads. The piers and the buttresses are connected through the flying buttresses and the cross vaults of the aisles. For sake of simplicity, the two lateral vaults are removed from the structure in the considered mechanisms and are replaced by the actions that they transmit to the piers and the buttresses.

11.4.3.1 The Local Failure Mechanisms of the Main Trusses

The local failure of the trusses placed on the top of the main nave can occur or with an upturn mechanism, named A1 (Fig. 11.16), or with a sliding mechanism, called A2 (Fig. 11.17).

The total weight of the truss of the roof, G_{tr}, is equal to:

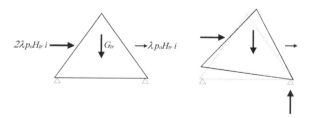

Fig. 11.16 The upturn failure mechanism of the truss of the roof, A1 in which p_o is the conventional unit wind pressure, assumed equal to 1 kN/m^2

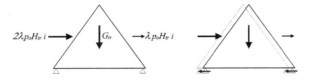

Fig. 11.17 The sliding failure mechanism of the truss of the roof, A2

$$G_{tr} = 2G_{cr} \approx 800 \text{ kN} \tag{11.17}$$

The overall horizontal wind force, considering the effects of the pressure on both the windward and the leeward sides of the roof, is:

$$S_w = \lambda \cdot 3p_o \cdot H_{tr} \cdot i \tag{11.18}$$

The upturn failure equation, in which λ_{A1} is the failure multiplier relative to the mechanism A1, gives the condition:

$$S_w = \lambda_{A1} \cdot 3p_o \cdot H_{tr} \cdot i = G_{tr} \frac{0.5(L_n + s_w)}{0.5H_{tr}} \tag{11.19}$$

that yields to $\lambda_{A1} = 2.43$.

Introducing the failure multiplier λ_{A2} relative to the mechanism A2, the sliding equilibrium equation can be written as (Fig. 11.16):

$$S_w - f \cdot G_{tr} = \lambda_{A2} \cdot 3p_o \cdot H_{tr} \cdot i - f \cdot G_{tr} = 0 \tag{11.20}$$

in which f represents the kinetic friction factor (wood over stone), taken equal to 0.4. The obtained value of the failure multiplier is equal to $\lambda_{A2} = 1.015$.

11.4.3.2 Failures of the Leeward Side of the Cathedral

The concurrent actions of the thrust of the main vault and of the wind load can produce a mechanism concerning the failure of the leeward side of the Cathedral. Two different failure mechanisms, named B and C, have been chosen to describe this type of failure (Fig. 11.18). In both the two mechanisms the truss of the roof is pulled out by the overturning of the piers and of the walls. Consequently the windward support of the truss slides over the head of the pier. The resistant work includes the plastic dissipation due to the sliding of the windward support of the truss and the terms due to the uplift of the various involved weights. The upper internal wall and a portion of the lower outward wall, belonging to the leeward side of the Cathedral, are affected by the wind pressure. In Fig. 11.17, the elements loaded by wind are painted in red and the ones involved in the mechanism in green.

11.4 Transverse Wind Strength

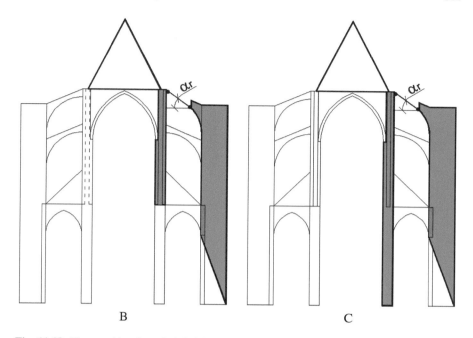

Fig. 11.18 The considered semi-global failure mechanisms involving the leeward side

The Mechanism B

Geometry of the Mechanism

The mechanism B considers that only the upper segment of the pier fails overturning around a hinge placed at the extrados of the lateral vault. The side buttress overturns with a diagonal crack that reaches the springing of the vault of the aisle. The lower triangular wedge of the buttress, the lower outward wall and the lower segment of the pier, are ineffective.

In the considered mechanism, the leeward side of the sectional modulus of the Cathedral moves to the right. A suitable number of hinges splits the sectional modulus into four segments, indicated in Fig. 11.19a as I, II, III and IV. Figure 11.19b shows the absolute centres of rotation of the pier and of the buttress, named C1 and C2 respectively, and the relative centres of rotation, called C13, C23, C14, C24, where the two numbers identify the two connected segments. A mechanism set is defined imposing the positions of these internal hinges. Each mechanism of the set is completely defined when the absolute centres of rotation of all the segments are determined.

Their positions are located taking into account the conditions of alignment between the various triplets of the absolute and relative centres of rotation: the position of the absolute centre C3 is obtained through the condition of intersection between the straight lines joining the two triplets C1, C13, C3 and C2, C23, C3

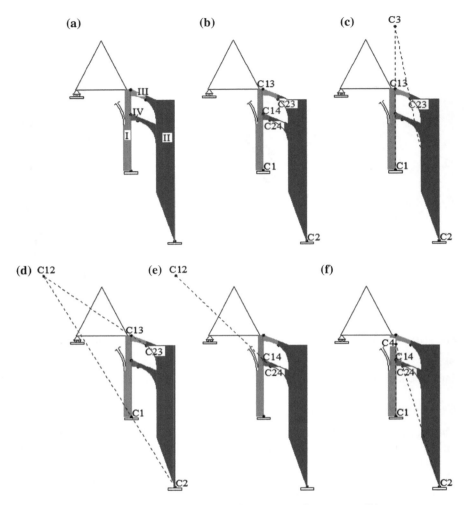

Fig. 11.19 Absolute and relative centres of rotation with the alignment conditions

(Fig. 11.19c); the position of the relative centre C12 is found through the condition of intersection between the straight lines joining the two triplets C12, C13, C23 and C1, C12, C2 (Fig. 11.19d); the position of the relative centre C24 is gotten through the condition of alignment between C14, C24 and C12, together with the condition that C24 is constrained to lie on the intrados curve of the flying buttress (Fig. 11.19e); the position of the absolute centre C4 is obtained through the condition of intersection between the straight lines joining the two triplets C1, C14, C4 and C2, C24, C4 (Fig. 11.19f).

The positions of the centres C13, C23 and C14 are independent variables.

Fig. 11.20 Limitations on the angle α_r

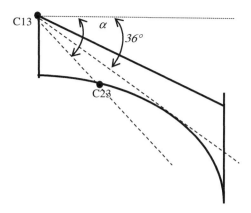

Once that all the positions of the centres of rotation have been obtained, the resulting deformation is defined by the rotation θ_p of the pier around its absolute centre C1.

The angle of inclination of the straight line joining the centres C13 and C23 to the horizontal axis is named α_r. Each choice of the position of the two previous hinges defines this angle (Fig. 11.20).

For the chosen geometry of the flying buttress, in order to obtain a kinematically admissible mechanism the value of the angle α_r cannot be less than 36°.

The relative centre C12 is defined for each position of the centres C13 and C23 and consequently, the centres C14 and C24 can be found as shown in Fig. 11.19e. The mechanism is then identified by only one independent variable, the angle α_r.

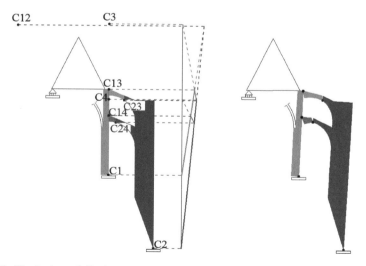

Fig. 11.21 The horizontal displacements occurring in the mechanism B

Figure 11.21 shows the horizontal displacements of the various structural components occurring in the mechanism. Despite the buttress rotates clockwise of an angle θ_b smaller than the one of the pier, named θ_p, detachments occur in the contact sections between the flying buttresses and the pier or the side buttress, respectively. This condition ensures that the mechanism B is kinematically admissible.

The Works in the Mechanism

The leeward wind loads, due to the suction effect, pull on the upper internal wall and the portion of the lower outward wall, while both the windward and leeward wind loads, acting on the roof, together with the main vault thrust, push on the whole leeward side. The weights of the various structural elements involved in the mechanism, together with the friction forces due to the sliding of the truss supports over the top section of the piers, produce the resistant work. While the leeward upper segment of the pier overturns in the mechanism, the roof trusses over the nave, pushed by the wind action, move to the right and their windward supports skid of $\Delta(\theta_p)$ on the head of the upper wall. Figure 11.22 shows the double rising of the right truss support occurring during the pier overturning.

The sliding force is fV_{left} where f is the *friction kinetic factor* and V_{left} is the vertical reaction of the windward truss support. Thus the energy dissipation occurring during the mechanism is

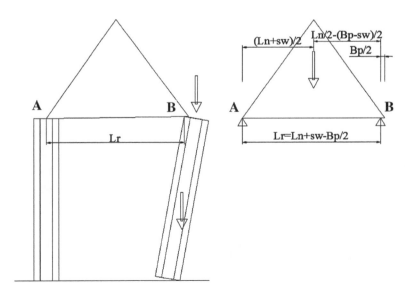

Fig. 11.22 The double rising of the right truss support

11.4 Transverse Wind Strength

$$D_{slid} = f \cdot V_{left} \cdot \Delta(\theta_p). \tag{11.21}$$

The equilibrium condition allows to calculate the kinematic multiplier λ of the wind pressure that causes the collapse:

$$D_{slid} + \langle g, v \rangle + \lambda \langle p_w, v \rangle = 0 \tag{11.22}$$

where the work of the vault thrust, depending only on the weight, has been included in the work $\langle g, v \rangle$ of the loads g.

We consider now the various loads involved in the mechanism.

The weight of the segment of the pier that moves upwards during the mechanism B is:

$$G_{pier} = \left(\pi \frac{B_p^2}{4}\right)(H_{cv} - H_{elv})\gamma_p + \left(\pi \frac{b_p^2}{4}\right)(H_p - H_{cv})\gamma_p = 1.17 \times 10^3 \text{ kN} \tag{11.23}$$

and the weight of the wall aligned to the pier is (Fig. 11.9):

$$G_{aw} = (H_p - H_{elv}) \cdot s_w \cdot (i - B_p) \cdot \psi_{gl} \cdot \gamma_b = 961.4 \text{ kN} \tag{11.24}$$

Then:

$$G_p = G_{pier} + G_{aw} = 2.132 \times 10^3 \text{ kN} \tag{11.25}$$

The buttress, due to the occurrence of the inclined crack, separates into two parts whose weights are (Fig. 11.23):

$$G_{b1} = s_b \cdot (H_b - H_{lv}) \cdot B_b \cdot \gamma_b = 4.94 \times 10^3 \text{ kN} \tag{11.26}$$

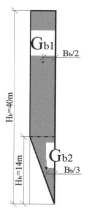

Fig. 11.23 The weights of the two fractions

Fig. 11.24 Evaluation of the roof truss reaction at its left support

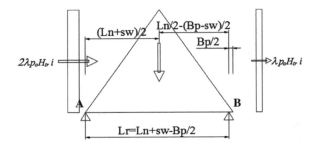

$$G_{b2} = \frac{1}{2} s_b \cdot H_{lv} \cdot B_b \cdot \gamma_b = 1.33 \times 10^3 \text{ kN} \tag{11.27}$$

The weight of the portion of the outward wall that works in the mechanism is:

$$G_{sw} = s_w \cdot (H_{elv} - H_{lv}) \cdot (i - s_b) \cdot \psi_{gl} \cdot \gamma_b = 250.8 \text{ kN} \tag{11.28}$$

The vertical reaction V_B of trusses at the right support B (Fig. 11.24) is equal to:

$$V_B(\lambda) = V_{B-G_{cr}} + \Delta V_B(\lambda) \tag{11.29}$$

where V_{B-Gr} depends on its weight $2G_{cr}$ and is equal to:

$$V_{B-G_{cr}} = \frac{2G_{cr} \frac{L_n + s_w}{2}}{L_r} \tag{11.30}$$

The increment of the vertical reaction of the leeward truss support ΔV_B due to the wind pressure is equal to:

$$\Delta V_B(\lambda) = \frac{3}{2} \lambda p_o \frac{H_{tr}^2}{L_r} \cdot i \tag{11.31}$$

With these assumptions, the works in Eq. (11.25) become:

$$\langle g, v \rangle = -\left(\left(\frac{G_p}{2} + V_{B-G_{cr}} + G_{cv}\right)B_p - S_{cv}(H_{cv} - H_{elv})\right)\theta_p$$
$$+ \left(\left(\frac{G_{b1}}{2} + \frac{G_{b2}}{3} + \frac{G_{lr}}{2} + G_{lv}\right)B_b + G_{sw}\left(B_b - \frac{s_w}{2}\right) - S_{lv}H_{lv} + G_{pi}\frac{B_{pin}}{2}\right)\theta_b \tag{11.32}$$

11.4 Transverse Wind Strength

$$\lambda \langle p_w, v \rangle = \lambda \left[\int_{H_{elv}}^{H_p} \left(\frac{p_o}{H} h^2 i \right) dh + \int_{H_p}^{H} \left(\frac{3p_o}{H} h i H_p \right) dh \right] \theta_p - \Delta V_B(\lambda) B_p \theta_p$$

$$+ \lambda \left[\int_{H_{lv}}^{H_{elv}} \left(\frac{p_o}{H} h^2 i \right) dh + \int_{H_b}^{H_b + H_{pin}} \left(\frac{p_o}{H} h^2 s_b \right) dh \right] \theta_b \quad (11.33)$$

The works of the thrusts S_{cv} and S_{lv} are included in the term $\langle g, v \rangle$.

The energy dissipation D_{slid} in Eq. (11.25), due to the sliding of the left truss support is:

$$D_{slid}(\lambda) = -f \cdot V_A \cdot \Delta \quad (11.34)$$

if Δ is the horizontal displacement of the truss involved in the mechanism and V_A the vertical reaction of the windward truss support (Fig. 11.23), equal to:

$$V_A(\lambda) = V_{A-G_{cr}} - \frac{3}{2} \lambda p_o \frac{H_{tr}^2}{L_r} \cdot i = \frac{2 G_{cr} \frac{L_n - (B_p - s_w)}{2}}{L_r} - \frac{3}{2} \lambda p_o \frac{H_{tr}^2}{L_r} \cdot i \quad (11.35)$$

where:

$$V_{A-G_{cr}} = \frac{2 G_{cr} \frac{L_n - (B_p - s_w)}{2}}{L_r} \quad (11.36)$$

is the reaction of the left support A due to the weight $2G_{cr}$. The dissipation of energy is then:

$$D_{slid}(\lambda) = -f \cdot V_A(\lambda) \cdot \Delta = -f \left(G_{cr} - \frac{3}{2} \lambda p_o \frac{H_{tr}^2}{L_n} \cdot i \right) (H_p - H_{elv}) \theta_p \quad (11.37)$$

where the kinetic friction factor f has been assumed equal to 0.4.

The relationship between the virtual rotation θ_p and θ_b of the pier and the buttress respectively depends on the assumed value of the angle α_r.

By varying the values of the inclination angle α_r we obtain the corresponding values of the kinematical multiplier λ. The minimum value of $\lambda(\alpha_r)$ is attained for $\alpha_r = 36°$ and is equal to:

$$\lambda_B = 0.969. \quad (11.38)$$

When the angle α_r reaches the value of 36°, the straight line connecting the centres C13 and C23 becomes tangent to the intrados of the flying buttress (Fig. 11.19). This situation is explained taking into account that the minimum value of the kinematical multiplier λ_B is also statically admissible. The condition $\alpha_r = 36°$, due to the chosen geometry of the flying buttress, assures that the two conditions of kinematic and static compatibility are both satisfied.

This minimum value corresponds to a value of the rotation θ_b equal to:

$$\theta_b = 0.67\theta_p \tag{11.39}$$

The corresponding collapse wind pressure acting on the leeward side is:

$$p_{collBl} = 0.969 \, \frac{\text{kN}}{\text{m}^2} \tag{11.40}$$

and the corresponding collapse wind pressure acting on the windward side is

$$p_{collBw} = 1.937 \, \frac{\text{kN}}{\text{m}^2} \tag{11.41}$$

11.4.3.3 The Mechanism C

The mechanism C assumes that both the leeward pier and the adjacent wall are involved along all their length in the overturning, in spite of the wind pushes only on the upper wall. The weight of the elements subjected to the uplift during the mechanism C is evaluated according to Eq. (11.6) and is equal to:

$$G_p = G_{pier} + G_{aw} = 3.639 \times 10^3 \text{ kN} \tag{11.42}$$

The dissipation of energy (Eq. 11.40) is evaluated taking into account that the rotation of the pier occurs at its base, and thus becomes:

$$D_{slid}(\lambda) = -f \cdot V_A(\lambda) \cdot \Delta = -f \left(G_{cr} - \frac{3}{2}\lambda p_o \frac{H_{tr}^2}{L_n} \cdot i \right) H_p \theta_p \tag{11.43}$$

With these assumptions, the work of the dead load is:

$$\langle g, v \rangle = -\left(\left(\frac{G_p}{2} + V_{B-G_{cr}} + G_{cv} \right) B_p - S_{cv}H_{cv} + S_{lv}H_{lv} \right)\theta_p +$$
$$- \left(\left(\frac{G_{b1}}{2} + \frac{G_{b2}}{3} + \frac{G_{rl}}{2} + G_{lv} \right) B_b + G_{sw}\left(B_b - \frac{s_w}{2} \right) - S_{lv}H_{lv} + G_{pi}\frac{B_{pin}}{2} \right)\theta_b$$
$$\tag{11.44}$$

By varying the angle α_r the search of the minimum value of λ gives the collapse load multiplier:

11.4 Transverse Wind Strength

$$\lambda_C = 1.501 \tag{11.45}$$

that is attained again for $\alpha_r = 36°$.

The corresponding collapse wind pressure acting on the leeward side of the Cathedral is:

$$p_{collCl} = 1.501 \, \frac{kN}{m^2} \tag{11.46}$$

and the corresponding collapse wind pressure acting on the windward side is

$$p_{collCw} = 3.002 \, \frac{kN}{m^2} \tag{11.47}$$

The minimum value of the load multiplier corresponds to a value of the rotation θ_b equal to:

$$\theta_b = 1.252 \theta_p \tag{11.48}$$

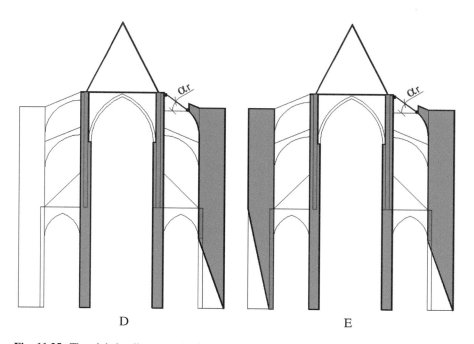

Fig. 11.25 The global collapse mechanisms D and E

11.4.3.4 The Global Failure

In these failure mechanisms, the piers and the upper walls flanking the main vault move together, without any variation of their distance since they are connected by the overlying roof trusses. The collapse of the piers is achieved with the formation of two hinges at their base. The two analyzed mechanisms are shown in Fig. 11.25.

The Mechanism D

For this mechanism, the works in Eq. (11.25) are:

$$\langle g, v \rangle = -\left(\left(\frac{2G_p}{2} + V_{B-G_{cr}} + \frac{V_{A-G_{cr}}}{2} + G_{cv} + \frac{G_{lr}}{2} + G_{lv}\right)B_p\right)\theta_p + $$
$$- \left(\left(\frac{G_{b1}}{2} + \frac{G_{b2}}{3} + \frac{G_{lr}}{2} + G_{lv}\right)B_b + G_{sw}\left(B_b - \frac{s_w}{2}\right) - S_{lv}H_{lv} + G_{pi}\frac{B_{pin}}{2}\right)\theta_b$$
(11.49)

$$\lambda \langle p_w, v \rangle = \lambda \left[\int_{H_{elv}}^{H_p} \left(\frac{p_o}{H}h^2 i\right) dh + \int_{H_{elv}}^{H_p} \left(\frac{2p_o}{H}h^2 i\right) dh + \int_{H_p}^{H} \left(\frac{3p_o}{H}hiH_p\right) dh\right]\theta_p +$$
$$- \Delta V_B(\lambda)B_p\theta_p + \Delta V_A(\lambda)0.5B_p\theta_p$$
$$+ \lambda \left(\int_{H_{lv}}^{H_{elv}} \left(\frac{p_o}{H}h^2 i\right) dh + \int_{H_b}^{H_b + H_{pin}} \left(\frac{p_o}{H}h^2 s_b\right) dh\right)\theta_b$$
(11.50)

Imposing a null energy dissipation in Eq. (11.25), the obtained collapse load multiplier is:

$$\lambda_D = 1.467 \qquad (11.51)$$

The corresponding collapse wind pressure acting on the windward side is:

$$p_{collDw} = 2.935 \, \frac{\text{kN}}{\text{m}^2} \qquad (11.52)$$

and the corresponding collapse wind pressure acting on the leeward side is:

11.4 Transverse Wind Strength

$$p_{collDl} = 1.461 \, \frac{\text{kN}}{\text{m}^2} \tag{11.53}$$

The Mechanism E

In this mechanism, the resisting and the pushing works are:

$$\langle g, v \rangle = - \left(\left(\frac{2G_p}{2} + V_{B-G_{cr}} + \frac{V_{A-G_{cr}}}{2} + G_{cv} + \frac{1}{2} G_{lr} + G_{lv} \right) B_p \right) \theta_p + \\
- \left(\left(\frac{G_{b1}}{2} + \frac{G_{b2}}{3} + \frac{G_{lr}}{2} + G_{lv} \right) B_b + G_{sw} \left(B_b - \frac{s_w}{2} \right) - S_{lv} H_{lv} + G_{pi} \frac{B_{pin}}{2} \right) \theta_b + \\
- \left(\left(\frac{G_{b1l}}{2} + \frac{G_{b2l}}{3} \right) B_b + G_w \frac{s_w}{2} + S_{lv} H_{lv} + G_{pi} \left(B_b - \frac{B_{pin}}{2} \right) \right) \theta_{bl} \tag{11.54}$$

$$\lambda \langle p_w, v \rangle = \lambda \left[\int_{H_{elv}}^{H_p} \left(\frac{p_o}{H} h^2 i \right) dh + \int_{H_{elv}}^{H_p} \left(\frac{2p_o}{H} h^2 i \right) dh + \int_{H_p}^{H} \left(\frac{3p_o}{H} h i H_p \right) dh \right] \theta_p + \\
- \Delta V_B(\lambda) B_p \theta_p + \Delta V_A(\lambda) 0.5 B_p \theta_p + \lambda \left(\int_{H_{lv}}^{H_{elv}} \left(\frac{p_o}{H} h^2 i \right) dh + \int_{H_b}^{H_b + H_{pin}} \left(\frac{p_o}{H} h^2 B_{pin} \right) dh \right) \theta_b \\
+ \lambda \left(\int_{0}^{H_{elv}} \left(\frac{2p_o}{H} h^2 i \right) dh + \int_{H_b}^{H_b + H_{pin}} \left(\frac{2p_o}{H} h^2 s_b \right) dh \right) \theta_{bl} \tag{11.55}$$

in which the weights of the two parts of the left buttress, identified by the inclined crack, are equal to:

$$G_{b1l} = s_b \cdot (H_b - H_{elv}) \cdot B_b \cdot \gamma_b = 3.8 \times 10^3 \, \text{kN} \tag{11.56}$$

$$G_{b2l} = \frac{1}{2} s_b \cdot H_{elv} \cdot B_b \cdot \gamma_b = 1.9 \times 10^3 \, \text{kN} \tag{11.57}$$

The relationship between the rotation of the windward pier and of the windward buttress is:

$$\theta_{bl} = \theta_p \tag{11.58}$$

The collapse load multiplier obtained for this mechanism is:

$$\lambda_E = 1.71 \tag{11.59}$$

The corresponding collapse wind pressure acting on the windward side of the Cathedral is:

$$p_{collEw} = 3.42 \, \frac{\text{kN}}{\text{m}^2} \tag{11.60}$$

and the corresponding collapse wind pressure acting on the leeward side is:

$$p_{collEl} = 1.71 \, \frac{\text{kN}}{\text{m}^2} \tag{11.61}$$

11.4.3.5 The Local Failure Mechanism of the Pinnacle

The local failure of the pinnacle placed on the top of the buttress can occur with an upturn mechanism. Calculations show that the load multiplier is very high.

11.4.3.6 The Wind Speeds at the Failure

For the considered case study, the smallest critical multiplier is attained in the mechanism B:

$$\lambda_{cr} = \lambda_B = 0.969 \tag{11.62}$$

The corresponding windward pressure at the height H of the top of the trusses spanning the nave is then:

$$p_{cr,cl} = 2\lambda_{cr} p_o = p_{collBw} = 1.937 \, \frac{\text{kN}}{\text{m}^2} \tag{11.63}$$

A simple estimate of the corresponding critical wind velocity is:

$$V_{cr}(H = 57\,\text{m}) = \sqrt{\frac{2 \cdot 1.937 \cdot 9.81}{1.25}} = 55.675 \, \frac{\text{m}}{\text{s}} \tag{11.64}$$

This velocity is attained at the height of 57 m above terrain. The corresponding velocity at the standard height of 10 m can be evaluated taking into account that:

$$V(57\,\text{m}) = V(10\,\text{m}) \left(\frac{57}{10}\right)^{0.35} = V(10\,\text{m}) \cdot 1.84 \tag{11.65}$$

and it reaches a value equal to:

11.4 Transverse Wind Strength

$$V_{cr-}(10\,\text{m}) = \frac{55.675}{1.84} = 30.277\,\frac{\text{m}}{\text{s}} \approx 109\,\frac{\text{km}}{\text{h}} \quad (11.66)$$

Concluding, the transverse failure of the studied model, similar to the Cathedral of Notre-Dame in Amiens, is reached through a semi-global mechanism when the wind action reaches a velocity at the height of 10 m above terrain equal to 109 km/h.

11.4.4 Conclusion

The map of strong winds in Europe gives the reference mean wind velocities at a height of 10 m above the terrain (with a roughness length $z_o = 0.05$), having an annual exceedance probability equal to 2 %, which corresponds to a return period of 50 years. Different values of the reference wind velocities are suggested by various Codes. Among these, the value of 26 m/s is one of the strongest and most frequent. Thus, summing up all the obtained results, we can conclude that the winds that can be sustained by ancient gothic cathedrals correspond to the reference wind intensities considered by actual Codes in Europe.

11.5 The Failure at Beauvais at the 1284

11.5.1 Introductory Notes

In many respects, the *Cathédrale Saint-Pierre de Beauvais* may be considered the most daring achievement of Gothic architecture; it is made up solely of a transept

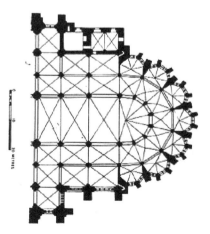

Fig. 11.26 Original plan of the choir and apse

Fig. 11.27 Plan with transept

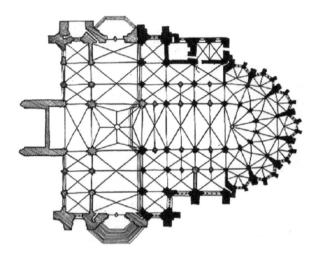

Fig. 11.28 Benouville's reconstruction

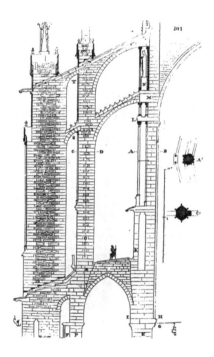

and choir apse with seven apse-chapels. The vaulting in the interior exceeds 48 m. in height, to make it the tallest cathedral in Europe (Fig. 11.26).

Construction of the choir and apse was begun in 1247 and finished in 1272. The work was interrupted in 1284 by the collapse of the choir vaulting. The work of reconstructing it spanned the following 50 years and included the addition of extra

11.5 The Failure at Beauvais at the 1284

Fig. 11.29 Cross section at the *"chevet"* (Viollet-le-Duc)

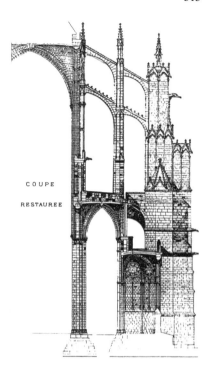

piers between the original ones of the choir, so that the bays were halved from about 9–4.5 m in breadth (Fig. 11.27a, b). It was in fact thought that the original pier spacing of the failed choir was too large. The choir was completely rebuilt by about 1337, but the work was interrupted for the next 150 years due to the Hundred Years War. It was not until 1500 that work on the transept was taken up again and brought to completion in 1548. In 1573 the collapse of the extremely high central tower halted the work once again. Various attempts were made to complete the cathedral, but by 1605 the decision was taken to consolidate the existing structure and abandon the enterprise, leaving it incomplete. Beauvais became what it is today, a choir and transept without a nave. A detailed history of the building has been provided by Branner (1962).

The 1284 failure was quite inexplicable: the cathedral had stood in good condition for 12 years and there are no historical accounts of earthquakes or wind storms occurring before the collapse. Figure 11.28 shows Benouville's (1891–1892) reconstruction of the state of Beauvais in the years 1272–1284 (Fig. 11.28). Figure 11.29 shows the details of the piers at the *chevet*, as illustrated by Viollet le Duc. The cause of the 1284 collapse has long been a matter of a great deal of speculation. Some *slow action* seems to be the most likely cause.

Two different conjectures on the identity of these slow-moving actions have been advanced: creep of the pier mortar and uneven foundation subsidence.

According to Viollet-le-Duc (1854–1868), the mortar creep triggered the transfer of loads from the masonry piers to the adjacent slender marble columns illustrated in Fig. 11.29. As a consequence, the columns buckled, producing rotation of the so-called *tas de charge* and consequent distortion of the adjacent flying buttresses with failure of the central vaults and piers.

Heyman (1995), instead, considered it more likely that uneven subsidence of piers foundations, due to soil consolidation, were responsible for the catastrophic collapse. Wolfe and Mark (1976) challenged this conjecture. Firstly, they noticed that there are no signs of any major differential subsidence in the existing building. Moreover, the cathedral was probably built on the site of a pre-existing building, near the walls of an old Roman precinct.

Determining for certain the cause of the fall of the original vaults of Beauvais cathedral is naturally a difficult, if not impossible, task, especially given that any and all documentary evidence on the failed construction has been long lost. Nevertheless, some insight into the issue can be gained from the following considerations. First of all, it should be noted that, while the distance between the centers of the main piers was 15 m—the same as many other cathedrals, such as Amiens, Chartres, etc.—the Beauvais piers were considerably higher. Moreover, the piers' cross sections were much smaller and the means by which they were stiffened, represented by the adjacent slender marble columns, seems very doubtful. It appears that no one ever performed a scientific analysis of the piers' buckling under the loads of the original quadripartite vaults, except perhaps a rather simplistic, conservative evaluation through application of Euler's elastic theory, as recently reported by Wolfe and Mark (1976). In this regard, however, in preceding sections we have seen the dramatic effects of load eccentricities on the buckling strength of masonry piers, whose behavior is very different from Eulerian columns.

Figures 11.28 and 11.29 show the complex geometry of the piers of the Beauvais cathedral, with all their offsets and misalignments, which produce eccentricities in the axial loads. In addition, we have also seen how mortar creep slowly increases

Fig. 11.30 The cross vault spanning the transept and the sliced web

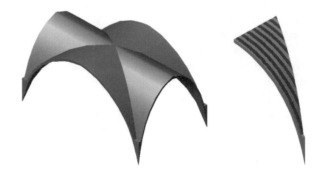

the destabilizing effects of eccentric axial loads. Accordingly, it seems natural to harbor strong suspicions that the piers' slenderness, combined with their geometrical irregularities and mortar creep, could have been responsible for the failure.

The next sections will delve more thoroughly into this hypothesis. However, to this end, it is first necessary to obtain a preliminary estimate of the order of magnitude of all the forces acting on the cathedral piers.

11.5.2 Thrust of the Cross Vault Spanning the Choir

With the aim of evaluating the various actions transmitted by the vault to the piers, the following provides a brief static analysis of both the cross vault spanning the choir and transept and the two adjacent flying buttresses (Fig. 11.30).

The cross vault rib is loaded by its weight and the vertical and horizontal forces conveyed by the sliced arches dividing the webs, each of which is assumed to have a thickness of 25 cm and made of strong bricks. The small inflexions of the vertical buttresses yield minimum thrust states in the sliced arches and cross ribs of the vault. The thrust has been evaluated by Magrì (2004–2005), who also took in account the presence of filling near the vault corners (Fig. 11.31).

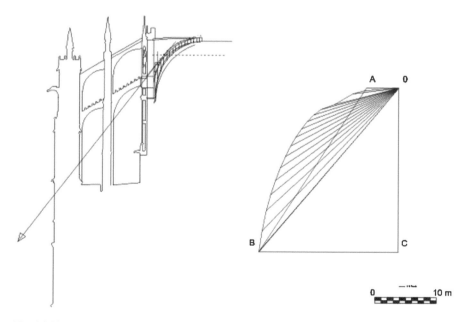

Fig. 11.31 Minimum thrust conveyed by diagonal cross ribs on a pier

Fig. 11.32 Lines of thrust of the lower and upper flying buttresses

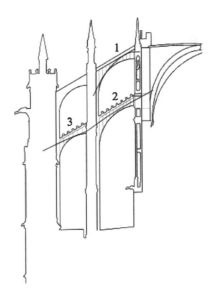

The thrust of the cross vault ribs is transmitted to the flying buttress across the *tas de charge,* the large marble block that joins the cross vault ribs and the lower flying buttress, as shown in Figs. 11.31 and 11.32.

According to calculations, the vertical and horizontal components V and H of force S transmitted by the vault to the pier are about 58 and 19 t, respectively (Fig. 11.22). The lower flying buttress intercepts this force S on the internal edge of the pier and transfers it through an intermediate pier to the large vertical buttress. The lines of thrust of the lower and upper flying buttresses have been traced. The thrusts of the higher and the lower fliers are respectively near the minimum and maximum thrust values. The lower flier effectively acts as a compressed prop.

11.5.3 Loads Acting on the Piers

The pier has a height of about 44.80 m, as measured from the foundation level to the extrados of the *tas de charge*. The pier is fixed here because of the presence of the flier, on the inner side, and the springing of the nave cross vault, on the other (Fig. 11.33).

Above the *tas de charge* the pier extends for about another 7 m to become part of the upper transverse walls. The total length of the pier is thus about 44.80 + 7.00 = 51.80 m, of which 3.80 m run from the foundation to the floor level.

The length of the pier under examination can be subdivided in two main parts (Fig. 11.33). The lower, 25 m-long section, extends from the foundation level to the extrados of the side vault, with its first 3.80 m built into the foundation; its circular cross section is about 1.60 m in diameter. The second part, of 18 m length, is

11.5 The Failure at Beauvais at the 1284

Fig. 11.33 Section of the Cathedral at the transept and two parts of the pier (lengths in m)

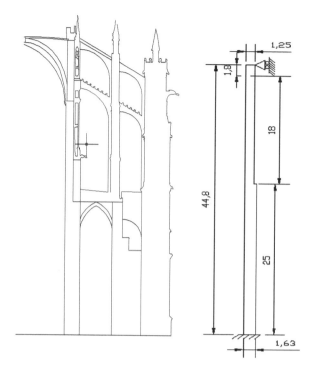

misaligned with respect to the first and reaches the extrados of the *tas de charge*. This section of the pier length, which is approximately circular in cross section with a diameter of 1.25 m, is strengthened by four adjacent small, slender marble columns with circular cross sections, each about 0.15 m in diameter. The center of these four columns is at a distance of about $0.625 + 0.35/2 = 0.80$ m from the axis of the adjacent pier's cross section (Fig. 11.24).

The lower flier takes the horizontal component of force S transmitted by the main vault, so only vertical loads act on the pier. In particular, these latter forces acting on the pier section located just at the vault springing are due to (Fig. 11.34):

(1) weight of the roof and the wooden trusses:
Considering an average weight of 500 kg/m² in plan, we have:
$W_{roof} = 9.00 \cdot 7.60 \cdot 0.5 = 34.2$ t, applied along the masonry pier axis, at a distance of 1.25 m² = 0.625 m from its internal edge (i.e. the edge towards the nave).

(2) weight of the transverse masonry walls:
Considering an average area of $8.50 \cdot 9.00$ m², a thickness of 0.5 m, a reduction factor of 0.65, to take into account the presence of the openings, a specific gravity of 2 tons/m³, we have, including the pier weight:

Fig. 11.34 Section of the upper pier with the four small columns (lengths in m)

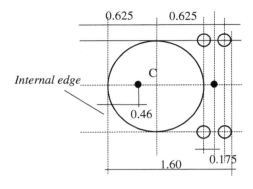

Fig. 11.35 All the forces transmitted across the *tas de charge*

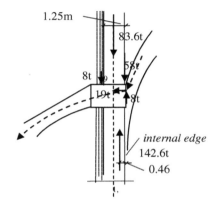

$$W_{sup.wall} = 9.00 \times 8.50 \times 0.5 \times 0.65 \times 2 = 49.4 \,\text{t}$$

This force is also applied along the pier axis, i.e., at a distance of 0.625 m from its internal edge.

(3) weight of the marble statue on the extrados of the *tas de charge*:
We have approximately: $W_{statue} = 4$ t, which force acts along the axis of the four small circular columns cross sections at a distance $e = 0.625 + 0.35/2 = 0.80$ m from the center of the pier cross section (Fig. 11.35).

(4) weight of the solid masonry block atop the marble columns, between the external edge of the pier and the upper flying buttress:

$$W = 0.80 \times 1.05 \times 1.80 \times 2.2 = 3.3 \text{ t}$$

(5) weight of the pinnacle:

$$W = \pi \cdot 1.0^2/(3 \cdot 4) \cdot 3.0 \cdot 2.2 = 1.7\,\text{t}.$$

Both of these last forces act along the axis of the four marble columns, situated at a distance of about $d = 0.625 + 0.35/2 = 0.80$ m from the center of the pier section.

(6) vertical actions conveyed by both the cross vault spanning the transept and the lower flying buttress.

In a minimum thrust state the vault conveys a vertical force of 58 t, while the vertical counterthrust of the lower flying buttress is 8 t. Consequently,

$$W_{vault,fl.buttr} = 58 - 8 = 50\,\text{t}.$$

This force acts along the *internal edge* of the upper pier and thus at a distance of —0.625 m from the center of the pier cross section.

The resultant summing up all the above forces:

$$W_{tot} = 83.6 + 4 + 3.3 + 1.7 + 50 = 142.6\,\text{t}.$$

By evaluating the moment of all these forces around the center of the pier section, we obtain the eccentricity of the resultant force:

$$83.6 \cdot 0 + 4 \cdot 0.80 + 3.3 \cdot 0.80 + 1.7 \cdot 0.80 - 50 \cdot 0.625 = 142.6\,\xi,$$

whence $\xi = -0.17$ m.

The distance of W_{tot} from the internal edge of the pier at the extrados of *the tas de charge* is $e = 0.625 - 0.17 = 0.46$ m

The center of the composite section made up of the sections of the pier and four columns is very near the center of the pier's section itself. The core of this composite section is about 15 cm in width. Consequently, the two external marble columns, placed on the outer side, would have to be *unloaded*. The eccentric axial load is for all intents and purposes sustained solely by the circular masonry cross section of the pier head.

Finally, at its springing the side vault spanning the aisle transmits to the pier the following vertical and horizontal actions:

$$V_2 = 21\,\text{t} \quad H_2 = 4.8\,\text{t}.$$

These forces are applied at the external edge of the pier section at its intersection with the side vault.

11.5.4 Creep Buckling of the Piers

A stability analysis of the pier has been conducted by applying the delayed modulus approach, according the results shown at Sect. 7.2.6.

The complex geometry of the Beauvais piers (Fig. 11.36) has been input into the program Athena in two-dimensional form (Di Carlo 2005–2006). The uppermost length of the pier, the marble stone of the *tas de charge*, exhibits the greatest strength, in both compression and tension.

Because of the eccentricity of the axial load, the small slender marble columns adjacent to the masonry pier have not been taken into account. The pier is loaded at its top section by an axial load P = 142 t, with eccentricity e = 0.46 m, as well as other loads applied along its length. The program automatically includes the weight of the various pier lengths.

All the loads are corrected via the parameter λ, which varies in the interval (0, 1), so that all the loads are effective when λ reaches unity. The horizontal displacement Δ of the pier is evaluated at the level of the side vault extrados, where the pier section changes dimensions.

The various axial load versus lateral displacement curves have been plotted by gradually increasing λ for different values of the delayed elastic modulus (Fig. 11.36).

Due to the long-term creep effects in old mortars, the delayed modulus can also be assumed to be four times lower than the initial value of E. Taking into account the order of magnitude of the initial pier masonry elastic modulus, whose value ranges from 10000 to 5000 Mpa, and the corresponding decay due to creep effects, the pier clearly results to be quite unstable.

It is in fact worth noting that when the modulus is near its initial value, the pier exhibits rather good axial strength and would fail only under loads much larger than the actual ones. For delayed modulus values that take into account the strong creep effect, on the other hand, the pier is prone to delayed failure under its ordinary loads. Figure 11.37 provide a representation of the different inflexions of the pier under the action of the full loads by gradually reducing the values of the delayed elastic masonry modulus (Di Carlo 2005–2006).

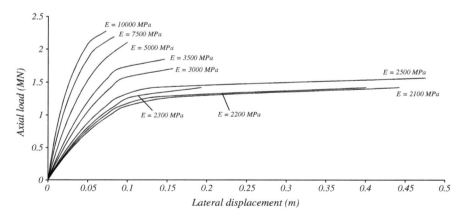

Fig. 11.36 Axial load—lateral displacement curves for different values of the delayed elastic modulus for the Beauvais pier

11.5 The Failure at Beauvais at the 1284

Fig. 11.37 Pier deformations corresponding to modulus values of 10000, 7500, 5000, 2100 MPa. Cracks distribution by gradually increasing load parameter λ in the case of E = 2100 MPa

11.5.5 Conclusions

The structural failure that struck the Beauvais Cathedral in 1284 can in all likelihood be attributed to the slenderness of its masonry piers, with their axis misalignments and eccentric loading, compounded by the inevitable effects of mortar creep. The loads on the lower length of the choir piers are highly eccentric due to the section change occurring at the extrados of the side vaults.

The pier is subject to lateral inflexion, which is exacerbated by the destabilizing effects of the axial loads due to their eccentricity and axis misalignments. The lower part of the pier bends and moves slowly towards the nave. Such displacement cannot be constrained by the side vault, which, as it bears a mechanism deformation and is in a minimum thrust state, follows the lateral inflexion of the pier. The upper part of the pier, fixed at the top by the presence of the fliers and the main cross vault, undergoes strong counter-flexing.

Despite all these deformations and the destabilizing effect of the axial loads, the pier equilibrium would turn out to be stable, if only its initial elastic no-tension response were considered. However, creep deformations of the medieval mortars has gradually aggravated both the pier inflexion and the destabilizing action of the axial loads. Thus, the pier slowly moved through ever more precarious equilibrium states up to the disastrous failure.

All told, these results allow for concluding that the 1284 collapse of the Beauvais Cathedral was likely due, in part, to delayed creep instability of the masonry piers.

References

Abruzzese, D., Como, M., & Lanni, G. (1990). *On the horizontal strength of the masonry Cathedrals*, Rapp. Sc. del Dipart. di Ingegneria Civile, n° 29, Univ. Di Roma Tor Vergata.

Benouville, L. (1891–1892). *Etude sur la Cathédrale de Beauvais, Encyclopédie d'Architecture*, S. 4,4, Paris, pp. 52–54, 60–62, 68–70.

Branner, R. (1962). *Le Maitre de la Cathédrale de Beauvais Art de France*, 2, Paris, pp. 77–92.

Cervenka, V.J. (2002). *Atena program documentation, User's manual for Atena 2D*. Prague.

Choisy, A. (1899). *Histoire del l'Architecture*. Paris.

Como, M. (2004). *Principi costruttivi e tecniche esecutive delle cattedrali gotiche*, in *Europa: Civiltà del costruire: dodici lezioni di cultura tecnologica dell'architettura*, a cura di G. Morabito, Gangemi editore, Roma.

Di Carlo, F. (2005–2006). *Verifica della stabilità dei pilastri non resistenti a trazione sotto carico di punta eccentrico: un'applicazione ai pilastri della cattedrale di Beauvais*, a.y. University of Rome Tor Vergata, Faculty of Engineering.

Frankl, P. (1962). *Gothic Architecture*. London, p. 101.

Heyman, J. (1971). Beauvais cathedral. *Transactions of the Newcomen Society, 40*, 15–36 (1967–1968) (London).

Heyman, J. (1995). *The Stone Skeleton*. Cambridge University Press.

Iannotti, A. (1995) *Approfondimenti sul Calcolo della spinta della volta a crociera della cattedrale di Beauvais*. Exercises in the Course in "Statics of masonry historic constructions (M. Como, Faculty of Engineering, Un. of Roma Tor Vergata, Rome (Italy).

Jessop, E. L., Shrive, N. G., & England, G. L. (1978). Elastic and Creep Properties of Masonry. In *Proceedings of the North American Masonry Conference*, The Masonry Society, pp. 12.1–12.17.

Magrì, G. (2004–2005). *Analisi delle volte a crociera della cattedrale di Beauvais*. Supervisor M. Como, a.y. University of Rome Tor Vergata, Faculty of Engineering.

Morabito, G. (2004). *Caratteri e tecniche del costruire nell'Europa del Medioevo*, in *Europa: Civiltà del costruire: dodici lezioni di cultura tecnologica dell'architettura*, a cura di G. Morabito, Gangemi editore, Roma.

Orsini, S. (2011–2012). *Resistenza al vento delle cattedrali*. Supervisors M. Como, S. Coccia, a.y. University of Rome Tor Vergata, Faculty of Engineering.

Pavone, A. (1994). *Calcolo della spinta della volta a crociera della cattedrale di Beauvais*. Exercises in the Course in "Statics of masonry historic constructions (M. Como, Faculty of Engineering, Un. of Roma Tor Vergata, Rome (Italy).

Sardoni, S. (2007–2008). *La resistenza della cattedrale gotica all'azione del vento*. Supervisor, M. Como, a.y. University of Rome Tor Vergata, Faculty of Engineering.

Shrive, N. G. & England, G. L. (1981). Elastic, creep and shrinkage behavior of masonry. *International Journal of Masonry Construction, 1*(3).

Viollet-le-Duc, E.E. (1854–1868) *Dictionnaire raisonné de l'architecture francaise du XIe au XVIe Siècle*, 10 vols. Paris.

Wolfe, M. I., & Mark, R. (1976). The collapse of the vaults of the Beauvais Cathedral in 1284. *Speculum, 51*(3), 462–476 (Medieval Academy of America).

Yokel, F. Y. (1971). Stability and load capacity of members with no tensile strength. *Journal of the Structural Division ASCE, 27*(ST7) Washington D.C.

Chapter 12
Masonry Buildings Under Seismic Actions

Abstract This last chapter deals with the study of the seismic behavior of historic masonry buildings. Five problems, particularly, are analyzed in the chapter:

- evaluation of the seismic response of the anelastic oscillator, representative of the masonry behavior
- evaluation of the horizontal strength and ductility of the masonry panel, archetype of the building structure
- definition of the seismic forces acting on a masonry building and analysis of their transmission across its structural components
- determination of the seismic out of plane strength of masonry walls
- determination of the seismic in-plane strength of masonry walls with openings.

All the results presented have been obtained in the framework of the Limit Analysis of masonry structures, according to the approach followed in the book. Numerical examples and comparisons with Code prescriptions are given.

12.1 Introduction

This chapter deals with the behavior of masonry buildings under horizontal loads, representing seismic forces, the most hazardous actions for such structures. Traditional masonry buildings were simply not built to offer any resistance to horizontal actions. The main function of their structures is instead to transmit vertical loads. Earthquakes cause movements of the soil beneath constructions, thereby imparting to it both horizontal and vertical accelerations. While the vertical component of seismic accelerations generally produce only moderate variations in the vertical loads to be borne by the structure, the horizontal accelerations bring about unprovided for horizontal forces. Consequently, traditional buildings experience earthquakes as exceptional events which they are ill-equipped to withstand. This is why most of the damage occurs in old historic centers, as well as why there is currently such a great demand to determine the most suitable means to reinforce them. The first such research in Italy was probably prompted by the earthquake that that ravaged its southern regions of Calabria and Sicily in 1783. Bourbon engineers surveyed sites to document the damage and prepare the earliest guidelines for repairing or rebuilding damaged structures. The importance of connections between the walls was grasped immediately and the earliest types of wall reinforcements were defined: inserting wooden frames into them. The first safety standards for building heights were moreover established. After the great earthquake in Messina (Sicily) in 1908, detailed standards were laid out for rebuilding the failed constructions.

In the early 20th century the field of Earthquake Engineering was aimed mostly at reinforced concrete and steel constructions and modern anti-seismic methods were developed for such structures in the USA and Japan, while research on the seismic strength of masonry buildings faded. Only after the quake of Skopje in 1976 was research in the field revived. The so-called POR method was introduced (Tomazevic 1978) when the collapse mechanism of masonry buildings was identified with the occurrence of shear failures of walls arising with diagonal cracking. Strength evaluations were based on tests results on small, two-story buildings, generally made of the rubble masonry so typical of the region. After the great earthquake that struck the Italian regions of Campania and Basilicata in 1980, research turned to analyzing the seismic behavior of masonry buildings with a larger number of stories and more complex plans. New approaches were developed, such as the Porflex method (Braga and Dolce 1982) and the method Vem (Fusier and Vignoli 1993),that added the so-called story-bands to POR analyzes and proposed a new failure model for walls under eccentric axial load. On a different front, in the wake of the Heyman model, Limit Analysis was applied to the collapse of masonry walls with openings (Como and Grimaldi 1985, etc.). Further developments followed with the introduction of the new concept of ductility and the introduction of new procedures, such as the SAM (Magenes and Fontana 1996), and the TREMURI method (Galasco et al. 2002). Although continuing research constantly furthers our knowledge of all the aspects involved in the seismic behavior of masonry structures, a universally accepted approach does not yet exist.

Currently, simplified criteria are used to represent building structures as loaded by dead loads and a gradually increasing distribution of horizontal forces, which are proportional to the mass of the building—the so-called *push-over* action. The *seismic strength* of the building is thus obtained by evaluating the intensity of these gradually increasing horizontal loads that produce failure of the building. The question is: for an assigned intensity level of a seismic action defined, for instance, by the maximum horizontal acceleration of the ground motion, what seismic strength is *required* of a building in order for it to survive the given action?

The answer to this question requires knowing the *ductility* of the structure, that is, its capacity to maintain its strength during the development of a failure mechanism, which, in brief, depends on the choice of the so-called strength reduction factor, also known as structure factor q. It is, in fact, impossible for any structure to sustain the actions of an intense earthquake without some structural damage. These issues are currently still the subject of debate, and the as yet partial answers furnished so far have come mainly from the study of the seismic behavior of reinforced concrete structures, which clearly cannot be transferred wholesale to masonry constructions.

The aim of this chapter is to discuss the main aspects of such issues in the context of historic masonry buildings while attempting to keep the analysis well-grounded in current theory. The basic assumption underlying such theory is once again the same considered in previous chapters of this book—the no-tension response of masonry material. The problems will thus be analyzed within the framework of the Limit Analysis of masonry structures.

After a preliminary analysis on the peculiar aspects of the seismic response of masonries, some fundamental aspects of the problem will be discussed: force transmission in building structures and evaluation of the involved strengths, with regards to the various possible failure mechanisms. A primary analysis will thus be developed to define the transmission chain that carries the forces from the building's mass, where they are generated, as far as the foundation. Some preliminary reinforcing systems can be conceived to guarantee full capacity of this transmission chain.

A masonry wall subjected to horizontal forces orthogonal to its plane is the most vulnerable element in the structural makeup of such buildings and, as a rule, requires suitable 'hook' reinforcements. The in-plane strength of masonry walls with openings represent, on the other hand, the actual strength of the building, provided that failure of the walls under actions orthogonal to their plane is avoided. These will be thoroughly investigated within the framework of the no-tension model of masonry materials.

12.2 The Masonry Panel Under Horizontal Forces: Strength and Ductility

12.2.1 Limit Strength in the Framework of the Rigid in Compression, no-Tension Model

A masonry panel is the simplest resistant model for representing the behavior of a masonry building. The essential features of such a panel's behavior are the same as the walls of a masonry building. The strengths of a masonry panel or a masonry building are thereby due essentially to their weight and geometry, both connected in the interplay between the pushing work of the horizontal forces and the resistant work required to raise their weights during a failure mechanism. The weight is thus of essence in such structures' resistance to lateral forces. Consequently, we assume that the compression stresses involved will prevail over masonry's tensile strength, as discussed in Chap. 1.

Fig. 12.1 The masonry panel

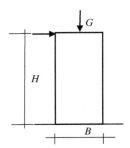

Fig. 12.2 Overturning mechanisms of the panel

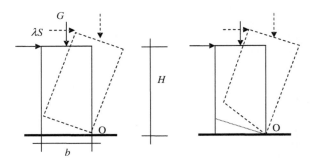

A panel, with geometry shown in Fig. 12.1, of height H, width B and unit thickness s, is loaded at its head section by the central weight G and the horizontal thrust λS, where λ is the load factor.

In the framework of the rigid in compression, no-tension masonry model, Fig. 12.2 shows the failure mechanism of the panel that, cracking at its base, rotates around its toe and opposes, by its raising, the overturning thrust.

As a rule, cracks would radiate upward, delimiting an inert masonry zone of roughly triangular shape. These mechanisms are not too relevant if the dominant vertical force is represented by the weight G applied at the head of the panel.

We may begin evaluating the collapse multiplier by applying the kinematic theorem. Thus, with reference to Fig. 12.3, the mechanism $\mathbf{u}^+$ is represented by the displacement field induced by rotation of the panel around the toe O. The resistant

Fig. 12.3 The pressure line of the panel at the collapse

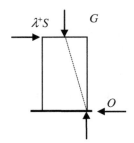

work $\langle \mathbf{g}, \mathbf{u}^+ \rangle$ is produced by the lifting of the central weight G located at the head of the panel. Thus

$$\langle \mathbf{g}, \mathbf{u}^+ \rangle = -G\theta B/2. \tag{12.1}$$

The pushing work is then

$$\lambda^+ \langle \mathbf{q}, \mathbf{u}^+ \rangle = \lambda^+ SH\theta \tag{12.2}$$

and an *upper* bound of the collapse thrust is

$$\lambda^+ S = G\frac{B}{2H} \tag{12.3}$$

The search for admissible stresses in equilibrium with the loads via the static theorem enables determining a lower bound $\lambda^- S$ of the collapse multiplier.

This solution reflects the actual resisting lower zone of the panel, which reduces to a *wedge*. In fact, at the limit equilibrium state, the panel is loaded at its head by a force R, the resultant of the weight G and the thrust, $\lambda^- S$, passing through the panel toe O and inclined by angle α with respect to the vertical, so that

$$tg\alpha = \frac{B}{2H}. \tag{12.4}$$

The horizontal and vertical components of R are thus $\lambda^- S$ and G. Operating in terms of stress *resultants*, the kinematically admissible thrust (12.3) will also be statically admissible.

The search for statically admissible states is more complicated in terms of stress components. In this case, we cannot refer to the usual stress fields corresponding to beam sections loaded by an eccentric axial force and shear. Admissible compression stresses in equilibrium with the loads can, on the contrary, be obtained using the stress distributions occurring inside a wedge loaded at its vertex and along its axis, as shown Fig. 12.4 (Como and Grimaldi 1983, 1985).

The resultant R of the weight G and the limit horizontal force $GB/2H$ passes along the wedge axis *a-a*. The stress state in the portion ABO of the panel, of unit thickness, is described by the radial stress field

$$\tau_{r\vartheta} = \sigma_\vartheta = 0 \qquad \sigma_r = -kR\frac{\cos \vartheta}{r} \tag{12.5}$$

by assuming negative the compressions (Timoshenko 1955). The constant k in the expression of the radial stress σ_r is obtained imposing equilibrium with the force R applied at the toe O.

Now, by evaluating along any cylindrical surface, the resultant of the radial compressions σ_r along the direction of the axis of R, we get:

Fig. 12.4 Statically admissible equilibrium states in a panel wedge

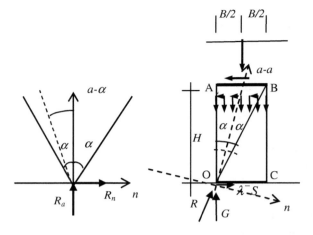

$$-2\int_0^\alpha \frac{kR\cos^2\vartheta}{r} r d\vartheta = -kR(\alpha + 1/2 \sin 2\alpha) = -R, \qquad (12.6)$$

whence

$$k = \frac{1}{r(\alpha + 1/2 \sin 2\alpha)}. \qquad (12.6')$$

The *radial compression state* acting in the portion ABO is thus represented by the stress components

$$\sigma_{r\vartheta} = \sigma_\vartheta = 0 \quad \sigma_r = -R\frac{\cos\vartheta}{r(\alpha + 1/2 \sin 2\alpha)}. \qquad (12.7)$$

This compression stress state is in equilibrium with the external loads, represented by the two forces R and $-R$, inclined by α from the vertical and respectively applied at the toe O and head of the panel. The stress field (12.7) is thus *statically admissible* and the thrust (12.3), both statically and kinematically admissible, represents the *collapse load* S_o of the panel, i.e. its lateral strength, where

$$S_o = \lambda_o S = G\frac{B}{2H}. \qquad (12.8)$$

These results can be generalized also to take into account the presence of an eccentricity e of the axial load at the head of the panel, We immediately have

$$S_o = G\frac{B}{2H}(1 - \frac{2e}{B}). \qquad (12.8')$$

12.2 The Masonry Panel Under Horizontal ...

Safety checks of the panel under the action of a horizontal force S thus requires

$$S \leq S_o. \tag{12.9}$$

12.2.1.1 Other Failure Mechanisms

Other different failure mechanisms are now taken into account: the so called partial mechanisms that consider the overturning of only some parts of the panel. In the first scheme of Fig. 12.5 a triangular band of the panel, having the same height H of the panel, fails for overturning; in the second scheme a triangular part of the panel, having height smaller than H, fails.

The head of the panel is loaded by vertical compressions g and horizontal shearing actions τ. With reference to Fig. 12.5, the limit equilibrium conditions of the failed panel band thus are

$$gx \cdot \frac{x}{2} = \tau_o x \cdot H \qquad gx \cdot \frac{x}{2} = \tau'_o x \cdot y \tag{12.10}$$

where τ_o and τ'_o are the limit shear in the two different considered cases.

In the first case we have

$$\tau_o = g \cdot \frac{x}{2H} \tag{12.11}$$

and in the second

$$\tau'_o = g \cdot \frac{x}{2y} \tag{12.11'}$$

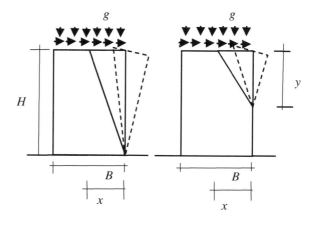

Fig. 12.5 partial mechanisms of the panel of triangular fragments of various height

Hence, with $y < H$ we have $\tau_o < \tau'_o$ and the partial failure of the panel, if can occur, engages the triangular half, having the same height of the panel. The whole width of the head of the panel is engaged,

Let us consider now the panel failure involving fragments of the same height but with different widths x (Fig. 9.6). As above we get

$$gB \cdot \frac{B}{2} = \tau_o B \cdot H \qquad gx \cdot \frac{x}{2} = \tau'_o x \cdot H \qquad (12.12)$$

and

$$\tau_o = g \cdot \frac{B}{2H} \qquad \tau'_o = g \cdot \frac{x}{2H} \qquad (12.12')$$

When $x < B$ we get $\tau_o > \tau'_o$. In this case tha panel fails splitting up into fragments.

Actually, panels want represent the piers of masonry walls (Fig. 12.6). These pierIn s are, as a rule connected at floor levels by steel ties, ring beams or other connection systems. The partial failure corresponding to the second scheme of Fig. 2.9 thus can be excluded.

At the same time panel that are tested in laboratories have their head connected by testing equipment and also in these cases partial splittings of the panel head cannot occur (Fig. 12.7).

The limit thrust given by (12.8) or (12.8') is based on the assumption of infinite compression strength, as per the no-tension masonry model. We will evaluate now the horizontal strength of the panel taking into account the effects of finite masonry compression strength. As discussed in Chap. 1, the thrust under which the stresses reach the masonry tensile strength along the diagonal panel section is sometimes deemed to represent the panel's lateral strength. Note that diagonal cracking occurs before collapse of the panel by overturning. Another interpretation of the panel

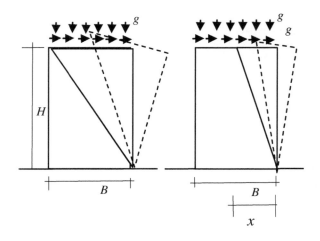

Fig. 12.6 Partial failurse of panel fragments of the same height but different widths

Fig. 12.7 Typical failure side mechanisms of a masonry wall of a building

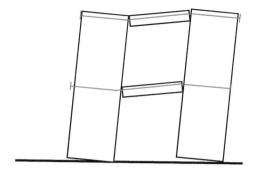

failure is frequently given within the framework of the Coulomb criterion (Contro and Nova 1982; Benedetti and Casella 1980; Binda 1983).

Note that the essential features of the lateral strength of a panel are very different from those corresponding to reaching the masonry tensile strength along the diagonal or to shear failure with internal friction. Indeed, even assuming zero masonry tensile strength, the panel possess lateral strength, which for the most part depends on the panel geometry alone. However, we can express the limit thrust (12.8′) by means of a fictitious limit shear stress ratio, τ_o/σ_m, as

$$\frac{\tau_o}{\sigma_m} = \frac{B}{2H}(1 - \frac{2e}{B}) \qquad (12.13)$$

where σ_m is the average compression stress acting on the entire panel section. Equation (12.10) indicates the geometrical and mechanical factors characterizing the panel lateral strength. Comparisons of the results predicted by Eq. (12.13) with numerous experimental results, including those of Yokel and Fattal (1976), and Murthy and Hendry (1966), have shown quite good agreement.

12.2.2 Panel Side Strength via Plastic Analysis

The research of the side strength of the panel can be also conducted in the framework of the theory of Plasticity. The limit staste of the panel is shown in Fig. 12.8. The panel has height H and width B and is composed by regular masonry, with bricks and horizontal mortar beds.

The resultant of the vertical dead load G and of the limit thrust S_o acts along the line CK joining the centre C of the head section of the panel with the point K, the centre of the segment BD, orthogonal to CK, along which the limit stresses $\sigma_{o<}$ are distributed. The limit compression stress acting at the toe of the panel is, in fact, the inclined compression strength $\sigma_{o<}$, analyzed at Sect. 1.11, because the mortar beds are horizontally directed. The rectangular triangle OBD is the compressed corner of

Fig. 12.8 Masonry plastic state at the toe of the panel

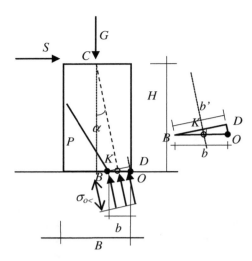

the panel. The horizontal segment OB, aligned along the panel base, is inclined of the angle α with BD and has the right end concurrent with the toe O.

The rotational equilibrium of the panel around the centre K gives

$$S_o = \frac{GB}{2H}\left(1 - \frac{b}{B}\right) \qquad (12.14)$$

On the other hand, taking into account of the equilibrium of the panel along the vertical direction we have

$$(\sigma_{o<} b's)\cos\alpha = G \qquad (12.15)$$

Likewise, considering the vertical equilkibrium condition at the panel head

$$\sigma_m Bs = G \qquad (12.15')$$

where σ_m is the uniform vertical compression at the panel head. Equality between (12.30) and (12.31) gives

$$(\sigma_{o<} b's)\cos\alpha = \sigma_m Bs \qquad (12.16)$$

Hence

$$\frac{b'}{B} = \frac{\sigma_m}{\sigma_{o<}\cos\alpha} \qquad (12.16')$$

But

12.2 The Masonry Panel Under Horizontal ...

$$b' = b \cdot \cos \alpha \tag{12.17}$$

and

$$\frac{b}{B} = \frac{\sigma_m}{\sigma_{o<} \cos^2 \alpha} \tag{12.17'}$$

Substituting (12.17') into (12.14) gives

$$S_o = \frac{GB}{2H}\left(1 - \frac{\sigma_m}{\sigma_{o<} \cos^2 \alpha}\right) \tag{12.18}$$

We have also

$$\frac{1}{\cos^2 \alpha} = 1 + tg^2 \alpha = 1 + \left(\frac{S_o}{G}\right)^2 \tag{12.19}$$

Substituting (12.37) into (12.36) gives the ratio S_o/G. In fact, with the positions

$$\Sigma_o = \frac{S_o}{G} \qquad \beta = \frac{B}{2H} \qquad \Phi = \frac{\sigma_m}{\sigma_{o<}} \tag{12.20}$$

we get

$$\Sigma_o = \frac{\sqrt{1 + 4\beta^2 \Phi(1 - \Phi)} - 1}{2\beta \Phi} \tag{12.21}$$

that, neglecting very small quantities, gives

$$\Sigma_o = \beta(1 - \Phi) \tag{12.22}$$

or

$$S_o = G\frac{B}{2H}\left(1 - \frac{\sigma_m}{\sigma_{o<}}\right) \tag{12.22'}$$

Equation (12.22') gives the failure thrust of the panel taking into account the reduced masonry strength at the corner of the panel due to the inclined compressions with the joints mortar. In this last aspect lies the effect of the shear on the limit strength of the panel. When the ratio

$$\sigma_m/\sigma_{o<} \qquad (12.23)$$

is negligible with respect to the unity, i.e. when

$$s_m/s_{o<} \rightarrow 0,$$

Equation (12.22′) finds again the previous limit thrust assessed acconding to the rigid compression assumption

$$S_o = G\frac{B}{2H}$$

On the contrary, when the ratio (12.23) cannot be neglected, in the assessment of the limit thrust we have to take into account that the limit strength depends on the ratio S_o/G. The limit thrust S_o can be thus obtained by using a trial and error procedure. In this case very useful can be the dependance (12.59) of $\sigma_{o<}$ as an appropriate function on the inclination angle α, (see Sect. 1.10).

12.2.3 Panel Ductility in the Framework of the Elastic-Plastic Model

Once the thrust S has reached its limit value S_o, the panel maintains its strength for a certain extent while plastic strains develop at its toe, as far as the crushing of the masonry doesn't occur (Fig. 12.9).

Thus the panel can rotate for a certain angle while the pushing force S_o keeps itself approximately constant. The question now arises: how far the panel is able to maintain its lateral limit strength S_0? This capacity defines the panel's *ductility*. The plastic ductility ratio can be defined as

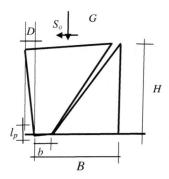

Fig. 12.9 The panel near the failure

12.2 The Masonry Panel Under Horizontal ...

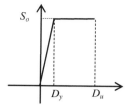

Fig. 12.10 The simplified elasto-plastic S—D diagram

$$\mu_p = \frac{D_u}{D_y} \qquad (12.24)$$

where D_u and D_y are the horizontal displacements of the panel head respectively at the failure and at the beginning of the plastic state. A first evaluation of the panel ductility can be performed within the framework of the elastic-plastic model (Fig. 12.10).

The plastic strains *localize* in a narrow band of length l_p along the compressed edge of the panel base. The order of magnitude of l_p can be assumed to be no larger than $B/4$, and the value of the ultimate strain ε_u along this band reaches about the value of 0.25 % (Fig. 12.11). The maximum plastic vertical shortening Δ_p of the toe is due to the plastic strains that occur, on average, along a suitable vertical length l_p of the toe. Thus we have

$$\Delta_p = \varepsilon_p l_p. \qquad (12.25)$$

Still referring to Fig. 12.11, the plastic rotation ϕ_p corresponding to the shortening Δ_p is given by

$$\phi_p = \frac{\Delta_p}{b} = \varepsilon_p \frac{l_p}{B} \frac{\sigma_o}{\sigma_m}, \qquad (12.26)$$

Fig. 12.11 The plastic strains occurring at the panel toe

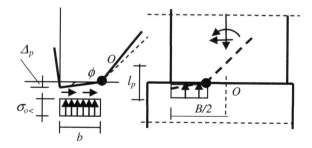

so that the limit rotation ϕ_u is

$$\phi_u = \varepsilon_u^* \frac{l_p}{B} \frac{\sigma_{o<}}{\sigma_m}. \qquad (12.26')$$

Similarly, the rotation ϕ_y at the onset of plastic strains is

$$\phi_y = \varepsilon_y^* \frac{l_p}{B} \frac{\sigma_{o<}}{\sigma_m} \qquad (12.26')$$

The head horizontal displacements D_u and D_y are respectively given by

$$D_u = H\phi_u \qquad D_y = H\phi_y \qquad (12.27)$$

and, consequently, the plastic ductility factor is

$$\mu_p = \frac{\varepsilon_u}{\varepsilon_y} \qquad (12.27')$$

According to the previous valuations of the ultimate and yield strains ε_u and ε_y (see Sect. 1.10), we can assume respectively $\varepsilon_u = 0.25\ \%$ and $\varepsilon_y = 0.1\ \%$ so that the plastic ductility ratio takes the value of 2.5. This valuation has been developed making reference to the case of regular masonry. In case of masonries of worse quality the plastic ductility factor takes a lower value. On the contrary, in case of solid masonry the value of $\mu_p = 2.5$ seems to be too small. In this case at the toe only local failures occur that round off the panel corner. A *rounded corner* enables the panel to continue rotating and preserves its capacity to oppose the thrust maintaining more or less the same lateral strength for a wider interval of rotations ϕ, as it will be shown in the next section.

12.2.4 Geometrical Ductility in the Framework of the no-Tension Panel Model

Once the thrust has reached the failure load, the panel, with rounded corner, passes through a sequence of rotated equilibrium configurations. As the rotation increases, the arm of the thrusting action rises while that of the resisting action falls (Fig. 12.12). Equilibrium of the panel thus requires reductions in the thrust as the panel rotation increases. Figure 12.13 shows a plot of the equilibrium thrust S versus the horizontal head displacement Δ of the panel for two different panel geometries: the decline of the thrust in the first is clearly greater than in the second.

As mentioned, the capacity of the panel to maintain its lateral strength during increasing rotation defines its ductility. As it depends solely on the panel geometry, it is therefore defined as *geometrical ductility*.

12.2 The Masonry Panel Under Horizontal …

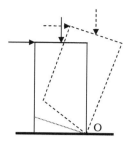

Fig. 12.12 Equilibrium of the panel in its rotated configurations

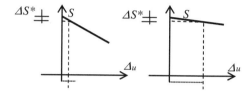

Fig. 12.13 Thrust—horizontal displacement Δ for various geometries of panels

Figure 12.14 shows the overturned configuration of the panel with the forces G and S at their displaced positions. The weight G is applied to the middle of the top section and the thrust S pushes at the left corner of this section. (We assume that special testing equipment is used to maintain these positions). The coordinates of the point B of application of force S at the initial vertical configuration of the panel, with respect to the reference system Oxy, are

$$x_B = -B \quad y_B = H \tag{12.28}$$

Similarly, the coordinates of the point of application of the weight G are

$$x_G = -B/2 \quad y_G = H. \tag{12.29}$$

Fig. 12.14 Equilibrium configuration of the panel

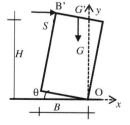

Let B' and G' indicate the position of the same point in the rotated configuration of the panel: they have coordinates

$$x_{B'} = -B\cos\theta + H\sin\theta \quad y_{B'} = B\sin\theta + H\cos\theta \quad (12.29')$$

$$x_{G'} = -B/2\cos\theta + H\sin\theta \quad y_{G'} = B/2\sin\theta + H\cos\theta. \quad (12.29')$$

The horizontal displacement Δ of point B, along which thrust S works, is

$$\Delta = x_{B'} - x_B = H\sin\theta + B(1 - \cos\theta), \quad (12.30)$$

while the vertical displacement of point B, along which the weight G works, is

$$V = y_{G'} - y_G = B/2\sin\theta + H\cos\theta - H = B/2\sin\theta + H(\cos\theta - 1). \quad (12.31)$$

The equilibrium configuration at the rotated position of the panel thus yields

$$S = G\frac{B/2\sin\theta + H(\cos\theta - 1)}{H\sin\theta + B(1 - \cos\theta)}, \quad (12.32)$$

which, with the position

$$\beta = \frac{B}{H}, \quad (12.33)$$

becomes

$$S = G\frac{\beta\sin\theta + \frac{2}{\beta}(\cos\theta - 1)}{2\sin\theta + \beta(1 - \cos\theta)}. \quad (12.33')$$

For $\theta \to 0$, Eq. (12.33') goes to $G\beta/2$, thereby once again yielding Eq. (12.8). Figure 12.15 shows the variation of the functions $S = S(\theta)$ and $S = S(\Delta)$ with varying θ and Δ, for the case of $\beta = 0.5$. From these diagrams it can be seen that the thrust falls by no more than 7 % for Δ/H ratios of up to about 5 %.

In brief, common panels, with geometries defined by a ratio $\beta = 0.5$, can maintain their lateral strength during increasing rotation, as long as the lateral displacement of their head sections is no greater than 7–8 % of the panel height.

Evaluation of the derivative $dS/d\Delta$ can be useful to obtain more information about panel ductility. We have

$$\frac{dS}{d\Delta} = \frac{dS}{d\theta}\frac{d\theta}{d\Delta}. \quad (12.34)$$

Thus, for small values of θ,

$$\frac{d(S/S_o)}{d(\Delta/B)} \approx -(1 + \frac{\beta^2}{2}), \quad (12.34')$$

12.2 The Masonry Panel Under Horizontal ...

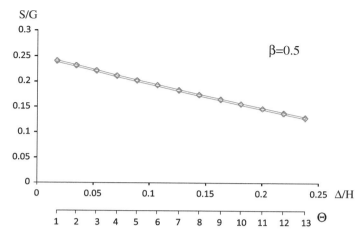

Fig. 12.15 Functions $S = S(\theta)$ and $S = S(\Delta)$ for $\beta = 0.5$

and therefore, the lower the ratio β (i.e., the ratio width/height of panel), the greater the panel ductility (i.e., the less is the decay $d(S/S_o)$ of the thrust ratio S/S_o versus Δ/B).

12.2.5 The Problem of the Comparison with Test Results

Many experimental tests have been conducted to obtain information about the lateral strength of masonry panels. It should be borne in mind, however, that the aim of such tests is to arrive at a good approximation of the actual behavior of the masonry wall of buildings.

The problem, in brief, is that the order of magnitude of the compression stresses at the base of masonry walls or piers is far greater than the adhesion strength between mortar and bricks or stones, or the mortar tensile strength itself. Such conditions have to be fulfilled by the test, otherwise the test results will represent only the strength of the tested panel. Only within this framework can the no-tension assumption or any panel strength evaluations according to (12.8), (12.15) and (12.21) be considered satisfactory. Figure 1.34 describes this type of panel failure.

We can define a proper factor able to establish whether or not the geometry and loading conditions of the panel satisfy these conditions. As previously highlighted, the no-tension assumption signifies that the compression stresses binding stones together will be much greater than the masonry tensile strength. It is thus important that the following condition hold for a test panel:

$$\chi = \frac{N}{\sigma_{ct} A} \gg 1, \qquad (12.35)$$

where:
N is the axial load applied at the panel head;
σ_{ct} the adhesion strength between mortar and bricks;
A the area of the panel base section.

Parameter χ is defined as the *similarity factor* with actual masonry walls. Condition (12.35) defines the category of test panels able to accurately represent the behavior of the walls or piers of masonry buildings. For example, any test panels for which

$$N = 50\,\text{t}, \ \sigma_{ct} = 1\,\text{kg/cm}^2, \ A = 2\,\text{m} \times 0.5 = 1\,\text{m}^2$$

or

$$N = 100\,\text{t}, \ \sigma_{ct} = 0.5\,\text{kg/cm}^2, \ A = 4\,\text{m} \times 0.5 = 2\,\text{m}^2,$$

and with adequate heights, can effectively represent the actual conditions of the masonry walls of a building. In the first case the similarity factor is $\chi = 5$ and in the second $\chi = 10$. On the contrary, for a panel with N = 200 kg, σ_{ct} = 10 kg/cm^2; A = 100 × 10 = 1000 cm^2, we have $\chi = 0.02$, and such panels could not thus be deemed representative of the behavior of masonry walls of buildings.

12.2.6 Behavior of the Panel Under Alternating Thrust Actions

It should be noted that for well-consolidated masonry, as in the case of brick masonry, the high compressions tighten up the bricks near the toes, so that a rounded corner will be present there. Under alternating lateral deformations, the panel will thus exhibit the S—Δ diagram shown in Fig. 12.16. Following this deformation, sketch (12.2) shows a slight fall in lateral strength during increasing panel rotation, while the sketch (12.21), to the contrary, shows the reverse response of the panel during back rotation.

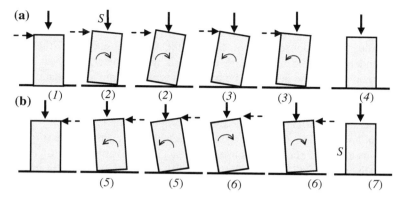

Fig. 12.16 Energy exchanges occurring during alternating motion of the panel

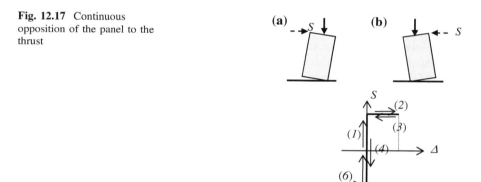

Fig. 12.17 Continuous opposition of the panel to the thrust

12.3 Some Recalls of Earthquake Engineering: The Elastic and the Elasto-Plastic Simple Oscillator

When a structure is subjected to ground motion in an earthquake and behaves elastically, the maximum response acceleration will depend on the structure's natural period of vibration and the magnitude of the damping present. Dynamic analyses of structures responding elastically to typical earthquakes records indicate the order of response acceleration that the structures may experience.

Very useful in Earthquake Engineering is thus the construction of elastic response spectra derived by dynamic analyses of a large number of single-degree-of-freedom oscillators to the specified earthquake motion. The maximum acceleration $S_e(T)$ is plotted as a function of the natural period T of vibration and the magnitude of damping, expressed as a percentage of the critical viscous damping. This diagram represents the elastic response spectrum corresponding to a given seismic event.

The value of response spectra lies in their condensation of the complex time-dependent dynamic response to a single key parameter, the peak response. This information can the generally be treated in terms of equivalent static response, simplifying design calculations. Response spectra for a defined level of strong motion shaking are commonly used to define peak structural response in terms of peak acceleration (PGA), directly usable in computing inertia forces.

The Italian Building Code furnishes the elastic spectrum $S_e(T)$ (Fig. 12.18), which represents the elastic response under a given seismic input of the elastic oscillator. Spectrum $S_e(T)$ defines the maximum acceleration occurring in the mass of an oscillator, with fundamental period T, whose base is shaken by the assigned seismic input, represented by a given accelerogram. In brief, the elastic oscillator, of stiffness k and mass m, that is, with period

$$T = 2\pi(m/k)^{1/2} \qquad (12.36)$$

is the ideal model representing the behavior of elastic structures. The maximum horizontal force F_e that the elastic oscillator will absorb is thus $mS_e(T)$. The corresponding maximum horizontal displacement is D_e (Fig. 12.19), whence we have.

$$F_e = kD_e. \qquad (12.37)$$

It is well known that it is generally uneconomic to design structures to respond to design-level earthquake in the elastic range. In regions of high seismicity elastic response may imply, in fact, lateral accelerations as high as 1.0 g.

If the strength of the lateral resisting structural system of the building has a level less than that corresponding to the acting earthquake, inelastic deformations result, involving, for instance, in reinforced concrete structures, yield of reinforcements and possible crushing of concrete.

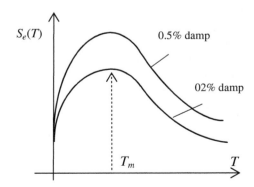

Fig. 12.18 The elastic response spectrum

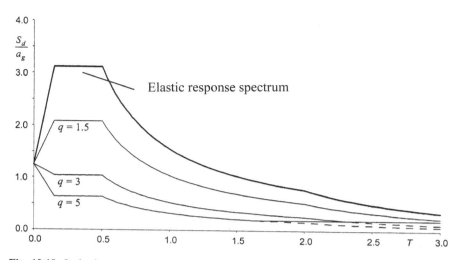

Fig. 12.19 Inelastic response spectra correspondig to various q (Da Ghersi e Lenza 2014)

Provided that the strength does not degrade as a result of inelastic action, acceptable response can be obtained. Displacements and damage must, however be controlled at acceptable levels.

In Earthquake Engineering it is well known that the same seismic input can be sustained by an *elasto-plastic oscillator* with the same period as the elastic oscillator, but having the horizontal plastic limit strength F_y, R times lower than $F_e = mS_a(T)$, *provided* that the oscillator is able to sustain the plastic horizontal displacement D_u, μ times larger than the limit elastic displacement:

$$D_y = F_y/k. \qquad (12.38)$$

The parameter

$$q = R = F_e/F_y \qquad (12.39)$$

is the *strength reduction factor R*, frequently called also the *structure factor q*. The parameter

$$\mu = D_u/D_y \qquad (12.40)$$

is the *ductility factor* of the oscillator: it controls the strength reduction factor R or the structure factor q. Knowing the function

$$q = q(\mu, T) \qquad (12.41)$$

is thus crucial to defining the design seismic strength of a building. Systematic numerical research studies on the behavior of elastic–plastic oscillators have defined function (12.41).

A fundamental reference parameter is the period T_m, the period where the elastic response attains its maximum (Fig. 12.18). According to the elasto-plasic response (Newmark and Hall 1982), it has been established that:

– For structures with period $T > T_m$, the maximum displacement D_u attainable by the inelastic oscillator is near the maximum displacement of the elastic system having the same initial period T as the inelastic system. Thus, in this case we have

$$D_u = D_e, \qquad (12.42)$$

whence we get

$$q = R = \frac{F_e}{F_y} = \frac{D_e}{D_y} = \frac{D_u}{D_y} = \mu \qquad (12.43)$$

– For structures with period $T < T_m$, the criterion of energy balance holds and we thus have

$$\frac{1}{2}F_e D_e = \frac{1}{2}F_y D_y + F_y(D_u - D_y) \qquad (12.44)$$

and

$$q = \sqrt{2\mu - 1}. \qquad (12.45)$$

– for very small periods, (T < 0.1 s)

$$q = 1. \qquad (12.46)$$

Based on these criteria, Italian Building Codes furnish the values of the corresponding strength reduction factor q (or the structure factor) for the various structural types of reinforced concrete structures, which exhibit different ductility factors μ.

Figure 12.19 shows various inelastic spectra corresponding to different values of the structure coefficient q of the Italian seismic Code (NTC2009). The diasgram corresponding to $q = 1$ is the elastic spectrum.

Masonry behaves very differently from reinforced concrete materials. and it is thus useful to search the corresponding values of the reduction factor q taking into account the peculiar behaviour of masonry, as it will be pursued in the next Section.

12.3.1 The Elastic Bilinear Oscillator Representative of the Masonry Behavior

Generally, standards for masonry buildings are expressed in terms of values of the the structure factor q, which originally refer to reinforced concrete structures (Fig. 12.20).

However, as seen in the foregoing, the behavior of the masonry panel under lateral oscillations is quite different from that of an elastic-plastic oscillator. The lack of dissipation would suggest that the relation between ductility factor μ and strength reduction factor q may differ significantly from relations (12.43) and (12.45).

The behavior of the masonry panel under alternating thrust actions has revealed a substantial bilinear elastic response.

It is thus reasonable to assume the *bilinear elastic oscillator* shown in Fig. 12.21 for masonry constructions in place of the classic elastic–plastic oscillator.

The oscillator is excited at its base by compatible accelerograms representing assigned seismic inputs. For the sake of simplicity, the assumed bilinear elastic oscillator does not take into account any deterioration in strength. Only very small lateral oscillations of the panel are thus considered. A valuable numerical calculation has been expressly developed in order to gather more information about the appropriate strength reduction factors to assume when dealing with masonry structures (Calderoni et al. 1991; Coccia and Como 2009).

In the scheme at the right of Fig. 12.22 the reference horizontal displacements D_y and D_u are also indicated required to link the behavior of the masonry panel to those of the elastic and elastic–plastic oscillators. Following the indications given in Fig. 12.14, the well-known equation of motion of such oscillators is expressed as

$$F(t) - k(D) \cdot D - c \cdot \dot{D} = m \cdot \ddot{D}. \tag{12.47}$$

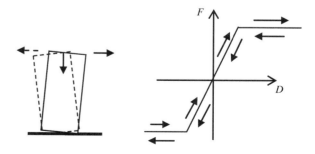

Fig. 12.20 The bilinear elastic oscillator typical of masonry systems

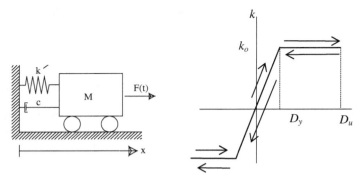

Fig. 12.21 Elastic, elastic-plastic and elastic-bilinear oscillators

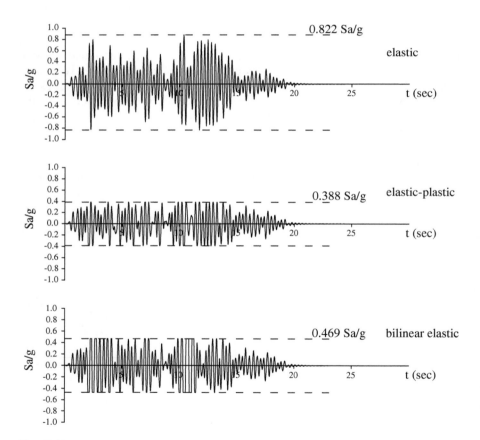

Fig. 12.22 Mass accelerations of elastic, elastic-plastic and bilinear elastic oscillators excited at their bases by an assigned seismic motion (Coccia and Como 2009)

12.3.2 The Structure Coefficient Q for Masonry Buildings

Figure 12.22 shows the results of the numerical integration of Eq. (12.47), for both the elastic-plastic and the bilinear elastic oscillator excited at their bases by the seismic motion corresponding to the spectral accelerogram shown in the upper diagram of the figure. The three diagrams show the variation over time of the mass acceleration of the elastic, elastic–plastic and the bilinear elastic oscillators.

Figure 12.23 shows the concluding results obtained for the bilinear elastic oscillator. Once again in this case, the peak period T_m of the elastic oscillator governs the oscillator response depending on its ductility.

For periods, $T > T_m$, of the bilinear oscillator, the strength reduction factor R depends only on the ductility coefficient μ defined by (12.40): it is substantially independent of the period T. With good approximation (Fig. 12.23) the relation,

$$q = \mu^{0.607}, \tag{12.48}$$

can replace the previous relation (12.43).

For smaller periods, that is, when $T < T_m$, the strength reduction factor depends on both T and μ. In place of (12.45), we can now take the relation (Fig. 12.23)

$$q = 1 + (\mu^{0.607} - 1)(\frac{T}{T_m})^{1.22} \tag{12.49}$$

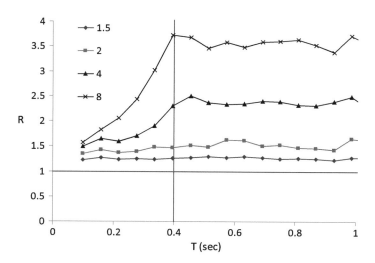

Fig. 12.23 Strength reduction factor R (or structure factor q) with varying ductility factor μ and period T, with $T_m = 0.4$ s (Coccia and Como 2009)

We develop now some examples.
Elasto–plastic Systems

$$T_m = 0.4 \text{ s}. \ T = 0 \text{ s} > T_m : q = \mu.$$

Let us assume $\mu = 3$. We have $q = 3$. This means that with a ductility factor $\mu = 3$, the strength of the elastic—plastic structure can be reduced by 3 times with respect to the strength of the elastic system.

Masonry systems
Bilinear elastic model. $T = 0.5 \text{ s} > T_m$
With ductility factor $\mu = 3$, we have $q = 3^{0.607} = 1.95$. The strength of the masonry structure can be reduced by 2 times with respect to that of the elastic system. Equations (12.48) and (12.49) are more conservative than the previous Eqs. (12.43) and (12.45): they require larger ductility, given the same reduction strength factor, or more strength, given the same reduction in ductility, as compared to the behavior of the elastic–plastic oscillator. The following examples serve to further illustrate the problem.

Elastic–plastic Systems
$T_m = 0.5 \text{ s}. \ T = 0.3 \text{ s} < T_m$ Let us assume $\mu = 3$. From (12.34) we obtain $R = 2.24$. With a ductility factor $\mu = 3$, the strength of the elastic—plastic structure can be reduced by 2.24 times with respect to the elastic system.

Masonry systems
Bilinear elastic model: $T = 0.5 \text{ s} > T_m$. From (12.37) we have

$$R = 1 + (3^{0.607} - 1)(0.3/0.5)^{1.22} = 1.51$$

With a ductility factor $\mu = 3$, the strength of the elastic—plastic structure can be reduced by 1.5 times with respect to the elastic system.

Concluding remarks
The reference seismic period T_m is generally not lower than 0.3 s so that for the more common masonry buildings, with a limited number of $T < T_m$. According to the elastic bilinear oscillator results, the Eq. (12.73) rules out the structure coefficients q. In this case the values q will reduce by decreasing the periods T. Also in presence of solid masonry, to which correspond higher values of the ductility factor μ, Eq. (12.73) gives lowvalues of q. For instance, in case of periods of the order of 0.2 s, with $\mu = 2.5$ the structure coefficient q is near to 1.5. According to the elasto-plastic oscillator in this case we have $q = 2.05$. For higher buildings, the bilinear oscillators gives higher values of the structure coefficient. Concluding, it is reasonnable to assume for the historic buildings values of the strcture coefficient not larger than q = 2. For the common constrctions with no more than two stories and built in poor masonry it is convenient to assume $q = 1.5$.

Such considerations only give qualitative indications regarding the ductility properties of masonry constructions. Despite their being too strongly tied to

12.3 Some Recalls of Earthquake Engineering ... 551

reinforced concrete structures, the standards put forth in various codes furnish generally quite conservative values of the strength reduction factors.

12.4 Seismic Resistant Structure of Masonry Buildings: Active and Inactive Walls

During an earthquake the seismic motion of the underlying soil produces horizontal forces that bear on a building and impart accelerations to its different masses. These accelerations are due to the coupling of the free oscillations of the building with the seismic motion at its base.

Figure 12.17 shows a scheme of a masonry building with the distribution of the horizontal forces. These forces impact the masses of the masonry walls and floors and tend to break up the connections between the various arrays of walls (Fig. 12.24).

The layout of a traditional masonry building usually involves a grid of two arrays of orthogonal walls (Fig. 12.18). Seismic actions, which can be considered directed along one of the two axes of the grid, determine two different roles for the d walls: *active* and *inactive* walls, depending on whether their planes are respectively parallel or orthogonal to the seismic action (Abruzzese et al. 1986).

The inactive walls, which are subjected to actions orthogonal to their planes, represent the weakest components of masonry buildings. Building codes assign the intensities of the conventional seismic forces that must be taken into account when checking the safety of a building structure. For a modern steel or reinforced concrete building, modal analysis can be used to obtain the maximum accelerations bearable by the building masses. Modal analysis provides information about the oscillation modes and their corresponding periods and thereby enables evaluating the maximum intensities of the horizontal forces that a building can bear. However, the situation for masonry buildings is quite different. Owing to an eventual random

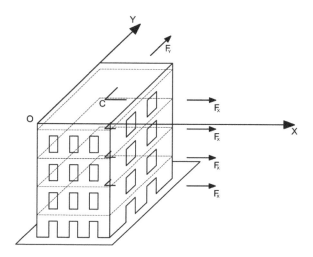

Fig. 12.24 Masonry building under seismic actions

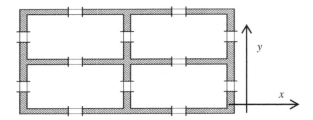

Fig. 12.25 Typical layout of a masonry building

distribution of cracks and the uncertain state of the connections between the walls, modal analysis becomes meaningless. In spite of the complexity and uncertainties involved in the problem, some simple devices can, on the other hand, be employed. The masses of the higher stories of a building are undoubtedly subjected to greater horizontal accelerations than the lower stories (Fig. 12.25).

A simplifying distribution of these horizontal forces along the height can thus be defined. These forces can represent both the total seismic loads impacting the overall building structure as well as the loads orthogonal to the plane of a single vertical band of masonry wall.

12.5 Distribution of Seismic Forces Along the Height

Seismic input has random features in both intensity and frequency distribution. The seismic strength of buildings is generally checked under conventional seismic forces established by the codes of various countries, which take into account the basic dynamic and ductility capacities of building structures.

Traditional masonry buildings generally have a limited number of stories: only in large urban areas do ever reach five or six stories. Now to look into the response of such structures to seismic actions, let W_k be the weights of the various masses of the building, lumped at story k, and let z_k be the height of the story with respect to level zero of the building, corresponding to the extrados plane of the foundation walls. The horizontal accelerations of the building masses excited by the seismic motion increase with their height. Consequently, the inertial forces acting on the building masses W_k/g, lumped at the various storys, will increase with the height as well. A simple, rough estimate of this effect can be made by recourse to a story distribution factor

$$\gamma_k = z_k \frac{\sum W_i}{\sum W_i z_i}, \qquad (12.50)$$

based on the assumption of horizontal acceleration increasing linearly with height.

The horiontal force acting at the k story of the building can be considered uniformly increasing with the multiplier λ so that

12.5 Distribution of Seismic Forces ...

$$F_k = \lambda \gamma_k W_k \qquad (12.51)$$

Thus, summing up the seismic forces on all the storys, for the entire structure we obtain the total thrust

$$F_{Tot} = \sum F_k = \lambda \sum W_k. \qquad (12.52)$$

because

$$\sum W_k \gamma_k = \sum W_k = W_{Tot} \qquad (12.53)$$

The same descrption of seismic forces can be given to the forces acting on the inactive walls. Let us first consider the masses involved in a span of the inactive wall between two adjacent active walls and two adjacent stories. With reference to Fig. 12.26 let:

- k be the number of the story;
- P_k, the weight of the span of inactive wall between story $(k+1)$ and k and between the two active walls;
- ς_k, the height of the midline of inter-story k, measured as the distance between level 0 and the midline of inter-story k.

Let us now evaluate the distribution factors γ_k (Fig. 12.26) defined similarly to (12.50):

$$\gamma_k = \varsigma_k \frac{\sum P_i}{\sum P_i \varsigma_i}, \qquad (12.54)$$

under the assumption, discussed above, of accelerations varying linearly with height.

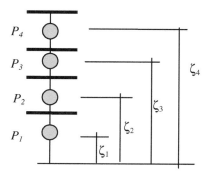

Fig. 12.26 Mass distribution of inactive walls at the interstory midlines

Fig. 12.27 Uniformly distributed loads applied at the inter-storys of inactive walls

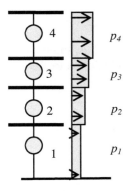

The corresponding seismic forces F_k applied to the masses of inactive walls are thus

$$F_k = \lambda \gamma_k P_k, \tag{12.55}$$

with the factor η defined by (12.55). The multiplier λ in Eq. (12.65′) has been added in order to describe a uniformly increasing distribution of seismic loads F_k:.

A sequence of uniformly distributed loads p_k can replace the point loads applied at the midline of the inter–stories (Fig. 12.27). Thus, we have

$$p_k = \lambda \frac{\gamma_k P_k}{h_k}. \tag{12.56}$$

The horizontal masonry bands running along the floor levels above the wall openings are also subjected to horizontal forces due to the floor masses (Fig. 12.28).

Figure 12.32 shows the assumed distribution of loads between the two supporting masonry bands. The corresponding seismic forces can thus be obtained as

$$S_k = \lambda \bar{\gamma}_k Q_k, \tag{12.57}$$

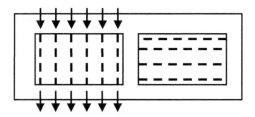

Fig. 12.28 Seismic actions due to the floor masses acting on the story bands

12.5 Distribution of Seismic Forces ...

where:

- Q_k is the weight of the half the mass of the corresponding floor;
- z_k, the height of inter-story k, measured as the distance from level 0 to the floor height of level k.

$\bar{\gamma}_k$, the distribution factor, which, by analogy with (12.50) is

$$\bar{\gamma}_k = z_k \frac{\sum Q_i}{\sum Q_i z_i} \qquad (12.58)$$

12.6 Design for Seismic Loadings

If the building structure sustains the seismic action elastically, the seismic forces F_k acting on the masses at various storys will thus be given by

$$F_{Ek} = \gamma_k \frac{W_k}{g} S_E(T) \qquad (12.59)$$

where, as defined above, $S_E(T)$ is the acceleration *elastic spectrum* of the assigned earthquake Θ. These represent the maximum thrusts that the building structure must be able to sustain *elastically* at the various storys. Equation (12.51) shows that the forces (12.51) depend moreover on the building's fundamental period T. The Italian Building Code provides for obtaining a rough, but simple estimate of the fundamental oscillation period of a building by applying the following:

$$T = C_1 H^{3/4} \qquad (12.60)$$

where H indicates the height of the building in meters, and C_1 is a factor that can be assumed equal to 0.05. In average, we have also

$$T \approx \frac{N_p}{10} \qquad (12.61)$$

where N_p is the number of stories. A common masonry building of two or three floors has an oscillation period of 0.2–0.3 s.

In accordance with the foregoing considerations, we can assume that the building structure can sustain the action of an assigned earthquake Θ with reduced lateral strength, but presenting inelastic displacements D_{Ak} of suitable amplitude. Under this assumption, the building structure would exhibit a distribution of story limit strengths reduced by *structure factor q* with respect to the elastic case

$$F_{Dk} = \gamma_k \cdot W_k \cdot S_D(T) \tag{12.62}$$

The quantity

$$S_D(T) = \frac{S_E(T)}{q} g \tag{12.63}$$

is the *demand* of the minimum horizontal average acceleration that the structure has to be able to sustain to be correlated to the chosen ultimate state. Generally, the reference ultimate state is the limit failure state. The Italian Code specifies required limit strengths for different seismic areas and for different soil types.

Thus, summing up the seismic forces on all the storys, for the entire structure we obtain the total thrust that the code requires to be absorbed by the construction

$$F_{totD\,Norm} = \sum F_k = S_D(T) \sum W_k. \tag{12.64}$$

where

$$\sum W_k \gamma_k = \sum W_k = W_{Tot} \tag{12.65}$$

is the total weight of the building. The quantity

$$S_D(T) \cdot g \tag{12.66}$$

is thus the demand of the minimum horizontal average acceleration that has to be sustained by the construction.

Limit Analysis will allow to obtain the distribution of the corresponding *resistant* limit horizontal forces a

$$F_{Rk} = \lambda_o W_k \gamma_k \tag{12.67}$$

so that the total limit resistance force of the building is

$$F_{R\,tot} = \lambda_o \sum W_k = \lambda_o W_{Tot} \tag{12.68}$$

The structure is charactarized by the collapse multiplier λ_o that defines its *resistant capacity*.

In this framework the collapse multiplier thus will represent the *average resistant horizontal acceleration* of the structure. The resistant total thrust cannot be lower than the demand of strength, i.e.

$$F_{R\,tot} \geq F_{totD\,Norm} \tag{12.69}$$

12.7 Seismic Failure Modes

The collapse of thre building may occur in compliance two different modes

- failure of walls under out of plane actions.
- failure of walls under inplane actions
- N

In the first case the failure mechanism is called *the first modc* mehanism and can affect a single wall, paricularly a facade wall of the building. In other word this failure is also defined as a *partial collapse* of the building if the failureconcerns only a limited number of walls.

The second case entails inplane failure mechanisms and is colled collapse mechanism of *second mode*. This type of collapse has as a rule more destructive features.

Only ther presence of rigid horizontal diaphragms, joined with the walls, can avoid the inplane failure of a single wall. Very seldom in traditional masonry buildings are, on the other hand present floors with rigid slabs connected with thge walls.

Generally the second mode failure occurs when the weakest wall fails in its plane. Figure 12.29 sahows the scheme of the collapse of the weakest wall.

The limit strength of the building is thus reached when a wall reaches its limit strength, i.e. when

$$\lambda_{oEDIF} = \underset{k}{Min}(\lambda_{owall,k}) \quad (12.70)$$

if $\lambda_{o\,wall,k}$ is the inplane collapse multiplier of the *knh* building wall.

The valuation of the partial failure multipliers $\lambda_{o,i}$ of all the single structural elements permits to check that early unexpected failures cannot occur. The condition

$$Min(\lambda^*_{oi,I°\text{mode}}) > \lambda_{obuilding} \quad (12.71)$$

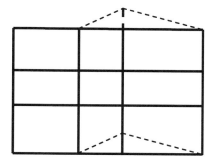

Fig. 12.29 The failure of the weakest wall of the building

Fig. 12.30 An historic centre destroyed by the Hirpinia Lucania quake of 1980

implies that the lowest first modes strengths will not larger than the building strength, defined according (12.96).

In the historic centres is generally impossible to find a single building but, on the contrary, a continuos aggregation of constructions. In these cases the first modes failures prevail. Figure 12.30 shows a global view of the serious damages suffered by an historic centre of the Irpinia during the quake of the 1980.

12.8 Out-of-Plane Strength of Inactive Walls

12.8.1 Unreinforced Wall. Collapse of the First Mode

The most frequent seismic failures occur in facade walls when they are subjected to horizontal out-of-plane seismic actions. Figure 12.31 shows typical seismic damage occurring at the corners of buildings. Figures 12.32 and 12.33 show the damage suffered by facades walls. It is identifiable the onset of the overturning detachment of the facades.

Figure 12.34 Shows the collapse of the whole corner of a building while Fig. 12.35 shows the breaching of the top wall of a facade wall. Further, Fig. 12.36 shows the out of plane onset of the breaching collapse of a facade that affect two

12.8 Out-of-Plane Strength ...

Fig. 12.31 Corners damagement

Fig. 12.32 Overturnig Detachment of facades

contiguous stories. We can therefore in this case undrrline the staving effect of the intermediate floor.

The above failures thus highlight the importance of studying the seismic refitting of old buildings in historic city centers (Di Pasquale 1982; Giuffrè 1986; Giangreco 1983; Pozzati 1986).

560   12   Masonry Buildings Under Seismic Actions

Fig. 12.33 Overturnig Detachment of facades

Fig. 12.34 Corner faikure

12.8 Out-of-Plane Strength ...

Fig. 12.35 Breaking of the tip of as facade wall

Fig. 12.36 Breaking of a facade involving two stories

Figure 12.37 shows a horrifying view of a street in Messina after the earthquake of 1908: all the building façade walls on the street failed, collapsed and dragged down the overlying roofs.

Fig. 12.37 A street of Messina destroyed by the earthquake of 1908

The simplest scheme of an unreinforced wall under horizontal out-of-plane forces is a cantilever wall. Such a wall fails by overturning. The geometry of the internal structure of the wall itself plays a fundamental role in its strength under out-of-plane actions. The transverse brick pattern, with *stretchers* and *headers*, considered in Chap. 1 in fact imparts unity and firmness to the wall (Giuffrè 1988). Headers, in particular, add strength to the wall.

Figure 12.38 shows the difference in the lateral strength of a wall with and without headers. Headers ensure that the two faces of a wall behave as a tightly connected unit, so that the wall's lateral resistance is that of a single full-width mass.

The lateral strength of a single isolated masonry wall of height H and width B can be easily calculated.. Assuming a triangular distribution of horizontal forces,

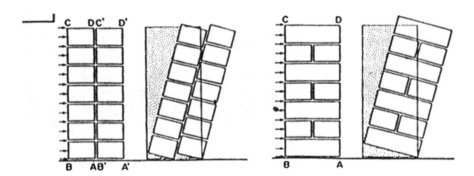

Fig. 12.38 Different wall lateral strength without (*left*) and with (*right*) headers (from Giuffrè 1988)

12.8 Out-of-Plane Strength ...

Fig. 12.39 Lateral failure of a simple wall under seismic action

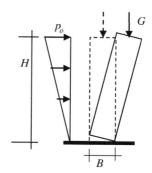

the strength of the wall, valuated as the ratio between the thrust S_o and the weight G, is given by (see Fig. 12.39):

$$\frac{S_o}{G} = \frac{3B}{4H}\left(1 - \frac{\sigma_m}{\sigma_k}\right), \qquad (12.72)$$

where S_o indicates the ultimate total thrust, equal to $p_o\,H/2$, and σ_m is the mean compression at the base under the vertical load G. The factor

$$\frac{3B}{4H}\left(1 - \frac{\sigma_m}{\sigma_k}\right) \qquad (12.73)$$

thus represents the mean failure acceleration of the wall.

For instance, with a height of 4.00 m and width B = 0.60 m, a wall made of good consistency masonry, yields about $S_o/G \approx 0.11$, while for poor masonry we obtain $S_o/G \approx 0.08$. Both these values are very low. In this regard, it should be stressed that isolated walls or, equivalently, walls with poor transverse constraints, are particularly vulnerable. The above Figs. 12.32 and 12.33 give examples of overturning failure mechanisms of walls. The presence of strong transverse connections is essential to impart strength to walls. It is thus easy to understand why it is so important that compact grid patterns be used for building walls in seismic areas (Fig. 12.40).

Fig. 12.40 Corner brick toothing

12.8.2 Walls Reinforced Against Out of Plane Actions

Also in the past numerous engineers have made various and interesting attempts to improve the out-of-plane resistance of building walls to seismic actions. Figure 12.38 shows a star plan of a building and Fig. 12.41 the corresponding section of an aseismic building from a 19th century design (Ruffolo 1912).

Note the design concept: concave facade walls are able to mobilize horizontal resisting arches that convey the thrusts to terminal buttresses. The floors of traditional buildings are usually made of wood or old steel beams connected transversally by horizontal masonry panels with a mixture of rubble and mortar. In such buildings, the floors thus provide only weak connections between the walls. They transmit vertical forces to the walls, but are incapable of transmitting horizontal forces.

12.8.3 Force Transmission from Inactive to Active Walls

The inactive walls of a building, subjected to actions orthogonal to their plane, must present strengths no lower than the active walls, otherwise partial failures could occur before the collapse of the main resisting structures of the building.

In order to evaluate the actual out-of-plane strength of walls it is necessary to accurately determine the force transmission from the inactive walls to the active ones and thereby define if *essential* reinforcements are called for.

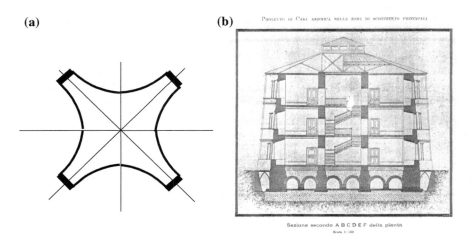

Fig. 12.41 Star plan and section of an aseismic masonry building, from a 19th-century design (Ruffolo 1912)

12.8 Out-of-Plane Strength ...

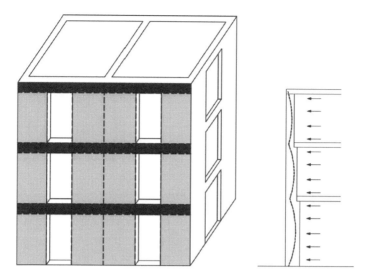

Fig. 12.42 Resistant wall system and vertical arches

To avoid that the last story wall works as a cantilever, the transmission of horizontal forces striking inactive walls has to occur by means of vertical resistant arch systems. These arches develop inside the continuous vertical masonry bands in the walls, laterally to the openings, between two adjacent stories (Fig. 12.42). In turn, the inter-story vertical bands link up with continuous horizontal bands running at floor levels and connected to the transverse lateral walls.

Efficiency of this transmission chain is fundamental to the lateral strength of the wall. However, the thrusts transmitted by the vertical arches at the top story must be adequately resisted. t intermediate stries thrusts balance each other. On the last story, this cannot occur., Only the weight of the roof sustainable by the wall contributes to withstanding the thrust of the vertical arch. This contribution is quite modest, so the walls of the last story are the most vulnerable to out-of-plane actions, as the old photograph in Fig. 12.37 illustrates.

We can valuate the order of magnitude of this thrust. The seismic horizontal load acting on the last story wall can be valuated as $p_o = g\gamma_{last} S_d(T)$ if g is the unit weght of the wall, γ_{last} the distribution factor for the last story, given by (12.59), $S_d(T)$ the design spectrum factor, valuated at the fundamental oscillation period T of the building. Assuming for $S_d(T)$ the maximum valuer given by the Italian Code, i.e. $S_d(T) = 2.5\, a_g/gq$, if a_g is the ground peak acceleration, g the the gravity and q the reduction or the structure factor, taken equal to 2. With a peak acceleration ratio a_g/g equal to 0.2, corresponding to a mean seismic intensity area and a structure factor $q = 2$, we have $S_d(T) = 2.5 \times 0.2/2 = 0.25$ and the corresponding seismic horizontal load acting on the last story wall is $p_o = g\gamma_{last} S_d(T) = g \times \gamma_{last} \times 0.25$. Assuming a thickness of the wall of 0.4 m, an unit weight of the masonry of 1.6 t/mc, the unit weight is 0.64 t/mq. With a distribution factor $\gamma_{last} = 2$, the load p_o equals $0.64 \times 2 \times 0.25 = 0.32$ t/mq and the thrust,

Fig. 12.43 Connection of the top ring beam to the wall through steel tie rods

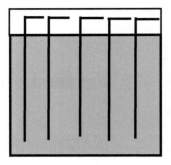

for unit length conveyed by the last story arch systems is $S = 0.32 \times 3.0^2 // 8 \times 0.4) = 0.9$ t/m. The weight of r.c. ring beam, equal to $0.4 \times 0.3 \times 2.5 \times 1.0 = 0.3$ t/m summed up with the weight of the horizontal masonry band equal to $0.4 \times 0.4 \times 1.6 \times 1.0 = 0.256$ t/m give a total weight of 0.556 t/m, scarcely larger than the half of the vertical thrust of the arch.

A simple reinforced concrete ring beam running along the top of the whole building walls and fixed firmly to them can ensure the necessary connection. Various reinforcement systems can be used. The case of the r.c. ring beam with anchoring systems is shown in Fig. 12.43. This system will be efficient only if the anchors have sufficient length. The realization in the masonry of the long vertical drillings with the insertion of the long steel bars and the injection of the connection mortar is a complex operation and cannot be convenient.

Alternatively, the reinforcement can be worked out wrapping vertical strips of FRP both on the outside and the inside face of the wall (Fig. 12.44). On the top of the wall the two FRP strips have to be suitably connected together or through the wall thickness. The r.c. ring beam, connected to the wall by the vertical steel sewings, can sustain both the vertical and the horizontal loads conveyed by the vertical arches. With the use of vertical FRP strips, on the contrary, only the vertical trhusting action can be opposed. In this last case other horizontal FRP strips will be suitably arranged in order to sustain the horizontal load conveyed on the horizontalm top band.

Fig. 12.44 Vertical FRP strips wrapped on the two faces of the wall

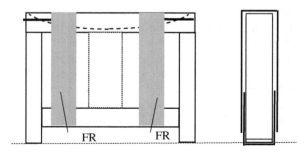

12.8 Out-of-Plane Strength ...

Fig. 12.45 Simple scheme of the steel ties fitted at floor levels to constrain walls in a traditional masonry building

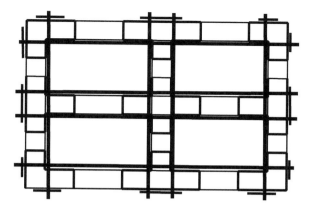

Between each story the vertical arches, mobilized within the walls, link into the horizontal bands at floor level. In modern masonry buildings the masonry walls at the toe and head are connected to reinforced concrete ring beams running along the walls at the floor levels. The reinforced concrete floors, whose slabs are connected to the ring beams, effect an efficient connection and the inactive walls are securely constrained to the transverse walls.

Traditional or historic masonry buildings lack these internal ring beams. However, if suitable *tie rods* are fitted, the continuous horizontal masonry bands running at floor levels could accomplish the required connection (Fig. 12.45).

12.8.4 Out of Plane Limit Strength of Fastened Walls

For the last story a reinforced concrete ring beam can be fitted at the top, along the wall edge. The aim of these reinforcements is to prevent the top sections of the walls from moving out of plane. Evaluating the breaching strength of such a reinforced wall is a complex problem, given the many geometric and mechanical parameters involved. However, such study can be simplified by using a one-dimensional wall model (Fig. 12.46). Study the failure of such models under orthogonal horizontal loads can furnish valuable information on the breaching strength of actual buildings fitted with the above essential reinforcements.

The first scheme on the left of Fig. 12.46 shows the wall under the action of its own weight. This wall has unit transverse width and its span corresponds to a generic inter-story. We will focus on the uppermost story because it is here that the seismic forces are strongest and the walls thinnest.

The wall, of height H, is constrained at its head by a horizontal strut, representing the ring beam fitted to prevent horizontal displacement of the wall's top section. In the left of Fig. 12.46 the wall is loaded only by vertical forces, i.e. by its weight g, uniformly distributed along the height, and by a vertical point force Q_W

Fig. 12.46 One-dimensional model of a reinforced wall under vertical and horizontal forces

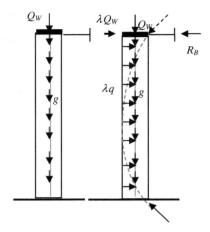

applied at the head. The right scheme instead shows, in addition to the vertical loads, the horizontal forces applied to the wall, which increase through the load multiplier λ. These forces are composed of a point load, λQ_W, applied at the top of the wall and a horizontal load distribution, λq, representing the forces specified by (12.65″) for the last story. We must now evaluate of the failure multiplier, λ_o, of such a load distribution.

Figure 12.47 shows a possible wall mechanism, CDB, composed of three hinges: one at base C; another at intermediate position D on the opposite side of edge CB at an unknown distance x from the base; and the last at B, at the head of the wall, where the horizontal strut prevents horizontal displacements. At actual failure, the position of the internal hinge will coincide with the position of the tangent point of

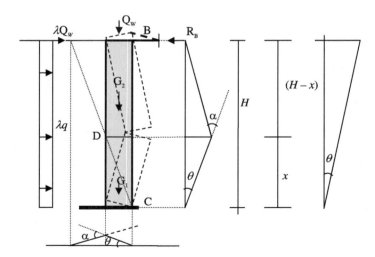

Fig. 12.47 Out-of-plane mechanism of the wall reinforced by a top ring beam

12.8 Out-of-Plane Strength ...

the pressure line with the wall edge, as shown in Fig. 12.50. The counter-rotation angle α and the rotation angle θ are linked by the following relation (Fig. 12.47)

$$H\theta = (H - x)\alpha, \qquad (12.74)$$

ensuring that the horizontal displacement vanishes at the head of the wall.
Thus, we obtain

$$\alpha = \frac{H}{(H-x)}\theta. \qquad (12.75)$$

At collapse, the pressure line is represented by the dotted curve in Fig. 12.45. The weight of the wall segment between the base and hinge D is G_1, while that of the remaining part is G_2, both proportional to the length of the corresponding segments. The weight of the wall per unit length is thus

$$g = G/H, \qquad (12.76)$$

and the total weight G is,

$$G = G_1 + G_2 = gH. \qquad (12.77)$$

By using the kinematic theorem of Limit Analysis, the work of the pushing loads is

$$L_{sp} = \lambda^+ q \frac{1}{2} H x \theta, \qquad (12.78)$$

and the resistant work due to the raising of the weights is

$$L_{res} = -Q_W \frac{B}{2} \theta - Q_W \frac{B}{2} \alpha - G_1 \frac{B}{2} \theta - G_2 \frac{B}{2} \theta - G_2 \frac{B}{2} \alpha \qquad (12.79)$$

or

$$\begin{aligned} L_{res} &= -(Q_W + G_1 + G_2) \frac{B}{2} \theta - \frac{B}{2} \alpha (Q_W + G_2) \\ &= -\frac{B}{2} \theta [(Q_W + G) + \frac{H}{H-x} Q_W + G]. \end{aligned} \qquad (12.79')$$

At the same time we have to consider the work made by the unthreading strength Q_{ao} of the vertical anchors placed at the head of the wall or, otherwise, the work made by the wrench strength of the FRP strips (Fig. 12.49).

We can assimilate this last work to an equivalent raising work considering the overall weight Q for unit wall length, applied at the wall head

$$Q = Q_W + Q_{o\,vert} \tag{12.80}$$

The vertical force Q is the force applied at the head of the wall in Figs. 12.46 and 12.47.

Let

$$R_o \tag{12.81}$$

be the unthreading strength crried on by the single anchor or by the single FRP strip. The overall strength Q_{oa} for unit length, inclusive of all the strengths, can be valuated considering the presence of N anchors or of N FRP strips, placed along the horizontal band.

Thus we have

$$NR_o = Q_{overt}L \tag{12.82}$$

and

$$Q_{overt} = \frac{NR_o}{L} \tag{12.83}$$

Figure 12.48 shows the opposition of the anchor or of the FRP strip in the out of plane failure mechanism of the wall. Thus, according to the kinematic theorem, we have

$$\lambda^+ q \frac{1}{2} Hx\theta = \frac{B}{2}\theta[(Q+G) + \frac{H}{H-x}Q + G], \tag{12.84}$$

and the kinematic multiplier is given by

$$\lambda^+ q = \frac{B}{Hx} Q[(1+2\psi) + \frac{H}{H-x}], \tag{12.85}$$

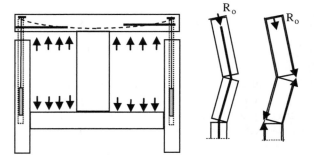

Fig. 12.48 Opposition worked out by the anchors or by the FRP strips in the failure mechanism of the wall

12.8 Out-of-Plane Strength ...

with position

$$\psi = G/Q \qquad (12.86)$$

equal to the ratio between weight Q, applied at the head of the wall, and the total wall weight G. Multiplier $\lambda^+(x)$ depends on the distance x of the internal hinge D from the base of the wall. The collapse multiplier equals the minimum attained by the function $\lambda^+(x)$ as the position of the internal hinge D varies, with $0 \leq x \leq H$. Function $\lambda^+(x)$ becomes unbounded for $x \to 0$ and $x \to H$. The search for the *minimum* of $\lambda^+(x)$ can thus be pursued by determining the abscissa $x = \bar{x}$ at which the derivative $d\lambda^+/dx$ vanishes. To this end, we have

$$\frac{d\lambda q}{dx} = -\frac{B}{Hx^2}Q[(1+2\psi) + \frac{H}{H-x}] + \frac{B}{Hx}Q\frac{H}{(H-x)^2} = 0, \qquad (12.87)$$

which yields the second degree algebraic equation

$$(1+2\psi)x^2 - 4(1+\psi)Hx + 2(1+\psi)H^2 = 0, \qquad (12.83')$$

whose roots are

$$x = \frac{4(1+\psi)H \pm \sqrt{8H^2(1+\psi)}}{2(1+2\psi)} \qquad (12.79')$$

The solution $\bar{x}$ cannot be larger than H. The root $\bar{x}$ is thus

$$\bar{x} = H\sqrt{2(1+\psi)}\frac{\sqrt{2(1+\psi)}-1}{(1+2\psi)} \qquad (12.88)$$

and the failure load multiplier is

$$\bar{\lambda}_o q = \frac{B}{H\bar{x}}Q[(1+2\psi) + \frac{H}{H-\bar{x}}]. \qquad (12.89)$$

The failure load multiplier is thus given

$$\bar{\lambda}_o = \frac{B}{H}\Lambda_{omin}(\psi) \qquad (12.90)$$

with

Fig. 12.49 Plot of factor $\Lambda_{\text{omin}}(\psi)$ versus ratio ψ

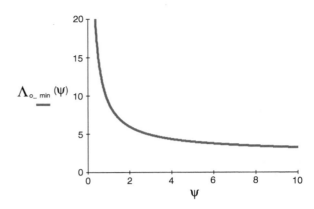

$$\Lambda_{\text{omin}}(\psi) = \frac{1}{\psi\sqrt{2(1+\psi)}} \frac{(1+2\psi)}{\sqrt{2(1+\psi)}-1} \\ [(1+2\psi) + \frac{(1+2\psi)}{(1+2\psi) - \sqrt{2(1+\psi)}[\sqrt{2(1+\psi)}-1]}] \quad (12.91)$$

The function $\Lambda_{\text{omin}}(\psi)$ is plotted in Fig. 12.49. A worthwhile exercise is to evaluate the increase in strength of the wall reinforced by the top ring beam with respect to the unreinforced wall, with free head, which fails as a vertical cantilever. In the latter case, whose corresponding mechanism is also diagrammed in Fig. 12.47, the pushing and resistant work components are respectively

$$L*_{sp} = \lambda q \frac{1}{2} H^2 \theta + \lambda Q H \theta = \lambda H Q (\frac{\psi}{2} + 1)\theta \quad (12.92)$$

$$L_{res} = (G+Q)\frac{B}{2}\theta = \lambda^* H Q(\frac{\psi}{2} + 1)\theta, \quad (12.93)$$

whence we obtain the corresponding load multiplier

$$\lambda* = \frac{(\psi + 1)}{(\frac{\psi}{2} + 1)} \frac{B}{2H}. \quad (12.94)$$

The ratio between the strength of the wall constrained at its head with the strength of the free wall is

$$\frac{\bar{\lambda}_o}{\lambda^*} = \Phi(\psi), \quad (12.95)$$

Fig. 12.50 Ratio $\Phi(\psi)$, between the lateral strengths of the head-constrained wall and the cantilever wall versus factor ψ

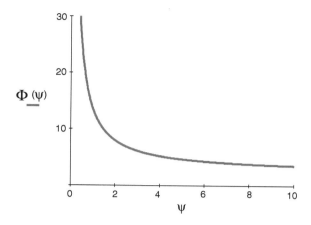

where the quantity $\Phi(\psi)$ is

$$\Phi(\psi) = \frac{(\psi+2)}{(\psi+1)} \frac{1}{\psi\sqrt{2(1+\psi)}} \frac{(1+2\psi)}{\sqrt{2(1+\psi)}-1}$$
$$[(1+2\psi) + \frac{(1+2\psi)}{(1+2\psi) - \sqrt{2(1+\psi)}[\sqrt{2(1+\psi)}-1]}]. \quad (12.96)$$

Figure 12.50 shows plot of function $\Phi(\psi)$. The weight Q acting at the head of the wall is essential for the formation of the resistant vertical arch. In fact, when $Q \to 0$, we have $\psi \to \infty$ and $\Phi(\psi) \to 1$. In this case, according to (12.88), $\bar{x} \to H$, and the internal hinge D in Fig. 12.47 moves to the head of the wall. The mechanism matches the overturning mechanism, and the internal resistant arch, needed to equilibrate load Q, vanishes.

Suitable anchors or equivalent reinforcing systems are thus required in the absence of an adequate load Q at the wall head.

The reaction R_B of the horizontal support fitted at the head of the wall is the transverse load acting on the ring beam and will be conveyed to the active walls. This load, R_B, is evaluated at failure by imposing that the resultant of all the forces acting on segment DB will pass through hinge D (Fig. 12.47). We thus have

$$(R_B - \bar{\lambda}_o Q)(H - \bar{x}) - QB/2 - \bar{\lambda}_o q(H-\bar{x})^2/2 - g(H-\bar{x})B/2 = 0 \quad (12.97)$$

and

$$R_B = Q\frac{B}{2(H-x)} + \bar{\lambda}_o[Q + q(H-x)/2] + gB/2. \quad (12.98)$$

This uniformly didtributed horizontal force is the limit transversal load $p_{\lim}$ acting along the top horizontal band that has to be conveyed to the active walls.

12.9 In-Plane Strength of Multi-storey Masonry Walls with Openings

12.9.1 The Different Models of Walls

Multi-story masonry walls are the most important structural components of any masonry building. A double array of multi-story walls with openings, for the most part weakly connected to wooden or iron floors, is the main resistant structure of a typical masonry building. Different systems of reinforcements can be fitted to improve their strength and, above all, prevent out-of-plane collapse of the external walls, as described in the previous sections.

Figure 12.51 shows a typical cracking pattern on the facade wall of a masonry building struck by an earthquake. The damage is due to considerable seismic forces acting in the plane of the wall. The X-shaped cracks suggest the occurrence of an inversion of the horizontal thrusts during the earthquake.

Multi-story masonry walls come in various geometrical arrangements. Here we will limit ourselves to considering only regular walls, with openings arranged in a regular pattern, both vertically and horizontally. It is thus possible to distinguish between vertical wall bands, termed "piers", and horizontal bands, called "architraves". Accorrding to the height of the architraves is large or small with respect o the height of the openings we can dist tinguish two different typologies of multi-story walls. We shall consider firstly the case of architraves of large height,—a very common arrangement (Fig. 12.52). The other tipology is frequently met in historic centres of many towns of the South of Italy.

Fig. 12.51 Damagement of a multi-storey wall under in-plane seismic actions

Fig. 12.52 Two common types of multi-story walls with window openings

12.9.2 Seismic Strength of Multi-storey Masonry Walls with Openings and Large Architraves

We consider the static behavior of the first typology of walls under the action of horizontal loads. One example of these walls is shown in Fig. 12.53. Horizontal steel ties at floor levels run in the masonry. Their presence is required to avoid partial detachment failure of single piers.

The study of horizontal strength of these walls has interested many researchers as, for instance, Calderoni ed altri (2009, 2010). Here the lateral strength of the wall will be obtained by means the Limit Analysis approach.

The common collapse mechanism of the wall is sketched in Fig. 12.54, distinguished by the simultaneous shear failure of the architraves and the overturnig of piers bands.

The shear failure of the architraves requires plastic stretchig of ties. In fact at the collapse the failure mechanism will develop under constant horizontal forces.

Suitable anchors are disposed at the ends of ties. Thus, in presence of too strong ties, the plastic re-entry of the anchors can replace the plastic stretching of steel ties.

Many anchors exhibit a ductile behavior as that shown in Fig. 12.55.

Other failure mechanisms could seem to be possible, in dependance of the geometry of the wall and strength of the ties (Fig. 12.56). But the shear failure of the architraves can be unlikely avoided because it is difficult to provide adequate shear strength ot the horizontal floor bands. These last mechanism must be thus excluded.

Fig. 12.53 The multi-storey masonry wall with openings and large architraves

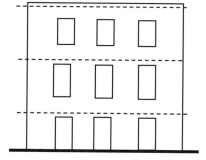

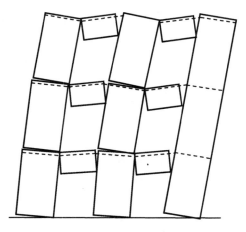

Fig. 12.54 Typical failure mechanism of a multi story wall with high architraves

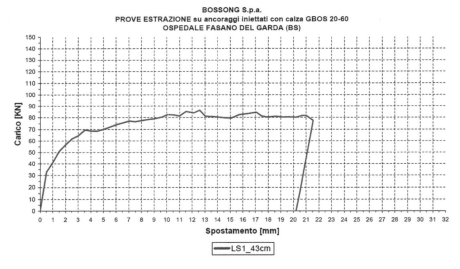

Fig. 12.55 Ductlile behaviour of the Bossong anchor in a pull out test (D.I.C.A.T.A., University of Brescia)

12.9.2.1 A Simple Model of Wall with Large Architraves

We will firstly consider a wall with two piers and two architraves, shown in Fig. 12.57.

The assumed mechanism of the wall is shown in Fig. 12.58.

12.9 In-Plane Strength of Multi-storey … 577

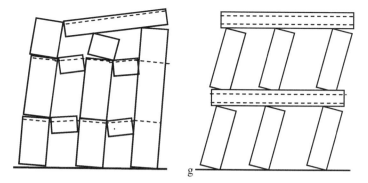

Fig. 12.56 Other mechanisms

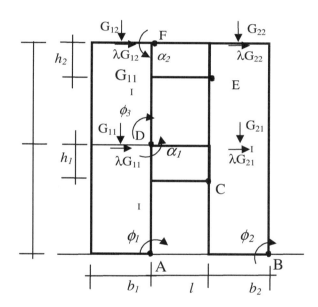

Fig. 12.57 A simple masonrt wall with larhe architraves

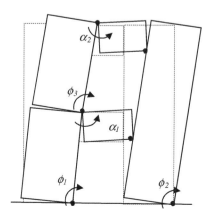

Fig. 12.58 The typical mechanism of the wall

Cracks occurring between panels and piers are due to the presence of panels of finite height. Consequently, different horizontal displacements will occur at the floor levels.

In the mechanism of Fig. 12.58 the wall subdivides into five parts: the whole right pier, that rotates around the hinge A of ϕ_1, the two panels and the two parts in which the left pier is necessarily partitioned. The left pier, is partitioned into two parts that rotate respectively around the hinges A and D of the angles ϕ_1 and ϕ_3 (Fig. 12.58).

The panels, in turn, rotate respectively with respect the piers of the angles α_1 and α_2. The basic parameter of the mechanism is the rotation ϕ_1 of the lower part of the left pier.

With refernce to Fig. 12.57 let us valuate the horizontal and vertical displacements u_C and v_C of the point C, the right lower corner of the first panel, considered solid with the left pier. These displacements have to be equal to the corresponding displacements of the same point C considered solid with the right pierThus we have the following equations

$$(H_1 - h_1)\phi_1 + \alpha_1 h_1 = (H_1 - h_1)\phi_2 \qquad (12.99)$$

$$l\phi_1 - l\alpha_1 = -b_2\phi_2 \qquad (12.100)$$

that solved give

$$\phi_2 = K\phi_1 \qquad (12.101)$$

with

$$K = \frac{\beta_2}{\beta_2(1-\rho_1) - \rho_1} \qquad (12.102)$$

and

$$\beta_2 = \frac{l}{b_2} \qquad \rho_1 = \frac{h_1}{H_1} \qquad (12.103)$$

Applying the same argument for the point E we get

$$(H_1 + H_2 - h_2)\phi_1 + \phi_3(H_2 - h_2) + \alpha_2 h_2 = K(H_1 + H_2 - h_2)\phi_1 \qquad (12.104)$$

$$l\phi_1 + l\phi_3 - \alpha_2 l = -Kb_2\phi_1 \qquad (12.105)$$

that give

$$\phi_3 = \delta\phi_1 \qquad (12.106)$$

12.9 In-Plane Strength of Multi-storey ...

where

$$\delta = [(K-1)\frac{H_1}{H_2} + K(1-\rho_2) - 1 - \rho_2\frac{K}{\beta_2}] \quad (12.107)$$

with the position

$$\rho_2 = \frac{h_2}{H_2} \quad (12.108)$$

Let us valuate the works of the vaious forces made during the development of the mechanism.

– *The resisting work of raising weights*

$$L_G = -G_{11}\frac{b_1}{2}\phi_1 - G_{12}\frac{b_1}{2}(1+\delta)\phi_1 - G_{21}\frac{b_2}{2}K\phi_1 - G_{22}K\frac{b_2}{2}\phi_1 \quad (12.109)$$

or

$$L_G = -\frac{1}{2}\{b_1[G_{11} + (1+\delta)G_{12}] + b_2K(G_{21} + G_{22})\}\phi_1 \quad (12.110)$$

– *The resisting work due to the plastic stretching of ties*
– Let T_o be the limit pull of the ties at the yielding. The plastic extensions Δ_1 and Δ_2 of the first and second level ties are

$$\Delta_1 = KH_1\phi_1 - H_1\phi_1 = H_1(K-1)\phi_1 \quad (12.111)$$

$$\begin{aligned}\Delta_2 &= K(H_1+H_2)\phi_1 - (H_1+H_2)\phi_1 - H_2\delta\phi_1 \\ &= [(H_1+H_2)(K-1) - \delta H_2]\phi_1\end{aligned} \quad (12.112)$$

and the plastic work is

$$L_{pl} = -T_o[2H_1(K-1) + H_2(K-1-\delta)]\phi_1 \quad (12.113)$$

– *The active work of the pushing horizontal loads*

$$L_{sp} = \lambda^+[H_1(G_{11} + KG_{21}) + H_2G_{12}\delta + (H_1+H_2)(G_{12} + KG_{22})]\phi_1 \quad (12.114)$$

The condition

$$L_G + L_{pl} + L_{sp} = 0 \qquad (12.115)$$

gives the kinematical multiplier of the pushing loads

$$\lambda^+ = \frac{b_1[G_{11} + G_{12}(1+\delta)] + b_2 K(G_{21} + G_{22}) + 2T_o[2H_1(K-1) + H_2(K-1-\delta)]}{2[H_1(G_{11} + KG_{21}) + H_2 G_{12}\delta + (H_1 + H_2)(G_{12} + KG_{22})]} \qquad (12.116)$$

In the simpler case of equal weights G, heights H and h we have

$$\lambda^+ = \frac{b}{2H} \frac{2(1+K) + \delta + 2t_o[3(K-1) - \delta]}{3(1+K) + \delta} \qquad (12.116')$$

with the position

$$t_o = \frac{T_o H}{Gb} \qquad (12.117)$$

If, at the limit, the height of panel vanishes, we have

$$K \to 1 \quad \delta \to 0 \qquad (12.118)$$

and

$$\lambda^+ \to \frac{b}{3H} \qquad (12.119)$$

The kinematical multiplier is the actual collapse multiplier is the stress state present in the wall, under the horizontal forces affected of the kinematical multiplier λ^+ is statically admissible.

12.9.2.2 The Vakuation of the Interaction Forced Between the Walls

To check this admissibility we have to valuate the interactions R_1 and R_2 that the piers transmit through the panels. as shown in Fig. 12.59. The problel is statically determined because the the pulls in the tiesare known, equal to T_o.

With refernce to Fig. 12.59 we have the following rotational equilibrium equations of the two piers around hinges D and A

$$-G_{12}\frac{b_1}{2} + (\lambda G_{12} + T_o - R_2 \cos\gamma_2)H_2 = 0 \qquad (12.120)$$

12.9 In-Plane Strength of Multi-storey ...

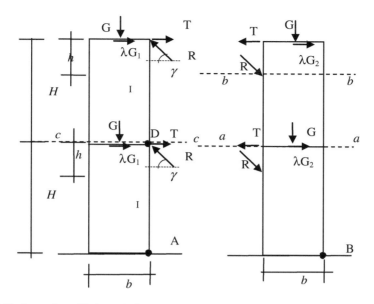

Fig. 12.59 Internal equilibrium of piers

$$-G_{11}\frac{b_1}{2} - G_{12}\frac{b_1}{2} + (\lambda G_{11} + T_o - R_1 \cos\gamma_1)H_1 \\ + (\lambda G_{12} + T_o - R_2 \cos\gamma_2)(H_1 + H_2) = 0 \quad (12.121)$$

with the positions

$$tg\gamma_1 = \frac{h_1}{l} \quad tg\gamma_2 = \frac{h_2}{l} \quad (12.122)$$

The third equilibrium equation, that of the rotational equilibrium of the right pier around the hinge B, is useless because the load multiplier is already known.

Solution of (12.120) gives

$$R_2 \cos\gamma_2 = [-G_{12}\frac{b_1}{2H_2} + (\lambda G_{12} + T_o)] \quad (12.123)$$

Substitution of (12.123) into (12.121) gives

$$R_1 H_1 \cos\gamma_1 = -G_{11}\frac{b_1}{2} + G_{12}\frac{b_1}{2}\frac{H_1}{H_2} + (\lambda G_{11} + T_o)H_1 \quad (12.124)$$

Particularly, in the simpler case of equal weights G, heights H and h we have

$$R_2 H \cos \gamma_2 = -G\frac{b}{2} + (\lambda G + T_o)H \quad R_1 \cos \gamma_1 = \lambda G + T_o \qquad (12.125)$$

or, with the positions

$$r_1 = \frac{R_1}{G} \quad r_2 = \frac{R_2}{G} \qquad (12.126)$$

we have

$$r_2 \cos \gamma_2 = -\frac{b}{2H} + (\lambda + \frac{T_o}{G}) \quad r_1 \cos \gamma_1 = \lambda + \frac{T_o}{G} \qquad (12.127)$$

12.9.2.3 Check of the Admissibility Conditions at the Kinematical State

First admissibility condition involving stresses in the panels

Efforts r_1 and r_2 have to determine compressions in the panels. Thus, from (12.127) we have

$$r_2 \cos \gamma_2 \geq 0 \rightarrow \frac{T_o}{G} \geq \frac{b}{2H} - \lambda^+ \qquad (12.128)$$

Admissibility conditions involving stresses in the right pier

The axial load acting at the line a–a, just over the level where the G_{21} is applied, has to be contained inside the section. The moment M_{a-a} of all the forces placed above the line a–a of all the forces placed above a–a valuated respect the centre of the section, and the axial load N_{a-a} are respectively

$$M_{a-a} = (\lambda G_{22} - T_o)H_2 + R_2(H_2 - h_2)\cos \gamma_2 - R_2 \frac{b_2}{2} \sin \gamma_2;$$
$$N_{a-a} = G_{22} + R_2 \sin \gamma_2 \qquad (12.129)$$

Hence, the eccentricità of N_{aa} is

$$e' = \frac{(\lambda G_{22} - T_o)H_2 + R_2(H_2 - h_2)\cos \gamma_2 - R_2 \frac{b_2}{2} \sin \gamma_2}{G_{22} + R_2 \sin \gamma_2} \qquad (12.130)$$

It is thus required that

$$-\frac{b_2}{2} \leq e\prime \leq \frac{b_2}{2} \qquad (12.131)$$

Likewise, in the section just under the line a–a eccentricity of the axial load is

12.9 In-Plane Strength of Multi-storey ...

$$e'' = \frac{(\lambda G_{22} - T_o)H_2 + R_2(H_2 - h_2)\cos\gamma_2 - R_2\frac{b_2}{2}\sin\gamma_2}{G_{22} + G_{21} + R_2\sin\gamma_2} \qquad (12.132)$$

and we have the third condition

$$-\frac{b_2}{2} \leq e'' \leq \frac{b_2}{2} \qquad (12.133)$$

But $e'' < e'$ and only condition la condizione (12.131) is effective.
Admissibility conditions involving stresses in the left pier
Head section of the pier
It is required that

$$G_{12} - R_2 \sin\gamma_2 \geq 0 \qquad (12.134)$$

or

$$R_2 \sin\gamma_2 \leq G_{12} \qquad (12.134')$$

The resultant of G_{12} and of $R_2 \sin\gamma_2$ is placed at the left of the centre of the section. The corresponding eccentricità is

$$e = -\frac{b_1}{2}\frac{R_2 \sin\gamma_2}{G_{12} - R_2 \sin\gamma_2} \qquad (12.135)$$

Thus conditiopn $e < b_1/2$ gives $G_{12} - R_2 \sin\gamma_2 > R_2 \sin\gamma_2$ or

$$R_2 \sin\gamma_2 < G_{12}/2 \qquad (12.136)$$

Section just over the line c–c
The axial load is

$$N_{c-c} = G_{12} - R_2 \sin\gamma_2 \qquad (12.137)$$

Previous condition (12.136) assures that N_{c-c} is positive, i.e. a compression. Further

$$M_{c-c} = (T_o + \lambda G_{12} - R_2 \cos\gamma_2)H_2 - R_2\frac{b_1}{2}\sin\gamma_2 \qquad (12.138)$$

and the eccentricità of N_{c-c} is

$$e' = \frac{(T_o + \lambda G_{12} - R_2 \cos\gamma_2)H_2 - R_2\frac{b_1}{2}\sin\gamma_2}{G_{12} - R_2 \sin\gamma_2} \qquad (12.139)$$

Comparing numerator of (12.139) with (12.123) we get $e' = b_1/2$ because an hinge is lacated at D.

Likewise, *just under the line c–c* the axial load is

$$N_{c-c} = G_{12} - R_2 \sin \gamma_2 + G_{11} - R_1 \sin \gamma_1 \qquad (12.140)$$

while the moment around the centre takes the value

$$M_{c-c} = (T_o + \lambda G_{12} - R_2 \cos \gamma_2)H_2 - R_2 \frac{b_1}{2} \sin \gamma_2 - R_1 \frac{b_1}{2} \sin \gamma_1 \qquad (12.141)$$

and the eccentricità e'' of N_{c-c} is

$$e'' = \frac{(T_o + \lambda G_{12} - R_2 \cos \gamma_2)H_2 - R_2 \frac{b_1}{2} \sin \gamma_2 - R_1 \frac{b_1}{2} \sin \gamma_1}{G_{12} - R_2 \sin \gamma_2 + G_{11} - R_1 \sin \gamma_1} \qquad (12.142)$$

and we have the condition

$$-\frac{b_2}{2} \leq e'' \leq \frac{b_2}{2} \qquad (12.143)$$

Le (12.143), (12.134), (12.133) e (12.131), that enforce restrictions on the magnitude of the limit force limite T_o, are the required admissibility conditions.

A Numerical Example

let us consider the regular wall with two piers and two levels with $H_2 = H_1 = H = 4.00$ m.; $h_2 = h_1 = h = 1.20$ m; $b_2 = b_1 = b = 2.00$ m, $l = 2.00$ m (Fig. 12.60). The thickness of the wall measures 0.80 m and the masonry has an

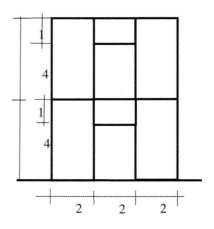

Fig. 12.60 The examined masonry wall

12.9 In-Plane Strength of Multi-storey ...

unit weight $\gamma = 1.6$ t/mc. Consequently $tg\gamma_1 = tg\gamma_2 = 1.2/2 = 0.60$; $\gamma = \gamma_1 = \gamma_2 = 30.964°$; $\sin \gamma = 0.514$; $\cos \gamma = 0.857$; $\rho = h/H = 0.30$; $\beta_1 = \beta_2 = 1$.

Kinematical quantities K and δ defined by (12.102) e (12.106) take the values: $K = 2.50$; $\delta = 1.50$. Hence $G = 10.24$ t. The yield sterss of the steel of ties is $\sigma_s = 2400$ kg/cmq. The kinematical multiplier, according to (12.116') is

$$\lambda^+ = \frac{8.5 + 6t_o}{48} = 0.177 + 0.125 t_o \tag{a}$$

TRhe corresponding interactions R_1 and R_2 between the piers, according to (12.127) take the expressions

$$r_2 = -0.08518 + 0.731 t_o \quad r_1 = 0.206 + 0.73 t_o \tag{a'}$$

We check the admissibility conditions to define the admissible strengths of the ties. From (12.128) we have, taking into account of (a') we have

$$t_o \geq 0.117 \tag{b}$$

Condition (12.143) requires

$$-\frac{b}{2} \leq \frac{(\lambda - T_o/G)H + r_2(H-h)\cos\gamma - r_2 \frac{b}{2}\sin\gamma}{1 + r_2 \sin\gamma} \leq \frac{b}{2} \tag{c}$$

i.e. with the previous calculations

$$-1 \leq \frac{0.547 - 0.121 t_o}{0.956 + 0.376 t_o} \leq 1 \tag{d}$$

that gives as $t_o \geq -5.91$ as $t_o \geq -0.823$ and therefore

$$t_o \geq -0.823 \tag{e}$$

Condition (12.136) requires $R_2 \sin \gamma \leq G/2$. Hence

$$t_o \leq 1.447 \tag{f}$$

while condition (12.143) finally gives

$$-\frac{1}{4} \leq \frac{(t_o/2 + \lambda - r_2 \cos\gamma) - r_2 \frac{1}{4}\sin\gamma - r_1 \frac{1}{4}\sin\gamma}{2 - r_2 \sin\gamma - r_1 \sin\gamma} \leq \frac{1}{4}$$

or

$$-1 \leq \frac{0.94 - 0.756 t_o}{1.938 - 0.752 t_o} \leq 1 \tag{h}$$

i.e.

$$t_o \leq 1.91 \tag{i}$$

Concluding, taking into account of (b), (e), (f) ed (i) we have

$$0.117 \leq t_o \leq 1.447 \tag{l}$$

With position (12.117) the yield strength T_o of each tie has to satisfy the condition

$$0.117 \cdot 5.12 \leq T_o \leq 1.447 \cdot 5.12, \text{ i.e } 0.60 \, t \leq T_o \leq 7.41 \, t.$$

We choose for the ties steel bars with diameter ϕ 12 in mild steel with yield stress $\sigma_{so} = 2400$ kg/cmq. Hence $T_o = 1.13 \cdot 2400 = 2.71 \, t$ and, taking into account of (12.117), we have $t_o = 0.529$. The collapse multiplier of the wall thus is, according to (a)

$$\lambda_o = 0.177 + 0.126 \cdot 0.529 = 0.243$$

From (a') we obtain also: $r_2 = -0.0852 + 0.731 \cdot 0.529 = 0.302$ and $R_2 = 3.09t$

Further $r_1 = 0.206 + 0.73 \cdot 0.529 = 0.592$ and $R_1 = 6.06t$

We proceed now to evaluate the eccentricity at various sections in order to determine the compressed zones in the piers.

Left pier
Head section
Condition (12.136) requires

$$R_2 \sin \gamma \leq G/2 \tag{m}$$

Considering the value just determined of R_2, condition (12.136) requires: $3.09 \cdot 0.514 = 1.59 < 10.24/2$, that is satisfied. The compression load acting at the head section is $N = G - R_2 \sin \gamma = 10.24 - 6.06 \cdot 0.514 = 7.12t$. This load acts at the left side of the section and has eccentricity

$$e = -\frac{b}{2}\frac{R_2 \sin \gamma}{G - R_2 \sin \gamma} = -\frac{2.0}{2}\frac{3.09 \cdot 0.514}{10.24 - 3.09 \cdot 0.514} = -\frac{1.588}{8.652} = -0.183m \tag{n}$$

contained inside the core of the section.

Section just over line c–c

$$N_{c-c} = G - R_2 \sin \gamma \tag{o}$$

Thus the previous condition (m) assicures that N_{c-c} is a compression load. Further

12.9 In-Plane Strength of Multi-storey ...

$$M_{c-c} = (T_o + \lambda G - R_2 \cos \gamma)H - R_2 \frac{b}{2} \sin \gamma \quad (p)$$

and the eccentricità of N_{c-c} is

$$e' = \frac{(T_o + \lambda G_{12} R_2 \cos \gamma)H - R_2 \frac{b}{2} \sin \gamma}{G - R_2 \sin \gamma} \quad (q)$$

Comparing the numerator of (q) we have $e' = b_1/2$ because an hinge of the mechanisn is located at D.

Likewise, *just under the line c–c* the axial load is

$$N_{c-c} = 2G - R_2 \sin \gamma - R_1 \sin \gamma = 2 \cdot 10.24 - (3.09 + 6.06) \cdot 0.514 = 15.78t \quad (r)$$

and the noment around the section centre is

$$M_{c-c} = (T_o + \lambda G - R_2 \cos \gamma)H - R_2 \frac{b}{2} \sin \gamma - R_1 \frac{b}{2} \sin \gamma =$$
$$= 10.20 - 1.588 - 3.115 = 5.497 \, \text{tm} \quad (s)$$

The corresponding eccebtricity e'' gives

$$e'' = \frac{5.497}{15.78} = 0.34 m$$

The axial load is applied just on the edge of the section core. The section can be considered fully compressed.

Right pier
Head section
$N = 10.24$ t applied at the section centre.
Section just over the line b–b

$$N = 10.24 \, \text{t}$$
$$M = -2.714 \times 1.20 + 0.243 \times 10.24 \times 1.20 - = -0.271 \, \text{tm}$$

eccentricity

$$e = -0.271/10.24 = -0.026 \, \text{m}$$

The section is fully compressed.
Section just under the line b–b

$$N = 10.24\,\text{t} + R_2\sin\gamma_2 \quad 10.24 + 3.09 \times 0.514 = 11.828\,\text{t}$$
$$M = -2.714 \times 1.20 + 0.243 \times 10.24 \times 1.20 - 3.09 \times 0.514 \times 1.0$$
$$= -1.86\,\text{tm}$$

eccentricity

$$e = -1.86/11.83 = -0.157\,\text{m}$$

The section is fully compressed.
Section just over the line a–a

$$M_{a-a} = (\lambda G - T_o)H + R_2(H - h)\cos\gamma - R_2\frac{b}{2}\sin\gamma, \text{ i.e.}$$
$$M_{a-a} = (0.244 \cdot 10.24 - 2.714) \cdot 4.0 + 3.09 \cdot (4.0 - 1.2) \cdot 0.857 - 3.09 \cdot 1 \cdot 0.514$$
$$= 4.964\,\text{tm}$$

The axial load takes the value

$$N_{a-a} = 10.24 + 3.09 \cdot 0.514 = 11.828\,\text{t}$$

and the ccentricity is

$$e\prime = \frac{4.964}{11.828} = 0.42\,\text{m}$$

The axial load is applied outside the core. The section is partially compressed.
Section just under the line a–a

$$e\prime\prime = \frac{(\lambda G - T_o)H + R_2(H - h)\cos\gamma - R_2\frac{b}{2}\sin\gamma}{G + G + R_2\sin\gamma}$$
$$M_{a-a} = 4.964\,\text{tm}$$
$$N_{a-a} = G + R_2\sin\gamma + G = 2 \cdot 10.24 + 3.09 \cdot 0.514 = 22.068\,\text{t}$$
$$e = 4.964/22.068 = 0.22\,\text{m}$$

The section is fully compressed.
Section just over the lower corner of the first level panel

$$N = 22.068\,t$$
$$M = (2.714 + 0.243 \cdot 10.24) \cdot (4.0 + 1.2) + 3.09 \cdot 0.857 \cdot 4.0$$
$$+ (-2.714 + 0.243 \cdot 10.24) \cdot 1.20 = 9.274\,\text{tm}$$
$$N = 22.068\,\text{t}$$

12.9 In-Plane Strength of Multi-storey …

Fig. 12.61 The masonry wall at the collapse with the compressed zones in piers and panels

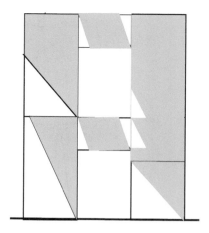

The eccentricity is e = 0.42 m.
The axil load is applied outside the core. The section is partially compressed.
Section just under the lower corner of the first level panel

$$M = 9.274 - 6.06 \cdot 0.514 \cdot 1.0 = 6.16 \text{ tm}$$
$$N = 22.068 + 6.06 \cdot 0.514 = 25.18 \text{ t}$$
$$e = 0.244 \text{ m}$$

The section is fully compressed.
The distribution of the compressed zones in the piers and panels of the masonry wall is traced in Fig. 12.61.

The General Case

Figure 12.62 shows the failure mechanism of a more general typology of masonry wall with openings and large architraves. The interfloor heights are respectively H_1, H_2, and the architraves heights are h_1, h_2, and so omn. All the geometrical quantities of the wall are indicated in figure.

In the figure are also indicated the rotations ϕ relative to piers and with α the rrelative rotations of the panels with respect to piers. Two indeces affect the rotations f and a: the first iundicates the reference pier and the second the reference floor. The contact joints pier/panel are indicated as C, D, E, F.

We proceed now to formulate the kinematical compatibility equations that will determine the connection among the various rotations a the basic one.
Equating horizontal and vertical displacements at the joint N.
Only one parameter puts in connection all the rotation ϕ eand α together.
The considered mechanism subdivides all the piers into a sequence of vertical panels except the last pier that, intact, rotates of the angle ϕ_N around its toe.

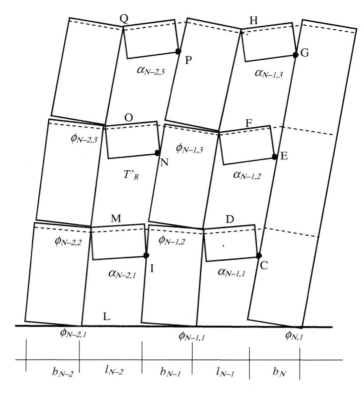

Fig. 12.62 The failure mechanism of the general sheme of wall

We equate the horizontal and the vertical displacements U and V of the joints C, I, E, N, G, P, considered once belonging to the left and the other belonging to to the right side of the single pier.

First level:
Equating horizontal and vertical displacements at the joint C

$$U_{Sin,C} = \phi_{N-1,1}(H_1 - h_1) + \alpha_{N-1,1}h_1 = U_{Des,C} = \phi_N(H_1 - h_1)$$
$$V_{Sin,C} = \phi_{N-1,1}l_{N-1} - \alpha_{N-1,1}l_{N-1} = V_{Des,C} = -\phi_N b_N$$

Equating horizontal and vertical displacements at the joint I

$$U_{Sin,I} = \phi_{N-2,1}(H_1 - h_1) + \alpha_{N-2,1}h_1 = U_{Des,I} = \phi_{N-1,1}(H_1 - h_1)$$
$$V_{Sin,I} = \phi_{N-2,1}l_{N-2} - \alpha_{N-2,1}l_{N-2} = V_{Des,I} = -\phi_{N-1,1}b_{N-1}$$

Equating horizontal and vertical displacements at the joint E

12.9 In-Plane Strength of Multi-storey ...

$$U_{Sin,E} = \phi_{N-1,1}(H_1 + H_2 - h_2) + \phi_{N-1,2}(H_2 - h_2) + \alpha_{N-1,2}h_2 = U_{Des,E}$$
$$= \phi_N(H_1 + H_2 - h_2)$$
$$V_{Sin,E} = \phi_{N-1,1}l_{N-1} + \phi_{N-1,2}l_{N-1} - \alpha_{N-1,2}l_{N-1} = V_{Des,E} = -\phi_N b_N$$
$$U_{Sin,N} = \phi_{N-2,1}(H_1 + H_2 - h_2) + \phi_{N-2,2}(H_2 - h_2) + \alpha_{N-2,2}h_2 =$$
$$U_{Des,N} = \phi_{N-1,1}(H_1 + H_2 - h_2) + \phi_{N-1,2}(H_2 - h_2)$$
$$V_{Sin,N} = \phi_{N-2,1}l_{N-2} + \phi_{N-2,2}l_{N-2} - \alpha_{N-2,2}l_{N-2} = V_{Des,N} = -\phi_{N-1,1}b_{N-1} - \phi_{N-1,2}b_{N-1}$$
$$(12.144)$$

Terzo livello:
Equating horizontal and vertical displacements at the joint G

$$U_{Sin,G} = \phi_{N-1,1}(H_1 + H_2 + H_3 - h_3) + \phi_{N-1,2}(H_2 + H_3 - h_3) + \phi_{N-1,3}(H_3 - h_3)$$
$$+ \alpha_{N-1,3}h_3 = U_{Dest,G} = \phi_N(H_1 + H_2 + H_3 - h_3)$$
$$\phi_{N-1,1}l_{N-1} + \phi_{N-1,2}l_{N-1} + \phi_{N-1,3}l_{N-1} - \alpha_{N-1,3}l_{N-1} = -\phi_N b_N$$
$$(12.145)$$

Equating horizontal and vertical displacements at the joint P

$$U_{Sin,P} = \phi_{N-2,1}(H_1 + H_2 + H_3 - h_3) + \phi_{N-2,2}(H_2 + H_3 - h_3)$$
$$+ \phi_{N-2,3}(H_3 - h_3) + \alpha_{N-2,3}h_3 = U_{Des,P} = \phi_{N-1,1}(H_1 + H_2 + H_3 - h_3)$$
$$+ \phi_{N-1,2}(H_2 + H_3 - h_3) + \phi_{N-1,3}(H_3 - h_3)$$
$$V_{Sin,P} = \phi_{N-2,1}l_{N-1} + \phi_{N-2,2}l_{N-1} + \phi_{N-2,3}l_{N-1} - \alpha_{N-2,3}l_{N-1} = V_{Des,P}$$
$$= -\phi_{N-1,1}b_{N-1} - \phi_{N-1,2}b_{N-1} - \phi_{N-1,3}b_{N-1}$$
$$(12.146)$$

The wall shown in figure has three piers and three floors. The number equations are in number of 12. The rotations ϕ and α are in number of 13. Equations (12.181) are thus able to express all the rotations in dependance of the rotation occurring at the toe of the third pier. We evaluate the plastic extension of the ties at various levels:

$$\Delta_1 = \phi_N H_1 - \phi_{N-2,1} H_1$$
$$\Delta_2 = \phi_N(H_1 + H_2) - \phi_{N-2,1}(H_1 + H_2) - \phi_{N-2,2} H_2$$
$$\Delta_3 = \phi_N(H_1 + H_2 + H_3) - \phi_{N-2,1}(H_1 + H_2 + H_3) - \phi_{N-2,2}(H_2 + H_3) - \phi_{N-2,3} H_3$$
$$\Delta_1 = \phi_N H_1 - \phi_{N-2,1} H_1$$
$$(12.147)$$

According this approach is thus possible operate as the the simple example firstly completely examined.

12.9.3 Seismic Strength of the Multi-storey Wall with Openings and Thin Architraves

12.9.3.1 Geometry of the Wall and of the Acting Loads

The wall to be examined has a regular array of openings, N_s stories and N_p piers (Fig. 12.63). The wall has been reinforced by a system of steel ties passing through the piers and running inside the floors with anchor plates at their heads. These, or other equivalent, reinforcements, fitted to increase the out-of-plane strength of the wall, establish longitudinal connections between the piers to prevent their detaching.

In the framework of the rigid in compression, no-tension model, the masonry architraves will prevent the occurrence of contractions along the horizontal lines at the floor levels. Further reinforcement of the wall can be also be achieved by inserting steel H beams above the openings and embedding them within the masonry piers. The steel beams on the same floor level are all the same and their strength is defined by a limit bending moment M_o.

The rigid in compression, no-tension assumption for masonry is the main assumption used to formulate a simplified model for in-plane wall collapse.

In evaluating the limit deformation of the masonry wall, usually composed of wide piers, the elastic strains produced in the piers under assigned horizontal forces can be considered negligible with respect to the detachments strains consequent to cracking. Thus, under the action of seismic horizontal forces, the wall piers will remain practically vertical as long as local overturning failure does not occur. The horizontal displacements of the failed piers will therefore be due solely to their rotations around their toes. The presence of horizontal connections, such as architraves and tie rods, will produce interactions between the piers up until a sideways mechanism affecting the entire wall develops. In analyzing these interactions, it is assumed that the steel H beams, lacking any anchoring, do not participate in these interactions and that the architraves, in plain masonry, will be able

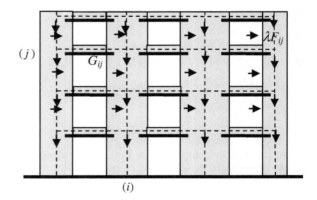

Fig. 12.63 Multi-story wall with openings and horizontal connections

12.9 In-Plane Strength of Multi-storey ...

to sustain only compressive forces. The steel ties, on the other hand, can withstand only tractions. The horizontal connections will thus exhibit so-called *unilateral* elastic behavior. Both the architraves and the ties will develop elastic strains whether they are in compression or in extension. The wall is loaded by constant vertical loads and linearly increasing horizontal forces. The problem thus becomes the evaluation of the in-plane strength of the wall under such loading conditions. To define the positions of the various loads, we can establish a reference grid represented by the axes of the horizontal bands above the openings and by the vertical piers' axes. The piers' widths may vary from story to story, and the grid could thus exhibit some misalignments between the floor levels.

The intersections points between piers and architraves axes are the grid nodes. Any given pier is denoted by the index i, while a generic story is identified by the index j (Fig. 12.64), where $1 \leq i \leq N_p, 1 \leq j \leq N_s$. A node (i, j) is thus the intersection of the axis of the pier i with story level j. At any node we can define the weight G_{ij} representing of masses assigned to node (i, j) and that produce vertical compressions along the pier.

To each node is also assigned horizontal seismic force, λF_{ij}, where λ is the load multiplier. These forces are due not only to the masses of the active wall connected to node (i,j), but also to the masses of the inactive walls connected to the node itself. Consequently, the thrusts λF_{ij} will be proportional to the force G'_{ij}, which includes the weights of both active and inactive walls bearing on node (i,j).

However, the assumption of forces increasing linearly with height requires applying the distribution factor γ_{ij} defined above. Thus, if z_j is the height of level j with respect to the wall base, the node thrusts can be thus expressed as

$$F_{ij} = \lambda \gamma_{ij} G'_{ij}, \qquad (12.148)$$

where γ_{ij} is the previously specified distribution factor (12.50). With reference to the generic regular plan of the building, two different choices are possible for defining the piers' sections. First, we can consider a reduced section, that is, only the rectangular section of the active wall in its plane.

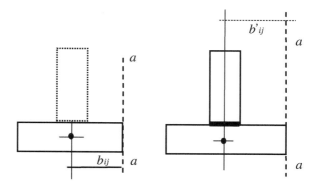

Fig. 12.64 Pier sections composed by active or by inactive and active wall portions

In this case, during development of the overturning collapse mechanism only the weight of the piers in this section, together with the corresponding floor loads, take part in the lifting work. The first scheme in Fig. 12.64 shows the reduced section of the second and third piers, together with the axis of rotation involved in the overturning mechanism, represented by a dotted line. This first scheme corresponds to the so called *disconnectedness assumption* between the two wall arrays. Conversely, the second option considers the full section of the piers obtained by joining the section of the adjacent *inactive* transverse wall to the rectangular part previously taken into account. In this case, the resistant lifting work performed during the overturning mechanism is generally larger than the first, in which case suitable connections would guarantee the shear capacity of the masonry in the contact area between the two section parts to transmit the forces involved. The second scheme in Fig. 12.65 illustrates this second choice: a bold line traces the contact area between the two parts of the section where shear transmission occurs. This second case corresponds to the *connection assumption* between the two arrays of walls.

Possible Failure Mechanisms

The collapse state of the wall, attained somewhere during loading, is characterized by a definite mechanism with horizontal displacements at the floor levels. There are various possible collapse mechanisms, some of which are shown in Fig. 12.66. The first can be called the *overall overturning* mechanism. Other mechanisms, by which only the top or some intermediate story of the wall collapses, can also occur. When a masonry wall has a regular distribution of openings and piers, the *overall* overturning mechanism, shown in the leftmost scheme in Fig. 12.66, is the most likely. The limit bending strength of the steel beams inserted into the piers above the openings also plays a relevant role in determining the failure mechanism. As will be

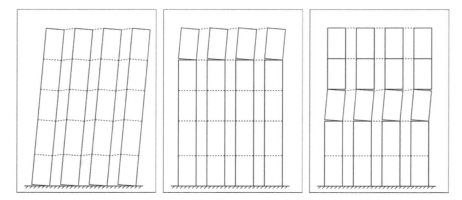

Fig. 12.65 Various wall collapse mechanisms

12.9 In-Plane Strength of Multi-storey …

Fig. 12.66 Overall overturning collapse mechanism of the wall

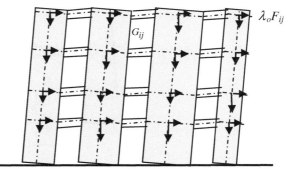

shown in the following, the overturning mechanism comes about when the steel beams are sufficiently weak.

We will now assume that the wall actually fails according to the overall overturning mechanism (Fig. 12.66). A check of the static admissibility of the wall stresses at collapse can demonstrate whether the failure has actually has been attained. Figure 12.54 shows in more detail the multi-story wall set in motion along the overall collapse mechanism under the thrusts $\lambda_o F_{ij}$, where λ_o is the failure multiplier.

Kinematic Multiplier in the Overall Overturning Mechanism

The collapse multiplier is obtained by applying the kinematic theorem, which equalizes the resistant and pushing work along the mechanism. The work of the pushing forces is

$$L_{thr} = \theta\lambda \sum_{i=1}^{N_p} \sum_{j=1}^{N_s} F_{ij} z_j = \theta\lambda \sum_{j=1}^{N_s} F_{tot_j} z_j \qquad (12.149)$$

summing up the thrusts on the same story. We thus consider the overall thrust on the given story:

$$F_{tot_j} = \sum_{i=1}^{N_m} F_{ij}. \qquad (12.150)$$

The first contribution to the resistant work is due to the *raising* of the weights G_{ij}, which is given by

$$L_{lift} = -\theta \sum_{i=1}^{N_m} \sum_{j=1}^{N_p} G_{ij} b_{ij}, \qquad (12.151)$$

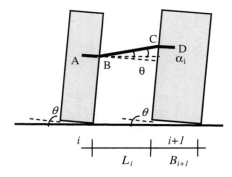

Fig. 12.67 Angular distortions of the steel platbands at their insertions into the walls

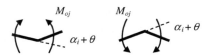

Fig. 12.68 Positive and negative limit bending moments at the platband insertion sections

where the quantities b_{ij} represent the lift arms of weights G_{ij}.

The second contribution is due to the *plastic dissipation* occurring in the steel beams inserted between the piers over the wall openings.

The end sections of the steel beams must bend plastically when the piers begin to rotate sideways. The plastic bending occurring at the end sections B and C of the steel beam over the opening to the right of pier i (Figs. 12.67 and 12.68) is due to rotation $(\theta + \alpha_i)$, where

$$\alpha_i = \frac{B_{i+1}}{L_i} \theta, \qquad (12.152)$$

and where B_{i+1} is the width of pier $i+1$ and L_i the span of the opening *subsequent to pier i*. All the steel beams of the same level present the same plastic limit bending moment, which will be denoted by M_{oj}. Figure 12.70 shows the limit plastic bending moments occurring at the left and right of the end sections of the steel beams. Taking (12.100) into account, the plastic work performed in the single steel beam at story level j, corresponding to pier i, is

$$D_{pi,i} = 2M_{oj}\left(\frac{B_{i+1}}{L_i} + 1\right)\theta. \qquad (12.153)$$

Summing up, the plastic work occurring along all piers i at the same level j, we get

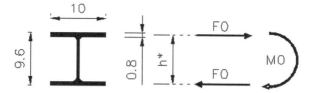

Fig. 12.69 Limit bending moment of the steel beam

$$D_{pl} = 2\sum_{j=1}^{N_s}\sum_{i=1}^{N_p-1} M_{oj}(\frac{B_{i+1}}{L_i}+1)\theta = 2\theta\sum_{j=1}^{N_s} M_{oj}\sum_{i=1}^{N_p-1}(\frac{B_{i+1}}{L_i}+1), \quad (12.154)$$

where the second sum is extended to all piers as far as the next to last. In fact, when $i = N_p-1$, the plastic work occurs in the steel beams overlying the last opening at level j. Figure 12.69 shows the evaluation scheme of the limit bending moment M_o for a section of an H beam.

In this case we have $F_o = 10 \times 0.8 \times 2400/1.15$ kg; $M_o = F_o \times h^*$; $h^* = 9.6$ cm.

Evaluation of the kinematic multiplier λ^+ of the horizontal forces along the assumed mechanism is thus performed by canceling the sum of all the resistant and pushing work along the mechanism, as follows:

$$D_p + L_{rais} + L_{thr} = 0 \quad (12.155)$$

$$-2\sum_{j=1}^{N_p} M_{oj}\sum_{i=1}^{N_m-1}(\frac{B_{i+1}}{L_i}+1)\theta - \theta\sum_{i=1}^{N_m}\sum_{j=1}^{N_p} G_{ij}b_{ij} + \theta\lambda^+\sum_{i=1}^{N_m}\sum_{j=1}^{N_p} F_{ij}z_j = 0.$$

$$(12.156)$$

The kinematic multiplier of thrusts λ^+ is thus given by

$$\lambda^+ = \frac{\sum_{i=1}^{N_m}\sum_{j=1}^{N_p} G_{ij}b_{ij} + 2\sum_{j=1}^{N_p} M_{oj}\sum_{i=1}^{N_m-1}(\frac{B_{i+1}}{L_i}+1)\theta}{\sum_{j=1}^{N_p} F_{totj}z_j} \quad (12.157)$$

and is independent of the elasticities of the horizontal constraints.

Multiplier λ^+ will represent the actual limit lateral strength of the wall for all possible mechanisms only if the stresses occurring in the wall under forces $\lambda^+ F_{ij}$ are *statically admissible*. Only in this case does multiplier λ^+ define the collapse λ_o multiplier of the thrusts. As can be seen from inspection of (12.155), depending on the values of the lift arms b_{ij}, the kinematic multiplier varies according to the direction of the thrusts, i.e. whether they act from left to right or vice versa. The seismic lateral strength of the wall will be the minimum of these two values.

Stresses in the Horizontal Constraints at the Limit Stat of the Wall:
Compatibility Conditions

The multiplier λ^+ is the actual collapse multiplier λ_o only if it is the smallest of the kinematic multipliers corresponding to all other possible mechanisms. On the other hand, it is possible to verify that multiplier λ^+ is the actual failure multiplier λ_o by checking the admissibility of the wall stresses.

This check calls for evaluating the stresses in the piers and the horizontal constraints of the wall under forces $\lambda^+ F_{ij}$. Steel ties cannot be compressed, masonry architraves cannot be stretched and the pressure lines running along the piers must never go beyond their edges. It is thus necessary to know the interactions between the horizontal constraints in order to carry out the above mentioned check.

Figure 12.70 shows the interactions occurring between the horizontal constraints: the internal interactions—those acting from the second pier to the next to last—have been distinguished from those acting on the first and last piers. The axial loads N_{ij} transmitted by the horizontal masonry platbands are located to the right of pier i and are thus denoted by

$$N_{1,j}, \quad N_{i,j} \quad (i = 2, 3, \ldots N_m - 1), \; N_{N_m - 1} \qquad (12.158)$$

Thus, axial load $N_{1,j}$ in the platband at level j is transmitted to the right of pier 1. Similarly, we have $N_{i,j}$ ($i = 2, \ldots N_m - 1$) for the other axial loads acting in all the platbands to right of pier i as far as pier ($N_m - 1$). The axial load $N_{Nm-1,j}$ instead acts to the left of the last pier N_m. Now, let us denote by

$$M_{oj} \qquad (12.159)$$

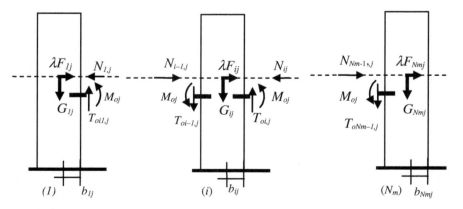

Fig. 12.70 Forces acting on the piers

12.9 In-Plane Strength of Multi-storey ...

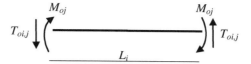

Fig. 12.71 Limit shear force in the platband

the limit bending moment at level j transmitted to the pier by the steel beam and

$$T_{o1,j}, \ T_{oi,j} \ (i = 2, 3, \ldots N_m - 1), \ T_{oN_m-1,j} \tag{12.160}$$

respectively the shear forces transmitted by each of the steel beams succeeding pier 1 as far as pier i ($i = 2, 3, \ldots N_m - 1$), where the last term denotes the shear preceding the last pier. In brief, the same notation used for the axial load has also been adopted for the shears (Fig. 12.71). From the equilibrium of the steel beams we have (Fig. 12.71).

$$T_{oi,j} = \frac{2M_{oj}}{L_i}. \tag{12.161}$$

Finally, tractions in the tie rods are denoted by

$$T_j. \tag{12.162}$$

Concerning the piers, the admissibility of the stresses can be expressed by means of the inequalities

$$-B_j/2 \leq e_j(z) \leq B_j/2, \tag{12.163}$$

where $e_j(z)$ is the eccentricity of the axial load on pier j at any given height z. In addition, the admissibility conditions for the masonry platbands and tie rods are

$$N_{i,j} \geq 0, \quad T_i \geq 0, \tag{12.164}$$

where both compression loads in the masonry platbands and tensile forces in the tie rods have been assumed *positive*. Checking the compatibility conditions requires preliminary evaluation of the stresses in the horizontal constraints at the limit equilibrium state of the wall. To this end, it is useful *to collect* all the unknowns regarding a given type of connection—tie rods or masonry platbands, etc., *by summing* their moments at the toe of each pier. All the axial loads on the masonry platbands and tie rods for a single pier are thus considered via the following positions

$$M_i^N = \sum_{i=1}^{N_s} N_{ij} z_j \tag{12.165}$$

$$M^T = \sum_{i=1}^{N_s} T_i z_i, \tag{12.166}$$

where:

- M_i^N is the moment at the toe of pier i of the axial loads transmitted by the masonry platbands concerning the span *following* pier i;
- M^T is the moment at the pier toe of the tensions in the tie rods.

Likewise, with reference to Fig. 12.59, we write

$$M_1^S = \sum_{j=1}^{N_p} G_{1j} b_{1j} + \sum_{j=1}^{N_p} M_{oj}, \; M_i^S = \sum_{j=1}^{N_p} G_{ij} b_{ij} + \sum_{j=1}^{N_p} 2M_{oj} + \sum_{j=1}^{N_p} T_{oi-1,j} B_i, \quad i=2,3,\ldots N_m-1$$

$$M_{N_m}^S = \sum_{j=1}^{N_p} G_{N_m j} b_{N_m j} + \sum_{j=1}^{N_p} M_{oj} + \sum_{j=1}^{N_p} T_{oN_m-1,j} B_i,$$

$$\tag{12.167}$$

$$M_i^R = \sum_{j=1}^{N_p} F_{ij} z_j \tag{12.168}$$

where:

- M_i^S is the *stabilizing* moment acting on pier i due to weights G_{ij} and the actions transmitted by the steel beams;
- M_i^R is the *overturning* moment of the thrusts evaluated with $\lambda = 1$.

By means of these positions, the equilibrium equations upon rotation at the toe of the single piers at wall *failure* can be obtained simply and quickly as follows:

- for the first pier,

$$-M_1^S + \lambda^+ M_1^R - M_1^N + M^T = 0 \tag{12.169}$$

- for pier i ($i = 2, \ldots, N_m - 1$),

$$-M_i^S + \lambda^+ M_i^R + M_{i-1}^N - M_i^N = 0 \tag{12.170}$$

12.9 In-Plane Strength of Multi-storey ...

- for the last pier,

$$-M_{N_m}^S + \lambda^+ M_{N_m}^R + M_{N_m-1}^N - M^T = 0 \qquad (12.171)$$

These conditions regulate the limit equilibrium state of single pier at wall failure. Some piers will be sustained by the others and other piers by means of the axial interactions of the platbands and tie rods.

Let us now define *the partial lateral strength of the single pier i*, by means of the *partial* failure multiplier λ_{oj} of the horizontal loads, which represents strength of the pier in absence of axial interactions with the other piers:

$$-M_i^S + \lambda_{oi} M_i^R = 0 \quad (j = 1.2..., N_m). \qquad (12.172)$$

The *partial* failure multiplier of a single pier is thus given by

$$\lambda_{oi} = \frac{M_i^S}{M_i^R}. \qquad (12.173)$$

The weakest pier is that corresponding to the smallest value among all the λ_{oi}. The overturning condition in the weakest pier is thus attained when multiplier λ of the horizontal loads reaches the value

$$\lambda_o^* = Min(\lambda_{oi}). \qquad (12.174)$$

Equations (12.120) and (12.121) can now be rewritten more simply using the definitions of the partial failure multipliers, as follows:

$$M_1^N - M^T = M_1^R(\lambda^+ - \lambda_{o1}) \qquad (12.175')$$

$$M_i^N - M_{i-1}^N = M_i^R(\lambda^+ - \lambda_{oi}) \qquad (12.176')$$

$$M^T - M_{N_m-1}^N = M_{N_m}^R(\lambda^+ - \lambda_{oN_m}). \qquad (12.177')$$

We can once again derive expression (12.155) for the kinematic multiplier of the horizontal loads by using Eqs. (12.175'), (12.176') and (12.177'). In fact, summing up the N_p equilibrium equations, the moments due to interactions cancel and we have

$$-\sum_{i=1}^{N_m} M_i^S + \lambda^+ \sum_{i=1}^{N_m} M_i^R = 0.$$

The kinematic multiplier of the horizontal forces acting on the wall is thus:

$$\lambda^+ = \sum_{i=1}^{N_m} M_i^S \Big/ \sum_{i=1}^{N_m} M_i^R,$$

and we obtain

$$\sum_{i=1}^{N_m} M_i^S = \left(\sum_{i=1}^{N_m} \sum_{j=1}^{N_p} G_{ij} b_{ij} \right) + \sum_{i=1}^{N_m-1} \sum_{j=1}^{N_p} 2 M_{oj} + \sum_{i=1}^{N_m-1} \left(\sum_{j=1}^{N_p} T_{oi,j} B_{i+1} \right)$$

$$= \left(\sum_{i=1}^{N_m} \sum_{j=1}^{N_p} G_{ij} b_{ij} \right) + \sum_{i=1}^{N_m-1} \sum_{j=1}^{N_p} 2 M_{oj} + \sum_{i=1}^{N_m-1} \left(\sum_{j=1}^{N_p} \frac{2 M_{oj}}{L_i} B_{i+1} \right)$$

$$= \sum_{i=1}^{N_m} \sum_{j=1}^{N_p} G_{ij} b_{ij} + \sum_{i=1}^{N_m-1} \sum_{j=1}^{N_p} 2 M_{oj} \left(1 + \frac{B_{i+1}}{L_i} \right).$$

Finally, taking into account expression (12.167) for M_i^S, we obtain, as in (12.157),

$$\lambda^+ = \frac{\sum_{i=1}^{N_m} \sum_{j=1}^{N_p} G_{ij} b_{ij} + \sum_{i=1}^{N_m-1} \sum_{j=1}^{N_p} 2 M_{oj}(1 + \frac{B_{i+1}}{L_i})}{\sum_{j=1}^{N_p} F_{ij} z_j} \qquad (12.178)$$

It should be noted that we have assumed masonry with unlimited compression strength. It is possible, on the other hand, to take the finite strength of masonry into account by correcting, through a rough estimate, Eq. (12.126) via the factor

$$\left(1 - \frac{\sigma_m}{\sigma_{o<}}\right), \qquad (12.179)$$

according to the previous specifications. Kinematic multiplier (12.175) thus becomes

$$\lambda^+ = \frac{\sum_{i=1}^{N_m} \sum_{j=1}^{N_p} G_{ij} b_{ij} + \sum_{i=1}^{N_m-1} \sum_{j=1}^{N_p} 2 M_{oj}(1 + \frac{B_{i+1}}{L_i})}{\sum_{j=1}^{N_p} F_{ij} z_j} \left(1 - \frac{\sigma_m}{\sigma_k}\right). \qquad (12.179')$$

Checking the Compatibility of the Limit Overturning State

The relevant equilibrium equations concern the vanishing of the overall moment at the piers' toes; there are thus N_m equations with the unknowns:

- the $(N_m - 1)$ moments M_j^N $(j = 1, 2, \ldots N_m - 1)$;
- the moment M^T;
- the multiplier λ^+.

The unknowns are thus $(N_m + 1)$, while the available equations are only N_m. We are now looking for the *missing last equation*. The following extra conditions must however be associated:

$$M_i^N \geq 0 \text{ (for } i = 1, \ldots, N_{m-1}); \quad M^T \geq 0. \tag{12.180}$$

These follow from the *unilateral character* of steel tie rods and masonry architraves. Equations (12.175′), (12.176′) and (12.177′) have in fact been defined for *compressed* masonry architraves and *stretched* the tie rods. At the same time, the pressure line in the piers cannot be never located outside the piers edges. It should now be noted that to solve Eqs. (12.175′), (12.176′) and (12.177′) associated to the inequalities in (12.177), at least one of the unknowns M_i^N and M^T must *cancel out* (Fig. 12.72).

In fact, if the tie rods at wall failure are stretched, and consequently forces T_j do not cancel out, it is not possible for *all* the horizontal architraves to be compressed. The existence of compressions in all the masonry architraves would contradict the existence of tensile stresses in the tie rods. Consequently, if the stresses in the tie rods do not cancel, the axial loads acting on at least one span between the piers would have to be equal to zero. Vice versa, if the masonry architraves are all compressed, the tie rods will be unloaded. Thus, in any event, the additional condition

$$M_1^N \cdot M_2^N \cdot \ldots M_{N_m-1}^N \cdot M^T = 0 \tag{12.181}$$

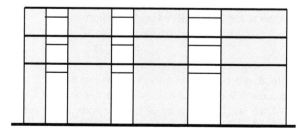

Fig. 12.72 Wall horizontal constraints: tie rods and masonry architraves

must hold. In short, at least one of the unknowns must equal zero. The number of the actual unknowns becomes N_m and the problem has an univocal solution. This result can be better illustrated by considering some examples.

The multi-story wall in Fig. 9.75 is loaded by horizontal forces acting from left to right. The masonry architraves in the figure are drawn with light lines. The weakest pier is the leftmost: under loading, it will be the first to rotate around its toe and lean towards the second pier. This latter will also overturn, followed then by the third, and all three will thus be sustained by the last pier. In this case, all the masonry architraves will be compressed and the tie rods will be unloaded.

On the contrary, if the horizontal forces act from right to left, when the first pier overturns, the tie rods will stretch and transmit to the last pier any excess thrust that the first pier is unable to sustain. It is thus possible that, by increasing the thrusts, the second pier will overturn, once again followed by third as well, with three transmitting onto the last pier the thrusts that they cannot sustain. When the last pier also reaches overturning, all the tie rods will be stretched and all the masonry architraves will be compressed, except those in the last span, between the next to last and the last pier. Depending upon the overturning strengths (12.173) of the single piers, it is thus possible to identify the unloaded constraints of the wall at failure. An algebraic approach can be followed in searching for the zero unknowns, taking into account the values of the partial multipliers (12.173) and directly using the system of Eqs. (12.114), (12.115) and (12.116'), which we rewrite here for the sake of convenience:

$$M_1^N - M^T = M_1^R(\lambda^+ - \lambda_{o1})$$
$$M_i^N - M_{i-1}^N = M_i^R(\lambda^+ - \lambda_{oi})$$
$$M^T - M_{N_m-1}^N = M_{N_m}^R(\lambda^+ - \lambda_{oN_m})$$

In fact, if the sign of the second member term in equation i is positive, unknown M_i^N cannot be zero, otherwise unknown M_{i-1}^N could be negative, contradicting condition (12.178). By examining the sign of the known terms, we can thus exclude at least one unknown. At this point it is worth remarking that the signs of the known terms in the second member of Eqs. (12.175'), (12.176') and (12.177') cannot all be the same, because if they were, then none of the unknowns could be equal to zero. At least one of the signs of the known terms will be different from the others because the collapse multiplier λ_o belongs to the set of partial multipliers λ_{oi}. Once the system of equations has been solved and the moments

$$M_1^N, M_2^N, \ldots, M_{N_m-1}^N, M^T \qquad (12.182)$$

have been determined, we can proceed to evaluating the forces in the tie rods and the masonry architraves. We can assume that the these forces will vary linearly along the height of the wall. Assuming all the tie rods in all the storys to be the same, we have

$$T_j = T_i \frac{z_j}{z_i}. \tag{12.183}$$

Thus, taking Eq. (12.166) into account, we get

$$T_j = M^T \frac{z_j}{\sum_{i=1}^{N_p} z_i^2}. \tag{12.184}$$

Forces (12.181) represent the second portion of the total forces acting on the tie rods in the active walls. The total force T_{jtot} at level j is given by adding to force (12.181) the other force T_y'', addressed in Sect. 9.4.2, due to the transmission of seismic forces from the inactive to active walls along the horizontal arch with reverse chains. Thus,

$$T_{jtot} = \lambda_o(V'' + S''\frac{b}{a}) + T_j. \tag{12.185}$$

Similar relations can be obtained for the axial forces in the masonry architraves.

12.9.3.2 Checking the Node's Capacity to Transmit Shear from Inactive to Active Walls

The participation of the inactive walls' masses to the resisting work can be very significant in defining the wall strength. It is thus important to check the actual capacity of the connection between active/inactive walls to transmit shear. Figure 12.61 shows a scheme of the active and inactive walls at a connection node between the two arrays of walls (Fig. 12.73).

The seismic action is indicated in the figure by a left-to-right arrow, and the active and inactive walls are represented by the rectangular sections ABCD and EFGH respectively. During the development of the overturning mechanism the pier rotates at its toe, BC, along the rotation axis a–a. All the weights G_{ij} of the active wall, with their lifting arms b_{ij}, aligned along the nodes of pier i from the first story

Fig. 12.73 Participation of the inactive wall masses to the resisting work

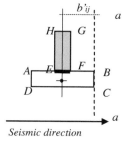

j as far as the last story $j = N_s$, will thus be raised as the mechanism arises. On the other hand, given the participation of the inactive walls' masses, the *total* weights G'_{ij}, with their lifting arms b'_{ij}, including the masses of both the inactive and active walls, *must* be taken into account in the lifting resistant work.

In brief, the entire weight of inactive wall EFGH, with the possible weights of the floors sustained by this wall, has to be raised during the overturning mechanism. The shear

$$V_i = \sum_{j=1}^{j=Ns} (G'_{ij} - G_{ij}) \qquad (12.186)$$

must thus be transmitted vertically along the contact section EF. If we indicate A_S as the total area of this vertical contact section, we can evaluate the average shear stress

$$\tau = \frac{\sum_{j=1}^{j=Ns} (G'_{ij} - G_{ij})}{A_S} \qquad (12.187)$$

to obtain an estimate of the order of magnitude of the shear action that occurs.

Analyzing the state of the connections between the two arrays of walls can suggest whether reinforcing works are necessary, possibly by inserting special steel ties. As a rule, the presence of uncracked masonry with regular blocks interpenetrating along the vertical contact section suggests that reinforcements can be avoided: an average limit shear value of 10 t/m² can usually be assumed.

12.9.3.3 Checking the Fixity of Steel Beams

The steel beams spanning the openings, with their ends fixed in the piers, produce strong interactions with the masonry during the development of a sideways mechanism (Fig. 12.62).

The masonry may fail by crushing around the steel beams' heads and unloose them, consequently reducing the wall strength. Checking the condition of the connections of the steel-beam ends to the wall is thus crucial. Figure 12.74 shows a length a of the steel beam built into the masonry, together with the internal equilibrium at the limit state of the wall. The limit bending moment M_o occurs at the insertion sections in the piers, and the shear

$$T_o = \frac{2M_o}{L} \qquad (12.188)$$

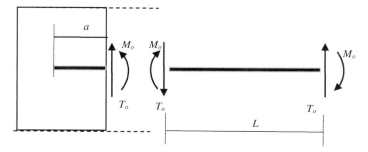

Fig. 12.74 Stress state in the steel beams at wall failure

is transmitted along the steel beam spanning the opening of width L. Figure 12.63 then shows the resultant stresses T_o and M_o acting at the beam section flush with the pier (Fig. 12.75).

The resultant reactions exerted by the masonry surrounding the length of beam are defined by the axial force N, applied at the center of the built-in length, and the moment M. The solid equilibrium of the built-in length of beam gives

$$M = T_o a/2 + M_o \qquad (12.189)$$

and

$$N = T_o. \qquad (12.190)$$

We assume that the masonry, at its ultimate state, bears the limit stress distribution around the built-in length of steel beam shown in Fig. 12.77, where σ_{oo} indicates the masonry crushing strength. This stress state is equivalent to the limit state of an elastic-plastic section under the action of axial load N and bending moment M. The interaction yield locus, drawn in Fig. 12.77, is thus given by the equation

$$M = M_{oo}[1 - (\frac{N}{N_{oo}})^2], \qquad (12.191)$$

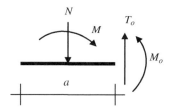

Fig. 12.75 Equilibrium of a steel beam end built into the masonry

Where

$$M_{oo} = \sigma_{oo}\frac{ba^2}{4} \qquad N_{oo} = \sigma_{oo}ba \qquad (12.192)$$

are respectively the limit moment and axial limit load, depending on the built-in length a.

The problem is thus to obtain the required built-in length a for a given masonry crushing strength σ_{oo} and for a given value of the ratio

$$tg\chi = \frac{M}{N} \qquad (12.193)$$

between the resultant moment M and the axial load N acting on length a (Fig. 12.76).

Taking into account the expressions for M and N, from (12.189) to (12.190), the ratio M/N becomes

$$tg\chi = \frac{L}{2}(1 + \frac{a}{L}). \qquad (12.194)$$

Ratio (12.194) defines the limit stress distribution around the built-in length a. Thus, stress point σ (Fig. 12.77) is located on the interaction locus, and taking Eq. (12.191) into account, we get

$$\frac{L}{2}(1 + \frac{a}{L}) = \frac{M}{N} = M_{oo}(\frac{1}{N} - \frac{N}{N_{oo}^2}). \qquad (12.195)$$

However, the resultant axial load N equals the shear force T_o in the steel beam, so condition (12.195) thus becomes

$$\frac{L}{2}(1 + \frac{a}{L}) = M_{oo}(\frac{1}{T_o} - \frac{T_o}{N_{oo}^2}). \qquad (12.196)$$

Fig. 12.76 Distribution of limit compressions σ_o around the beam end

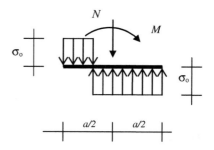

12.9 In-Plane Strength of Multi-storey ...

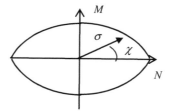

Fig. 12.77 Interaction yield locus as a function of M and N

Taking into account expression (12.192) for the limit moment M_{oo} and the axial force N_{oo}, we also obtain

$$\frac{L}{2}\left(1+\frac{a}{L}\right) = \frac{\sigma_{oo}ba^2}{4T_o}\left(1 - \frac{T_o^2}{\sigma_{oo}^2 b^2 a^2}\right). \tag{12.197}$$

With the positions

$$x = \frac{\sigma_{oo}bL^2}{2M_o} \quad y = \frac{a}{L}, \tag{12.198}$$

representing the masonry strength and the built-in length ratios, we obtain the equation linking factors x and y

$$2(1+y) = xy^2\left(1 - \frac{y^2}{4}\right), \tag{12.199}$$

Taking into account that quantity $y^2/4$ is negligible with respect to unity gives

$$2(1+y) = xy^2 \tag{12.200}$$

and

$$y = \frac{1 + \sqrt{1+2x}}{x}. \tag{12.201}$$

12.9.3.4 A Numerical Example of the Inplane Strength Valuation of a Masonry Wall with Thin Openings

Figure 12.66 shows the floor plan of a simple masonry building made up of two longitudinal and four short transverse walls. As discussed in Sect. 8.5.2, two different choices are available for defining the piers' sections. First, we can consider the so-called *disconnectedness assumption* between the two wall arrays and thereby address only the rectangular section of the active wall in its plane. Conversely, the

Fig. 12.78 Example masonry wall

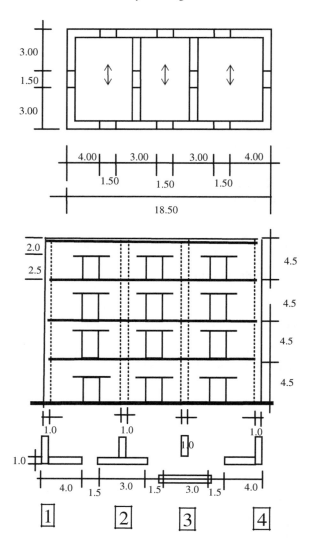

second option considers the *connection assumption* between the two arrays of walls and therefore regards the section of the adjacent *inactive* transverse wall as well (Fig. 12.78).

Two different values of the weights will thus be evaluated: the first G_{ij}, pertaining to the first choice; and G'_{ij} to the second. These values, with the corresponding values of the lifting arms, b_{ij} and b'_{ij}, are given in Tables 8.1 through 8.5 (Tables 12.1, 12.2, 12.3, 12.4 and 12.5).

12.9 In-Plane Strength of Multi-storey ...

Table 12.1 Weights G_{ij} in tons

39.98	34.61	34.61	39.98
37.40	32.49	32.49	37.40
34.84	30.39	30.39	34.84
31.72	27.75	27.75	31.72

Table 12.2 Lifting arms b_{ij} in meters

1.82	1.50	1.50	2.18
1.81	1.50	1.50	2.19
1.79	1.50	1.50	2.21
1.77	1.50	1.50	2.23

Table 12.3 Weights G'_{ij} in tons

56.78	51.41	51.41	56.78
53.17	48.26	48.26	53.17
49.43	44.98	48.26	49.43
44.99	41.02	41.02	44.99

Table 12.4 Lifting arms b'_{ij} in meters

2.32	1.50	1.50	1.68
2.32	1.50	1.50	1.68
2.33	1.50	1.50	1.67
2.33	1.50	1.50	1.67

Table 12.5 Horizontal forces at node λF_{ij} in tons

$\lambda\cdot 23.85$	$\lambda\cdot 21.06$	$\lambda\cdot 21.59$	$\lambda\cdot 23.85$
$\lambda\cdot 44.13$	$\lambda\cdot 40.05$	$\lambda\cdot 40.05$	$\lambda\cdot 44.13$
$\lambda\cdot 61.79$	$\lambda\cdot 55.36$	$\lambda\cdot 55.36$	$\lambda\cdot 61.79$
$\lambda\cdot 74.68$	$\lambda\cdot 68.09$	$\lambda\cdot 68.09$	$\lambda\cdot 74.68$

The distribution factors are

$$\gamma_1 = 0.42;\ \gamma_2 = 0.83;\ \gamma_3 = 1.25;\ \gamma_4 = 1.66$$

Let us on now evaluate the partial strengths multipliers of the single piers according to (12.173) under the assumption of disconnectedness between the two arrays of walls. We obtain

$$\lambda_{o1} = \frac{259.09}{2683.1} = 0.0966 \qquad \lambda_{o2} = \frac{187.86}{2442.6} = 0.0769$$

$$\lambda_{o3} = \frac{187.86}{2442.6} = 0.0769 \qquad \lambda_{o4} = \frac{316.68}{2683.1} = 0.1180$$

According to (12.115), the corresponding strength multiplier of the wall is

$$\lambda^+ = \frac{259.09 + 187.87 + 187.87 + 316.68}{2(2683.1 + 2442.6)} = 0.0928$$

Conversely, the partial strength multipliers under the assumption of connections between the two arrays of walls are given by

$$\lambda_{o1} = \frac{494.38}{2683.1} = 0.1843 \qquad \lambda_{o2} = \frac{395.74}{2442.6} = 0.1620$$

$$\lambda_{o3} = \frac{395.74}{2442,6} = 0.1620 \qquad \lambda_{o4} = \frac{466.40}{2683.1} = 0.1738$$

and the corresponding strength multiplier of the wall is

$$\lambda^+ = \frac{494.38 + 38.9 + 395.74 + 395.74 + 466.40}{2(2683.1 + 2442.6)} = 0.1709$$

So far the presence of the steel beams over the wall openings and inserted into the piers has not yet been taken in account. We now assume that *two* HEA 120 beams made of Fe 360 steel run over each of the openings. Let us calculate the contribution of the steel beams to the kinematic multiplier for the overturning mechanism of the wall.

Mechanical properties of the section HEA 120

Flange section: A_s = 12 cm × 0.8 = 9.6 cm^2;
Distance between flange centers: h* = 9.8 cm + 0.8 cm == 10.6 cm
Yield strength of the Fe360 steel: σ_{sy} = 2400 kg/cm^2
Ultimate bending moment
M_{tH} = 2400 × 9.6 × 10.6 = 2442 kgm; 2 M_{02H} = 4884 kgm

Let us now evaluate the kinematic multiplier under the disconnectedness assumption:

$$\lambda^+ = \frac{266.09 + 187.76 + 187.76 + 302.36 + 364.8}{2(2682.9 + 2428.2)} = 0.128$$

Conversely, under the connection assumption, the partial failure multipliers are first given by

$$\lambda_{o1} = \frac{502.7}{2682.9} = 0.187 \qquad \lambda_{o2} = \frac{387.94}{2428.20} = 0.1598$$

$$\lambda_{o3} = \frac{387.94}{2428.20} = 0.1598 \qquad \lambda_{o4} = \frac{502.7}{2682.9} = 0.187$$

and the corresponding strength multiplier of the wall is

$$\lambda^+ = \frac{502.7 + 38.9 + 387.9 + 447.5 + 364.8}{2(2682.9 + 2428.2)} = 0.169$$

Note the gradual increase in the value of the kinematic multiplier when the contributions of the inactive walls and then the steel beams are included.

Stresses in the horizontal connections

To evaluate these stresses we start by applying Eqs. (12.175′), (12.176′) and (12.177′). The evaluation considers both the presence of the inactive walls, according to the connection assumption, and the participation of the HEA 120 steel beams inserted in the piers. By using expressions (12.173) for the partial failure multipliers, we have

$$M_1^N - M^T = M_1^R(\lambda^+ - \lambda_{01}) \tag{a}$$

$$M_2^N - M_1^N = M_2^R(\lambda^+ - \lambda_{02}) \tag{b}$$

$$M_3^N - M_2^N = M_3^R(\lambda^+ - \lambda_{03}) \tag{c}$$

$$M^T - M_3^N = M_4^R(\lambda^+ - \lambda_{04}) \tag{d}$$

where the overturning moments due to the horizontal forces evaluated according to the values given in Table 8.5 are

$$M_1^R = 2683.10\,\text{tm}; \quad M_2^R = 2442.260\,\text{tm}$$
$$M_3^R = 2442.60\,\text{tm} \quad M_4^R = 2683.10\,\text{tm}$$

From (12.124), the partial failure multipliers of the single piers under the connection assumption in the presence of HEA 120 steel beams inserted into the piers are

$$\lambda_{01} = 0.1843 \quad \lambda_{02} = 0.1620 \quad \lambda_{03} = 0.15620 \quad \lambda_{04} = 0.1738$$

while the corresponding overall kinematic multiplier is $\lambda^+ = 0.169$. We have

$$(\lambda^+ - \lambda_{01}) < 0 \quad (\lambda^+ - \lambda_{02}) > 0 \quad (\lambda^+ - \lambda_{03}) > 0 \quad (\lambda^+ - \lambda_{04}) > 0 \tag{e}$$

Thus, from Eqs. (a), (b), (c) and (d), we obtain

$$M_1^N - M^T < 0 \quad M_2^N - M_1^N > 0 \quad M_3^N - M_2^N > 0 \quad M^T - M_3^N < 0$$

All the unknowns M_1^N, M_2^N, M_3^N and M^T must be positive and, at the same time, satisfy the condition $M_1^N \cdot M_2^N \cdot M_3^N \cdot M^T = 0$ and at least one of the unknowns has to cancel out. Thus, from the previous inequalities, it is a simple matter to prove that each of the three possibilities

$$M_2^N = 0 \quad M_3^N = 0 \quad M^T = 0$$

is inconsistent, and that only the condition

$$M_1^N = 0$$

is, on the contrary, compatible. This outcome can be explained simply by mechanical considerations. Under the seismic forces acting from left to right, piers 2 and 3, the weakest, are the first to fail and are thus sustained by pier 4. Then when, as the thrusts increase, pier 4 also fails, it pushes on pier 1 through the tie rods. In this state all the tie rods are stretched and the architraves to the right of piers 2 and 3 are compressed. The architrave located just after pier 1 is, on the contrary, unloaded. In short, at global failure the masonry architrave after pier 1 will be unloaded and we have $M_1^N = 0$. Given this last result, the first and second equations, (a) and (b), become

$$M^T = M_1^R(\lambda_{01} - \lambda^+) \quad M_2^N = M_2^R(\lambda^+ - \lambda_{02})$$

we can first calculate the values

$$M^T = M_{R1}(\lambda_{01} - \lambda^+) = M_{R1}(0.1843 - 0.1709) = 35.76 \text{ tm}$$

$$M_2^N = M_{R2}(\lambda^+ - \lambda_{02}) = M_{R2}(0.1709 - 0.1620) = 21.77 \text{ tm}$$

Whence

$$M_3^N = M_3^R(\lambda^+ - \lambda_{03}) + M_2^N = 2442.60(0.1709 - 0.1620) + 21.77 = 43.53 \text{ tm}$$
$$M_1^N = 0.$$

We can now go on to evaluate the actual stresses in the tie rods and compressed architraves. Taking into account the values of the floor heights, we have

$$z_1 = 4.50 \text{ m} \quad z_2 = 9.00 \text{ m} \quad z_3 = 13.50 \text{ m} \quad z_4 = 18.00 \text{ m}$$

$$\sum_{i=1}^{4} z_i^2 = 4.5^2 + 9.0^2 + 13.5^2 + 18.0^2 = 644.50 \, m^2$$

Hence, from (12.184) the forces in tie rods are

12.9 In-Plane Strength of Multi-storey ...

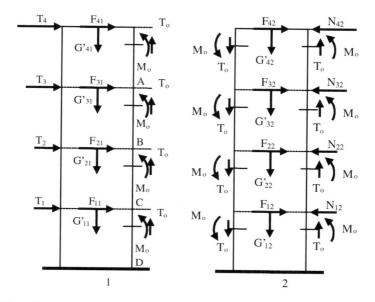

Fig. 12.79 Actions transmitted by the horizontal connections to pier 1 and 2

$$T_1 = 0.265\,t \quad T_2 = 0.530\,t \quad T_3 = 0,795\,t \quad T_4 = 1.059\,t$$

Likewise, for the masonry architraves we get

$$N_{i1} = 0 \ (i = 1, 2, 3, 4)$$

$$N_{12} = 0.161\,t \quad N_{22} = 0.322\,t \quad N_{32} = 0.4684\,t \quad N_{42} = 0.645\,t$$
$$N_{13} = 0.322\,t \quad N_{23} = 0.645\,t \quad N_{33} = 0.967\,t \quad N_{43} = 1.290\,t$$

Knowing the stresses in the horizontal connections enables checking the static admissibility of the limit state of the wall under the thrusts $\lambda^+ F_{ij}$.

Checking the admissibility of the limit horizontal loads $\lambda^+ F_{ij}$
Figures 12.79 and 12.80 show the piers' walls under all the actions transmitted to them by tie rods, masonry architraves and steel beams according to the results above (Fig. 12.81).

Checking node capacity to transmit shear from inactive to active walls
By way of example of such a check, let us consider pier 2, for which we evaluate the shear force V that must be transmitted through it. From Tables 03 and 01, we have

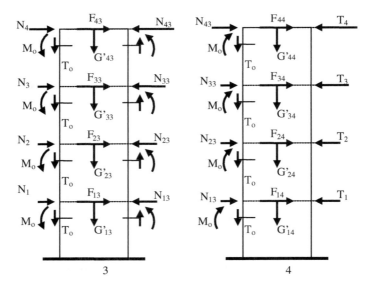

Fig. 12.80 Actions transmitted by the horizontal connections to piers 3 and 4

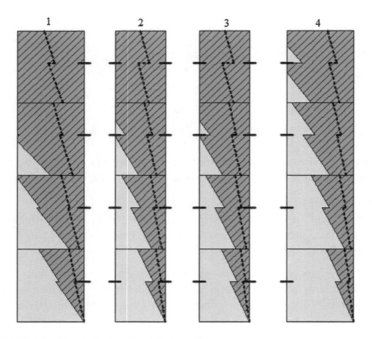

Fig. 12.81 Resistant bands in the piers (*dark gray*)

$$V_2 = \sum_{j=1}^{j=Ns} (G'_{2j} - G_{2j}) = 60.43 t,$$

where evaluation of the shear resistant vertical section gives

$$A_t = 4.50 \times (1.0 + 0.9 + 0.8 + 0.7) = 15.3 \, m^2,$$

and the average shear is thus $\tau = 60.43/15.3 = 3.9 \, t/m^2$

This value is low considering uncracked masonry and the presence of interpenetrating blocks at the nodes. The check furnishes a positive result.

Checking the fixity of the steel beam ends
Two Fe 360 steel HEA 120 beams are inserted into the piers over each wall opening, hence, $M_0 = 2 \times M_{0tr} = 2 \times 2.28 \, tm = 4.56 \, tm$
  $b = 2 \times 12 \, cm = 24 \, cm$ $L = 150 \, cm$. We assume $\sigma_{oo} = 50 \, kg/cm^2$. Consequently, according to (12.198) and (12.201) we get

$$x = \frac{50 \cdot 2 \cdot 12 \cdot 150^2}{2 \cdot 488400} = 27.641 \quad y = \frac{1 + \sqrt{1 + 2 \cdot 27.641}}{27.641} = 0.308$$

$$y = \frac{a}{150} = 0.308 \quad a = 46 \, cm$$

In brief, two HEA 120 steel beams are to be fitted over each opening of span 1.50 with an insertion length of 35 cm. The masonry around the ends of the steel beams must be firmly consolidated.

Concluding remarks
The lateral strength of such a building wall is equal to 0.169 of its weight. Taking approximately into account the effect of the finite compression strength, consideringf for the reduction factor $(1 - \sigma_m/\sigma_{o<})$, equal to about 0.85, the wall lateral strength now becomes $0.169 \times 0.85 = 0.14$ of its weight.

References

Abruzzese, D., Como, M., & Grimaldi, A. (1986). *Analisi limite degli edifici murari sotto spinte orizzontali*. Rome: Atti Dipartimento di Ingegneria Civile, Università di Roma Tor Vergata.

Abruzzese, D., Como, M., & Lanni, G. (1992). On the lateral strength of multistory masonry walls with openings and horizontal reinforcing connections. In *Earthquake Engineering, Proceedings of the Tenth World Conference* (pp. 4525–4530). Rotterdam: Balkema.

Benedetti, D., & Casella, M.L. (1980). Shear strength of masonry piers. In *7th World Conference on Earthquake Engineering*, Istanbul.

Benedetti, D., Binda, L., Carabelli, E., Contro, R., Corradi Dell'Acqua, L., Franchi, A., Genna, F., Gioda, G., Macchi, G., Nova, R., Peano, A., & Rossi, P.P. (1982). Comportamento statico e sismico delle strutture murarie. In G. Sacchi Landriani & R. Riccioni (Eds.), Milan: CLUP.

Binda, L. (1983). *Metodi statici di stima della capacità portante di strutture murarie*. Milan: da Comportamento Statico e Sismico delle strutture Murarie, CLUP.

Braga, F., & Dolce, M. (1982). A method for analysis of antiseismic masonry multistory buildings. In *6th International Brick Masonry Conference*.

Carbone, I. V., Fiore, A., & Pistone, G. (2001). *Le costruzioni in muratura* (pp. 58–59). Milan: Hoepli.

Chopra, A.K., & Goel, R.K. (2002). A modal pushover analysis for estimating seismic demands for building, *Earthquake Engineering and Structural Dynamics*, 31.

Coccia, S., & Como, M. (2009). Sull'Analisi sismica delle costruzioni in muratura. In *WONDER masonry, Proceedings of the Workshop on Design and Rehabilitation of Masonry Structures, Ischia*. Florence: Polistampa.

Como, M., & Lanni,G. (Eds.). (1981). *Elementi di Costruzioni Antisismiche*. Cremonese: Rome.

Como, M., & Grimaldi, A. (1983). Analisi limite di pareti murarie sotto spinta. In *Atti Istituto di Tecnica delle costruzioni*, n. 546. Naples: Università di Napoli.

Como, M., & Grimaldi, A. (1985). An unilateral model for the limit Analysis of masonry walls. In *Unilateral problems in Structural Analysis*, Proceedings of the *2nd Meeting in Unilateral Problems in Structural Analysis*. New York: Springer (Ravello, 22–24 Sept. 1983, CISM Courses and Lectures, 288).

Como, M., Lanni, G., & Sacco, E. (1991). Sul calcolo delle catene di rinforzo negli edifici in muratura soggetti ad azione sismica. *V° Conf. Naz.le "L'Ingegneria sismica in italia"*, ANIDIS, Facoltà di Ingegneria, Università di Palermo.

Como, M., Grimaldi, A., & Lanni, G. (1998). New results on the strength evaluation of masonry buildings and monuments. *9th World Conference on earthquake Engineering*. Tokyo: Balkema.

Como, M. (2006). Modellazioni semplici per l'analisi della resistenza sismica degli edifici in muratura, Atti del Workshop *WONDERMasonry 2006*, Dipartimento di Ingegneria Civile, Università di Firenze, Edizioni Polistampa, Florence.

Contro, R., & Nova, R. (1982). *Modello fisico e matematico del legame sforzi e deformazioni del comportamento a rottura della muratura*. ISMES, Bergamo: Corso I.P. sul comportamento statico e sismico delle Strutture Murarie.

Di Pasquale, S. (1982). Architettura e Terremoti, *Restauro*, 59–61.

Fajfar, P. (1999). Capacity Spectrum method based on inelastic demand spectra. *Earthquake Engineering and Structural Dynamics*, 28.

Galasco, A., Lagomarsino, S., & Penna, A., (2002). TREMURI Program: Seismic Analyzer of 3D masonry Program, Università di Genova.

Galasco, A., Lagomarsino, S., & Penna, A. (2006). On the use of pushover analysis for existing masonry buildings. *1st ECEES*, Geneva.

Giangreco, E. (1983). La normativa sismica. Tappe e prospettive, In *Fondamenti di Ingegneria Sismica*, Bologna.

Fusier, F., & Vignoli, A. (1993). *Proposta di un metodo di calcolo per edifici in muratura sottoposti ad azioni orizzontali*. Anno X: Ingegneria sismica. 1.

Giuffrè, A. (1986). *La meccanica nell'architettura*. Rome: Nuova Italia Scientifica.

Giuffrè, A. (1988). Monumenti e Terremoti, aspetti statici del restauro. *Scuola di Specializzazione per lo Studio ed il Restauro dei Monumenti*, Multigrafica Editrice, Rome.

Magenes, G., & Della Fontana, A. (1998). Simplified Non linear Seismic Analysis of Masonry Buildings. In *Proceedings of the British Masonry Society*, 8.

Murthy, C. K., & Hendry, A. W. (1966). *Model experiments in load bearing brickworks*. *Building Science* (Vol. 1). London: Pergamon Press.

Newmark, N., & Hall, W. (1982). *Earthquake Spectra and Design, Monograph, Earthquake Engineering Research institute, Oakland*. USA: Calif.

Norme Tecniche DM 16.01.1996, come integrate dalla CM n.65 del 10.04.1997.

Norme tecniche OPCM n. 3431 del 03.05.2005.

Norme tecniche DM 14.01.2008, e relative istruzioni per l'applicazione.

Petrini, L., Pinho, R., & Calvi, G.M. (2006). Criteri *di progettazione antisismica degli edifici*, IUSS Press.

Priestley, M.J.N., Calvi, G.M., & Kowalsky, M.J. (2007). Displacement—Based Seismic Design of Structures, IUSS Press.

Pozzati, P. (1986). *Processo di approfondimento delle conoscenze tecniche: inquietanti tendenze del nostro tempo*. Ingegneri, Architetti, Costruttori: INARCOS.

Ruffolo, F. (1912). La stabilità sismica dei fabbricati, Casa editrice "L'Elettricista", Rome.

Timoshenko, S. (1955). *Theory of Elasticity*. New York: McGraw Hill, Book Company.

Tomazevic, M. (1978). The Computer Program POR, Report ZRMK. In *Institute for Testing and Research in Materials and Structures*, Ljubljana.

Yokel, Y., & Fattal, S.G. (1976). Failures hypothesis for masonry walls. *ASCE, Structural Division*.

Theses (University of Rome Tor Vergata, Fac. of Engineering).

Ambrogi Leonard Diego, (a.a. 2008–2009) *Resistenza e duttilità delle pareti murarie multipiano*, Supervisors: Coccia S., Como M.

Iannotti Antonio (a.a. 2006–2007) *Sulla resistenza sismica degli edifici in muratura a pareti regolari multipiano*, Supervisors: Como, M.; Fabiani, F.M.

Erratum to: Masonry Vaults: Cross and Cloister Vaults

Mario Como

Erratum to:
Chapter 7 in: M. Como, *Statics of Historic Masonry Constructions*, Springer Series in Solid and Structural Mechanics, DOI 10.1007/978-3-319-24569-0_7

The original version of the chapter 7 was inadvertently published with partially processed figures 7.36 and 7.37. The figures have been replaced in the original chapter. The erratum chapter have been updated with the changes.

The updated original online version for this chapter can be found at
DOI 10.1007/978-3-319-24569-0_7

M. Como (✉)
Department of Civil Engineering, University of Rome Tor Vergata, Rome, Italy
e-mail: como@ing.uniroma2.it

© Springer International Publishing Switzerland 2016
M. Como, *Statics of Historic Masonry Constructions*,
Springer Series in Solid and Structural Mechanics 5,
DOI 10.1007/978-3-319-24569-0_13